C. Ludwig S. Hellweg S. Stucki

Municipal Solid Waste Management

Springer-Verlag Berlin Heidelberg GmbH

Christian Ludwig
Stefanie Hellweg
Samuel Stucki
Editors

Municipal Solid Waste Management

Strategies and Technologies for Sustainable Solutions

with 163 Figures

 Springer

DR. CHRISTIAN LUDWIG
Paul Scherrer Institut
5232 Villigen PSI
Switzerland
Email: *christian.ludwig@psi.ch*

DR. STEFANIE HELLWEG
Safety and Environmental Technology Group
ETH Hoenggerberg
HCI G143
8093 Zurich
Switzerland
Email: *hellweg@tech.chem.ethz.ch*

DR. SAMUEL STUCKI
Paul Scherrer Institut
5232 Villigen PSI
Switzerland
Email: *samuel.stucki@psi.ch*

ISBN 978-3-642-62898-6
Library of Congress Cataloging-in-Publication Data Applied For

Die Deutsche Bibliothek – CIP Einheitsaufnahme
Municipal solid waste management : strategies and technologies for
sustainable solutions / Christian Ludwig ... (ed.). - Berlin ; Heidelberg ;
New York; Hong Kong ; London ; Milan ; Paris ; Tokyo :
Springer, 2003
  ISBN 978-3-642-62898-6   ISBN 978-3-642-55636-4 (eBook)
  DOI 10.1007/978-3-642-55636-4

springeronline.com

© Springer-Verlag Berlin Heidelberg 2003
Originally published by Springer-Verlag Berlin Heidelberg in 2003
Softcover reprint of the hardcover 1st edition 2003

Camera ready by editors
Cover design: E. Kirchner, Heidelberg
Printed on acid-free paper SPIN 10976395 30/3111 5 4 3 2 1

Preface

Motivation

The other day I was waiting at the station for my train. Next to me a young lady was nonchalantly leaning against the wall. Suddenly, she took a cigarette pack out of her handbag, pulled out the last cigarette, put it between her lips, crushed the empty pack, threw it on the ground and hedonistically lit the cigarette. I thought to myself, "What a behavior?!". The nearest trashcan was just five meters away. So I bent down, took the crushed pack and gave it back to her, saying that she had lost it. She looked at me in a rather deranged way, but she said nothing and brought the piece of waste to the trashcan.

Often people are not aware of the waste they produce. They get rid of it and that's it. As soon as the charming lady dropped the cigarette pack, the problem was solved for her. The pack was on the ground and it suddenly no longer belonged to her. It is taken for granted that somebody else will do the cleaning up.

There is a saying that nature does not produce waste. For long as humans obtained the goods they needed from the ground where they lived, the waste that was produced could be handled by nature. This has drastically changed due to urbanization and waste produced by human activities has become a severe burden. The amount of waste production increased enormously in the last century in conjunction with increasing prosperity and, at the same time, an increase in toxic pollutants, which have been enriched in soil, water and air. All of this puts enormous stress on the environment. Moreover, pollutants are not only potentially harmful for nature, but also for human health.

The main focus of this book is on *Municipal Solid Waste* (MSW) management. Think of a waste bin at your home or office, or of a public trashcan. You might remember having seen pictures with huge mountains of waste on TV or in newspapers. Astonishingly, not many people have ever been to a landfill themselves. Once confronted with the real situation, most people would be deeply impressed by the huge amounts of waste, the dirt, and the off awful odors. Standing at such a site, one might begin to think that waste potentially causes some severe problems. Research and development has shown that wastes produced by our contemporary society are not wrongly found to be hazardous to environment and health.

People are much more aware of the "waste problem" if they directly suffer from the impacts, e.g. from the bad odors, dust, and traffic due to a landfill in close proximity to their home. This is more likely if space is very limited. Therefore, it is not astonishing that a densely populated country such as Switzerland begins to think about waste management guidelines at quite an early stage.

The Guidelines, set up in 1986 include political, technical, and economic principles and objectives that have had a strong influence on waste management practices and legislation. From the scientific and technical point of view, the following long-sighted objectives are still an issue after more than 15 years of discussion (quotation):

"Waste disposal systems should generate only two groups of materials from waste, namely those which can be recycled and those which are suitable for deposing on final disposal sites, i.e. sites where flows of materials into the environment (air, water, and soil) are environmentally compatible both in the short- and long-term without the need for additional treatment."

"Waste treatment methods should be designed to ensure that environmentally hazardous materials occur in their most highly concentrated form and environmentally compatible materials in their purest form (i.e. similar to earth or soil)."   *(Guidelines on Waste Management, BUWAL, Switzerland, 1986)*

Established by the Federal Parliament in 1991, the Swiss Priority Program Environment (SPPE) was set up by the National Science Foundation. It was designed to explore and help solve the urgent problems facing the environment on both a national and global level. Within the scope of SPPE, the Integrated Project Waste (IP Waste, 1993-2001) was an interdisciplinary framework in which emphasis was placed on the improvement of waste management systems, technologies, procedures, and waste products. The IP Waste involved collaborative projects from various disciplines, including academia, consulting firms, and industry. The results of IP Waste were by no means complete. With this book the editors have aimed to achieve a synthesis of the fundamental and practical findings obtained by IP Waste in the last decade and to put them into an international context. The knowledge and experience is complemented by other views from different countries.

Synopsis

Current state-of-the-art waste management systems and treatment technologies are presented and discussed along with potential future systems and technologies. More than 40 authors, many from outside of Switzerland, have contributed to this book. Along with an outline given by the editors, a complementary set of contributions has been written in order to produce a consistent and uniform book.

We have tried to achieve a writing style that is easily understandable and accessible to a rather large and interdisciplinary audience. The content has been kept simple and application oriented. The synthesis texts at the beginning of each main chapter should function as guides to the more detailed contributions and reflect the general content in a very concise manner. These synthesis texts, contrasted from the other texts by a gray shaded background, were written by the editorial team and do not necessarily reflect the opinion of the authors who contributed to the corresponding chapters. The editors acknowledge Helmut Brandl, who wrote the synthesis to chapter 4.

Chapter 1 provides a general introduction to the various aspects of the "waste problem" and information about the historical trends of waste.

The composition of waste basically determines the toxic potential of an MSW landfill. Major concerns are emissions to soil, water, and atmosphere. Chapter 2 supplies information on the composition of MSW and the emission problems of MSW landfills. This chapter explains why such landfills are not sustainable and provides motivation for considering better solutions.

Better solutions are available and already in practical use today (chapter 3 to 5). The approaches and possibilities to solve the waste problems strongly vary from one country to another. Differences between developing and developed countries are particularly evident in this regard. There is a huge potential for future economic growth in developing countries, which is always accompanied by a strong increase in the MSW amounts and also a change in composition of MSW towards more toxicity. The implementation of new management strategies and technologies will be crucial.

Landfills are a major source of greenhouse gases, such as methane. Methane collection right at the site is a rather inexpensive way to reduce the amounts of methane released to the atmosphere; furthermore, the gas can be used for the production of electricity. Methane collection is a step in the right direction and may be a real improvement in many countries where waste is not treated at all. However, methane collection is no longer considered a state-of-the-art-option for developed countries. In developed countries there is a strong push to stop landfilling, i.e. to stop MSW disposal. Chapter 3 discusses various options which are already in use today. Using modern *recycling technologies,* the *zero waste* approach is very promising. However, as long as the product design for any existing product is not optimized for the reuse of all materials, even the best recycling method will fail. Therefore, as long as MSW is produced, *end-of the-pipe* solutions have to be considered. Many countries have started to develop incineration technologies. Although these technologies are able to destroy many toxic substances, they still produce harmful products. New technologies effectively clean the off gases of an incineration plant; however, toxic filter and bottom ashes are still produced. Of major concern are the heavy metals contained in these residues.

Future MSW treatment processes will focus on two goals: a) proper treatment of waste by total detoxification of the residues producing secondary raw materials, and b) production of heat and power from the biomass and organics contained in waste. Biological processes may have a high potential, specifically because they are associated with low costs. Chapter 4 provides a short overview of the current state-of-the-art methods using biological and bio-mechanical processes.

In the last 15 years research and development has made substantial progress in thermal treatment technologies. They can efficiently remove toxic substances whether they are organic (e.g. dioxins) or inorganic (e.g. heavy metals). Without claiming to have composed a complete list, chapter 5 presents some of the most innovative of these technologies, considering energy and/or material recovery.

Choosing the appropriate technology is often a complex task. Different factors have to be taken into consideration, such as the geographic, economic, social and cultural conditions. Assessment of such conditions is of particular relevance. Life Cycle Analyses (LCA) is a powerful tool in the environmental assessment of waste treatment systems. Chapter 6 offers insight into the uses and functions of this tool by applying it in the assessment of some of the technologies presented in this book. Long-term effects from landfills and their possible evaluation in LCA are discussed.

Independent of technological feasibility and environmental soundness, humans are the most crucial factor in establishing sustainable behavior and new technologies. In democratic systems the people, finally, make the decisions. Therefore, solutions must be socially acceptable and compatible. Dialogue between the various stakeholders is essential for finding the proper regional solution. Various case studies presented in Chapter 7 offer insight into the methodology of mediation and the development of a new assessment tool for evaluating social acceptance.

This book presents current and past developments in waste management. Modern and new technologies for waste treatment and the recycling and recovery of materials from wastes are discussed. Options and technologies for directing material flows for the production of secondary raw materials and for the use of waste in the production of heat and electricity are considered. In the final chapter, roadmaps toward integrated waste management are outlined. The most central conclusions of the book have been summarized in chapter 9; here, future challenges regarding technological, economic, ecological and social aspects are discussed.

Christian Ludwig
Villigen PSI, June 14, 2002

Acknowledgements

List of Guest Reviewers

The editors thank the following persons for their review work:

PD Dr. H. Brandl, University of Zurich, Switzerland

Prof. Dr. P.H. Brunner, Vienna University of Technology, Austria

Prof. Dr. K. Gruiz, Budapest University of Technology and Economics, Budapest, Hungary

Dr. N. Patel, AEA Technology Environment, Culham, Abingdon, Oxfordshire, UK

M. Schaub, CT Environment, Winterthur, Switzerland

W. Snow, Envision New Zealand, Takapuna, Auckland, New Zealand

Dr. A. Tukker, TNO (Netherlands Organisation for Applied Scientific Research), Delft, The Netherlands

Dr. J. Vehlow, Forschungszentrum Karlsruhe GmbH, Germany

Dr. H.J.M. Visser, Netherlands Energy Foundation ECN, Petten, The Netherlands

Prof. Dr. R. Zevenhoven, Helsinki University of Technology, Finland

Prof. Dr. H. Zillessen, Carl von Ossietzky University of Oldenburg, Oldenburg, Germany

Other anonymous reviewers

Language Editing

The editors would like to acknowledge Scott Loren, M.A. (University of Zurich) for his excellent English language corrections.

Further Acknowledgments

The editors acknowledge the Swiss National Science Foundation for its financial support of this book project (projects 5001-044556 and 5001-058291) and the support by all institutions represented by the many authors.

C. Rootes acknowledges the support by Contract no. ENV4-CT96-0239 from the European Commission (DG XII - Science, Research and Development).

After many ups and downs this project could finally be completed. Without the full activity of G. Herlein, Ch. Kunz, H. Schröder, Dr. R. Struis, PD Dr. H. Brandl, and J. Wochele the project could not have been realized.

The editors thank Jürg Leuzinger for drawing the cartoons shown in the synthesis text of each main chapter.

V. Carabias thanks J.V. Holmes (La Force, France) for English language corrections.

M. Beckmann thanks Ch. Swanson (Clausthal-Zellerfeld, Germany) for English language translation.

Further we like to acknowledge the following persons and/or institutions for valuable discussions and for supplying figures and/or data

- Mr. M. Schaub, **CT Environment Ltd.**, Switzerland
- Mr. M. Stengele, **IMMARK Ltd.**, Regensdorf, Switzerland
- Dr. D. Chambaz, **Service de Gestion des Déchets du Canton de Genève**, formerly **SAEFL**, Switzerland
- Dr. Graham Shields, James Cook University, Queensland, Australia, formerly at **CNRS Strasbourg**
- Dr. Thierry Advocat, **CEA Marcoule**, France
- Mr. Hugues Siegenthaler, Pully, Switzerland
- **Primeverre**, Montpellier, France
- Dr. L. Stenmark, **Chematur Engineering AG**, Karlskoga, Sweden
- M. Asbury, R. Steer, and R. Whale, all **University of Kent at Canterbury**, UK
- Prof. Dr. A. Wokaun, head **General Energy Research Department at Paul Scherrer Institut**, Switzerland

Table of Contents

List of Contributors

Richard V. Anthony, Principal, Richard Anthony Associates, 3891 Kendall St., San Diego 92109, CA, USA, tel:++1-858-2722905, fax:++1-858-2723709, e-mail: RicAnthony@aol.com, *contributions to chapters 3 and 7*

Prof. Dr. Michael Beckmann, Lehrstuhl für Verfahren und Umwelt, Bauhaus Universität Weimar, Coudraystr. 13c, D-99423 Weimar, Germany, tel:++49-3643-584675, fax:++49-3643-584679, e-mail: michael.beckmann@bauing.uni-weimar.de, *contributions to chapters 3 and 5*

Marcel A.J. van Berlo, Head of Process Management & Control Group, GDA / AVI Amsterdam, Australiëhavenweg 21, 1058 EN Amsterdam, The Netherlands, tel: ++31-20-5876282, fax:++31-20-5876200 e-mail: berlo@avi.amsterdam.nl, *contribution to chapter 5*

Prof. Dr.-Ing. habil. Bernd Bilitewski, Technische Universität Dresden, Institut für Abfallwirtschaft und Altlasten, Pratzschwitzer Str. 15, D-01796 Pirna, Germany, tel:++49-3501-530030, fax:++49-3501-530022, e-mail: abfall@rcs.urz.tu-dresden.de, *contribution to chapter 3*

Dr. Serge Biollaz, Head, Group of Thermal Process Engineering, Paul Scherrer Institut, CH-5232 Villigen PSI, tel:++41-56-3102923, fax:++41-56-3102199, e-mail: serge.biollaz@psi.ch, *contribution to chapter 5*

PD Dr. Helmut Brandl, Lecturer and Senior Researcher, University of Zurich, Institute of Environmental Sciences, Winterthurerstrasse 190, CH-8057 Zurich, Switzerland, tel:++41-1-6356125, fax:++41-1-6355711, e-mail: hbrandl@uwinst.unizh.ch, *contributions to chapters 2 and 4*

Prof. Dr. Rainer Bunge, Head of Institute, University of Applied Science Rapperswil, Institut for Applied Environmental Technology (Umtech), Oberseestr. 10, 8640 Rapperswil, Switzerland, tel:++41-55-2224862, fax:++41-55-2224861, e-mail: rainer.bunge@hsr.ch, *contribution to chapter 5*

Vicente Carabias, M. Sc., Lecturer, Zurich University of Applied Sciences Winterthur, Center for Social-Ecology, P.O. Box 805, CH-8401 Winterthur, Switzerland, tel:++41-52-2677674, fax:++41-52-2677473, e-mail:crb@zhwin.ch, *contribution to chapter 7*

Marc Chardonnens, Swiss Agency for the Environment, Forests and Landscape, Waste Management, CH-3003 Berne, Switzerland, tel: ++41-31-3226956, fax:++41-31-3230369, e-mail: marc.chardonnens@buwal.admin.ch, *contribution to chapter 5*

Gabor Doka, M. Sc., Consultant in LCA, Stationsstr. 32, CH-8003 Zurich, Switzerland, tel:++41-1-4631608, e-mail: doka@unite.ch, *contribution to chapter 6*

Dr. Göran C.B. Finnveden, Research Leader, Environmental Strategies Research Group (fms), Swedish Defence Research Laboratory, PO Box 2142, SE 103 14 Stockholm, Sweden, tel:++46-8-4023827, fax:++46-8-4023801, finnveden@fms.ecology.su.se, *contribution to chapter 6*

Dr. Stefanie Hellweg, Senior Research Associate, Swiss Federal Institute of Technology Zurich (ETH), Chemical Engineering Department, Safety and Environmental Technology Group, ETH Hoenggerberg, HCl, CH-8093 Zurich, Switzerland, tel:++41-1-6334337, fax:++41-1-6321189, e-mail: hellweg@tech.chem.ethz.ch, editor and *contribution to chapter 6*

Peter Hofer, dipl. Eng. ETH, Senior Consultant, GEO Partner AG Umwelt Management, Baumackerstrasse 24, 8050 Zürich, Switzerland, tel:++41-1-3112728, fax:++41-1-3112807, e-mail: hofer@geopartner.ch, *contribution to chapter 8*

Martin Horeni, dipl. Eng., Lehrstuhl für Verfahren und Umwelt, Bauhaus Universität Weimar, Coudraystr. 13c, D-99423 Weimar, Germany, tel:++49-3643-584669, fax:++49-3643-584679, e-mail: martin.horeni@bauing.uni-weimar.de, *contribution to chapter 5*

Prof. Dr. Konrad Hungerbühler, Head of Safety and Environmental Technology Group, Swiss Federal Institute of Technology Zurich (ETH), Chemical Engineering Department, ETH Hoenggerberg, HCl, CH-8093 Zurich, Switzerland tel: ++41-1-6326098 fax: +41 1 632 11 89, e-mail: hungerbuehler@tech.chem.ethz.ch, *contribution to chapter 6*

Dr. Frank Jacobs, Consultant, Technical Research and Consulting on Cement and Concrete (TFB), Lindenstrasse 10, 5103 Wildegg, Switzerland, tel:++41-62-8877332, fax:++41-62-887-7270, e-mail: jacobs@tfb.ch, *contribution to chapter 3*

Prof. Dr. Walter Joos, Vice-Director of the Department for Applied Languages and Cultural Sciences, Zurich University of Applied Sciences Winterthur, Center for Social-Ecology, P.O. Box 805, CH-8401 Winterthur, Switzerland, tel:++41-52-2677210, fax:++41-52-2677482, e-mail: joo@zhwin.ch, *contribution to chapter 7*

Dr. Hans Kastenholz, Head Interdepartmental Discourse Unit, Center for Technology Assessment in Baden-Würtenberg, Industriestr. 5, 70565 Stuttgart, Germany, tel:++49-711-9063166, fax:++49-711-9063175, e-mail: hans.kastenholz@ta-akademie.de, *contribution to chapter 7*

Dr. Charles Keller, Consultant, Inchema Consulting AG, Lavaterstrasse 66, CH-8002 Zürich, Switzerland, tel:++41-1-2017707, fax:++41-1-2017714, e-mail: inchema.consulting@switzerland.org, *contribution to chapter 5*

Dr. Christian Ludwig, Head, Group of Chemical Processes and Materials, Paul Scherrer Institut, General Energy Research Department, CH-5232 Villigen PSI, Switzerland, tel:++41-56-3102696, fax:++41-56-3102199, e-mail: christian.ludwig@psi.ch, editor and *contributions to chapters, 1,2,3,5, and 9*

Dr. Harald Lutz, Paul Scherrer Institut, General Energy Research Department, CH-5232 Villigen PSI, Switzerland, tel:++41-56-3102663, fax:++41-56-3102199 *contribution to chapter 5*

PD Dr. Urs Mäder, University of Berne, Institute of Geological Science, Baltzerstrasse 1, CH-3012 Berne, Switzerland, tel :++41-31-6314563, fax : ++41-31-6314843, e-mail: urs@geo.unibe.ch, *contribution to chapter 5*

Prof. Dr. Saburo Matsui, Kyoto University, Graduate School of Global Environmental Studies, Yoshida, Sakyoku, Kyoto City, Japan 606-8501, tel:++81-75-7535151, fax:++81-75-7533335, e-mail: matsui@eden.env.kyoto-u.ac.jp, *contribution to chapter 4*

Dr. Leo S. Morf, Senior Consultant, GEO Partner AG, Umweltmanagement, Baumackerstrasse 24, 8050 Zurich, tel:++-1-3112728, fax:++41-1-3112807, e-mail: morf@geopartner.ch, *contribution to chapter 8*

Dr. Didier Perret, head of the analytical lab, Ecole Polytechnique Fédérale de Lausanne (EPFL), Laboratoire de Pédologie, ENAC-ISTE-LPE CH-1015 Lausanne, Switzerland, tel: ++41-21-6933772, fax :++41-21-6935670, e-mail: didier.perret@epfl.ch, *contribution to chapter 5*

Sebastian Plickert, Research fellow, University of Potsdam, Working Party on Ecological Technology, Park Babelsberg 14, D-14482 Potsdam, Germany, tel:++49-331-9774445, fax:++49-331-9774410, e-mail:plickert@rz.uni-potsdam.de, *contribution to chapter 4*

Prof. Dr. Ortwin Renn, Chair of the Board of Directors, Center for Technology Assessment in Baden-Württemberg, Industriestr. 5, 70565 Stuttgart, Germany, tel:++49-711-9063166, fax:++49-711-9063175, e-mail: ortwin.renn@ta-akademie.de, *contribution to chapter 7*

Christopher Rootes, Reader in Political Sociology and Environmental Politics, Director, Centre for the Study of Social and Political Movements, School of Social Policy, Sociology and Social Research, Darwin College, University of Kent at Canterbury, Canterbury, Kent CT2 7NY, UK, tel:++44-1227-823374, fax:++44-1227-827005, e-mail: c.a.rootes@ukc.ac.uk, *contributions to chapter 7*

Dr. Shin-ichi Sakai, Director, Research Center for Material Cycles and Waste Management, National Institute for Environmental Studies, Onogawa 16-2, Tsukuba, Ibaraki 305-8506, Japan, tel:++81-298-502806, fax:++81-298-502808, e-mail: sakai@nies.go.jp, *contribution to chapter 5*

Dr. Kaarina Schenk, Swiss Agency for the Environment, Forests and Landscape, Waste Management, CH-3003 Berne, Switzerland, tel: ++41-31-3244603, fax: ++41-31-3230369, e-mail: kaarina.schenk@buwal.admin.ch, *contribution to chapter 5*

Dr. Konrad Schleiss, Consultant, Umwelt und Kompostberatung, Eschenweg 4, CH-6340 Baar, Switzerland, tel:++41-41-7612432, fax:++41-41-7612413, e-mail: k.schleiss@bluewin.ch, *contribution to chapter 4*

Elke Schneider, Research Associate, Interdepartmental Discourse Unit, Center for Technology Assessment in Baden-Württemberg, Industriestr. 5, 70565 Stuttgart, Germany, tel:++49-711-9063166, fax:++49-711-9063175, *contribution to chapter 7*

Prof. Dr. Reinhard Scholz, Technische Universität Clausthal, Institut für Energiever-fahrenstechnik und Brennstofftechnik, Agricolastraße 4, 38678 Clausthal-Zellerfeld, Germany, tel:++49-5323-722032, fax:++49-5323-723155, e-mail: scholz@ievb.tu-clausthal.de, *contributions to chapters 3 and 5*

Prof. Dr. Kai Sipilä, VTT Energy, New Energy technologies, Biologinkuja 5, Espoo, PO Box 1601, 02044 VTT, Finland, tel:++358-9-4565440, fax:++358-9-460493, e-mail: kai.sipila@vtt.fi, *contribution to chapter 5*

Dr. Konrad Soyez, Senior Researcher, University of Potsdam, Working Party on Ecological Technology, Park Babelsberg 14, D-14482, Germany, tel:++49-331-9774693, fax:++49-331-9774410, e-mail: soyez@rz.uni-potsdam.de, *contribution to chapter 4*

Dr. Peter Stille, ULP Ecole et Obervatoire des Sciences de la Terre CNRS, Centre de Géochimie de la Surface UMR7517, Cycles Géochimiques et Impacts Anthropiques, 1, rue Blessig, F-67084 Strasbourg Cedex, France, tel :++33-0390240434, fax : ++33-0388367235, *contribution to chapter 5*

Dr. Rudolf P.W.J. Struis, Senior Scientist, Paul Scherrer Institut, General Energy Research Department, CH-5232 Villigen PSI, Switzerland tel:++41-56-3104169, fax:++41-56-3102199, e-mail: rudolf.struis@psi.ch, *contribution to chapter 5*

Dr. Samuel Stucki, Head, Laboratory for Energy and Materials Cycles, Paul Scherrer Institut, General Energy Research Department, CH-5232 Villigen PSI, Switzerland, tel:++41-56-3104154, fax:++41-56-3102199, e-mail: samuel.stucki@psi.ch, editor and *contributions to chapters 1,2,5 and 9*

Dr. Frédéric Vogel, Senior Scientist, Paul Scherrer Institut, General Energy Research Department, CH-5232 Villigen PSI, Switzerland tel:++41-56-3102135, fax:++41-56-3102199, e-mail: frederic.vogel@psi.ch, *contribution to chapter 5*

Jörn Wandschneider, dipl. Eng., Director, W+G, Kattunbleiche 18, D-22041 Hamburg, Germany, tel:++49-40-51312900, fax:++49-40-51312903, e-mail: joern.wandschneider@t-online.nl, *contribution to chapter 5*

Herbert Winistörfer, M. Sc., Lecturer, Zurich University of Applied Sciences Winthertur, Center for Social-Ecology, PO Box 805, CH-8401 Winterthur, Switzerland, tel:++41-52-2677675, fax:++41-52-2677473, e-mail:win@zhwin.ch, *contribution to chapter 7*

Regula Winzeler, dipl. Eng. NDS FH, Senior Consultant, GEO Partner AG Umwelt Management, Dufourstrasse 5,. CH-4052 Basel, Switzerland, tel:++41-61-2066525, fax: ++41-61-2066599, e-mail: rw@jsag.ch, *contribution to chapter 8*

Jörg Wochele, Senior Scientist, Paul Scherrer Institut, General Energy Research Department, CH-5232 Villigen PSI, Switzerland, tel:++41-56-3102626, fax:++41-56-3102199, e-mail: joerg.wochele@psi.ch, *contributions to chapters 2 and 5*

Prof. Dr. Zhao Youcai, National Laboratory of Pollution Control and Resource Reuse, Tongji University, Shanghai 200092, China, tel: ++86-21-6592684, fax:++86-21-65980041, e-mail: zylmk@online.sh.cn, *contributions to chapters 2 and 8*

1 Introduction

Samuel Stucki and Christian Ludwig

Today, nearly half of the world's growing population lives in urban areas, causing enormous pressure on the local environment. Particularly in the large agglomerations of the developing countries, inadequate waste management is the cause of serious urban pollution and health hazards. Affluent industrialized economies are facing an ever-increasing load of wastes and declining landfill space to dispose of these materials. Sustainable management of waste with the overall goal of minimizing its impact on the environment in an economically and socially acceptable way is a challenge for the coming decades.

Different countries have adopted different strategies for reaching their goals, be it by applying advanced environmental technologies, by extending recycling and re-use, or by reducing the material intensity of their economy. Sustainable waste management will have to consider all possible options for the reduction of the negative impacts of consumption. Waste is a source of pollution as long as we don't learn how to better use the material and energy resources contained in waste: "Don't waste waste!"

1.1 The Problem with Waste

When a useful material good, such as a car or a newspaper or a computer reaches the end of its life cycle, it loses its economic value and turns into waste material. The car may turn into waste because its engine breaks down through wear and material failure, or because it crashes into a tree. Today's newspaper will turn into waste after it has been read and/or because the information it carries will be obsolete tomorrow. The computer might turn into waste if it breaks down, or, what is more likely nowadays, because its technology is outdated by the development of new and more powerful machines.

The transformation of a useful product into waste strongly depends on the function it has for the owner and its economic value. Waste is, so to say, in the eye of the beholder: the used car might still serve as a shelter for a homeless person, yesterday's newspaper might be useful for wrapping up some fish-and-chips and the outdated computer might still be sufficient for use in a primary school class. Waste is not just waste! It can be transformed back into a valuable material by a different user and be useful for another life cycle. When the car starts to get so rusty that it can no longer keep the rain out, it will be abandoned by the homeless person. When the fish-and-chips are eaten up, the newspaper will have turned into a greasy and soggy bit of paper that will be dropped on the pavement by the thoughtless consumer.

As it will take years for the newspaper to transform into topsoil and as it will soon become evident that the rusting car body pollutes the roadside with corrosion products and poisonous liquids, there will be citizens who will object. They will want to have the car removed from the side of the road and the newspaper picked up from the pavement because litter interferes with their esthetic preferences, or because they are concerned about the ecological consequences of potentially dangerous emissions.

At this stage waste clearly reaches the state in which it has a negative value, e.g. it will give rise to costs because it will have to be removed and treated so that it does not cause additional costs in the future. *Waste management* is about all the options society has to manage the transition of the value of goods and materials from positive to negative. Ideally, waste management will ultimately turn waste into a zero-value good (e.g. appropriately treated residues which can be left in a safe landfill for indefinite durations) or recycle it by transforming it physically and/or chemically so that it becomes valuable again as a raw material for new products.

From the above, we can conclude the following:

1. Waste management is inextricably linked to economy, as waste is defined by its relative economic value;
2. Waste management is likewise linked to ecology, as, left on its own waste is likely to affect the environment;
3. Waste management is a social issue, as waste is mainly a social construct (what is perceived as waste depends to a large part on life-style and social rank) and it raises the questions about the responsibilities of individuals towards society.

The short introduction should illustrate that all three aspects of sustainability (economic, environmental and social) have to be considered in waste management.

1.1.1 Economic Aspects of Waste

The economic problems of waste are due to the fact that, by definition, waste is material with no value. Classical economic mechanisms of supply and demand controlling the flow of goods, therefore, fail for waste materials. As a consequence, they tend to accumulate in the natural environment if no countermeasures are taken by authorities. Countermeasures include regulations prohibiting the uncontrolled disposal of waste and prescribing minimum standards for treatment and deposit. Controlled disposal or recycling involves costs, i.e. waste materials are assigned a negative economic value in the form of a disposal fee. In exchange for the disposal fee, the economic value "absence of pollution" is created. The central economic problem is allocating the costs for a clean environment to the stakeholders. Waste management used to be the responsibility of the public domain and financed by taxpayers, with little or no incentive for the consumer to diminish the rate of waste production. In order to create incentives for waste reduction, the "polluter pays" principle has been introduced and increasingly used.

Mining of ores and fossil fuels leads to the depletion of resources and to irreversible (at least in human, as opposed to geological time spans) dissipation of material (entropy increase). One would expect that the dissipation of concentrated mineral resources will eventually lead to increased economic value of the materials in question (increased prices for a given raw material) and thus to an efficient reduction in their use, or their substitution. So far, however, the economic value of most, if not all, mined raw materials has decreased over the past centuries and there is hardly any indication of a reverse trend. Although the concentrations of, for example, Cu in mined ores has continually decreased, the copper industry has been successfully dealing with this trend by employing new extraction technologies (and improving existing ones), even with declining prices. While the primary world production of copper increased from less than 8.8 million tons to more than 13.3 million tons between 1976 and 1996, the average price (corrected by the consumer price index) dropped by 50% within the same period [7]. The percentage of recycling to total copper consumption has leveled close to 40% in the same time span. Of this 40%, about half has been recycling of manufacturing waste. This means that about 20% of the total copper consumption is recovered from post-consumer waste. It is quite clear that this share of recycling will not be increased as long as prices from primary production sources keep declining. Clearly, the motivation for tapping municipal solid waste (MSW) as a source of mineral raw materials is not given at the time being. The currently proved reserves of copper will last for another 40 years, the resources for another 105 years [7]. Imposing taxes on non-renewable resources to encourage more efficient management of such resources or their substitution has been proposed (see Chapter 8).

1.1.2 Ecological Aspects of Waste

Industrial civilization relies on materials that are mined from enriched deposits (ores and fossil fuels) and are hence very different in composition from the resources that natural systems (i.e. the biosphere) draw on. Producing, using and discarding these materials will ultimately lead to their dissipation and accumulation in specific compartments of the environment with potentially harmful consequences. The waste resulting from the high carbon intensity of our industrial civilization has led to the well-known surge in CO_2 concentrations in the atmosphere with potentially disastrous consequences for the world climate. Metallurgical activities of humans have left footprints in the global environment as well: the analysis of Greenland ice cores has revealed that human activities have resulted in enhanced concentrations of, for example, Pb in the precipitation already more than 2000 years ago, with the spread of Graeco-Roman metallurgy exploiting lead containing silver ores [9].

Our industrialized civilization clearly leaves chemical footprints in the environment, i.e. it changes the chemical composition of some of the sensitive ecological compartments (such as air, water, soil, etc.). Whether or not an increase in the concentration of a given compound or chemical element in an environmental compartment is ecologically harmful or not, depends very much on the activity it exhibits with living systems. Many of the trace deposits of heavy metals found in the different environmental compartments are detectable but need not be harmful; others clearly have an impact globally or locally due to their quantities and or quality. The assessment of contemporary ecological impact and, in the long-term, of the irreversible dissipation of materials to the environment is the subject of Chapter 6.

Materials (compounds, elements), which are not found in nature, can cause particularly harmful effects on ecosystems and human health. Such xenobiotic materials can be damaging at very low concentrations. Examples from domains of chemistry, physics and biology include the following:

- Xenobiotic chemicals: persistent chemical compounds entering the food chain, such as the infamous pesticide DDT turning up in unexpected places and concentrations – or the release of very stable chloro-fluoro-carbons (CFCs) leading to the destruction of the stratospheric ozone layer in polar regions.
- Radio-nucleids: radioactive isotopes formed by artificially induced nuclear reactions (atomic bombs, reprocessing of nuclear fuels, etc.) with unknown and/or difficult to assess hazards to the environment and human health.
- Bio-hazardous material: there is increasing concern about the release of traces of bio-active substances, such as hormones and/or antibiotics, mainly via the water cycle. These substances can be biologically active in very low concentrations. The unknown risk associated with the release of transgenic organisms is the most recent example of potential dangers arising from the interaction of the "artificial" with the "natural" environment.

From an environmental perspective, sustainable development means that the chemical footprints of civilization should be reduced to a level where they are not

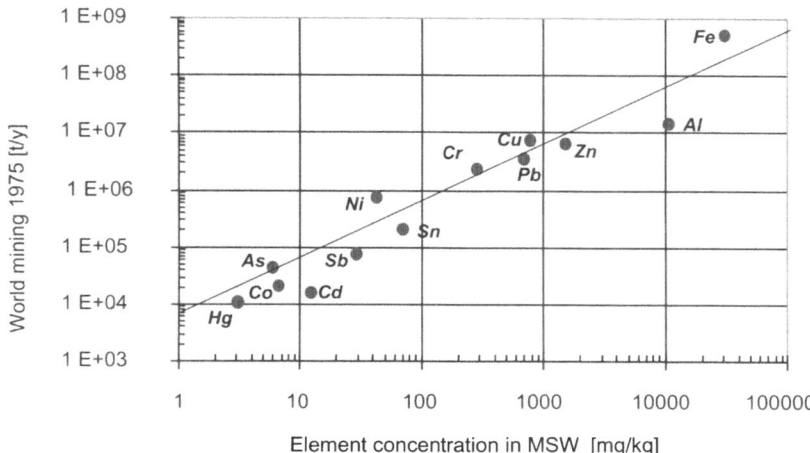

Fig. 1.1. World mining productivity plotted vs. the average concentration of elements in MSW. Concentrations reveal a correlation to the production rates for a number of metals

harmful for future generations. The elemental composition of average solid discards from households and local small business (MSW) differs from the background composition ("earth crust") by an order of magnitude for a number of heavy metals. Clearly, the solid residues of MSW, if discarded, will show an environmental footprint, the extension of which depends on the rate of dissipation into sensitive compartments such as aquifers. A number of chapters and case studies in this book focus on the recovery of metals from MSW treatment plants, with emphasis on avoiding any long-term pollution of aquifers by toxic metals.

The elemental composition of MSW reflects the spectrum of products that enter the consumption chain. This is illustrated in Fig. 1.1, which shows the correlation between the production rate of mining products and the mean concentration of elements in MSW. This means that, at least for a number of elements, MSW is a representative sample of what is being consumed. In order to devise efficient means of guiding the flow of materials and/or avoiding hazardous pollution to the environment, the flows of materials need to be analyzed in the entire system. Material flow analysis, i.e. methods used to establish the material balances of the inputs and outputs of human settlements ("the anthroposphere"), has contributed to our understanding of the material flows and material accumulations in human-made systems [1]. This analysis is essential in assessing the relative importance of the material flows and the suitability of waste management options available for influencing these flows. For an example of such an analysis, see section 8.3.2.

1.1.3 The Social Problem of Waste

The minimum requirement that solid waste management has to fulfill is the removal of solid discards from the immediate vicinity of settlements. This task is being solved in more-or-less efficient ways in most societies. As soon as waste is out of sight it is no longer perceived as a problem. This attitude becomes more of a problem, the better the consumers are shielded from the consequences of their consumption. According to [10], the transition to consumer society after World War II has been connected with a high degree of division of labor and an increasing separation of production and consumption. "The price for this separation has been a loss of immediate perception and control of the individual over the material and energy fluxes associated with his or her activities, and a growing difficulty to keep track of what is going on in the world." [10]. The relative prices for the raw materials and energy have come down tremendously during the same period. Consumption has become more and more effortless as a consequence. In order to provide unspoilt consumption pleasure, the chores of dealing with the resulting wastes have also been taken away from the consumer. Efficient waste management systems which deal with the material output at the end of the pipe are in place: waste water is dealt with from the moment you push the flush button; solid waste is removed efficiently from the side of the street by public waste collection workers, and the further processing of this material has little or no implications for our daily lives.

Therefore, as long as waste management operations do not lead to emissions which immediately interfere with the pleasures of consumer life we are enjoying, the consequences of the highly increased material output of our consumer society are not generally perceived. The better the management of wastes, the less the public is aware of our high-throughput economy. The vision of a zero-waste society can therefore only become a reality if the immediate material flux consequences of consumerism can be made transparent or tangible again.

Current waste management practices are mostly based on centralized operations: waste is collected and brought to a landfill or a central processing plant (e.g. incinerator). Social acceptance problems arise mainly from local opposition to the erection of such plants because of real or feared negative impacts, such as noise, odor, air pollution, traffic (see Ch 7).

1.2 History of Waste Management

Waste has been an issue of public concern ever since humans started to live in towns, i.e. in an area which was smaller than the land area needed to sustain their food production. Most European countries introduced legal regulations regarding waste management only in the second half of the 19th century [3]. It is no coincidence that this time coincides with the population explosion in central Europe of the same century, which led to the growth of big cities. It was also during this time that growing scientific evidence showed that epidemics related to overpopulation, most prominently cholera (e.g. the 1832 epidemic of London), were related to hygienic problems. Waste management was, therefore, first of all a hygienic issue: preventing the spread of diseases through rotting waste.

In industrialized countries the decades after the end of the 2nd World War were characterized by a sharp increase in the per capita consumption of energy and material resources. Between 1950 and 1990, as an example for a European country, the consumption of fossil fuels in Switzerland rose by a factor of 5, the production of MSW by a factor of 4.3, and the number of cars by a factor of 19 – with a population growth of 44% for the same period [10]. By the end of the 1970s, this unprecedented out-of-control surge in material throughput had fuelled fears that we may not be able to master the associated waste avalanche. The publication of the "Limits of Growth" [12] by the Club of Rome in 1975 made clear that the resources the consumer age depended on were not unlimited and that therefore waste management had to change its focus from "efficient removal" to waste avoidance, minimization, and recycling as options with higher priority. New guidelines had to be developed to manage the material flows of consumer society. For example, in Switzerland the Swiss Guidelines for Waste Management were published in 1985 [5], as a highly visionary set of guidelines, which have since influenced the Swiss waste management policies (see Box). To enforce these policies, environmental legislation and specific regulations regarding waste management were drawn up as a result. In European countries, most of the actual waste management regulations came into force in the early eighties [4]. Different countries have adopted different strategies for reaching their goals of reducing the ecological and economic burden of wastes.

Summary of Swiss Guidelines for Waste management

Scientific and technical guidelines:

- Waste disposal systems should generate materials which can be recycled or deposited in a final disposal site.
- Hazardous substances must be concentrated, not diluted.
- Organic substances are not compatible with final disposal sites.

Political guidelines:

- Waste management is guided by the objectives of the environmental protection laws.
- Waste disposal systems must be environmentally compatible.
- Waste should be disposed of within Switzerland.
- Regional responsibility for planning of landfill sites is applicable.
- Public authorities play a subsidiary role in waste management.

Economic guidelines:

- Public authorities should not subsidize waste disposal systems.
- The *polluter pays* principle is to be adhered to.

> • Waste should be recycled if the result is less environmental pollution than disposal and production from virgin materials. Recycling must be profitable.

1.3 Directing Material Flows

Waste is an inherent result of economic activity, as it is of any metabolic system. The ultimate goal of waste management is the absence of waste, i.e. to get rid of it, to use it as a resource, or not to have it in the first place. Preventing the production of waste materials is not usually considered a part of waste management in its strict sense. The term 'Integrated Waste Management' (IWM) has been coined to include front-end measures such as design for recycling, exclusion of problematic materials in products, etc. as integral parts of waste management (see Ch. 8).

Waste management in the narrower sense directs the flows of materials so that their impact on the environment, the depletion of resources, and the resulting costs are minimized. Various strategies to achieve these conflicting goals have been adopted by different countries. Strategies include technical fixes 'at the end of the pipe' as well as recycling schemes and prevention of waste at the source, i.e. on the drawing boards of production. According to Buclet and Godard [4], three main "myths" underlie most of the strategies adopted in Europe, and probably in developed countries all over the world. These principles are referred to as "myths" because, implemented individually, they will not work.

1. **The myth of a dematerialized post-modern society** maintains that the material intensity of the consumer society can be drastically reduced so that fewer material goods end up in the waste management cycle. Consumption shifts from using and discarding material goods to the consumption of immaterial "services."

   While it is true that an increasing share of the economic throughput today is made with immaterial goods, the overall material throughput itself has not declined, only its relative value (i.e. the contribution of material products to the GDP). The much praised substitution of the flows of material goods by flows of information ("moving bytes instead of tons") [14] has thus far not been successful. Information and communication technologies have not done away with travel – mobility demands are steadily increasing. The computer has not brought us the paper-less office; it has rather enabled us to increase our printer outputs tremendously. It is, however, undisputed that the material efficiency (and energy intensity) of the economy could be increased by factors [15].

2. **The myth of perpetual material cycles** maintains that the material cycles of the "artificial" human civilization can be closed completely, as can the materials cycles in "natural" ecosystems. Any waste material can be reused or recycled. Dissipation of materials from the "artificial" to the "natural" can be neglected and has therefore no environmental impact ("zero waste" vision).

The zero waste vision has been inspired by the fact that, apparently, in natural ecosystems there is no waste, i.e. all refuse is recycled by specialized organisms in an ecosystem. Industrial ecology has been dreamed up as an artificial equivalent to natural systems [8]. However, no ecosystem can persist without a throughput of materials and energy drawn from and discarded to sufficiently large reservoirs. Complete recycling is not possible, as any material cycle will include irreversible losses (entropy production) which cannot be prevented without excessive energy input. Energy use and materials recycling ultimately need to be put into their right balance in order to optimize the life cycle performance of products.

3. **The myth of mastery and containment** maintains that waste can be treated and deposited safely; that landfilling can be mastered so that any spilling of pollutants into the environment can be prevented in any timescale, or is environmentally irrelevant.

   Potential leakage from a landfill can be minimized by adopting appropriate technology of treatment and landfill construction. However, no one can guarantee the long-term safety of landfills with respect to emissions. Incinerators greatly contribute to reducing the risk of groundwater contamination by landfills; nevertheless, the products of incineration still contain toxic elements unless subjected to further treatment. Purely technical solutions to the waste problem easily run into acceptance problems with respect to costs and operational risks ("not in my back yard!").

The "three myths" discussed above aim at reducing the flow of materials on different levels in the life cycle of goods (see Fig. 1.2): prevention at the front end; recycling at the end of life of a useful product, before disposal; and finally end-of-pipe treatment and safe disposal (avoiding any uncontrolled flows to the environment). In integrated waste management (IWM), one assumes that there is not simply one unique way of dealing with material flows in a sustainable way, but that there should be control mechanisms on all levels. Realizing that any material flows and conversion processes involve potential emissions to the environment (see horizontal arrows in Fig. 1.2), it is clear that control measures further upstream are more efficient in preventing environmental damage and in saving resources. Avoided waste is the cheapest, the most efficient and most effective method of 'disposal'! This has lead to the hierarchy of integrated resource management options that has been adopted as a guideline in many countries: prevention > reduction > recycling > disposal. A pragmatic way of setting priorities is to minimize the total impact of waste on the environment, i.e. the impacts on water, air, soil have to be minimized as well as the consumption of energy, materials and landfill space.

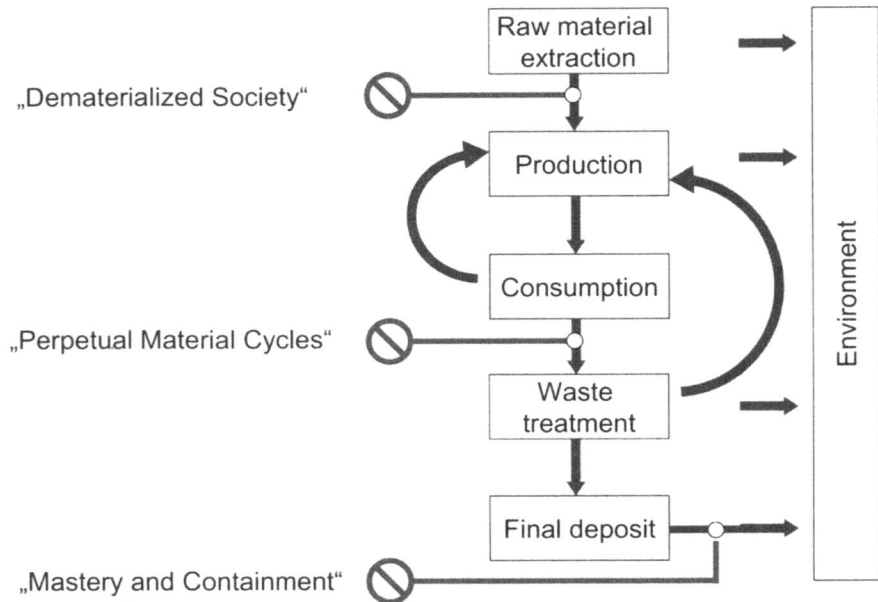

Fig. 1.2. From the cradle to the grave: product life cycles and the "myths" of controlling material flows

1.3.1 Preventing and Reducing Waste

Waste can only be prevented at the front end of the material cycles by changing the way goods are produced and consumed. The rapid rise in material and energy throughput of western economies is clearly not sustainable ("… it is simply impossible for the world as a whole to sustain a Western level of consumption for all. In fact, if 7 billion people were to consume as much energy and resources as we do in the West today we would need 10 worlds, not one, to satisfy all our needs" [2]). This means that reducing the throughput for a given set of services is a primary goal of sustainable production and consumption. Realizing that the ecological efficiency of the way goods are produced and consumed today is far below what it could be, some leading industrialists have come to the conclusion that the efficiency of material usage has to be improved along the life cycle of products ("doing more with less"). 'Eco-efficiency', a way of balancing ecological and economic efficiency, has been coined by the World Business Council for Sustainable Development as a strategy for advancing production towards sustainable goals [13]. The eco-efficiency strategy has tried to achieve an economically competitive advantage by reducing the material intensity of products and production processes, which, at the same time, lead to an overall reduction of waste production. The ideas of Schmidheini and his colleagues have been successfully implemented by a broad range of companies and have contributed to a general "greening" of industry over the past 10 years, although they have been criticized as doing little more than slowing down the current trends and perpetuating the paradigm of consumerism [11]. Clearly, new concepts for the development of products and services need to be discussed and put to work if waste prevention is to become more than just another myth. While this is essential, a thorough discussion of sustainable production and consumption is beyond the scope of this book. Important strategies that can be adopted in this area are the subject of a recent book [6].

1.3.2 The Great R's: Re-use, Recycle, Recover

Recycling is the second level in the hierarchy of waste management options. The myth of closed material cycles relies on using all waste as a raw material for production. Waste can be used as a resource for a great number of materials. Separate collection schemes have been very successful in enabling cost effective recycling of materials such as paper and cardboard, glass, aluminum, etc. (see chapter 3.). Recycling can be done at different levels of material complexity:

- Re-using products or parts of products for the same application (e.g. re-using parts of a machine after disassembly), or for a different application (e.g. using a newspaper as packing material);
- Recycling of complex materials (e.g. producing recycled paper from waste paper, recycled plastic from waste plastic);
- Recovering raw materials and energy from waste (e.g. metals from scrap, monomers from polymers, heat from combustibles).

The higher the level of recycling, the lower the technical complexity of the recycling process usually is and the lower the required energy input. At the same time, the recycling of products and of complex materials leads to the deterioration of the resulting recycled product with each recycling loop. This phenomenon, which limits the number of potential recycle loops, has been referred to as "downcycling," which means that the quality of the secondary material or product is usually inferior.

The situation is different in the case of raw material recovering, which can be done in a way that does not impair the product quality, because the material is broken down to the level of its building blocks (i.e. metals or metal salts from inorganic residues, or synthesis gas from organic material). The recovery of raw materials usually involves thermal processing steps (smelting and metallurgical processing; gasification; purification steps), which are intrinsically more energy intensive. The energy input economically limits the degree to which raw materials can be recovered.

The Swiss Guidelines for Waste Management [5] state that recycling should be favored only if the recycling process leads to less environmental damage overall than production from virgin raw materials. This means that decisions regarding recycling should be based on a full life cycle analysis of the underlying processes, including energy consumption.

1.3.3 Recovering Materials and Energy from Waste

However much waste prevention there is at the front end, and however many improvements in the cycles of consumption and production, waste managers will never become redundant: At the end of the cycle there always remains a material flow which needs to be taken care of. In its classical sense, waste management is about dealing with those materials and, accordingly, the main part of the present book deals with end-of-pipe technologies and their implications.

MSW contains high fractions of organic compounds, which, in a thermodynamic sense, are not in equilibrium. The organic matter in MSW will eventually get transformed in the landfill by biological and chemical processes. The rate of these transformations depends on conditions in the landfill. The chemistry of these processes is very unpredictable and so are potential emissions resulting from such a large chemical reactor (2.2). MSW landfills have been recognized as potential hazards to aquifers and therefore need long-term monitoring and emissions management (treatment of leachate water and collection and use of landfill gas).

Incineration of MSW gets rid of the organic carbon in MSW (via the flue gas, in the form of CO_2) and produces solid residual materials (bottom ash and fly ash), the chemistry of which is more predictable. The environmental performance of early incinerators was highly questionable, as untreated flue gases contain pollutants (SO_2, NOx, dust, heavy metals, dioxins, ...) which might be more damaging than the potential emissions resulting from the slow decay of the same mix of materials in a landfill. This particular drawback of incineration has, however, been overcome by applying highly sophisticated filtering technology. Modern air pollution control technology removes air pollutants of waste-to-energy plants to levels that are comparable with equivalent power plants fuelled by fossil fuels (see Ch. 6). The age of environmental technology, starting with the eighties, has solved the negative impact of incinerators on the local environment, and a modern MSW incinerator can be regarded as a sink, not a source of pollutants (see chapters 3, 5).

The inorganic solid residues from incinerators and from flue gas treatment equipment (fly ash, scrubber residues) are a mixture of materials with a composition which differs from average rock materials ("earth crust") for a number of elements, notably for the heavy metals. Much of recent research and development has tried to separate minority components, such as heavy metals, from ashes by thermal, mechanical or even biological treatment, or to optimize the incineration processes with respect to its separation efficiency. The goal of these new developments has been to produce residues which can either be used as secondary raw materials, or deposited safely in a long-term landfill. The processes usually involve one stage at temperatures above the melting point of the ashes and produce glassy residues with much improved leaching stability. The major competitors in the environmental technology business have developed new processes with greatly improved residue quality over the past 10 years. They are being introduced to the market in Japan (5.3).

1.4 Conclusions

Waste is an inherent product of economic activity. The surge in productivity and connected consumption in the second half of the past century has led to a massive increase in material flows in the anthroposphere, creating environmental problems of sinks and (re-)sources. Solid waste management is relevant in directing the flow of some of the materials, especially heavy metals as essential ingredients of modern civilization, and as persistent toxic elements. There are various strategies to control the flow of materials through the economy. The overall goal of IWM must be the reduction of the total environmental impact resulting from waste, its handling recycling, treatment or final disposal. Reducing the input to the waste management system by increasing the material efficiency of the economy is the preferred option in the long term. Recycling of materials can substantially contribute to minimize the amount of material, which needs to be deposited. At the end of the pipe there is, and there will be no alternative to treating and depositing the remains. Technologies for doing this in a way that does not charge the burdens to our children and grandchildren are available. They take a good fraction of the following chapters.

References

1. Baccini P, Brunner P (1991) Metabolism of the Anthroposphere. Springer, Berlin Heidelberg New York
2. Brundtland GH (1994) The Challenge of Sustainable Production and Consumption Patterns. Symposium on Sustainable Consumption, Oslo
3. Buclet N, Godard O (2000) Municipal Waste Management in Europe. Kluwer Academic Publishers, Dordrecht Boston London
4. Buclet N, Godard O (2000) Municipal Waste Management in Europe: A Comparison of National Regimes. In: Buclet N. G, O. (ed) Municipal Waste Management in Europe. Kluwer, Dordrecht, pp 203-224
5. BUWAL SAftEFaLS (1986) Guidelines on Swiss Waste Management. Schriftenreihe Umweltschutz 51
6. Charter M, Tischner U (2001) Sustainable Solutions. Greenleaf, Sheffield
7. Dzioubinski O, Chipman R (1999) Trends in Consumption and Production: Selected Minerals. DESA Discussion Paper ST/ESA/1999/DP.5. United Nations
8. Graedel TE, Allenby BR (1995) Industrial Ecology. Prentice Hall, Englewood Cliffs, NJ
9. Hong S, Candelone JP, Boutron CF (1994) Greenland Ice History of the Pollution of the Atmosphere of the Northern Hemisphere for Lead during the last three Millennia. Analysis 22:38-40
10. Kaufmann R (2001) Increasing Private Consumption, or, Why "the King" Lost Control. In: Sitter-Liver B, Baechler G, Berlinger-Staub A (eds) Supporting Life on Earth. Council of the Swiss Scientific Academies, Bern, pp 27-31
11. McDonough W, Braungart M (2001) The Next Industrial Revolution. In: Charter MT, U. (ed) Sustainable Solutions. Greenleaf, Sheffield, pp 139-150

12. Meadows DH, Meadows DL (1972) The Limits to Growth: a Report for the Project on the Predicament of Mankind. Universe Books, New York
13. Schmidheini S (1992) Changing Course: A Global Perspective on Development and the Environment. MIT Press, Cambridge MA
14. Stahel WR (2001) Sustainability and Services. In: Charter MT, U. (ed) Sustainable Solutions. Greenleaf, Sheffield
15. von Weizsäcker E-U, Lovins AB, Lovins LH (1997) Factor Four:: Doubling Wealth, Halving Resource Use. Earthscan, London

2 Waste Disposal: What are the Impacts?

Contributions by Samuel Stucki, Jörg Wochele, Christian Ludwig, Helmut Brandl, and Zhao Youcai

Take 500 g of wet biomass, add 1200g of combustibles (paper, plastics, wood and textiles) and some 140g of minerals, season with some salts and top with 170 g of metals and you get the mix of roughly 2 kg of municipal solid waste (MSW) that an average American produces per day. The recipe is different in different parts of the world: If you reduce the amount of combustibles and metals in the American waste by roughly a factor of 10 you will end up with the composition of MSW in a Chinese City (section 3.5). MSW contents and quantities are a mirror of the material turnover of a society and reflect the consumption habits of the population. There is a clear correlation between the Gross Domestic Product of a country (GDP) and the amounts of waste it produces (section 2.1). Waste management in China and in particular in Shanghai faces the problems that are typical of an economy in transition (section 2.3). The development in Shanghai confronts the local authorities with the difficult task to adapt existing and to invest into new infrastructure in order to cope with the rapidly changing quantities and qualities of MSW.

Common to MSW of any origin is that it contains high proportions of organic compounds that are more or less easily bio-degradable. Normal practice throughout the world is to pile up the above cocktail of wastes in more or less organized landfills, or to just dump it wherever suitable. A MSW landfill is an uncontrolled bio-chemical reactor. The number of chemicals found to be released by landfills to the atmosphere and/or to the hydrosphere is huge (section 2.3). The ecological consequences of these emissions have local as well as global character. Emissions of polluted water from landfills to soil, surface and ground water are local, but can persist for centuries. With the potent greenhouse gas methane as the main component, gaseous emissions from landfills have a strong impact on a global scale. Although modern landfills attempt to collect, clean and use the methane resulting from anaerobic fermentation as a fuel, in most cases it is released to the atmosphere. Of the total global emissions of methane, estimated in 1999 at 535 million tons annually, 375 million tons are the immediate result of human activities, and 18% of those come from waste disposal. Methane emissions from landfills can be avoided if MSW is incinerated.

Figure 2.2 in section 2.1 shows that the composition of MSW deviates considerably from the composition of the geological formations it is discharged to ("average earth crust") for a number of elements. Next to carbon, chlorine and sulfur, associated mainly with food and vegetable wastes, the heavy metals, notably Zn, Cu, Cd, Pb, differ by one to two orders of magnitude from background. The release of these materials to the environment cannot be prevented unless they are efficiently separated and recycled (chapters 3 to 5). Even prohibition of certain toxic substances will have only an effect on the MSW composition on the mid- to long-term scale, because society has built up huge reservoirs of toxic objects, which will eventually become waste.

2.1. The Diversity of Municipal Solid Waste (MSW)

Samuel Stucki, Christian Ludwig, and Jörg Wochele

Municipal solid waste (MSW) includes the solids discarded by the end consumers, i.e. private households, small business and public areas, and typically collected by public authorities for disposal. Normally, separately collected waste for recycling, such as paper, metals, aluminium, glass, etc. is included in the MSW quantities given. MSW refers specifically to that part of MSW which is sent to landfill, incineration, or other final treatment [6]. MSW is only a relatively small fraction of all the solid waste that is generated in an advanced economy. According to the OECD Environmental Outlook [5] the total solid waste generated in the OECD countries reached 4 billion tons in 1997, of which 14% were classified as MSW. Table 2.1 shows that the major sources of solid wastes are in primary production (agriculture, forestry and mining) and in manufacturing.

Most of the waste streams generated in primary production are dealt with locally on the site where they are generated (e.g. agricultural and forestry wastes are generally used as fertilizers or as fuels, most other production waste is being disposed of locally or recycled).

Table 2.1. Percent share of solid waste in OECD countries; Total amount: 4 billion tons in 1997

Manufacturing	25
Agriculture & Forestry	21
Mining & Quarrying	14
MSW	14
Construction & Demolition	14
Energy Production	4
Water Purification	2
Others	6

The present book is dealing with post-consumer waste, its prevention, treatment and disposal, i.e. specifically with the 14% of MSW and some of the wastes arising from building sites (Construction and Demolition).

2.1.1. Quantities of MSW Collected

MSW production in developed economies has grown continually, very much in line with economic growth. MSW production has increased by 40% between 1980 and 2000, matching very nearly the increase in Gross Domestic Product (GDP) (50 %) over the same time span [5], and illustrating the fact that so far the increase in prosperity has been linked with an increase in material throughput. Annual MSW production has reached an average of 500 kg/cap in OECD countries. Figure 2.1 shows that there are marked differences in the specific per capita waste production of different countries. The correlation with GDP in the same countries confirms the strong link between affluence and MSW quantities. A similar correlation is also seen in a comparison of waste quantities with GDP for different cities in China (section 3.5.1).

The collection and assessment of MSW data in developing countries is much more difficult, as rural areas of these regions are hardly connected to an organized waste management infrastructure. Even in the big cities of the developing world, especially in Asia, only a fraction of the population is connected to regular waste collection services. Much of the waste there is dealt with informally, i.e. it is dumped in an uncontrolled way, and/or recycled very efficiently by scavengers and waste pickers. Table 2.2, taken from data published in World Resources 1996-97, shows the MSW generation for a number of Indian cities, together with an estimated collection efficiency (% waste collected). A detailed analysis of the evolution of waste quantities and compositions in China, and in particular in the booming city of Shanghai is given in section 2.3 of this book.

As mentioned above, the amounts of waste are expected to rise further with increased economic development and very likely a near 1:1 correlation of the increases of MSW and GDP has to be expected. In fact, a growth of 50% MSW production is expected in the period 1995-2020 for OECD countries, and of 100% in non OECD countries. Some of this growth is expected to be offset by more efficient recycling [5].

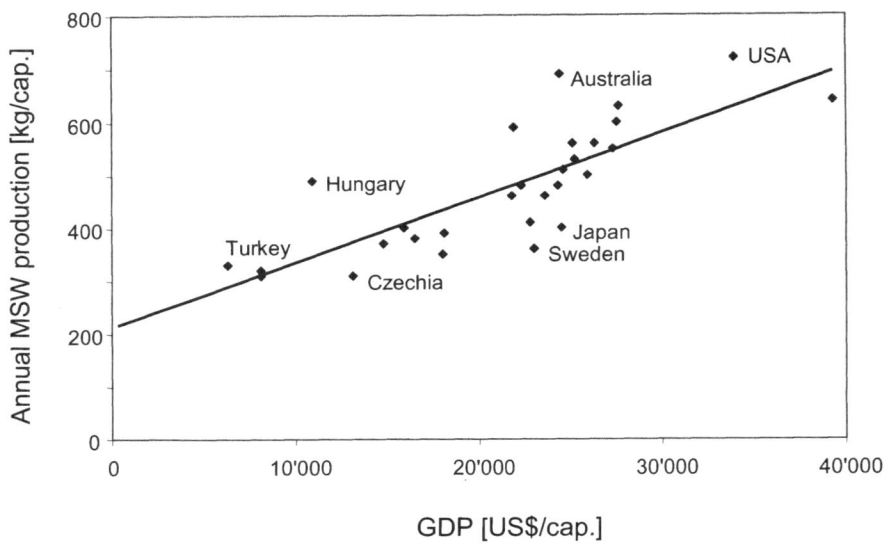

Fig. 2.1. Production of per capita MSW in OECD countries as a function of per capita Gross Domestic Product (GDP) [4].

Table 2.2. Waste collected in selected cities in India [35]

	MSW [kg/cap]	% collected
Bombay (India)	200	90
Delhi (India)	440	77
Bhiwandi (India)	100	40

2.1.2. The Composition of MSW

The composition of MSW is a mirror of consumption patterns, eating habits, social structure etc. of the societies producing the waste. Whereas in low income areas the main components of MSW are readily bio-degradable (food waste), this fraction is strongly reduced in highly developed cities. In section 2.3 it is shown that, in some of the less developed Chinese cities, food waste and coal ashes account for over 80% of MSW, whereas a significant increase of plastics and paper is seen in the cities of Shanghai and Beijing, which are rapidly developing towards western consumption standards. This trend is also reflected by the data given for the USA (Table 2.3). The shares of paper and especially of plastics increased over the period of 30 years between 1960 and 1990 in the USA, while, over the same period, the percentage od food waste declined drastically, although, in absolute numbers, it stayed constant.

Table 2.3. Development of average composition of MSW in USA (1960 to 1990). In million tons per year [19], and in %.

	Paper	Glass	Metals	Plastics, Rubber, Leather	Textiles	Wood	Food Waste	Garden waste	Total MSW
1960	29.90	6.70	10.50	2.40	1.70	3.00	12.20	20.00	86.4
1970	44.20	12.70	14.10	6.30	2.00	4.00	12.80	23.20	119.3
1980	54.70	15.00	14.50	11.20	2.60	6.70	13.20	27.50	145.4
1990	73.30	13.20	16.20	20.80	5.60	12.30	13.20	35.00	189.6
in %									
1960	34.61	7.75	12.15	2.78	1.97	3.47	14.12	23.15	100
1970	37.05	10.65	11.82	5.28	1.68	3.35	10.73	19.45	100
1980	37.62	10.32	9.97	7.70	1.79	4.61	9.08	18.91	100
1990	38.66	6.96	8.54	10.97	2.95	6.49	6.96	18.46	100

Table 2.4. Composition of MSW to incineration (separately collected fractions not included) in Switzerland (1992/93) [11]

	[%]
Paper and cardboard	28
Vegetable matter	23
Plastics	14
Mineral materials	6
Wood, leather, bones,…	5
Composite materials	8
Composite packaging	3
Metals	3
Glass	3
Textiles	3
Fines (< 8 mm)	4

Table 2.4 shows the average MSW compositions as sent to incineration in Switzerland [11]. The numbers do not include separately collected recyclable fractions such as paper, glass, metals etc. for which a separate collection system has been introduced. The average composition of MSW delivered to incineration, of incinerator ashes, and of the earth crust are given in Table 2.5. The data for the elemental composition of waste is taken from a representative study carried out in Switzerland in1993 [8].

The composition is typical for waste from an affluent society. Figure 2.2 shows the relative concentrations of elements, normalized with the average earth crust composition as reference. It is striking to see that significant deviations of concentrations in MSW from average background concentrations exist for C, Cl and S and for heavy metals, notably Pb, Zn, Cu, Cd, Hg.

Table 2.5. Average composition of average dry MSW, incinerator bottom ash, air pollution control (APC) residues [8] and average earth crust (mg/kg) [28]

	MSW (CH) (Belevi)	Bottom ash (Belevi)	APC residues (Belevi)	Earth crust (Reimann)
C	334000	20000		200
H	40000			
O	257000			
N	3120			
S	1120	2000	30000	260
Cl	6870	4000	259000	130
P	890	3000	3600	1100
Si	48500	190000	53000	280000
Fe	30000	100000	8000	56000
Ca	14000	120000	184000	41000
Al	12400	50000	48000	82000
Na	5140	25000	40000	24000
Mg	3380	16000	8200	23000
K	2060	10000	29000	21000
Zn	1310	3000	22000	70
Pb	500	1600	6900	13
Cu	1200	2000	690	60
Cd	12	7	360	0.20
Hg	2	0.04	130	0.08
Cr	315	1100	760	100
Ni	107	190	50	80
Co	2	16	12	

A large fraction of MSW is actually organic material and water, which will eventually disappear, be it by fast mineralization in an incinerator, be it by s low mineralization in a landfill. The inorganic fractions, notably the ashes left over after incineration, contain concentrations of heavy metals which exceed background concentrations by two to three orders of magnitude. Clearly, the deposition of untreated MSW in a landfill causes the enrichment of potentially harmful substances in the landfill, which will eventually lead to emissions to water, soil or air. The emissions resulting from the deposition of untreated MSW are described in the following section.

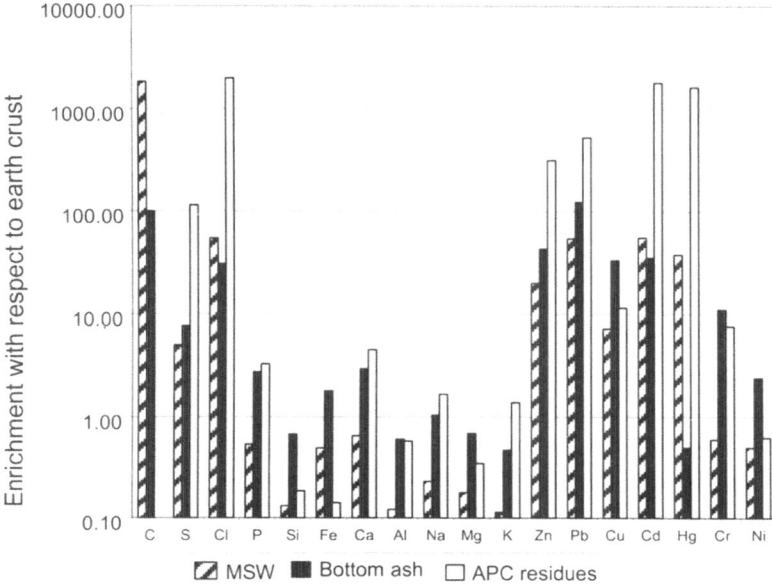

Fig. 2.2. Enrichment factors of chemical elements in MSW, incinerator bottom ash and air polution control (APC) residues, relative to the average composition of the earth crust, drawn from data in Table 2.5. Factor 1 = no enrichment.

2.2 Emissions from Municipal Solid Waste Landfills

Helmut Brandl

2.2.1 Emission to the Atmosphere

In principal, municipal solid waste (MSW) landfills represent anoxic environments with oxic conditions and aerobic processes occurring only at the surface [17]. MSW contains usually a large fraction of organic matter which can be metabolized by microorganisms. During biological mineralization of the organic material, electrons originating from the degradation have to be transferred to a terminal electron acceptor. Generally, electron acceptors have to be present in a sufficiently high concentration and thermodynamically favorable in such a way as to enable the microorganisms to gain energy for growth from the process. The development of a landfill system from an oxidizing to a reducing state is characterized by a typical series of electron acceptors which are microbially utilized in a typical sequential order (redox sequence): oxygen, nitrate, manganese, iron, sulfate, and carbon dioxide [31]. Under oxic conditions, oxygen offers the highest energy yield. After its depletion, anoxic conditions start to prevail and electron acceptors with lower energy yields are utilized. Additionally, a wide variety of fermentative microorganisms are active using organic metabolic intermediates as electron acceptors. Consequently, microbial communities and their activities are subject to gradual changes during the degradation of organic matter.

A general "classical" pattern of the degradation of organic matter is presented in Figure 2.3. Organic material is degraded in the presence of oxygen by fungi and heterotrophic bacteria resulting in the formation of simple organic compounds. As example, glucose is formed by the degradation of cellulose. These compounds can be further utilized under oxic conditions by heterotrophic microorganisms and carbon dioxide is formed as a final product, entering a "carbon dioxide pool" which is supplied by other metabolic reactions as well. In the absence of oxygen, denitrifying microorganisms can metabolize carbon compounds to form carbon dioxide and reduced nitrogen compounds such as dinitrogen or ammonium. In the absence of nitrate as electron acceptor, fermentation processes lead generally to the formation of acetate, other volatile fatty acids (such as e.g. propionate or butyrate), simple carbon compounds, carbon dioxide, and hydrogen. Although hydrogen is a main product in anaerobic mineralization, it can only be detected in significant amounts in anoxic systems in rare cases. It is only possible to detect high levels of hydrogen in a phase where fermentative microorganisms and processes are dominant. Hydrogen can be used as electron donor by a series of microorganisms (nitrate, iron, and sulfate reducers as well as autotrophic methanogens).

Generally, organic material is only completely degraded below a certain partial pressure of hydrogen due to thermodynamic reasons. High hydrogen concentrations can inhibit fermentation reactions (product inhibition). The inhibition can only be overcome by the activity of hydrogen scavenging microorganisms. In natural ecosystems (sediments, sewage, sludge, rumen) sulfate reducing and methanogenic bacteria keep the hydrogen partial pressure at values of <10 Pa ($<10^{-4}$ atm). Partial pressures of 10 to 100 Pa inhibit fermentative reactions which can be confirmed either by thermodynamic calculations or by experimental approaches [18]. However, in some cases the formation of microbial flocs or biofilms allows the functioning of fermentative reactions even at elevated hydrogen concentrations. Flocs and biofilms can reduce gas diffusion through these structures.

Volatile fatty acids (VFA) can be utilized by iron and manganese reducing bacteria. Again, carbon dioxide is formed as final product along with reduced iron and manganese. Sulfate is reduced by sulfate reducing bacteria (using e.g. acetate as carbon source) leading to the formation of hydrogen sulfide whereas in the absence of sulfate VFAs are converted to acetate by acetogenic bacteria. Acetate is used by acetoclastic methanogenic microorganisms to form methane (and also carbon dioxide) as product of the anaerobic mineralization. Additionally, methane can be formed by autotrophic methanogens using carbon dioxide from the "pool" and hydrogen from fermentation processes. As a consequence, methane and carbon dioxide are the main terminal products resulting from the degradation of organic matter in an ecosystem.

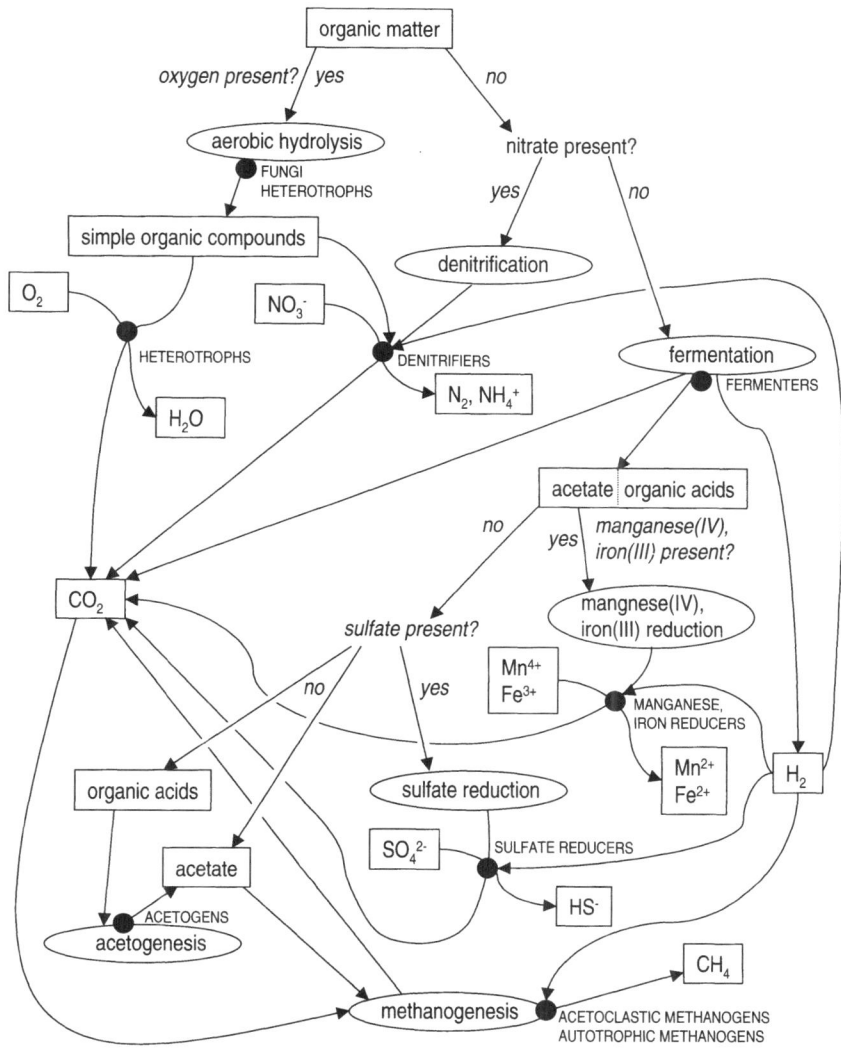

Fig.2.3. Schematic view of microbial degradation of organic matter in a landfill (adapted from Humphreys et al. [22]). Texts in boxes represent specific compounds (educts or products of the microbial metabolism). Texts in ellipses represent microbial processes. Solid symbols represent specific main functional groups of microorganisms.

Gas formation is one of the most important processes occurring in landfills. Methane and carbon dioxide are formed by microorganisms as terminal product during the anaerobic degradation of organic material. However, it is known that organic materials such as paper, cardboard, food and garden wastes are easily degraded under laboratory conditions or in compost heaps, whereas in landfills the process can take quite a long time [17].

Gaseous emissions from MSW landfills are characterized by their complex composition [16, 32]. Generally, landfill gas contains 40 to 60% methane (CH_4) and 60 to 40% carbon dioxide (CO_2) which is formed in the process of microorganisms degrading organic material [1]. Both of them are important greenhouse gases with methane possessing a global warming potential which is about 25 times greater than carbon dioxide. Regarding greenhouse gases, the concentration of nitrous oxide (N_2O) found in landfill gas is negligible. Methane is emitted at typical annual emission rates of about 10 m^3 of gas per ton of deposited waste, finally resulting in 150 to 300 m^3 of landfill gas [23, 32]. Factors affecting biological decomposition of MSW and landfill gas emission include presence and spatial distribution of microorganisms, moisture content of the waste, pH, temperature, redox potential, nutrient concentration as well as physical dimension of the landfill site, type and particle size of deposited waste, age of the waste, waste compaction, coverage, capping and so forth [2, 32].

Landfill gas formation is characterized by four to eight phases, depending on the point of view [12, 14]:

1. An oxidative phase dominated by oxic conditions where oxygen is gradually consumed by microbial activities and carbon dioxide is formed. Nitrogen concentration remain more or less constant.

2. Start of anoxic conditions after oxygen depletion where electron acceptors such as nitrate, iron, or sulfate are used instead of oxygen. Gaseous products are carbon dioxide and hydrogen. A series of short chain alkanoic acids are formed which are finally converted to acetate (acetogenesis). Nitrogen is displaced.

3. Start of methane formation where acetate, hydrogen and part of the carbon dioxide is consumed by methanogenic bacteria ("unstable" phase of methane formation). Methane concentrations gradually increase. The duration of the first three phases ranges from 180 to 500 days [14].

4. Methane and carbon dioxide are formed at a relatively constant rate ("stable" phase of methane formation) resulting in a constant gas composition over a certain period of time. This phase can last relatively long in comparison to the first three phases.

5. Gas formation rate starts to decrease. A significant portion of carbon dioxide is dissolved in the leachate. Ambient air start to intrude into the landfill body.

6. In this phase methane is aerobically oxidized resulting in the consumption of the intruding oxygen. Nitrogen concentrations increase along with carbon dioxide concentration originating from methane oxidation.

7. Methane oxidation terminates and oxygen concentration increase.

8. This phase is the final phase in which the waste has been fully degraded. Landfill gas more or less resembles interstitial air present in soil.

A schematic overview of the gas formation kinetics is shown in Figure 2.4. It has been stated that the duration and the relative amount of gases formed are affected by a variety of factors mentioned above [14]. Whereas for the first phases data are available, the later phases are only speculative [12].

Besides the major compounds methane and carbon dioxide, a wide variety of minor constituents such as hydrocarbons, halogenated hydrocarbons, alcohols, aldehydes, ketones, esters, ethers, and organosulfur compounds can be detected, sometimes in concentrations that are of toxicological significance [32]. However, a differentiation between their biotic and abiotic (purely chemical) formation is almost impossible. For instance, ethane can be abioticly produced by the reductive dehalogenation of chlorinated solvents, but a microbiological formation is also possible by the hydration of ethylene, which itself is microbially produced from sugars or ethanol. However, the source and microbial metabolism of non-methane hydrocarbons is not fully understood.

A series of compounds in landfill gas are already present in the MSW in their original form (e.g. propellant gases escaping from cans), whereas others are formed by chemical reactions. Alternatively, certain gaseous compounds from the sulfur and nitrogen cycle, such as e.g. H_2S, NH_3; $(CH_3)SH$; $(CH_3)_2S$, $(CH_3)NH_2$, or $(CH_3)_2NH$, can be microbially produced [7]. These substances are formed by the reduction and methylation of oxidized sulfur or nitrogen compounds. Elements other than sulfur and nitrogen can also be methylated by the metabolic activities of microorganisms [24]: A wide series of metals and metalloids are known to occur as methyl compounds which are characterized by a high volatility and mobility. In particular, volatile species of antimony, arsenic, bismuth, bromine, iodine, lead, mercury, silicon, tellurium, vanadium, and tin have been detected in gases released from domestic waste deposits in a concentration range of 0.1 ng to 10 µg per m^3 of gas [20]. It has been demonstrated that at least some of these compounds are formed under anoxic conditions by methanogenic, sulfate-reducing, and peptolytic bacteria [27]. This might also be the case in MSW landfills. Metals can, therefore, be emitted from MSW landfills not only as ionic, water soluble compounds, but also in gaseous forms [25].

Several models have been developed to simulate gas or leachate composition from MSW landfill emissions [36]. For example, net emissions of methane at individual landfill sites follow the following simplified equation [15]:

net methane formation = Σ(methane emission + lateral migration + methane

recovery + methane oxidation + methane storage)

A more complex model was used to simulate gas and leachate emissions [36]: waste composition, size and shape of site, water input, waste pretreatment, temperature, moisture level, pH, redox potential, bacterial population, and solute concentrations were used as input parameter for a mathematical model combining a number of subsystems. It has been stated that "all the mechanisms involved in regulating landfill degradation interact and cannot ultimately be considered in isolation from each other" [36].

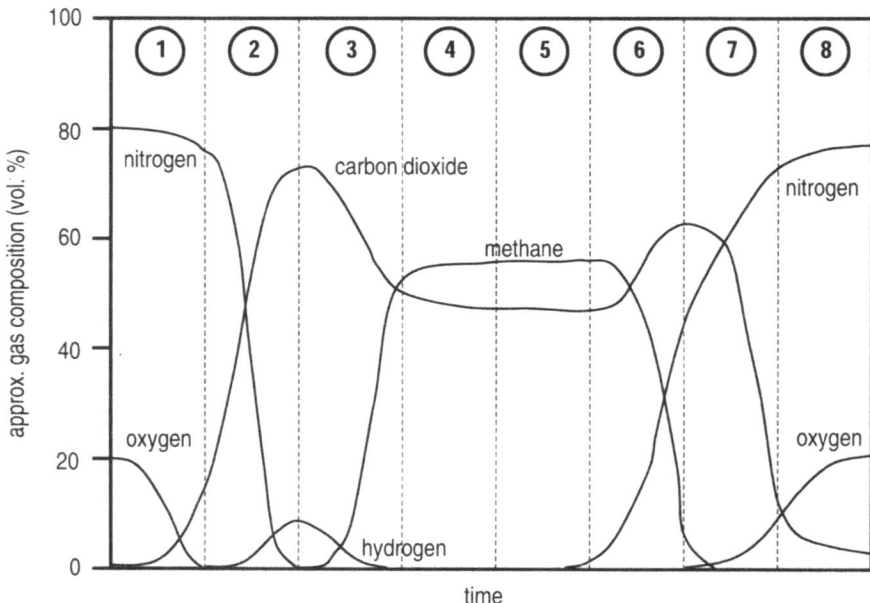

Fig.2.4. Kinetics of gas formation in a landfill (modified after Christensen & Kjeldsen [12]). Numbers represent distinct phases of gas formation (see text for explanation).

In a case study, a landfill composed of waste containing per m³ 150 kg of carbohydrates, 300 kg of protein, 20 kg of fat, and 450 kg of non-degradable materials was established [36]. Initial dissolved and gaseous concentrations of methane, hydrogen, oxygen, hydrogen sulfide, ammonia, acetate, carboxylic acids, alcohols, and glucose were taken as zero in an atmosphere of 20% carbon dioxide and 80% nitrogen assuming that all the oxygen has been rapidly consumed in the aerobic phase and that the anaerobic phase has just been started. The biomass of methanogenic bacteria was assumed as 10 mg per m³. A 14-month simulation showed a pH decrease in the leachate from approximately 7.3 to 6.5 after 30 weeks, followed by a sharp increase to 7.5, and after 38 weeks a more or less constant value of 7.3. Correspondingly, the concentration of acetic acid increased to 1.3 g per liter after 30 weeks followed by a complete consumption within 8 weeks. The population of autotrophic methanogenic bacteria consuming hydrogen and carbon dioxide was large enough after 16 weeks to completely utilize hydrogen which had been formed in the early stages. At week 30, acetoclastic methanogens (consuming acetic acid) with a modeled biomass of 55 mg per liter started to use acetic acid until its full depletion after week 38. Consequently, biomass decreased to approximately 15 mg per liter after 14 months. Maximum methane concentrations of 70% could be found at around week 38 decreasing to approximately 50% after 40 weeks. Finally, it was concluded that methanogens eventually reach steady state conditions where substrates are utilized at the same rate as products are formed [36]. Waste is degraded to carbon dioxide and methane in a 1:1 molar ratio.

2.2.2 Emissions to the Pedosphere and Hydrosphere

In addition to uncontrolled gas release from landfills to the atmosphere, emissions can also occur affecting the pedo- and hydrosphere. Theses emissions are mostly in the form of liquid leachates which are generated by percolating rainwater and contain run-off of organic and inorganic compounds resulting in the contamination of soil, surface and groundwater. It has been estimated that groundwater pollution originating from landfills may be a risk even after several centuries [3, 21]: At medium rates of leachate formation (e.g. 200 mm per year), 300 years are needed until the final storage quality is reached and the leachate can be released into the hydrosphere without risk. With coverage systems allowing leachate formation at rates of only 100 mm per year or less, it is evident that, even for a very long time period, the quality criteria are not met.

Four groups of pollutants are characteristic for landfill leachates [13]:

1. dissolved organic matter, expressed as Chemical Oxygen Demand (COD) or Total Organic Carbon (TOC), also including methane and volatile fatty acids;

2. inorganic macro-compounds such as calcium, magnesium, sodium, potassium, ammonium, iron, manganese chloride, sulfate, and carbonate;

3. heavy metals such as cadmium, copper, chromium, lead, nickel, and zinc;

4. xenobiotic organic compounds such as aromatic hydrocarbons, phenols, and halogenated aliphatics.

Leachate composition varies depending on the waste type, rainfall conditions, landfill design and operation, and landfill age [32].

A major survey of landfill leachates showed that over longer periods of time ammonium in concentrations of up to 2.5 grams per liter has the highest hazardous potential to affect surface or ground water [13, 32]. By contrast, heavy metal concentrations (e.g. cadmium, chromium, copper, lead, mercury, nickel, or zinc) in leachates are generally low and present in amounts that are below those usually detected in household sewage. Recently, the biogeochemistry of landfill leachates affected aquifers has been critically reviewed. It was demonstrated that most contamination plumes are relatively narrow [13]. A spatial heterogeneity of leachate composition and concentrations can be observed with areas showing relatively low concentrations and some "hot spots" of high concentrations. Depending on climatic conditions (rainfall), temporal variations of compounds found in landfill leachates also occur [13]. In general, it has been found that natural physical, physico-chemical, chemical, and microbial attenuation processes such as dilution, sorption, ion exchange, precipitation, redox reactions, and degradation processes significantly contribute to natural remediation resulting in effects of the leachate in a distance from the landfill that does not exceed 1 kilometer [13].

With a leachate plume originating from a landfill, distinct redox zones are present with specific oxidation/reduction regimes. These zones are typical for certain microbiologcal metabolic activities and can also be found in other anoxic environments , e.g. in aquatic ecosystems such as freshwater lake sediments [10]. The typical redox sequence is present starting with a zone of methane formation closely located to the landfill body. Downgradient from this location, zones of sulfate reduction, iron reduction, manganese reduction, and nitrate reduction (denitrification) can be found, sometimes overlapping to a certain extent. Finally, at the edge region of the plume (furthest away from the landfill), oxic conditions prevail where aerobic processes occur. Additionally, fermentative reactions are possible where the electron acceptor is of organic nature. Fermentation can basically occur in the whole anoxic zone of the leachate plume. This sequence strongly depends on the presence (type, concentration) of terminal electron acceptors and the thermodynamic energy yield available for the microorganisms from each redox reaction [18].

The behavior of microbial communities involved in biogeochemical processes in each zone can be deducted from chemical thermodynamics [18]. Each redox zone in a leachate plume is the habitat of specific and typical microorganisms: It was shown that groundwater aquifers contaminated with landfill leachate are dominated by the presence of bacteria (eubacteria and archea) and that protozoae are absent [26]. Over a distance of approximately 300 meters, a total of 10^7 to 10^8 bacterial cells (determined by acridine orange direct counts) per gram of dry aquifer solids have been found [13]. Methane-forming bacteria were restricted to the most polluted part, closest to the landfill, showing the most reduced conditions. Around 10^5 cells per gram have been detected. On the other hand, highest cell numbers of nitrate-reducing bacteria (10^6 to 10^7 cells per gram) have been found at a distance of approximately 80 meters away from the landfill. On the basis of specific biomarkers (phospholipid fatty acids, PLFA), a decrease on viable microbial biomass as well as shifts of microbial community composition were detected along a horizontal gradient with increasing distance from the landfill body.

Biochemical and molecular techniques have been used to investigate the composition and the physiological capabilities of microbial communities present in aquifers contaminated by landfill leachates [29]. Anaerobic community-level physiological profiles (by BIOLOG multi-well plates) and DNA fragment analysis (by denaturing gradient gel electrophoresis) were applied to groundwater collected near a landfill site. With both techniques it was possible to differentiate microbial communities from the aquifer underneath the landfill as compared to sampling locations up- or downstream the aquifer. It was demonstrated that functional diversity of microbial populations regarding the range of metabolizable substrates was significantly enhanced in the plume of pollution resulting from the landfill [29]. Degradation of organic compounds occurred in the plume under iron-reducing conditions, whereas upstream of the landfill, nitrate reduction (denitrification) was the most important process [30]. Iron reduction was related to the presence of members of the family *Geobacteraceae* which strongly contributed to the microbial communities. Microorganisms of the class β-proteobacteria were dominating upstream of the landfill. Beneath the landfill, however, this group was not found and gram-positive microorganisms were mostly present. A profound effect of landfill effluents rich in organic matter on the chemistry and microbiology of aquatic environments underlying the landfill was clearly shown.

Table 2.6. Emissions of hazardous substances from landfills and their environmental impact (adapted from [15, 32]). COD = chemical oxygen demand.

path	group	compound	environmental impact	importance
gas	volatiles	methane	global climate change. explosive, asphyxia	high contribution of landfills to overall emissions
		carbon dioxide	global climate change	minor contribution of landfills to overall emissions
		hydrogen sulfide	odors, corrosion	minor impact due to fast oxidation in the presence of oxygen
		halogenated organics	human toxicity, cancerogeneity, ozone depletion	important for employees and local communities
		organics	human toxicity, cancerogeneity, nuisance	important for employees and local communities
		alkylated metals	human toxicity	importance unknown
leachate	salt	e.g. chloride	ecotoxicity	high contribution from landfill waste water treatment
	nitrogen	e.g. ammonia	eutrophication	important, due to local contamination of surface and groundwater
	metals	Cd, Ni	human toxicity, cancerogeneity	less important, small contribution to total emissions
		Cu, Hg, Pb, Zn	ecotoxicity	less important, small contribution to total emissions
	carbon	COD	eutrophication	less important, small contribution to total emissions from waste water treatment

In summary, Table 2.6 shows major emissions from landfills into the atmosphere, hydrosphere, and pedosphere and their corresponding impact on the environment [32]. At global levels, is has been estimated that methane emissions can contribute for approximately 18% to of total methane emissions. Regarding leachates, chloride is quantitatively the most significant compound. Approximately 2% of chloride discharged to the environment by waste water treatment systems originates from landfill leachates.

Case Study: Landfill 'Ritzer' Near the City of Aarau, Switzerland

An area near the city of Aarau (Switzerland) in the Jura Mountains was used until 1921 as a quarry for the production of raw materials (carbonate rocks) utilized in the cement industry. As early as 1959, landfilling of a variety of wastes already began to include the disposal of household, hospital, and industrial wastes, sludge from neutralization processes (iron chloride, calcium chloride), bitumen, soil excavated from gas works or from spills of chemicals or oil, and foundry sands. The volume of the landfill is approximately 360'000 m^3. After the opening of a waste incineration plant nearby in 1974, the landfill was closed and covered.

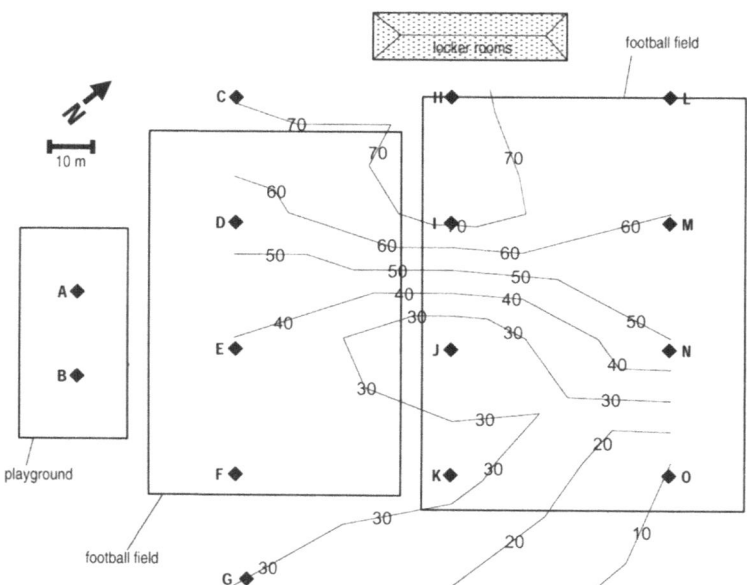

Fig. 2.5. Methane isopleths (% v/v) in a landfill which has been used as football field and playground since the closure of the landfill in 1974. Lines represent locations of equal methane concentrations. Diamonds (labelled with letters A to O) represent sampling wells.

Two years later, playground and football fields were established. After a certain time period, waste materials began to settle resulting in several depressions in the football fields which increased yearly by approximately ten centimeters. After rainfalls, water was retained in these depressions, preventing all sports activities. In addition, it was observed that from few spots in the field gas was emitting and that, therefore, the nearby locker rooms were endangered by possible explosion due to the gas penetrating the building.

During an investigation in 1998, physico-chemical characteristics, such as the composition of gas as well as leachate originating from the landfill, were determined (Eberhard & Partner AG, Aarau, Switzerland; personal communication). No microbiological studies were performed. Table 2.7 gives an overview of a series of landfill gas constituents. Methane concentrations of up to 75% (v/v) were determined to be highest in the area where the locker rooms are located (Fig. 2.5). Areas with the deepest depressions showed the highest methane concentrations. In addition to methane, butane was found as an important constituent of the trace compounds. Unfortunately, the high methane concentrations prevented a quantitative determination of ethane and propane. Benzene was detected as the main compound in the group of aromatic hydrocarbons (Tab. 2). Collected landfill gas showed elevated concentrations of halogenated hydrocarbons, mostly trichloroethene (TCE). Since a typical TCE profile could be determined, it was suggested that this compound was emitted from a point source, possibly from chemical solvents disposed in the landfill (Fig. 2.6).

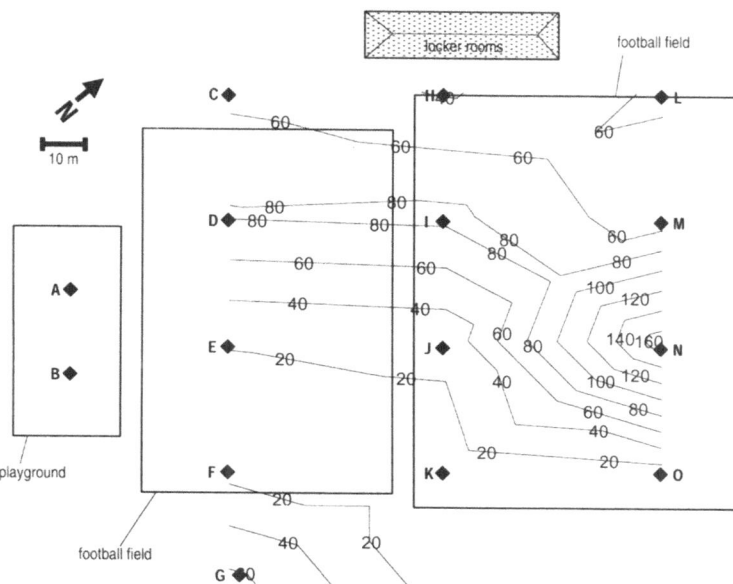

Fig. 2.6. Trichloroethene isopleths ($\mu g/m^3$) in a landfill which has been used as football field and playground since the closure of the landfill in 1974. Lines represent locations of equal trichloroethene concentrations. Diamonds (labelled with letters A to O) represent sampling wells.

Concentrations of heavy metals found in the landfill leachate were below the limit values of legal regulations. In one of the samples slightly elevated concentrations of sodium, potassium, nitrite, ammonium, sulfide, and boron were found. Additionally, traces of hydrocarbons were detected, suggesting the presence of residues originating from oil degradation. In contrast to the landfill gas, no trichloroethene was found. As a result from the investigation, it was calculated that between 22 and 54 million m^3 of methane could be formed in the landfill and that the gas formation would last up to twenty years until the waste in the landfill is consolidated (Eberhard & Partner AG, Aarau, Switzerland; personal communication). Important environmental effects resulting from gas emissions were identified, namely the damage of plant roots in the area due to oxygen depletion in the soil, the architectural instability of the locker room building due to the formations of depressions in the ground, and the danger of explosion due to gas/air mixtures containing high amounts of methane. As the primary measure to enhance the outgassing of the landfill, an active pumping and collection of the gas was suggested. The gas would serve as additional energy source for a nearby school.

Table 2.7. Composition of gas collected from a closed landfill site near the city of Aarau, Switzerland. For sampling locations A to O (figures 2.5 and 2.6). n.d. = not determined; n.a.=not applicable, because high methane concentrations matched ethane and propane; *=below detection limit. [1] Aliphatic hydrocarbon. [2] Aromatic hydrocarbon. [3] Halogenated hydrocarbon.

	unit	A	B	C	D	E	F	G	H	I	J	K	L	M	N	O
		Sampling location														
Drilling depth	m	2.4	2.4	3.5	3.5	3.5	2.5	2.5	2.5	2.5	2.5	2.5	2.5	2.5	2.5	2.5
Sampling depth	m	0.7	1.5	3.3	3.3	3.3	2.0	2.3	2.3	2.3	2.3	2.3	2.3	2.3	2.3	2.3
Main compounds:																
Methane	%	0.005	0.04	75	n.d.	40	n.d.	30	70	70	20	35	60	60	50	10
Carbon dioxide	%	n.d.	n.d.	18	n.d.	20	5	16	20	20	15	16	30	25	23	18
Trace compounds:																
Ethane [1]	ppm	*	*	n.a.	n.a.	n.a.	n.a.	n.a.	n.a.	n.a.	n.a.	n.a.	n.a.	n.a.	n.a.	n.a.
Propane [1]	ppm	*	*	n.a.	n.a.	n.a.	n.a.	n.a.	n.a.	n.a.	n.a.	n.a.	n.a.	n.a.	n.a.	n.a.
Butane [1]	ppm	1.0	0.4	2.1	5.0	3.5	1.1	2.4	2.2	0.8	0.8	2.0	1.4	2.3	1.1	0.5
Pentane [1]	ppm	*	*	1.6	2.0	1.4	0.9	1.0	1.1	2.2	*	1.5	0.4	0.8	0.6	*
Hexane [1]	ppm	*	*	0.5	0.5	0.3	0.6	0.2	0.5	0.7	*	0.4	0.4	*	0.5	*
Heptane [1]	ppm	*	*	0.8	*	*	*	*	0.3	0.1	*	*	0.2	0.1	0.3	*
N-octane [1]	ppm	*	*	0.4	0.3	*	*	*	0.8	*	0.7	0.2	0.4	0.2	0.4	*
Iso-octane [1]	ppm	*	*	0.6	*	*	*	0.4	*	*	*	*	*	*	*	*
Nonane [1]	ppm	*	*	0.3	0.5	0.2	*	*	0.4	*	*	*	0.3	*	0.5	*
Decane [1]	ppm	*	*	0.3	0.1	*	*	0.3	0.4	*	*	0.3	*	0.1	*	*
Benzene [2]	ppm	*	*	2.5	2.5	1.4	0.3	1.7	2.6	1.6	0.6	2.9	1.0	1.1	1.9	*
Toluene [2]	ppm	*	*	*	*	*	*	*	*	*	*	*	*	*	2.1	*
Ethylbenzene [2]	ppm	*	*	0.8	*	*	*	0.3	0.8	0.3	*	*	0.5	0.2	*	*
Ortho-xylene [2]	ppm	*	*	*	*	*	*	*	0.2	*	*	*	*	*	*	*
Propylbenzene [2]	ppm	*	*	*	*	*	*	*	0.4	*	*	*	*	*	*	*
1,2,4-trimethyl benzene [2]	ppm	*	*	*	*	*	*	*	0.4	*	*	*	*	*	*	*
Trichloroethene [3]	µg/m³	*	*	55	82	21	*	67	38	85	25	*	63	47	171	20
Tetrachloroethene [3]	µg/m³	*	*	*	*	*	*	*	*	*	*	*	*	*	208	35

2.2.3 Problems in Predicting the Long-Term Behavior of Landfills

Biotests for Toxicity Assessment

A series of ecotoxicological tests have been applied to determine the impact of MSW leachates on natural ecosystems. Under the term 'biotests' specific – mainly standardized - techniques are summarized which apply biological systems such as bacteria, protozoae, microalgae, small invertebrates, or fish to assess the impact of a sample or compound based on specific physiological reactions. Biotests find a wide application especially in aquatic ecotoxicology [9]. In general, acute and chronic toxicity of a sample or compound can be determined. Table 2.8 summarizes a selection of bacterial biotests which have been developed. A very popular and simple test is the determination of light emission inhibition by luminescent bacteria (*Vibrio fischeri*) after exposure to different amounts of an aqueous solution of the compound.

Biotests based on physiological reactions of higher organisms have been applied to investigate the toxicity of MSW leachates containing a variety of different organic and inorganic chemicals [34]. Invertebrates such as *Ceriodaphnia dubia* were used to test acute toxicity regarding the suppression of feeding activity by toxic compounds. It was suggested that toxicity of the leachates obtained form three sites was due mostly to organic compounds [34].

Zebrafish (*Brachydanio rerio*) were used as test organisms to study acute toxicity of MSW leachates in Brazil [33]. Leachates were differently treated (addition of EDTA or aluminium sulfate, aeration) to compare toxicities. It was found that the leachate was a highly toxic effluent potentially affecting aquatic species after discharge to aquatic ecosystems. The addition of aluminium sulfate significantly reduced toxicity.

Table 2.8. Selection of bacterial biotests for toxicity assessment

Type of assessment	Inhibition test	Standard norm	Organism	Physiological reaction	Duration (h)
acute toxicity	oxygen utilization	DIN 38412L27	*Pseudomonas putida*	substrate oxidation	0.5
acute toxicity	light emission	DIN 38412L34	luminescent bacteria (*Vibrio fischeri*)	energy metabolism	0.5
acute toxicity	respiration	OECD 209	sewage sludge culture	oxygen utilization	3
chronic	cell replication	DIN 38412L8	*Pseudomonas putida*	cell replication	16

2.2.4 Conclusions

Most of the problems related to landfill emissions are due to the amount of organic compounds in the waste which is microbially degraded, leading to soluble and volatile degradation products [32]. It has been suggested that proper landfill management (e.g. operational practices, controlling the waste type accepted for landfilling, appropriate leachate treatment prior to discharge) might reduce problems associated with landfills [32].

2.3 MSW Management and Technology in China

Zhao Youcai

China is one of the largest nations in the world, encompassing a vast area, with diversified nationalities and cultures, and a very large population. It is also the largest developing country and has relatively poor infrastructures and underdeveloped industry. From the viewpoint of MSW management, the country might be divided geographically into at least two sections, roughly the north and the south, with an approximate boundary along the Yangtze River. In the north of China, the weather is dry and cold for most seasons of the year, with a fragile ecological environment and a vast area of desert and high plateau. By contrast, it is humid and hot for nearly the whole year in the south of China, especially in the provinces along the East China Sea and South China Sea.

The south of China is densely populated and the available land seems to be very limited as nearly every inch of land has been used for agricultural, industrial, and living purposes. More land reserves might be available in the north, except for in the proximity of several big cities such as Tianjin, located in the Great Northern China Plain.

The agricultural and industrial sectors are also greatly varied from north to south. Rice is the main crop in the south, grain and corn in the north. Most heavy industries are located in the north, while the light industries are located in the south, though this situation is gradually changing.

There are differences in living and eating habits as well. While in the south, many kinds of soups are consumed because the weather is always hot, the food in the north is relatively dry. As a result, the MSW in the north and south differ in terms of humidity, composition, odor, etc. The humidity in MSW in the north is around 30-50%, compared with that in the south, which is around 40-60%. The humidity of MSW in the north disappears rapidly because of the dry weather, without strong odor or severe corruption. By contrast, the MSW in the south may degrade and corrupt very quickly, producing strong odor and leachate water. In the north, a great deal of coal is used for heat generation in the winter. Consequently, the proportion of coal ash in the MSW can be as high as 70%. Currently, the number of cities using natural gas or coal gas as fuel is increasing in the north (as well as in the southern cities), and the coal ash content is decreasing as a consequence.

The selection of treatment and disposal technologies of MSW in China should be flexible and adapted to local economical, social, geological and even cultural and historical conditions. Direct mechanical separation practices for the MSW in the south have proven difficult because of the high humidity. However, it is feasible in the north, where the MSW is relatively dry.

Around 120 million tons of municipal solid wastes were collected by the city authorities in China in 1999. Up to now, the MSW generated in rural areas has not been collected and is rather dumped into any available sites. About one quarter of the Chinese population are living in 700 cities and over 30,000 towns. The scale of these cities in population may vary from 0.1 to 9 million. It should be pointed out that, especially in small cities and towns, the service area for organized MSW collection by the relevant authorities usually covers only a small central part of the cities. The MSW generated in the suburbs and small towns is often not collected at all.

Table 2.9 shows the average composition of MSW in three typical cities in China. As mentioned above, the situation may vary greatly from one city to another. Generally speaking, the contents of plastics and papers are gradually increasing, while those of coal ash are decreasing. The construction and demolition wastes have increased in recent years as many families move into new housing. Food and ash wastes contents in Dalian are much higher than in the two other cities. Based on the *in situ* investigation, it was determined that the coal ash content in Dalian was still quite high in 1998.

The contents of recyclable wastes such as papers, plastics, metals, etc., are low. In fact, most of these wastes are collected and recovered by scavengers. The composition of MSW is determined *in situ* in landfills or dumping sites, not in the generation sites of the MSW. In large cities such as Shanghai and Beijing, the recyclable wastes are well recovered, including cans, cardboard, big pieces of woods, TV sets, nearly all kinds of plastics, and glass bottles. However, used batteries, lamps, thermometers, etc., have not been collected separately and are being mixed with MSW, ultimately entering the landfills or dumping sites.

In addition, the moisture in the MSW may vary from 30 to 60% in weight, depending on the seasons and locations. In the rainy seasons in the south of China, the moisture is so high (over 60%) that landfill operations become unacceptable .

Table 2.9. Average MSW composition in typical cities in China in 1998 determined in landfills [wt %]

Cities	Food wastes and ashes	Papers	Glass	Metals	Plastics	Textiles	Slag
Beijing	59.6	11.7	3.8	1.7	12.6	2.8	8.2
Shanghai	65.7	6.7	4.0	2.0	11.8	2.3	7.5
Dalian	82.1	3.4	2.6	0.5	5.7	1.6	4.1

The MSW yield per capita in China is shown in Table 2.10. Obviously, the values are relatively low for urban areas, as only 1.16 kg per day/person are generated. Lower yield in small cities (such as Maanshan in the Anhui Province in the south of China and Anshan in the Liaoning Province in the north of China) may be due to the relatively undeveloped economy and better recovery rates of recyclable wastes. The highest yield is found in Shenzhen, a newly developed city near Hong Kong. Table 2.11 presents the relationship between the GDP and the MSW generated in the large-scale cities, as reported in 1995. It can be found that the higher the GDP, the more the MSW, with nearly linear correlation between MSW quantity and the GDP, exclusive of the situation in Shenyang, a large city with a number of heavy industries (compare section 2.1). From this viewpoint, one may argue that the MSW yield per capita and the total quantity should increase as the economy in China develops. In this case, more and more treatment facilities for MSW will have to be planned and constructed.

Table 2.10. MSW yield [kg/day/person] in typical cities in China in 1996 based on investigations *in situ* in landfills or dumping sites

City	Beijing	Tianjin	Shanghai	Shenyang	Dalian
Yield	1.20	0.99	1.23	1.02	1.03
City	Hang-zhou	Shenzhen	Guangzhou	Maanshan	Anshan
Yield	0.92	2.62	1.20	0.66	0.76

There are three main alternative treatment methods for MSW in China; these are landfill, incineration, and composting. Landfill is the predominant method in China, while large-scale composting is limited and incineration is still being developed. The first large-scale incineration plant in Shengzheng was constructed in 1985. Meanwhile, an incinerator capacity of 2000 ton/d in Shanghai, 1000 ton/d in Ningbo, 600 ton/d in Zhuhai, 300 ton/d in Xiameng, and various capacities in Shengyang, Shengzheng, Tianjing, etc., will be constructed in the coming years. With increasing economic development, many of the existing landfills will soon be reaching their design capacity, and finding new landfill locations is becoming increasingly difficult. Although there are many treatment technologies being developed and applied in the world, the incineration technologies will be a method with priority for final MSW disposal in the coming years in China, especially in the more quickly developing big cities.

Presently, most MSW is dumped in the dumping sites around the cities, which results in serious environmental problems. Most so-called landfills have to be classified as dumping sites, but they can be restructured into sanitary landfills in the coming years. The statistic data shows that, over all, less than 10 % of MSW is treated in sanitary landfills, composting and incineration plants at present.

China is a developing country, not only economically but also in science and technology. However, for solid waste management in China, technology does not seem to be the key limiting factor in the obstruction of the development of solid waste management, although it should be improved in the future. In past years, a great deal of feasible and cost-effective technologies have been developed, and some of them have been applied in completed and ongoing projects.

Table 2.11. Quantitative relationship between GDP in a city and its MSW quantity in 1995 in China

City	City population [million]	Total population (incl. suburbs) [million]	GDP [billion Chinese Yuan]	MSW [million tons]
Maanshan	0.38	0.50	8.78	0.094
Anshan	1.44	3.31	39.5	0.401
Dalian	2.53	5.37	73.3	0.715
Shenyang	4.20	6.71	77.1	1.569
Hangzhou	1.96	6.03	90.6	0.660
Shenzhen	0.78	1.03	95.0	0.754
Tianjin	5.13	8.98	110.2	1.853
Guangzhou	4.03	6.56	144.5	1.764
Beijing	7.10	10.78	161.5	3.110
Shanghai	9.32	13.04	290.2	4.182

In China, almost all investments in solid waste management and operational costs are financed by local governments, which is usually not the case in industrialized countries. This situation has stifled the advancement of MSW treatment in contrast to the economic development. Moreover, many Chinese still think that solid waste management is the duty of the government. Action for *in situ* sorting and separation for MSW at home is also difficult to put into practice. Hence, the public environmental responsibility should be brought to task.

MSW Management in a Fast Developing Chinese City: Shanghai

MSW generated in Shanghai is increasing, with a total quantity of 11,620 tons per day in 2000. It is estimated that the MSW quantity in 2005 will reach 14,850 tons. In addition, the number of used TV sets, furniture, refrigerators, washing machines, bicycles, etc., as wastes have increased greatly in recent years. Table 2.12 shows the quantity of products sold in Shanghai in 1997. Theoretically, these products are expected to become bulky wastes in subsequent years. Nevertheless, most of this kind of waste has been reused or collected by the recycling plants as secondary materials when the users discard them. It is rare to see such wastes arrive at the landfills.

Used batteries and fluorescent lamps have not been collected and treated separately. At least 100 million pieces of small batteries are being used in Shanghai every year. None of them are treated properly. The main reason may be economical, in comparison to the primary raw materials. Investigations into recycling technologies for used batteries, relying on experience in Switzerland, USA, Germany, and France, have been conducted by the engineers, scientists and governmental officials in China, but implementing efficient recycling plants seems difficult. Currently, many private sectors have interests in the collection and treatment of used batteries, but progress is slow and there is the danger of potential secondary pollution.

Table 2.12. Quantity of products sold in Shanghai in 1997

Type	Used furniture	TV (set)	Refrigerator (set)	Washing machine (set)	Bicycle (set)
Quantity [ton]	440'000	390'000	240'000	235'000	475'000

There are two landfills in Shanghai. Liming Refuse Landfill, is a relatively small operation mainly used for the deposition of MSW collected in Pudong New Area. The large landfill, Shanghai Refuse Landfill Laogang, has been built and extended over the past 12 years, along the shore of East China Sea. Currently, there are around 6 km² of filling area available. An extension is planned which will increase the total area of the landfill to 12 km². Shanghai Refuse Landfill is not an ideal site. It was selected because there was no better site under consideration. Being on the shore of the sea, it is affected by the tides: It was found that the liners of the landfill are destroyed by the up and down motion of the tides. This problem has not been solved. In addition, the landfilling height is only 4 m, and the subsequent large area of placement leads to very high costs for liners. In addition to the landfills, there are still 12 large scale dumping sites in suburban Shanghai, of which 10 sites have been closed and 2 are still in use. Several million tons of refuse are stored in these sites.

Two incinerator plants are currently under construction. One is located in West Shanghai, another in East Shanghai. The key equipment was imported from Spain and France, with a loan from the foreign governments. It is claimed that the flue gases are treated at EU standard. 2000 tons of refuse can be incinerated, with a total investment of 0.75 billion Chinese Yuan for each plant.

There had once been a large scale composting plant in Shanghai. Unfortunately, it had to be closed because, on one hand, there was no market for the compost, and on the other, the composition of the waste became difficult to handle, with a large proportions of plastics, broken glass, textiles, etc. Currently, there is no composting plant in Shanghai.

Table 2.13 summarizes the current flows of waste materials in Shanghai. The MSW generated downtown, around 6,840 tons/day, is dumped in two controlled landfills and two dumping sites. The MSW in the suburb, around 4,010 tons/day, is simply dumped in the dumping sites without any pollution control measures. Planning for future MSW treatment facilities in Shanghai encompasses 'Integrated Treatment Plants', which consist of mechanical separation and sorting systems, composting systems, baling systems, perhaps also drying and compressing systems, and landfilling for the non-recoverable fraction. Shanghai is so large that reasonable planning for an economically feasible and technically viable MSW treatment system is quite difficult.

Table 2.13. Mass balances for Solid Wastes treatment in Shanghai in 1999

Items	Classified items	Quantity collected [ton]	Treatment methods
MSW	Total weight	6'840 in down-town	4500 tons in Laogang and Liming Landfills (with daily cover and drainage, and treatment of leachate, without liners)
			2'340 tons in two dumping sites
		4'010 in suburb	All in the various dumping sites without any pollution control facilities (leachate is directed to the sewage treatment plants in some sites)
	Bulky items (furniture, TV etc.)	230	Mostly recycled. Remnants are broken and deposited in landfills or dumping sites
	Plastics	990	Around 1/3 recycled, the remainder deposited in landfills or dumping sites
	Toxic waste, such as used batteries	3.4	Mixed with MSW and placed in landfills
Food origin wastes		1'100	Used as feed for 260,000 pigs until 1999. Prohibited since June 2000. Currently placed in landfills after dewatering, which makes the operation difficult, as the moisture is high.
Demolition and construction wastes		29'700	Mostly balanced in situ and partly recycled as feed for cement production, the remainder is deposited in the slag dumping sites
Human excrement		7'130	Mostly sewage, some recycled as organic fertilizer after digestion
Sludge		4'384	No way to go until 2002

Two viewpoints are always encountered when discussing the planning of MSW treatment establishments: centralized vs. decentralized facilities, which have their own individual advantages and disadvantages. According to the current political system in Shanghai, it is possible to establish a centralized facility. However, the problems may be the financial sources and sites selection. No Administration District or county wants to let the facility be constructed on its own land. In this regard, the Shanghai Government has to let every District construct its own 'Integrated Treatment Plant', in which mechanical separation, small scale landfill, composting, and perhaps incineration, should be located together at one site.

According to the experiences gained in the other large cities, such as in Guangzhou, a centralized treatment facility seems to be feasible if landfill space is available. In Shanghai, every inch of land has been used to a full extent and it has become increasingly difficult to find a sufficiently large site to host all the MSW.

Significant investments are required to construct the needed infrastructure mentioned above. In total, an estimated 4767.85 million Chinese Yuan (1 US dollar = 8.3 Chinese Yuan in 2001) in 6 years is required so that all the MSW in Shanghai can be treated to EU standards. However, it is impossible to get such a huge investment from the Shanghai Government. Currently, MSW in Shanghai (as in other Chinese cities) is collected, transported and treated by the Shanghai Environmental Sanitation Bureau, which acts as a company and an administration bureau. At the beginning of the year, the bureau receives all the funds from the Shanghai Government, based on the total expenses of the previous year. Generally, this fund can maintain only the lowest standard for MSW collection and treatment. If a new project is expected to be constructed, additional application must be presented and approved, which may take anywhere between some months and several years, depending on the scale of the investment.

Shanghai Government is the single investment source for MSW management. Nevertheless, the most important thing for local governments seems to be economic development, hence, the investment for environmental protection, including MSW and other waste treatment is usually put aside. Fortunately, many public and private companies are willing to invest in the treatment of MSW. Certainly, some profit should be guaranteed for these companies. One of most reliable financial sources is to collect payments from the MSW generators, including companies, households, and institutions from public and private sectors, etc. So far, in most cities in China such action has not been put into practice, as the local governments fear opposition from the households, especially ones with low income.

Many suggestions have been proposed for MSW management. For example, all the MSW facilities that are constructed by the governments can be rented to private companies, while relevant governmental organizations just act as regulator or supervisor. The governments should, of course, pay reasonable treatment fees to the companies. Landfills, incineration plants, waste water treatment plants, etc., can be sold and bought among interested customers. Currently, MSW collection and transportation operations still tend to be owned and operated by the local governments.

Private companies can construct their own treatment companies, and have the local governments pay the treatment fees. The prices can be negotiated. The dilemma facing the China cities is that everything is changing rapidly. Local governments are always reluctant to make any promises to private companies. It is very difficult to get a payment contract from a local government if one wants to treat MSW for the local community.

Hence, the Chinese government should speed up its reforms concerning MSW, including refuse fee collection, regulations for construction and operation treatment facilities.

Conclusions

The MSW management mechanism in China is basically centralized; local governments are responsible for MSW collection, transportation, treatment, facilities investment and construction, recruitment of all staff, and often affect ineffective productivity and heavy bureaucracy. The investment for MSW facilities construction is very constrained, as the governments are the single investors. Possibilities for the trading of MSW-related companies are under discussion. Private and public companies will be encouraged to invest and manage the MSW treatment facilities. Refuse tax may be levied in the future. Considering the advantages and disadvantages for the individual technologies, such as incineration, composting, landfill, the concept 'Integrated Treatment' may be adopted. It attempts to combine all the available technologies together at an optimum mode, in order to solve the difficulty of site selection and to facilitate the recycling of resources. The MSW management in Shanghai, in fact, involves a series of complex issues, e.g., adoption of centralized or decentralized manners, trading of facilities, maturing and developing of competitive and qualified companies and labors, etc. A significant amount of investment is required to create new facilities and upgrade the old ones. As most cities do not possess any modern treatment facilities for MSW, according to the experiences gained in recent years, landfilling seems to be the favored alternative for the rapid improvement of city sanitation, as the duration of construction of landfills is usually relatively short and the investment and operational costs are relatively low, provided that qualified liners be installed and leachate be properly treated. Nevertheless, the remediation of closed and functioning dumping sites should gain more attention from the public in general and from the local governments in particular, as the adverse long-term impacts the dumping sites have on human health and the environment have been clearly proven.

References

1.   Anonymous (1998) Option to Reduce Methane Emissions (final report). AEA Technology Environment, AEAT-3773: Issue 3. Canadian Agri-Food Research Council
2.   Anonymous (1999) Emission Estimation Technique Manual for Municipal Solid Waste Landfills. United Nations Institute for Training & Research, Geneva
3.   Anonymous (2000) Environment in the European Union at the Turn of the Century. Chapter 3.7. Waste Generation and Management. European Environment Agency, pp 203-226
4.   Anonymous (2000) OECD in Figures; Statistics of the Member Countries. OECD Publications, Paris
5.   Anonymous (2001) OECD Environmental Outlook, Paris
6.   Baccini P, Brunner P (1991) Metabolism of the Anthroposphere. Springer, Berlin Heidelberg New York
7.   Bachofen R (1991) Biogeochemische Zyklen, Mikroorganismen und atmosphärische Spurengase. Eco-Informations: Dechets - atmosphere

8. Belevi H (1998) Environmental Engineering of Municipal Solid Waste Incineration. Habilitation. vdf, Dübendorf, p 148
9. Blaise C (1991) Microbiotests in Aquatic Ecotoxicology: Characterstics, Utility, and Prospects. Environmental Toxicology & Water Quality 6:145-156
10. Brandl H, Hanselmann KW, Bachofen R, Piccard J (1993) Small-Scale Patchiness in the Chemistry and Microbiology of Sediments in Lake Geneva, Switzerland. Journal of General Microbiology 139:2271-2275
11. BUWAL SAftEFaLS (1993) Zusammensetzung der Siedlungsabfälle der Schweiz 1992/93. BUWAL Schriftenreihe Umwelt 248
12. Christensen TH, Kjeldsen P (1995) Landfill Emissions and Environmental Impact: An Introduction. In: Christensen TH, Cossu R, Stegmann R (eds) Fifth International Landfill Symposium. SARDINIA 95. Vol. 3. Environmental Sanitary Engineering Centre, Caglari, pp 3-12
13. Christensen TH, Kjeldsen P, Bjerg PL, Jensen DL, Christensen JB, Baun A, Albrechtsen HJ, Heron C (2001) Biogeochemistry of Landfill Leachate Plumes. Applied Geochemistry 16:659-718
14. Farquhar GJ, Rovers FA (1973) Gas Production during Refuse Decomposition. Water, Air, and Soil Pollution 2:483-495
15. Fischer C, Maurice C, Lagerkvist A (1999) Gas Emissions from Landfills. AFR-Report 264. Swedish Environmental Protection Agency, Stockholm
16. Gendebien A, Pauwels M, Constant M, Ledrut-Damanet MJ, Nyns EJ, Willumsen HC, Butson J, Fabry R, Ferrero GL (1992) Landfill Gas: From Environment to Energy. Commission of the European Communities, Luxembourg
17. Hamilton JD, Reinert KH, Hagan JV, Lord WV (1995) Polymers as Solid Waste in Municipal Landfills. Journal of the Air & Waste Management Association 45:247-251
18. Hanselmann KW (1991) Microbial Energetics applied to Waste Repositories. Experientia 47:645-687
19. Harrison P, Pearce F (2000) AAA Atlas of Population and Environment, Population, Waste and Chemicals. University of California Press, Berkeley, Los Angeles, London
20. Hirner AV, Feldmann J, Goguel R, Rapsomanikis S, Fischer R, Andreae MO (1994) Volatile Metal and Metalloid Species in Gases from Municipal Waste Deposits. Applied Organometallic Chemistry 8:65-69
21. Hjelmar O, Andersen L, Hansen JB (2000) Leachate Emissions from Landfills. AFR-Report 265. Swedish Environmental Protection Agency, Stockholm
22. Humphreys P, McGarry R, Hoffmann A, Binks P (1997) DRINK: a Biogeochemical Source Term Model for Low Level Radioactive Waste Disposal Sites. FEMS Microbiology Reviews 20:557-571
23. Köhler M, Völsgen F (1998) Geomikrobiologie. Wiley-VCH, Weinheim
24. Krishnamurthy S (1992) Biomethylation and Environmental Transport of Metals. Journal of Chemical Education 69:347-350
25. Lindberg SE, Wallschlager D, Prestbo EM, Bloom NS, Price J, Reinhart D (2001) Methylated Mercury Species in Municipal Waste Landfill Gas sampled in Florida, USA. Atmospheric Environment 35:4011-4015
26. Ludvigsen L, Albrechtsen HJ, Ringelberg DB, Ekelund F, Christensen TH (1999) Distribution and Composition of Microbial Populations in Landfill Leachate Contaminated Aquifer (Grindsted, Denmark). Microbial Ecology 37:197-207

27. Michalke K, Wickenheiser EB, Mehring M, Hirner AV, Hensel R (2000) Production of Volatile Derivatives of Metal(loid)s by Microflora Involved in Anaerobic Digestion of Sewage Sludge. Applied and Environmental Microbiology 66:2791-2796
28. Reimann DO (1994) Beiheft. Müll und Abfall 31
29. Röling WFM, van Breukelen BM, Braster M, van Verseveld HW (2000) Linking Microbial Community to Pollution: Biolog-Substrate Utilization in and near a Landfill Leachate Plume. Water Science and Technology 41:47-53
30. Röling WFM, van Breukelen BM, Braster M, Lin B, van Verseveld HW (2001) Relationships between Microbial Community Structure and Hydrochemistry in a Landfill Leachate-Polluted Aquifer. Applied and Environmental Microbiology 67:4619-4629
31. Sand W (2001) Microbial Corrosion and its Inhibition. In: Rehm HJ, Reed G, Pühler A, Stadler P (eds) Biotechnology. Vol. 10. Special Processes. Wiley-VCH, Weinheim, pp 265-316
32. Schmid J, Elser A, Ströbel R, Crowe M (2000) Dangerous Substances in Waste. Technical Report No. 38. European Environment Agency, Copenhagen
33. Sisinno CLS, Oliveira EC, Dufrayer MC, Moreira JC, Paumgartten FJR (2000) Toxicity Evaluation of a Municipal Dump Leachate using Zebrafish Acute Test. Environmental Contamination & Toxicology 64:107-113
34. Ward M, Bitton G, Townsend T (2000) Toxicity Testing of Municipal Solid Waste Leachates with CerioFAST. Environmental Contamination & Toxicology 64:100-106
35. World Resources Institute (1996) World Resources 1996-97. Oxford University Press, Oxford
36. Young A (1995) Mathematical Modeling of the Methanogenic Ecosystem. In: Senior E (ed) Microbiology of Landfill Sites, 2nd (edn). Lewis Publishers, Boca Roton, pp 67-89

3 Recycling, Thermal Treatment and Recovery

Contributions by Richard V. Anthony, Bernd Bilitewski, Michael Beckmann, Reinhard Scholz, Christian Ludwig, Jörg Wochele, and Frank Jacobs

Waste paper, cardboard, waste wood, plastic and glass bottles, tins, not to speak of broken appliances etc. hardly appear in waste statistics of developing countries, although they are being used and discarded by the minority of the population who can afford them. The reason is that all these categories of materials represent the often exclusive income for a host of informal waste pickers and recyclers who collect them and turn them into marketable goods by making e.g. roof tiles from spent cans, fuel briquettes from cardboard, second-hand goods from broken and repaired appliances etc. These kind of "zero waste" systems are driven by the needs of the poor and do not need to be encouraged by appeals to save resources and to close materials cycles, nor does it require the complex legislation that has been issued to encourage recycling in industrialized countries. The informal scavenging is hard work for a small income and exposes those involved to great occupational health risks. The waste picking business only functions, if sufficient material is thrown away that can be recycled, i.e. if there is a social class above who can afford to waste. It relies on large social differences and, form this perspective, is hardly sustainable.

It is probably no coincidence that the strong zero waste vision proclaimed in section 3.1 of this chapter originates from California, the wealthiest part of the richest country of the world! The zero waste vision has been inspired by the fact that in natural ecosystems there is no waste, i.e. all refuse is recycled by specialized organisms in an ecosystem. While true zero waste is not literally possible, the idea of closed material cycles is convincing and an absolute need to counterbalance the impacts of over-consumption. Technologies are available for producing quality products from waste materials by mechanical sorting and processing of the retrieved fractions with a high degree of automation (section 3.2).

Recycling is not automatically the best solution in all conceivable cases. Recycling processes shoud always be assessed with respect to their ecological impact. As a rule, recycling is only advisable, if it produces less damage than disposal and production from virgin raw material. Limits to material recycling exist in particular for organic materials such as plastics, paper and biomass, especially if they are composites, physically mixed with each other and/or contaminated with toxic substances. For such materials, the energy content is the most valuable and accessible resource to be recovered. Recovering the energy from combustible waste requires the application of highly sophisticated technology to avoid negative environmental impacts of the combustion process. The contribution 3.3 covers the fundamental control options one has to optimize the conversion of the combustible fraction of waste on the traditional grate furnace and to minimize the formation of air pollution by primary measures in the combustion chamber. Together with the highly perfected air pollution control systems (electrostatic precipitators, dry or wet scrubbers, catalytic nitrogen oxide converters and charcoal filters) these primary measures have contributed to the present advanced state of incinerator technology, which can truly be regarded as a sink and not a source of pollutants.

It has been wide-spread practice to use the ashes from incinerators as a filling material in e.g. road construction, especially the bottom ash. 250 kg of bottom ash are on average produced from 1 ton of MSW and it seems to make sense to recycle this material in construction applications. The composition of bottom ash corresponds very closely to the composition of cement and traditional additives to cements (section 3.5). However, in contact with water, the ashes are not inert materials. Contribution 3.4 shows that especially the fly ashes, and to a lesser degree bottom ash, contain soluble heavy metal compounds, which can be leached out and contaminate the aquifers, if landfills are not managed in a way that collects leachate water for appropriate treatment. The investigations reported in section 3.5 show that further thermally treated ashes resulting from advanced incineration plants could be used as substitute construction materials for applications that are not too much demanding.

It shows one feature of recycling that must be kept in mind for all recycling operations: due to their heterogeneous origin, recycled materials tend to have inferior properties compared to materials produced from virgin raw materials. This drawback limits the market acceptance of certain recycled products. This is in particular the case, if used objects, such as for example plastic bottles or newspapers, are used for reproduction of the same product. The residence time of the goods in the anthroposphere could be substantially increased using modern recycling methods, but "down-cycling" produces waste, which ultimately cannot be avoided and needs further treatment. Today, the most common alternative to MSW landfills is incineration (section 3.3). "Down-cycling" is less of a problem for the recovery of raw materials. New advanced technologies, which aim to recover the values of the materials at the end of a "down-cycling stream", are presented in chapters 4 and 5.

3.1 Reduce, Reuse, Recycle: The Zero Waste Approach

Richard V. Anthony

The notion of zero waste is as much as a principal of survival for the human species as it is a matter of fact in nature. A close examination of natural systems reveals that there is very little waste in nature. Everything is connected to each other. Every discard is an other's feedstock. When the planet is seen as a finite sphere in space, there can be no *away* on planet earth. Everything that is sent away must go some place.

Zero waste references can be found throughout past and current mythology and religion. References to burning fires of garbage analogous to Hell are found in the proverbs of the Old and New Testament. There are many Biblical references calling for stewardship of the land and resources. The Native American Chief Seattle said it best: "We do not inherit the earth from our forefathers, we borrow it from our children."

Waste is a concept of privilege, for only the truly rich can afford to waste. If all the people of the world consumed and wasted resources like the upper 20 percent of the world population currently does, we would need several planets to supply those needs. The average inhabitant of the earth cannot afford to be wasteful.

The genesis of the zero waste movement comes from the realization that discarded materials are resources. These resources have been manufactured from a raw state with energy and labor. In the cases of metal and oil they are irreplaceable. The value of that energy and labor is still in the commodity, even after the user has discarded it.

A zero waste system is a resource management system. The population of the world at the turn of the new millennium has reached six billion. Futurists project that even with zero waste and world cooperation in the production and distribution of food and survival commodities, the planet can support a maximum of ten billion people [107].

As the world population increases, world resources generally increase as well. The impact of this increasing demand on the remainder of the planet's finite resources like petroleum, metals, wild animals, birds and trees is leading to their depletion and in some cases, extinction.

The process of wasting resources is against nature. For humanity to survive over time on the planet, a balance of supply and demand must be attained that considers the whole biosphere. Extracting finite resources and destroying them after one use, while the population continues to grow upsets that balance. Therefore the ultimate option is to control population and recycle resources in order to survive.

In a zero waste system every thing has a place before, during and after use. In an ideal system, dismantling or de-manufacturing would be designed into the product. A society supporting a zero waste system would not subsidize wasting by paying for systems that destroy resources and create new sources of air and water pollution. Disposal funds would be used to collect and process resources. Existing disposal sites and systems would be used in transition but new facilities would be built for reuse, recycling and composting.

The system of extraction, manufacturing, use, and disposal to incinerators or landfill would be replaced by systems that capture the material and recycle them into a closed loop system of reuse, repair, recycle and redesign. Raw materials would be used as reserves.

The principle of the closed loop system requires that the consumer and the manufacturer buy recycled materials as well as recycle the product at its end use. A zero waste society would tax virgin and raw material use and give tax breaks for industries that use recycled materials. It would force the loop to be closed by making environmental dumping illegal or expensive.

Even though disposal by burning and landfill may be cost effective by today's standards, a legacy of depletion and pollution for our children will provide the basis for new standards. These new standards will take the future into much greater consideration concerning the planets' resources and will discourage waste.

3.1.1 Zero Waste Theory

A system design for the handling of discards from a household, business, institution, or city with a zero waste goal is theoretical. Recognizing that, scientifically, absolute zero is improbable, close counts. Any system that reduces wasting to ten percent of current generation rates would make a significant step toward sustainability and would be close to zero (90% successful). Pollution is measured quantitatively and a reduction of 90% is significant. In the area of discarded resources the remaining 10% may have to be dealt with through legislation, innovation and ultimately producer responsibility.

In a zero waste system everything has to go somewhere. Discards create jobs and products. In the cycle of use, reuse, repair and recycling close the loop to wasting. Discard management plans involve the transport and transition of discarded resources to a place where they will be used again.

In a perfect world this would happen logically. But the fact is that in many cases, the status quo of wasting is protected and encouraged by law. In many places in the world, tax laws and government funding for the collection and disposal of discarded materials encourage wasting over recovery.

Governments have the power to finance and implement discard management systems. Some systems are financed in such a way that they discourage the growth of recycling and composting industries. From a grassroots perspective, the individual has some influence on this process and can insist that the government include convenient recycling and composting programs in the disposal system.

If the natural forces of supply and demand were not subverted by subsidies for wasting, the consumer would place value on a product based on its reusability, reparability and recyclability, when making purchasing decisions.

When recycling and composting are considered a part of the disposal system in the analysis of costs, the revenue is used to reduce public subsidy for the proper handling of the discards. When adding the cost of long term disposal monitoring and considering the economic cost of taking land out of production forever, destroying ancient forests and polluting air and water downstream, planning a system to reduce or eliminate these future costs makes economic sense.

Zero waste discard management systems have political as well as technical considerations to take into account. The way each human being accounts for their daily generation of discards, and the way those discards are managed in a community are political and technical decisions. The future sustainability of the planet depends on people at the point of purchase, choosing products that can be ultimately composted or recycled and at the point of discard making a decision as to where in the system the discarded material should go. Aside from product design decisions and separation requirements, most of the decisions in the process are technical.

The consumer and the manufacturer are both responsible for the proper disposal of a product before, during and after purchase. The relationship between the consumer and the manufacturer should not be confused by government subsidies for the disposal of wasted resources. It is not fair that the entire population pay for the disposal of products enjoyed by only a few. The consumer should encourage the manufacturer to design products that can be repaired and eventually recycled or composted.

Manufacturers have a responsibility to the community to produce products that are designed for recyclability and/or compost-ability. Manufacturers should be encouraged to use recycled materials in their products. Products should be produced in a way that they can be repaired where wear and tear occurs and dismantled and recycled into new products when they cannot be repaired.

There is a hierarchy of use of materials that involves the highest and best use of materials in the areas of energy and resources [3, 4, 34, 66, 69, 92]: Source Reduction, Reuse, Repair, Recycling, Composting, Transformation and Landfill. In California in the early eighties, the "Three R's" were used to teach pollution prevention. The first of the "Three R's" (reduce, reuse and repair) refers to source reduction, or the area of discard management that addresses packaging and single use products. The 3 "R's" have been taught as a means to demonstrate how product design can lead to decreasing waste. Consumers have been encouraged to consider buying products that can be reused and repaired.

To many, the key component in a zero waste disposal system is reducing the amounts of discards at the source. Products designed for a single use are used as examples of unnecessary over consumption. The debate over paper versus cloth diapers for babies addresses reusable washable cloth diapers as opposed to introducing fecal mater to landfill via paper diapers. Some other reuse issues include cotton versus paper napkins, double sided versus single sided copies, and refillable versus single use products.

Repair-ability in a product is another design criteria that can be omitted when subsidized wasting occurs. The notion that the manufacturer owns the product and rents it to the customer is becoming popular in the carpet and computer industries. As the product wears out or becomes obsolete the product is returned to the manufacturer for dismantling and recycling. Automobiles, appliances and computers are primary examples of situations where landfill and incineration are unacceptable disposal options. The materials used in these products can pollute water and air. The proper disposal hierarchy for these items is reuse, repair and recycle.

After reuse and repair, the next step on the hierarchy is to separate materials at the source for recycling or composting. Sorting discards into recyclable and compostable categories assists in a zero waste system by, for example, keeping wet materials from contaminating paper products. Plastic and glass are contaminants that are hard to remove from finished compost. Wet organics are contaminants to paper recycling. Material processing facilities (MFR) are used to classify materials into specific categories. These categories include reusable products, paper, metal, polymers, glass, ceramics, vegetative debris, putrescibles, soils, wood, textiles, and chemicals.

The hierarchy places mixed waste (not source separated) composting, transformation processes and sanitary landfill at the bottom. In a zero waste system, items that cannot eventually be composted or recycled are returned to the manufacturer through advance recycling fees or product deposits. The price of recovery and recycling is added to the product cost to cover the infrastructure needed to reintegrate this material back into use.

Thus zero waste theory calls for disposal systems that place disposal cost responsibility on the manufacturers, influencing them to redesign products for recycle ability. The discard management service provider, whether government or private contractor, is mandated to collect source separated material from clearly labeled and conveniently located storage containers and deliver them to processing centers that will sort, process and reintroduce these materials back into the use system.

3.1.2 Zero Waste Analysis

A waste stream assessment is used to determine the quantity, source and composition of the discard stream. This information is needed to make planning, design, contractual, financial and regulatory decisions. Managing resources means managing discards.

Quantity is measured in weight and volume. Materials are sold by weight; however, storage and trucking capacity are measured in volume. Knowing the annual and seasonal generation rates for the targeted populations' discards will aid in assessing capital, labor, operation and maintenance costs for the system. Equipment and structures include collection vehicles, storage bins, processing equipment, and buildings. Estimating quantities by discard type provides the basis for the estimation of potential revenue.

Source refers to the generation point at which the discards are aggregated. Sources include residential, commercial, municipal service, institutional, industrial, and agricultural generators. Residential sources include single and multifamily dwellings. Commercial sources include offices, retail stores, entertainment centers, restaurants, hotels and motels, and service stations. Municipal services include demolition and construction, street cleaning, landscaping, storm drains, parks and beaches, and wastewater treatment bio-solids. Institutions include schools, hospitals and prisons. Knowing the source of discards allows for the routing of specific collection vehicles to areas and which are appropriate to the type of discards generated.

Discarded materials can be segregated into twelve categories. The Twelve Master Categories as defined by Urban Ore of Berkeley California, a non-government agency (NGO) that developed this system, sort and aggregate discarded materials into market based categories. Items like TV sets are placed in the reusable category. After reuse the next step for the TV is dismantling for recycling. The goal is complete recovery through design and source separation; however, during the transformation to a zero waste system some material will be incinerated and land filled at the real cost through the current infrastructure. The following is an outline of the twelve categories.

1. *Reusable Goods* are discarded materials that are useful in their present form. Examples are doors, windows, furniture, lighting, household goods, clothing, bricks, live plants, etc. Reuse operators need enclosed or covered space and enough room to organize, display, and sell all reuse items coming to the facility. They will also need to dismantle, clean, upgrade and store unsaleable merchandise for recycling.
2. *Paper*, one of the largest commodity sub-flows, comes in many forms, from newsprint to cardboard, all valuable for their fiber content. Paper collection and processing requires warehousing and sorting facilities, a baler, a forklift, and trucking.
3. *Plant debris* is another large sub-flow; it includes tree limbs and tree rounds, brush, weeds, grass clippings, and leaves. Plant debris operators need room to store green materials until they are dry enough to be fed into a grinding process. After grinding, plant debris may be screened, windrowed, turned, watered, and eventually blended with other nutrients and minerals into various types and grades of soil amendments. Composting plant debris and tilling it into soil is a carbon sink, a potential remedy to global warming.
4. *Putrescibles* are similar chemically to plant debris, but differ in their high nutrient value, which makes them a magnet for scavenger species of birds, mammals, and insects. Special handling requirements may include rapid mixing and dispersing with plant debris, containerizing for aerobic or anaerobic decomposition and odor control.
5. *Wood* may initially be divided into three streams: reusable/resalable, recyclable/unpainted, and painted. Reusable wood includes, doors, cabinets, dimensional lumber, furniture and plywood. Recyclable wood is usually chipped or ground, manufactured into particleboard, or blended with other ingredients

into compost. Painted and treated wood may require special handling due to entrained metals and other toxins.

6. *Ceramics* are hard, brittle materials such as stone, concrete, china tile and asphalt.

7. *Soils* are generated by road and foundation construction and by dredging. Clean soils can be sold for fill or added to compost blends to produce a more mineralized product. Soils contaminated by petrochemicals can often be cleaned up through bioremediation.

8. *Metals* have been recycled for thousands of years; the metals recycling industry recognizes hundreds of subcategories, most based on complex alloys of two or more elemental metals such as iron, aluminum and copper. Metals have a very large and varied reuse component. Metals are also recycled extensively: most new steel, for example, is recycled from old steel.

9. *Glass* comes to disposal facilities in two major sub flows: plate glass and container glass. Plate glass may be used as is, if unbroken, or recycled into fiberglass or sand. Container glass may be color sorted, then ground up and made into new containers or simply made into sand.

10. *Polymers* are carbon-based compounds manufactured into films or rigid forms such as containers or computer cases. In comparison with other master categories, polymer recycling is a very young industry experiencing multiple growing pains. Resin complexity and incompatibility, contamination, and "heat" history are primary limiting factors.

11. *Textiles* are fabrics woven from natural or synthetic fibers into objects such as clothing, bedding, carpeting, draperies, and upholstery. The textile reuse and recycling industry is very old and well developed, with worldwide markets for everything from old Levis to wiping cloths and paper.

12. *Chemicals* include unused paints, used oils and solvents, cleaners, acids and bases and the like. Deemed safe for their designated used, they become major pollutants when land filled or burned. Reuse is a preferred disposal option for many chemicals. Recycling requires filtration, distillation, mixing, or other refining operations to produce useful products.

The following table displays the 12 Master Categories in a waste composition study for Del Norte County, a small rural community with 32'000 inhabitants living in just over 9'100 households on the coast of northern California. This information is required by law for cities and counties in California to be published in an Integrated Waste Management Plan for each County that is reviewed by the State of California on a regular basis.

Certain conditions can impact the quantities and types of materials discarded, including seasonal variations in weather and tourism, demographic differences in age and wealth, state of the economy, laws (container deposits, mandatory recycling rules) and natural or unforeseen catastrophes.

The United States Environmental Protection Agency publishes a waste characterization protocol to provide a standardized procedure for measurement of the types and quantities of discards from an established area. The process of defining the characteristics of a discard stream will assist in determining costs, and identifying key generators.

Table 3.1. Discard Composition Analysis (based on data from [50])

Categories	Discarded [tons/year]	Discarded %
1. Reusables	1,014	5.7
2. Paper		21.2
3. Plant Debris		2.6
4. Putrescibles		
Sludge		4.9
Other		21.2
5. Wood		1.8
6. Ceramics		9.9
7. Soils		5.9
8. Metals		9.3
9. Glass		3.8
10. Polymers		9.4
11. Textiles		2.8
12. Chemicals		1.3
Total		*100.0*

In a zero waste system all discards have a place. A zero waste characterization study that sorts discards for the above-described "clean dozen" categories is recommended. Definitions for each category should be established before sampling begins to ensure consistency.

3.1.3 Storage and Collection

Collection systems for recovering resources have matured in the last twenty years. Mandatory recycling rules in states and nations have encouraged the development of technology to meet the new rules. Collection systems that started out with the customer doing sorting for plastic, paper, metal, flint and colored glass, and organic material are being challenged by collection systems that collect material separated by wet and dry distinctions. Materials Recovery Facilities (MRF) use a combination of technology (magnets, balers) and hand picking off conveyor belts to create recycling materials with high market value. Wet materials like putresibles, vegetative debris and food dirty papers are collected in sealed compaction vehicles and talken to composting facilities to be processed into soil amendment. New trucks have been introduced that collect wet and dry materials in the same vehicle by divided compartments within the truck.

Storage Containers are important for keeping materials separated. In the first phases buckets, baskets or barrels work. Today, automatic collection trucks pick up 60 – 90 gallon (227 – 341 liter) carts with wheels that are color-coded based on material type. The typical system uses blue for recyclable, green for organic and currently black for trash. Capturing the correct materials in the right container is most effective when containers are consistent in color and signage, convenient to the discard area, well marked as to what they are collecting, and large enough to handle all the designated discards between collection times.

3.1.4 Processing, Storage and Marketing

Just as the collection system feeds the processing plant, the commodity markets dictate the necessary processing systems. There exists a well-developed secondary materials market and there are uses and the corresponding demands for all 12 commodity types. International secondary fiber markets buy baled old newspaper, corrugated cardboard, high grade writing (ledger) paper, and magazines sorted grades and mixed paper. Metal smelters buy baled aluminum and non-ferrous, as well as ferrous metals (iron and steel). Sorted polymers can be formed into all kinds of product shapes and find markets. Glass must be sorted in order to have high value, but can also be used in aggregate. Organically composted materials are important feed-stocks for the agriculture industry as well as for urban gardeners.

Most of the transportation to market is done by truck and trailer rigs pulling load limits. Even if using rail, which typically is more economical for a large load, densification of the commodity is desirable. In the scrap industry, the baler is the primary technology used to densify loads for transport. Sometimes grinding is used to get more density in the bales. Still, most metal and fiber markets prefer the baled to the shredded material. Organic, ceramics and glass materials use grinding technologies to achieve transport densities as well as processing for feed-stock materials.

The processing area should have space to store load limit amounts of baled or ground material. Materials last longer under covered space. Ground up material can be stored in bunkers and loaded with a bucket-loading tractor.

Materials Recovery Facilities (MRF) are designed to recover source separated co-mingled materials. Trying to recover materials from mixed discards is problematic. The decomposing organic mater damages paper's recyclability. Source separating materials by designated recyclables, papers and containers, or wet and dry materials will enhance the quality of the baled sorted materials.

At the MRFs the collector dumps the load on acement pad. The load is checked for hazardous materials or bulky items and then pushed onto a conveyor where somewhere along the line magnets retrieve the ferrous metals. Sophisticated systems use automatic sorting technologies (section 3.1.5), but manual separation is still the best. In the USA, these jobs offer better than minimum wage and include health insurance.

Most large cities have Independent Processing Centers that reclaim fiber, resins, textiles, metals, rock and aggregate, as well as humus. Communities, institutions and individuals can sell discards to these Buy Back Centers.

Whether they are called Scrap Yards, Paper Yards, or Reuse Yards, these commodity brokers guarantee the mill, smelter, or plant, the type and the purity of the material. They know the commodity specifications and value and are essential to the system as they are the market.

3.1.5 Appropriate Technology

Source Separation is the first step in reuse and recycling. The technology of using people to sort discards is known as low technology. There is a historical presumption that almost any material can be reused or recycled if kept free of contaminants and where inexpensive hand sorting for recovery is the dominant technology. After establishing a yard debris composting facility which is publicly available and reliable, banning such materials from landfill disposal reduces reliance on disposal while directing recoverable materials to appropriate facilities. Thus keeping both material and the related jobs in the local economy.

The following schematic diagram illustrates the current use of materials in our society:

As shown in the above figure 3.1, we do not "consume" materials; we merely use them and ultimately return them, often in an altered state, to the environment. The production of useful goods for eventual use by those people called "consumers" requires an input of materials. These materials can come from one of three sources: raw materials, which are mined from the earth and used for the manufacturing of products; scrap materials produced in the manufacturing operation; and materials recovered after the product has been used.

The industrial operations are not totally efficient, producing some byproducts, which are either disposed of or used as raw material in other processes. The resulting processed goods are sold to the consumers. After the product is used, there are three options: to reuse the material for the same or a different purpose without remanufacture; to collect the material in sufficient quantities for use in energy production or to recycle it back into a manufacturing process; or to dispose of this material.

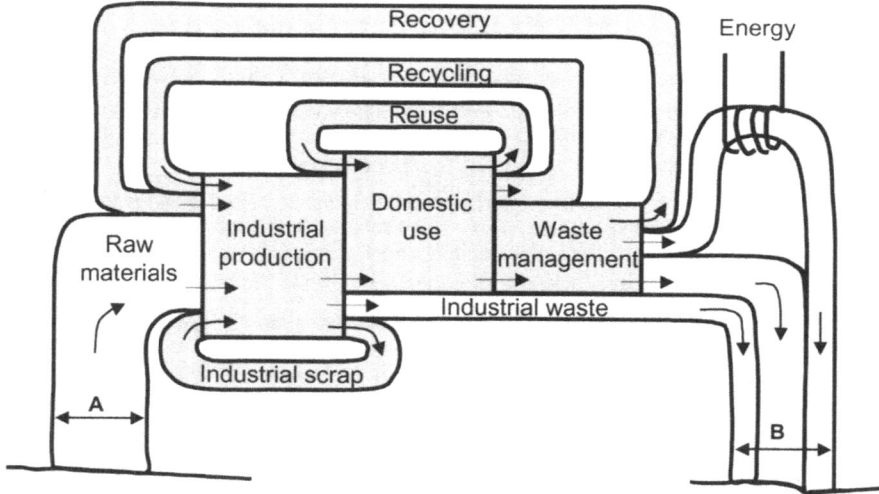

Fig. 3.1. Flow of material through society [106]

As shown in the figure, this is a closed system, with only one input and one output, emphasizing again the finite nature of our world. At a steady state, the amount of raw materials introduced into the process must equal the materials' disposal back into the environment. The closer the achievement of the zero waste goal, the less material is disposed of, thus raw material use is reduced.

The key to achieving the zero waste goal is having the appropriate source reduction, reuse, recycling and recovery programs. These programs must be supported by the applicable technology. In addition the programs must be accepted and used by the manufacturers and consumers.

Source reduction can be achieved in three basic ways:

- reducing the quantity of material used per product without sacrificing the utility of that product;
- increasing the useful life of a product;
- eliminating the need for the product.

As shown in Figure 3.1 source reduction results in a reduction in the size of the domestic use box and thus a similar reduction in the industrial production box.

Technology plays a role in each of these three options. For example, from 1972 to 2000 the amount of aluminum in an aluminum can significantly decreased. In 1972 an aluminum can weighed 0.74 ounces; by 2000 the same can weighed only 0.48 ounces or about 1/3 less. Another example is the advances in battery technology that have resulted in longer life battery. Thus one battery now lasts as long as two or three older batteries.

The preferred source reduction option is to eliminate the need for the product. This approach has been appropriately called voluntary simplicity. For example, when buying a single item at a store, it is usually not necessary to have the item placed in a bag.

Recycling involves taking material that has been used and sending it back to industrial production. The most important factor in the success of a recycling program is consumer participation. The centrally located recycling drop-off center is a low technological recycling solution which engages the public by having them do something about the wastefulness of contemporary society. A fixture of early (1970) recycling efforts to reach residential communities, these programs are quite appropriate for institutional, multi-family residential and small community recovery programs.

While a drop off site has low operational costs, it usually also results in low participation rate. Rarely do these programs engage more than 15% of the population unless the site is in a rural community and is near or adjacent to the dominant disposal site. The consumer has to be motivated to bring material to this site instead of disposing of the material in the garbage. In addition, the consumer must also sort the recyclables into various different categories.

A variation of a drop off site is a buy back site. At these sites consumers are paid for their recyclables. Because of the economic incentive, these sites have a higher participation rate. Curbside collection of residential recyclables is another option. In many cases recyclables are placed into containers, one for bottles and cans and one for paper products. Today's technology allows for paper and containers in one container. The key is to keep the paper dry. In addition, green waste, food and food dirty paper can be placed in a separate container.

Once the material is collected it is transported to a MRF. MRFs employ many different technologies for the sorting of materials. Many rely on hand sorting of recyclables. Others use automated equipment (see below). Once the material has been separated it is prepared for shipping.

Automatic sorting has been tried for mixed residential and commercial materials. In Italy there is a law that prohibits human beings from sorting mixed waste; they have therefore invented and assembled equipment for the sorting of mixed materials [9]. Currently, Germany has what is called the green dot take back program; they have assembled equipment that will sort source separated packaging materials [29]. Several generalities can be made regarding automatic sorting:

- Shredded materials make the size uniform so that other sorting techniques work better. Early experience in the US with shredders and municipal mixed materials resulted in explosions when sparks from the hammers ignited flammable materials.
- Magnets are reliable for ferrous metals. Eddy current magnets that send electricity through metals can effectively recover the non-ferrous conductible metals.
- Gravity is often used. A magnet hung high will allow non-ferrous metals to fall on the belt. Air knives, air classifiers, centrifuges, water and other kinds of liquid media are used to separate lights from heavies. Trommels are long cylinders with different size holes that allow separation by size when shedders are not used.
- Bounce adherence was demonstrated in the early nineties to effectively separate cylinders from papers. This piece of equipment, used in agriculture to separate tobacco leaf from the stems and seeds. The sytem uses a vertical conveyor under a trommel to bounce round things off on to another belt while the paper and plastic film adhears to the conveyor falls on another belt and is sent to a sorting area
- Bag breakers have been a problem. Plastic and broken glass are contaminants to compost. Spikes inside the trommels at the beginning of the line are effective in breaking the bags.
- Optical sorting with lasers is new and costly.
- Balers are state of the art in compacting materials for road load limits.

This equipment, it turns out, is very effective in sorting mixed materials that are in the same categories. Where compost from mixed municipal materials is contaminated with glass and plastic, and paper for recycling is contaminated by the food and vegetative debris in mixed material sorting programs, source separated materials enjoy the benefits of these technologies in conjunction with manual sorting. Most source separated material recovery programs use magnets and balers, while most source separated material composting programs use tub grinders and trommels.

Mixed waste technologies go against the natural flow of material use. In areas with no collection service recyclers divert discards by separating the twelve different categories of materials at the source into destination point clusters. An individual wanting not to waste would separate papers and containers (paper, metal, polymers, glass), load them into a private means of transport and take them to a buy back or drop off recycling center. Food, vegetative debris, and food dirty paper is collected in the home and taken to a composting area by the generator (resident) to an area which is usually on the property (sometimes near the vegetable or flower garden. Old broken furniture, appliances and other discards are loaded into a transportation vehicle and taken to a flea market or given to charity organizations. Chemicals, drain oil, and paint are stored and taken to the Household Hazardous Waste (HHW) roundup or center as needed. The destination points are the recycling center, compost heap, thrift store and HHW center.

Therefore an ideal system would have the public discarding the 12 categories of materials into destination clusters. These are: Paper and Containers, Organics (food, vegetative debris, food dirty paper), Discarded items (furniture, appliances, clothing, toys, tools, etc.), and Special discards (Chemicals, Construction and Demolition materials). An organized diversion program will use these clusters for developing processing centers and collection systems.

1. Co-mingled paper and containers clusters would be best processed when separated from other discards.
2. An organic materials processing center could be at the landfill or at a privately owned site.
3. A centralized Resource Recovery Park (RR park) can receive source separated recyclables, organics, construction and demolition debris and reusable goods to process, reuse, recycle and sell.
4. Household Hazardous Waste discards and construction and demolition materials are generated either in small amounts or at a specific time or event.

Two conceptual case studies are presented. They are the Resource Recovery Park for Reuse Repair and Dismantling for Recycling with a Paper and Container Recovery area and, the Organics Processing Facility. These examples are from the study for Del Norte County, California The earlier discussion on waste characteristics and the 12 categories shown are from this study. The County is still involved in siting and permitting the facility. Residuals will be shipped to a distant landfill so that the more diversion there is, the less the hauling and disposal costs will be.

Case Study: Resource Recovery Park for Reuse Repair and Dismantling for Recycling with a Paper and Container Recovery Area

In managing the Resource Recovery Park, it is important to let the market determine the details of where, how, and to whom materials move. Source separation principles should preside along with convenience, cleanliness, and satisfying the customer.

Table 3.2. Master categories, clusters and processing centers.

Twelve Master Categories of discarded materials	Clusters	Processing Centers
1.Reusable 2.Paper 3.Plant Debris 4.Putrescibles 5.Wood 6.Ceramics 7.Soils 8.Metals 9.Glass 10.Polymers 11.Textiles 12.Chemicals	*Paper and Containers* Paper, metals, glass, polymers *Organics* Food, vegetative debris, food dirty paper, paper, plant debris, putrescibles, wood *Discarded items* Furniture, appliances, clothing, toys, tools, reusable goods, textiles *Special discards* Chemicals, construction and demolition materials, wood, ceramics, soils,	*Recyclables:* Papers, plastic, glass and metal containers *Organics*: Food, vegetative debris, and food paper, putrescibles, untreated wood and sheetrock *Reuse & Repair:* Reuse, repair, dismantling, reconditioning, remanufacturing and resale of furniture, appliances, electronics, textiles, toys, tools, metal and ceramic plumbing fixtures, lighting, lumber and other used building materials *Metals*: scrap metals and auto bodies *Inert*: Rock, soils, concrete, asphalt, brick, land clearing debris, and mixed construction and demolition materials) *Household Hazardous Wastes*: Used motor oil, paint, pesticides, cleaners, and other chemicals

The Park is conceptualized as a central area where the public drops off discarded materials. The facility will have the capability and technology to 1. Sort discards for reuse, repair and/or dismantle for recycling; 2. process separately collected commingled paper and containers; 3. store bins for organics to be transferred to composting facilities; 4. store separated HHW materials for reuse and disposal; and 5. collect separated construction and demolition materials and transport if necessary (Table 3.3).

Types of services related to businesses located in a Resource Recovery Park include the following: drop-off or buyback recycling centers, food banks, repair services and retail store for resale of reuse items. Repair and reuse services include white goods (washers, dryers, refrigerators), brown goods (e.g., computers, TVs, electronics, and other small appliances), furniture, clothing, latex paint, vintage clothing, consignment, household item thrift shop, stove and porcelain refinisher, an antique restoration firm, Eco-artist, black smith, glass blower, etc. The Park would have collection points for the Composting Facility, C & D storage and/or Processing Area, and HHW storage and reuse.

Types of products recovered include the following: pipe, conduit, grills, gates, appliances, fasteners, patio furniture, tools, computer casings, toys, furniture, planter pots, discard collection receptacles, doors, fencing, furniture, cabinets, dishes, windows, lenses, glass blocks, lamps, toilets, sinks, dishes, plant pots, brick, block, stone, car parts, white goods, industrial scrap, dismantled structures, couches and mattresses, textiles, steel beams, equipment parts, metal fencing, metal building parts, and recovered dimensional lumber

HHW businesses include HHW material transporters and converters. Products and materials include left over paint, pool acids, fertilizers, pesticides, solvents and, other household chemicals.

Table 3.3. Process, Feedstock and Products

Process	Feedstock	Products
1. Sort discards for reuse, repair and/or dismantle for recycling	Discards	Materials, commodities
2. Process separately collected comingled paper and containers	Paper, containers	Newspaper, cardboard, ledgers, glass, steel, HDPE, PET, aluminum
3. Store bins for organics to be transferred to composting facility and	Food, paper, yard wastes, wood	Mulch, compost, wood
4. Store separated HHW materials for reuse and disposal	HHW	Reusable chemicals and paint, HHW
5. Collect separated construction and demolition materials	C&D	Wood, ceramics, inert, soil

Case Study: The Organics Processing Facility

Currently, more than half of the material disposed of in landfill is biodegradable or compostable. Paper is the largest quantity of organic material typically discarded for land filling. Mixed paper, when separated into paper stock grades, has a value as recyclable fiber for papermaking. The recovery of paper for recycling is discussed above. Only food contaminated paper (food paper) is considered as a potential feedstock for the Organic Management Program. Food paper includes food in wrappers, paper cups, waxed boxes, molded paper and pizza boxes. If aggregated, all these organics could be processed into mulch. The mulch could be composted by itself or with other organic nutrients into a high quality soil amendment.

There will be a large container at the Resource Recovery Park for organic material collection. This container will be transferred to the Organic Processing Center.

Businesses clustered around organics include collection and processing services for yard trimmings, food scraps, food-contaminated paper, wood, soils, and other putrescibles.

There are several levels of markets for organic material: lumber and wood recovery for reuse; live plants for reuse; mulched leafy green material for roadsides, garden beds and erosion control; mulched dense woody material for bio-mass based fuel burners; clean green and food compost; clean green and bio-solid mixed compost; and vermicompost, or worm castings, as a high grade potting soil.

Reusable lumber and furniture could be recovered and sold at the Resource Recovery Park. Mulched material can be used for roadside erosion control, or used as landfill cover. Clean green organic compost has a value for commercial farming and landscaping. Nitrogen enhanced organic compost has a higher value. There are local agricultural and nursery uses for soil amendment.

Construction and Demolition materials (C&D) can be separated at the source and brought to a materials yard and processed for resale. Businesses include collecting and processing C&D debris, deconstruction or dismantling. Products would include ceramics, concrete and asphalt, roofing materials, bricks, and mixed demolition debris.

Major potential end-uses for tires include producing crumb rubber for use in molded rubber products or rubberized asphalt.

3.1.6 Local Ordinances

In a zero waste system, rules need to be established, publicly accessible, and adhered to. Households and institutions can develop their own operating rules based on efficiency. The government or agency may by ordinance require the separation of designated recyclable and organic material just as hazardous or bulky materials require separate service. Health agencies can ban certain materials from burning and burying. It seems clear that metals, just based on recycle values for example, should never be buried or burned.

Where the government requires mandatory source separation, discarded material collection fees can be based on the types of materials collected (recyclable, organics, reusables). These charges are based on the collection and processing costs and resale value. Separate collection charges based on destination point make a lot of sense for both public and private collectors. The collector would ticket a container that is in violation of the separation ordinance. A member of the city would visit the generator and informs them of the rule violation and explain the separation procedure. A second violation can lead to a monetary fine.

Some governments require that packaging and durable products like electronic products be taken back by manufacturers. These programs require that the manufacturer add a deposit to the cost of the product that can be redeemed by a recycler when returned, or where the manufacturer pays a recycling fee to the government or a private agency to subsidize the operation of the recovery program [1, 3, 4, 34].

European law includes the Council Directive 94/62/EC of 15 December 1994 on packaging and packaging waste (Official Journal L 365, 31.12.1994). This directive covers all packaging placed on the market in the European Union (EU) and all packaging waste. The directive dictates that the Member States take measures to prevent the formation of packaging waste, which may include national programs and may encourage the reuse of packaging. The Member States must introduce systems for the return and/or collection of used packaging to attain a recovery target of 50% to 60% and a recycling target of 25% to 45%, with a minimum of 15% by weight for each packaging material. These targets are to be revised every 5 years (however, the first revision is still in process). The directive also regulates maximum concentrations of heavy metals in packages.

The Member States are to report regularly to the Commission on the application of the Directive. According to the reports from 1999, Germany, Scandinavia and Austria have been most successful concerning prevention/reuse. All member states met the overall recycling target of 25% of packaging material, the targets per material were met for all materials but plastic (these were only met in Germany, Finland, and Austria). Austria, Belgium, the Netherlands, Germany, Sweden, Denmark, and Finland have the highest recycling rates (around 45% or more).

The Directive on batteries 91/157EWG, Amendment 98/101/EG, regulates the use of heavy metals in batteries (especially mercury), thus making them more attractive for recycling. Several other directives are underway, e.g. concerning product responsibility for all electronic goods (take-back, obligation of member states to provide recycling infrastructure, recycling targets of between 60 and 80 % by weight, regulation of dangerous substances) (Directive on used electric and electronic goods 2000/0158 COD).

German law is the most advanced in the EU with regard to recycling and includes the German Packaging Ordinance [3], which regulates the waste hierarchy (reduce, reuse, recycle...) and product responsibility for all packages. This law obliges all manufacturers and distributors of packages to take back and recycle their used transport, secondary and sales packaging (recycling quota targets are established). According to the central recovery company, Dual System, the use of packaging declined from 96 to 83 kg per year and capita. A large fraction of this amount (78 kg per year and capita) is currently collected and most of it recycled [30]. The Ordinance was amended in 1998, reinforcing prevention and recycling targets and implementing the European Package Directive.

The Recycling and Waste Managing Act (Kreislaufwirtschafts- und Abfallgesetz) [4] introduced product responsibility for all goods circulated in Germany. However, this general law needs to be specified in further regulations. So far this has been done for old cars, electronic scrap, and batteries (full product responsibility: take back free of charge, recycling quota). A new law for old wood is on the way.

Industry established voluntary commitments for the take back of old cars and recycling targets (85% till 2002, 95% till 2015). The textile and carpet industry established their own reuse recycling system as a consequence of the Waste Managing Act. A large share of textiles are currently reused (40%) or recycled (50%), numerous carpet recycling facilities have been installed [31].

Producer responsibility take back laws are prevalent in parts of *Asia*. In Island Banking and Industrial Countries like Japan, Hong Kong and Taiwan, space is at a premium. Take back laws and recycling systems are prevalent in Taiwan and Japan. Taiwan has a series of take back laws that instruct manufacturers to residuals from the Island. In Japan, about 50 % of the discards are recovered for recycling, another 25% is burned for energy and volume reduction, and the residuals are buried. Much of the feedstock for the industries in these areas is recycled metal, paper or plastic. In Hong Kong, the government subsidizes wasting through a series of landfills and no disposal fees.

Several states in the *USA* have mandatory recycling laws. The States of New Jersey, Rhode Island and Pennsylvania have laws requiring the provision of recycling services. Some cites, such as Chula Vista and Poway California, require separation of designated recyclables from mixed discards. In California there are laws requiring that newsprint and plastics have recycled content, beverage containers are to have deposits and redemption centers, state agencies and cities are to have recycling plans with a 50 % reduction of waste generated, and advance recycling fees for oil, and tires. Currently the State legislature is contemplating a zero waste goal and take back laws for electronics.

United States federal law "The Resource Conservation and Recovery Act" is over twenty five year old and covers hazardous material collection, disposal and clean up, and sets basic standards for sanitary landfill operation. The Clean Air Act and the Clean Water Act control air and water emissions that are associated with typical disposal practices. The Federal government educates and funds programs that prevent pollution and promote recycling as a force against climate change.

3.1.7 Participant Education

Zero waste programs utilize source separation as a method to reclaim materials accurately and cost effectively. Involving the generator of the material in the program assures quality control at the point of separation. Program planners should include participants in the program planning process.

Whether the source of the discarded materials to be recycled is an entire city, an institution, office or home, participants in the generation of the discards need to be part of the process. In the city or office, a task force on zero waste could help provide direction to those in charge of implementing the program. Local coordinators in neighborhoods or at the business management level can work out collection location details and arrange for special pickups.

Promotion and education programs should be designed according to the needs of each generating group and maintained through out the year. Simple things like a quarterly newsletter tracking reclamation tonnage and explaining program details have a positive impact on those participating. Regular on time picks ups are the most important statement the program operators can make about their side of the program.

Planning for public education and involvement requires that program planners understand their audience, preparing formal plans, and establishing a method for evaluating programs. Using a task force made up of managers from collection, markets, and involved departments, the program goals and mission can thus be formed. Based on the goals, objectives to reach the goals for the short and long term are set. The plan is then put together on how to achieve the goals. A zero waste program is easy to monitor. Compare wasting both before and after.

Delivering an educational message, maintaining program participation, and funding activities are key challenges in making the program work. In a zero waste program the message is to eliminate waste by directing discards to reuse, recycling and composting programs. Participation in the program will vary based on the clarity of the message (What materials do you want?), and the participant's ability (Convenience) to take part (Where should I put them?).

One way to encourage participation is to explain how much wasting costs versus the costs of a zero waste program. The participant has a right and responsibility to understand the costs and liabilities of managing the discards they produce. In most cases, the citizen or employee has figured it out already. When the program is implemented , many will say "its about time."

Involving the participant in the planning, execution and evaluation of the program will help gather the data needed to make program evaluations, modifications and improvements. Participant involvement is necessary in the planning, education, execution and evaluation. Block leaders and program coordinators can provide value input as to types of material, generating areas, the necessary transport vehicles and competing interests.

When planning a Participant Education Program, a written document (plan) should be prepared using input from participants and program staff. The plan should identify the following:

- Main issues or challenges to be addressed,
- Short and long term goals to attain,
- Activities and events needed to accomplish each goal,
- Resources available for each activity and event,
- Timelines that coordinate public information with program implementation and take into account seasonal activities and events,
- A preset program to monitor and evaluate activities.

Most people are concerned about resources and the environment. In a volunteer program, participation can be anywhere from 15% to 80% depending on convenience. A drop-off recycling center will draw 15%, while a well promoted regular curbside pickup program will involve more than 80% participation.

Participation is always in the 90% range when the programs are (legally) required. Relatively few people refuse to participate in source separation programs, especially if it is the law (e.g., the packaging collection quota in Germany is currently 94% [30]. The usual reasons given for not participating in a required program include the following: inadequate knowledge, memory loss, being new in town, or temporary insanity.

3.1.8 Determining Costs and Benefits

The bottom line for most projects is the determining factor toward implementation. The bottom line must include the cost of wasting as well as the cost of recovery. Thus the cost of discard disposal starts when the material is source separated and includes the following: the cost of planning, storage, collection, and processing, as well as the reuse, recycling and composting revenue, and the cost of residual disposal in transformation facilities and landfill. The cost of monitoring and pollution mitigation at the transformation site and landfill must be included in the transformation and landfill disposal cost.

Households (numbers), container types, per unit costs (cost per ton), and land use categories, are all common denominators that can be used to calculate the cost of service or service fees. Annual costs are broken down into amortized capital (equipment and land) and operations (labor and materials). Revenue is broken down into annual material sales and service fees. In a cost benefit analysis, savings of land, resources and energy is factored in as a long-term benefit (revenue).

Using the twelve master categories, the following is a Cost/Benefit analysis of the 50-ton per day Discard System shown earlier in the waste characterization section and discussed in the recycling technology section. The clusters are based on destination points determined by markets and processing systems discussed earlier. The material is source separated and self-hauled to the center and placed into appropriate areas at the site.

Case Study: Cost/Benefit Analysis Resource Recovery Park by Cluster

Using the concept of a Resource Recovery Park as the processing centers and cost data developed by Urban Ore for The State of West Virginia to develop a reuse, recycle and organic Resource Recovery Park, the capital, operations and maintenance, revenue and total costs are estimated (Table 3.4). Note that avoided costs or disposal savings justify the project.

Table 3.4. Cost/Benefit Analysis of Resource Recovery Park by Cluster

Cluster	Capital [$/a*]	O&M [$/a]	Annual Costs	Trans/ Disp.** Savings	Sales [$/a]	t/a	Benefits (Costs) [$/Ton]
Reuse	34'817	432'311	467'128	106'425	413'700	1'419	+37
Recycling	58'475	169'928	228'403	323'925	108'410	4'319	+47
Organics	79'113	158'928	238'041	509'000	74'040	6'796	+51
Total	*172'405*	*761'167*	*933'572*	*940'050*	*596'150*	*12'534*	*+48*

* Amortization; 20 years land and structures, 6 years equipment and fixtures
** $75 dollar per ton savings from avoided transfer and disposal.

3.1.9 Measuring Diversion

One way to evaluate the success of a program is to measure the amount of discards generated. In order to measure the result of the prevention program source, reduction programs are quantified. For example, if before the prevention program there were hundreds of acres of lawns for public parks, golf courses and cemeteries that contributed lawn clippings, the prevention program implements a grass cycling policy. This policy would require mulching mowers that leave mulched grass clippings on the lawn instead of collecting them for disposal. The savings in material disposed is estimated and added to the diversion measurements as source reduced.

Other source reduction measurements include reducing paper use through two side copying and office e-mail, permanent drink cups for all employees, chemical diversion programs, etc. A substantial list can be obtained from the California Integrated Waste Management Board. Credit for reducing waste at the source provides support for prevention activities.

Transformation (burning, distillation, gasification etc) can also be calculated as diversion. Certain discards like forest slash, agricultural and hospital residue are best transformed by heat into a gas and/or inert materials. However, for the most part, the highest and best use for most discards is material recycling and composting.

The problem with a goal like "50% diversion" is, what do we do with the other 50%? If diversion of 50% of the discards is the goal for recycling, is the goal for wasting also 50%? Planners look 20 years into the future to estimate landfill needs. If 50% of the discarded resources are diverted and wasted resource generation increases at a positive number each year, in how many years will we need a new landfill?

A zero waste analysis would examine all materials currently discarded and then find a place for them. To the planner, the 20-year forecast will be zero waste. The existing capacity of the landfill is used as the zero target is approached. Materials that do not have a market must be sent back to the manufacturer with instruction that these materials are their responsibility. Ultimately violators will have their sales banned in localities where no recycling or composting options exist.

3.2 Mechanical Sorting Processes and Material Recycling

Bernd Bilitewski

The basic precondition of recycling is the separate collection of production wastes, used products and organic waste from residual waste. The form of collection and the logistics of the collected material have significant effects on the quality of the secondary materials, as well as on the economics and the environmental burdens.

- Primary material and primary energy resources can be replaced by secondary materials and secondary fuel only if the product and fuel quality is maintained.

The mechanical sorting process of separately collected waste fractions has, in spite of the development of new mechanical separation methods for MSW, still not succeeded in effectively replacing the method of handpicking. Because of this, the sorting plants still prefer hand picking. But it is a labor intensive and, therefore, expensive as well as slow method. Sorting machines have been developed to aid in handpicking efficiency and effectiveness.

The first automatic sorting plants for packaging and paper are in practical use. Because of their high investment capital needed and their large through-put, their applicability is limited to highly dense populated areas.

3.2.1 Glass Recycling

Waste container glass can be repeatedly melted without any quality loss. Separate glass collection, which began in 1992, has a recycling quota which amounts to approximately 85 % today. To achieve high glass quality, glass factories brought forth demands of low impurities and color sorted collection. The following impurities and glass qualities have been defined:

glass impurities
- glass from windows, lamps and special glass products are not suitable for the production of container glass
- wrong/mixed colors

impurities
- Ceramics and minerals
- top openings, cork, labels
- content matter

Ferrous and nonferrous metals disturb the melting process. Ceramics and minerals also have to be limited. Particles often cannot dissolve during the melting of glass, leading to inclusions, which make the recycled glasses more fragile.

The particle size distribution of shredded glass has an impact on heat transfer in the melting process. Table 3.6 shows the target demand of the glass manufacturing industries and the size distribution of two different types of shredder.

Figure 3.2 illustrates that semi-automatic glass separation has been the most frequently used method in the last twenty years [19].

To meet the new standards of color separated glass it is necessary to use an automatic separation unit to segregate glass into different colors and separate ceramics and minerals as well.

The automatic glass sorting machines – figure 3.3 –use opacity to separate glass into different colors and from fragments of non-glass material. The sensor used to distinguish the cullet functions using either infrared light or opto-electrical signals. As soon as the sensor identifies the cullets to be sorted out, a magnetic valve opens an air jet which blows off the particle(s).

Table 3.5. Impurities and quality demand by glass manufactures over time for container glass [20]

Year	1977	1992	1994
impurities max.	[g/Mg]	[g/Mg]	[g/Mg]
Ceramic, stones, porcelain	100	50	25
Nonferrous metals	15		
Aluminum		5	5
Lead		1	1
Ferrous metals	5	5	5
Organic matter	500	500	500
impurities max.	[%]	[%]	[%]
Water content	-	2	2
White glass:			
impurities of green	0.005	1	1
impurities of brown	0.01	2	2
impurities of colored white	0.02	-	-
Brown glass:			
impurities of green	10	3	3
impurities of white	5	5	5
Green glass:			
impurities of brown	15	10	10
impurities of white	10	15	15

Table 3.6. Particle size distribution of impact and hammer mill in comparison of the demanded target [59]

Particle size [mm]	Target [%]	Impact [%]	Hammer mill [%]
0 – 1.0	3	5	3
1.0 – 3.15	7	21	10
3.15 – 8.0	22	41	40
8.0 – 16.0	45	26	44
16.0 – 25.0	23	7	3

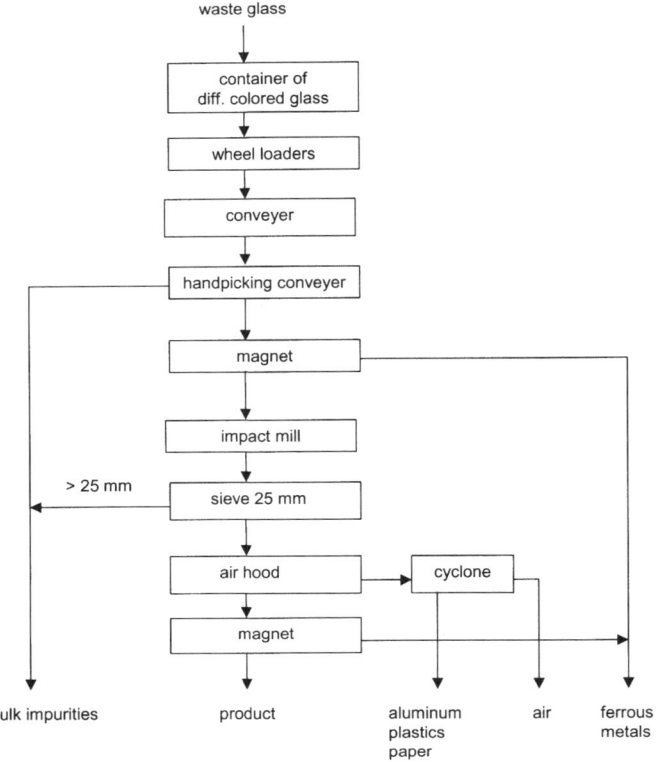

Fig. 3.2. Hazemags-separation plant for waste glass [98]

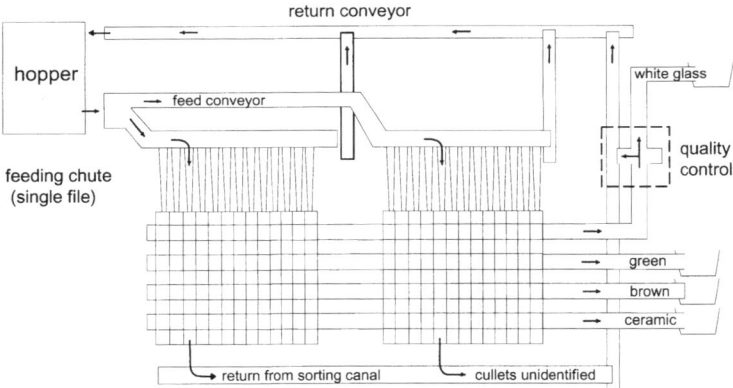

Fig. 3.3. Schematic picture of an automatic glass sorting machine [19]

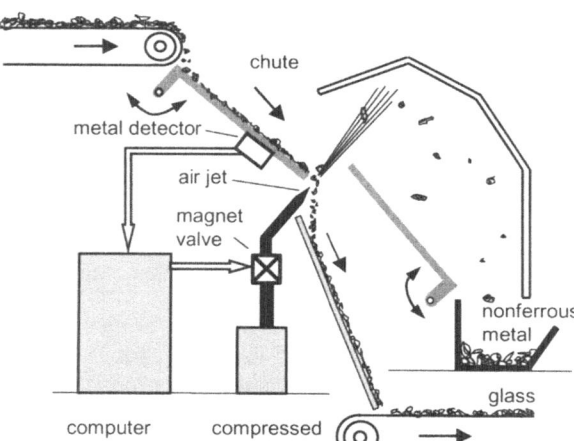

glass + nonferrous metal

Fig. 3.4. Principle of eddy current separator [20]

The efficiency of the automatic glass sorting depends on:

– amount of impurities to be separated;
– particle size distribution (very small cullet are problematic)
– an exact feed in single file to the optical sensor
– successful (effectively sorted by color) collection at drop-off or curb side locations

The purity of an automatic separation unit for white glass is approx 99,7 % - 99,8 % [7]. To meet the target standard, the glass has to pass a second unit:

Eddy current separation for non-ferrous metal can be combined with the glass separation process. Eddy current separators have been used in the recycling processes of sorting packaging, electronic scrap and automobile shredder plants. Figure 3.4 shows the schematic principles of an eddy current separator of nonferrous metal. The eddy current induces a strong repelling current which can be combined with an air jet to make sure that the separation efficiency is almost 100 %.

3.2.2 Recycling of Paper and Cardboard

In the year 2000 in Germany, approx. 60 % of the input of produced paper, a total of 10,921 Mio tons, is secondary fiber. Table 3.7 shows the development of different recycling quotes of the main paper products. Roughly 85 % of the graphical paper and 75 % of the packaging paper used in German households are separately collected. To achieve these recycling quotas, it is necessary to collect paper in as clean a state as possible and to improve the separation process.

Table 3.7. Breakdown of the waste paper input of the paper production (1990 and 2000) [38]

Main paper products	waste paper reused [ton/y]		production [ton/y]		recycling quota [%]	
	1990	2000	1990	2000	1990	2000
packaging paper/cardboard	3'847'000	6'101'000	4'166'000	6'733'000	92.3	90.6
graphic paper	1'046'000	3'411'000	5'784'000	9'125'000	18.1	37.4
hygienic paper	455'000	750'000	828'000	1'017'000	54.9	73.7
technical paper	422'000	659'000	1'095'000	1'307'000	38.6	50.4
sum	5'770'000	10'921'000	11'873'000	18'182'000	48.6	60.1

The main separation process for collected mixed waste paper is sieving and handpicking. In a normal drop-off container, 35 % of the waste paper will be cardboard, 60 % deinking quality and approx. 5 % impurities. The productivity of handpicking is 0,7 – 1,2 Mg per worker and hour [19]

As soon as graphical paper and packaging paper are collected in separate drop-off containers, the quality content improves dramatically, so that graphical paper can be accepted without any sorting process, as shown in figure 3.5.

The used technology is based on a series of air classifiers, which will be used in combination with a shredder, producing a uniform particle size and making automatic separation possible.

Table 3.8 shows the performance of an air classifier in correspondence to the particle size in weight, thickness, form and area.

The deinking quality of the recovered paper has less than 2,5 % impurities. All metal, glass particle, sand, adhesive backs of books and waste components are separated. The second fraction consists of big cardboard pieces and the third fraction of a mixture of cardboard and (approx 4 %) of deinking quality paper of in respect of the input.

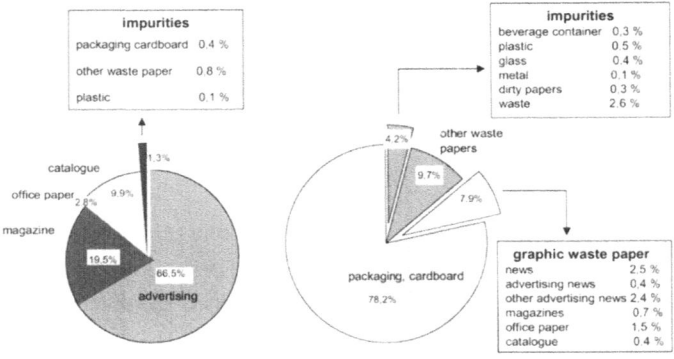

Fig. 3.5. Quality of waste paper containers by using separate collection systems for graphical paper deinking quality and card boards in Dresden [38]

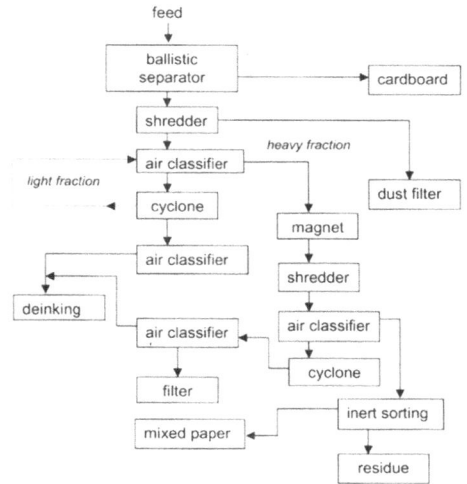

Fig. 3.6. Automatic separation plan for paper in deinking quality [73]

Table 3.8. Separation air velocity of card board particles in an air-classifier [73]

particle	form	weight	area	specific weight	thickness	volume	separation air velocity
		[g]	[mm²]	[g/m²]	[mm]	[mm³]	[m/s])
cardboard from deep frozen food	multi-corner	0.253	621	407.41	0.50	310.5	3.70
Tetra-Pack	trapeze	0.074	172	430.23	0.50	86.0	3.75
cardboard with torn side	trapeze	0.017	84	202.38	0.30	25.2	2.70
carton	trapeze	0.033	127	259.84	0.50	63.5	3.00

Deinking Technology For Recovered Paper

The flotation deinking technology for the removal of ink from a recycled pulp slurry was first established in Europe 1959. Since than the annual growth of production of DIP (De-Inked Pulp) has been 15 % p.a. Newsprint in Germany is only produced using secondary collected waste paper with a recovered paper utilization rate of 116 % [79].

Nevertheless, the utilization of DIP is a permanent challenge in the production of a constant quality, due to the variations of impurities and unwanted paper components in recovered paper deliveries, which are, for this purpose, still a mixture of newspapers and magazines mainly recovered from private households.

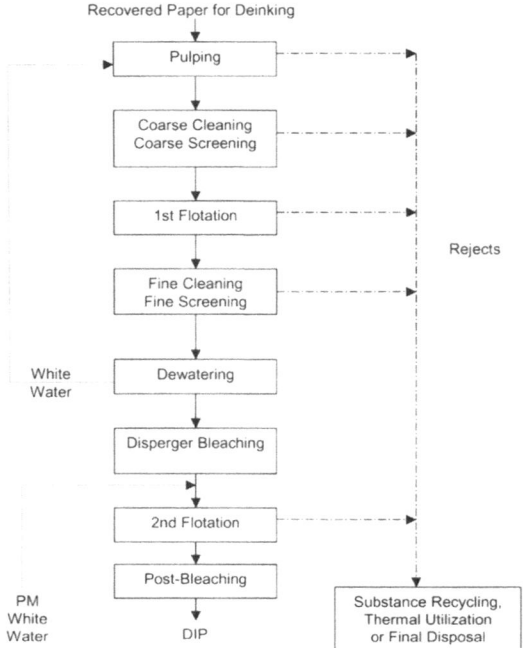

Fig. 3.7. Stock Preparation for Graphic Papers [79]

Finally, the composition of the recovered paper for deinking has changed in terms of the variety of printed matter; e.g. waterborne flexo inks, offset inks with natural binder systems, digitally printed papers, and adhesives such as self sealing envelopes, post-it notes and others.

Today´s flotation cells are mainly of a cylindrical shape, using injector aeration. For standard newsprint made from 100 % recycled fibers, today's deinking plants are commonly equipped with double flotation and a dispersion stage (Fig. 3.7). Processing based on double flotation and dispersion is intended to remove ink particles released from the fibers during pulping in the first flotation stage. For ink particles not detached from the fibers and for particles too large to be floated, the dispersing stage contributes both to the release and the break-down of smaller particles. In the second flotation stage these ink particles can be efficiently removed from the pulp.

In general, double flotation results in an improved DIP quality, not only in terms of brightness but also in residual ink content and visible ink specks. Additionally, variations resulting from the incoming recovered paper can be controlled better. In large stock preparation plants for newsprint manufacture up to 1'000 t/a DIP in one line with double flotation, the specific electric energy amounts to 350 to 500 kWh/t at a yield in the range of 75 to 82 % [79].

In general, the DIP processes need further improvements by increasing efficiency in terms of yield. This is only possible if the ink separation processes, flotation and washing, become more selective. Additionally, an advanced control of ash and stickies would be favorable for the utilization of DIP in all paper grades. The economic disposal of the rejects, energy recovery from them or their use for material recycling are pre-requisites for further DIP application in paper products. Finally, the composition of used recovered paper has to be controlled in a more sophisticated way in order to guarantee a more homogeneous raw material quality, resulting in smaller variations of important characteristics of the incoming stock as well as of the final DIP [79].

3.2.3 Recycling of Light Weight Packaging with the Green Dot System

The German business community founded the Dual System Deutschland (DSD) as a private organization. Under the DSD system, manufacturers apply for and pay DSD a fee to place their symbol, a green dot, on their packages to ensure that DSD will collect and recycle their packaging. DSD is not responsible for the actual recycling of materials; it is responsible for guaranteeing that recycling targets are met. Any product with a green dot signifies to the consumer that the package should be collected by DSD, and not returned to the retail outlet. At present, around 250 sorting plants in Germany are involved in sorting lightweight packaging collected by the Dual System.

The fully automatic SORTEC 3.0 process is divided into three steps: Dry mechanical pre-sorting, wet mechanical preparation and plastics processing (Fig. 3.8). The yellow bags are first opened mechanically. Then the lightweight packaging passes through various sieves which sort the waste according to size. After this, the fractions are transported past an air separator which blows out light plastic films and pieces of paper. This so-called lightweight fraction is forwarded directly to the central hydra-pulping step. The heavy fractions are transported past a magnet separator, which lifts out ferrous metals such as tin cans. PET bottles and beverage cartons are identified by means of near-infrared spectroscopy. All shapes, sizes and colors of beverage cartons and PET bottles are identified with the aid of spectral analysis. The unit establishes their position on the conveyor belt and compressed air valves blow them into appropriate collection containers.

The remainder of the heavy fractions is conveyed to a second pulper. Any adhering dirt and paper fibres are suspended in water in this unit and the plastics and packaging containing aluminum are washed clean. During the water treatment step, the paper fibres are removed and any pollutants are flocculated before being discharged as solid sludge. The purified water is returned to the hydrapulping process, thus forming a closed cycle.

The remaining material, primarily plastic and aluminum packaging, is subsequently shredded. Then it is routed to two series-connected sorting centrifuges containing water or a saline solution as separating medium. These separate the plastics according to their specific density. This method permits the recovery of basically homogeneous polystrene as well as polyolefine fractions [74]. The heavy material that has sunk to the bottom is subjected to eddy current separation for re-

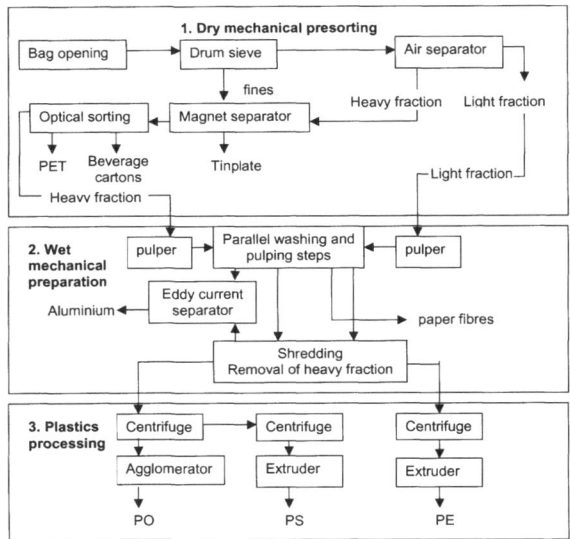

Fig. 3.8. Flow sheet of the SORTEC 3.0 [74]

moval of the aluminum components. In the subsequent plastic processing step, the homogeneous plastic fractions are melted in extruders and processed into granulate (Fig. 3.8). The material balance is shown in Table 3.9.

Table 3.9. Secondary raw material and residue in a material balance [74]

Tinplate	23.5 %
Beverage cartons	5.0 %
Paper fibres	8.0 %
Aluminum	4.0 %
PE granulate	13.0 %
PS granulate	3.5 %
PET	1.5 %
PO-Agglomerate	23.0 %
Residue (wood, textiles, stones)	18.5 %
Total	100.0 %

New Optical Sorting of Household Waste with Optibag

The contemporary concept of waste collection is the multi-bin, one used for organic waste, one for paper and cardboard, one for packaging; in large housing areas, three bins are use for glass of different colors.

The Optibag-System uses plastic bags with a thickness of al least 35 µm and a different color for each fraction. The sorting system is based on a color sensor that recognizes the color of the bag. Each color corresponds to a fraction. The system

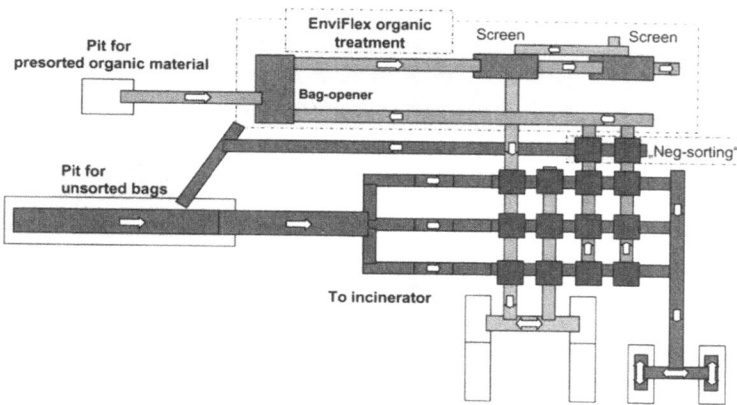

Fig. 3.9. Optibag plant in Sweden

has a capacity to handle up to 6-7 fractions. The bags are dumped into existing waste chutes or containers. A single truck collects the colored bags.

In the recycling plant, the optical system identifies the color of the bags and distributes them onto a receiving conveyer. If a bag does not match any of the color criteria, it will continue on to the end of the conveyer, together with material that is not contained in any bag and will be sent to landfill or incineration.

The sorted bags are delivered to bag openers and containers or directly to an incineration plant or biogas plant, and one or several fractions of recyclable material to sorting plants.

The Optibag-System ensures that separation at the source takes place and that transportation costs and environmental burdens are minimized. In Europe, 16 plants are in operation connecting 1.4 million people with system.

3.2.4 RDF-Production from Household Waste

Up to now, the market for substitute fuels has developed slowly. It is essentially influenced by three participating groups: the producer of substitute fuels, the users of substitute fuels and the approving authorities. The existing dissent in interests between these groups is clearly reflected in the discussion about quality standards for substitute fuels. While the authority represents primarily ecological interests and therefore judges quality standards by the emission standards and the transfer into different environmental compartments, the producers of substitute fuels maintain the role of waste processors. They define quality standards on the basis of waste as a raw material and its associated pollutant load. For the user, rather technical aspects are at the centre of interest, especially concerning operational reliability and aspects of product quality.

Table 3.10. Survey of quality standards for substitute fuels

Element	LAGA (proposal) [a] LAGA, [55] mg/MJ	BUWAL [b] BUWAL, [24] mg/MJ	BGS (proposal) [c] Püchel, [78] mg/MJ
As	1.9	0.6	0.5
Be	0.13	0.2	0.13
Cd	0.3	0.1	0.5
Co	1.2	0.8	0.75
Cr	3.7	4.0	15.09
Cu	3.7	4	35.21
Hg	0.02	0.02	0.05
Ni	3.5	4	7.55
Pb	n.a.	8	12.58
Sb	0.07	0.2	3.02
Se	0.2	0.2	0.5
Sn	0.4	0.4	7.55
Te	0.04	n.a.	0.4
Tl	0.15	0.12	0.15
V	6.7	4	1.51
Zn	8	16	n.a.
Cl	1 % by weight	n.a.	n.a.

n.a. not available. Basis for conversion of mg/kg(dry) into mg/MJ is LHV(dry) [d] equals 18'000 kJ/kg.
[a] Working Group on Waste by the German Bundesländer: Criteria for energetic utilisation in cement kilns.
[b] Federal Office for Environment, Forest and Landscape: Swiss guideline of waste disposal in cement kilns.
[c] German Quality Association for Secondary Fuels: Quality standards for substitute fuels.
[d] LHV denotes Lower Heating Value.

These contrary interests are also reflected in the ongoing discussion about the assurance of production quality and use of substitute fuels. From the different positions, standards for substitute fuels are presently suggested.

Mechanical processing techniques must be sufficiently selective to produce defined fuels. Different RDF-production facilities were investigated with respect to separation efficiency and their mass balances were determined for each element. Therefore, the input and output streams were sorted and the mass distribution of waste fractions connected with the fraction-specific concentrations. This mathematical model was useful for the understanding of the relative contribution to each output stream. Sankey diagrams offer a descriptive method for displaying material and energy flow balances. Additionally, Figure 3.10 shows the impact of the considered elements by individual waste fractions combined into four differently shaded groups. The grey tones represent the classification into

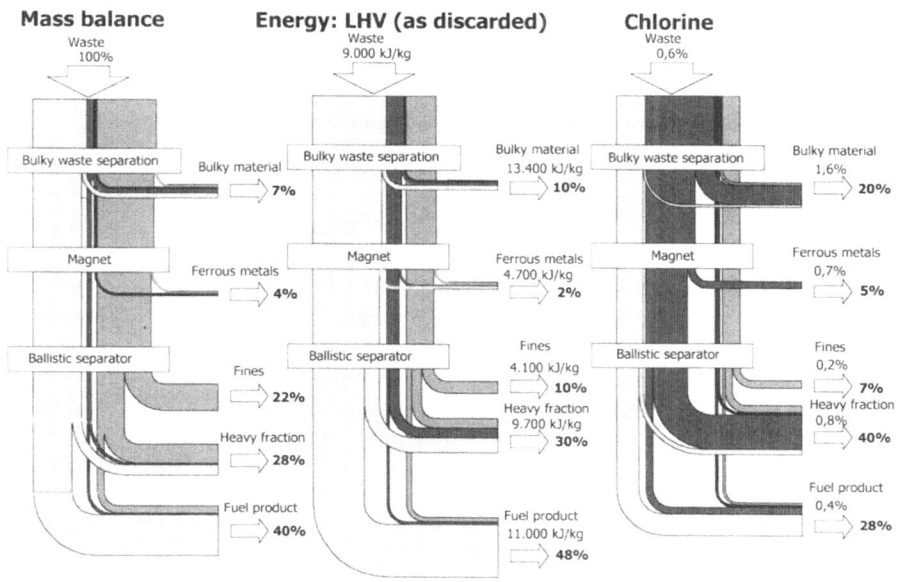

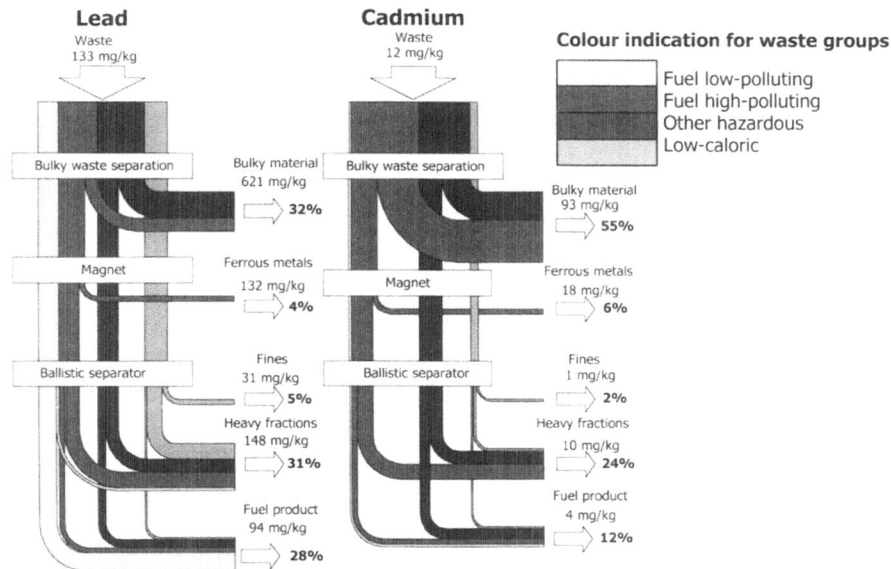

Fig. 3.10. Energy- and material flow balance of mechanical waste processing by ballistic separation, displayed as Sankey diagram; raw material is waste from urban areas [52]

- high-caloric fractions of small pollution impact (wood, paper/cardboard, plastic foils and - packaging);
- high-caloric fractions of high pollution impact (non-packaging plastic products, shoes, leather, rubber, other composites);
- low-caloric or inert fractions (organic waste, minerals, metals, fines);
- hazardous fractions (batteries, luminous tubes, electronic waste, household chemicals etc.).

From the examined mechanical processing of household waste (e.g. sizing, air classifying, automatic plastic detection), only the separation in the ballistic separator is used to ensure a sufficient recovery of paper and cardboard. The desired selectivity of the separation process was indicated by the reduction in concentration for the three regarded pollutants, while the caloric values increased compared to the input. Only small proportions of fuel with low pollutant contents in the household waste are lost with the heavy fraction group. In the heavy group selective pollutants are enriched. The example clearly indicates another pollutant sink. The high proportion of bulky waste in the household waste at the practical test was notable. The material balances show that an effective separation of bulky fractions is an important part in the production of quality-secured fuels [52].

3.3 Conventional Thermal Treatment Methods

Michael Beckmann and Reinhard Scholz

3.3.1 Introduction

Various processes, which integrate pyrolysis, gasification and combustion fundamental structural units, are currently being applied and tested in the field of the thermal treatment of municipal waste and similar industrial waste. The main thermal processes can be broken down into two units:

- First, for the conversion of the solid and pasty waste
- Second, for the treatment of the gas, flue dust or pyrolysis coke produced in the first unit (Table 3.11).

The conventional thermal treatment of residual waste, also often called "classical combustion of residual waste," represents a tried technology. The processes for thermal waste treatment can be classified into the so-called main thermal processes and processes or plants for flue gas purification, energy conversion, ash treatment, production of supplementary agents, etc. Classical waste combustion can be viewed as a combustion-post-combustion process with a grate system in the first unit and a combustion chamber system in the second unit. In relation to the classification in Table 3.11, Figure 3.11 shows the profile of a complete process as an example.

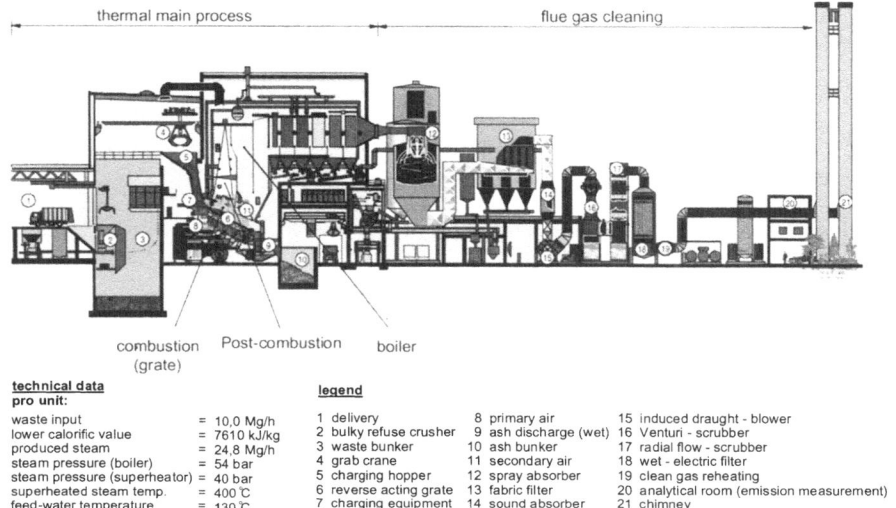

combustion   Post-combustion   boiler
(grate)

**technical data
pro unit:**

waste input	= 10,0 Mg/h	
lower calorific value	= 7610 kJ/kg	
produced steam	= 24,8 Mg/h	
steam pressure (boiler)	= 54 bar	
steam pressure (superheator)	= 40 bar	
superheated steam temp.	= 400 ℃	
feed-water temperature	= 130 ℃	

legend

1 delivery	8 primary air	15 induced draught - blower
2 bulky refuse crusher	9 ash discharge (wet)	16 Venturi - scrubber
3 waste bunker	10 ash bunker	17 radial flow - scrubber
4 grab crane	11 secondary air	18 wet - electric filter
5 charging hopper	12 spray absorber	19 clean gas reheating
6 reverse acting grate	13 fabric filter	20 analytical room (emission measurement)
7 charging equipment	14 sound absorber	21 chimney

Fig. 3.11. Schematic representation of a waste power plant (classical municipal solid waste combustion plant) [100]

Table 3.11. Systematic description of main thermal processes [87]

	Unit 1	Unit 2	Processes and Examples
A	combustion[1]	combustion	combustion - post-combustion-process (e.g. standard waste incineration) [27, 51, 56, 67, 77, 81 83, 89, 100]
B	thermolysis[2]	combustion	thermolysis - post-combustion-process (e.g. Schwel-Brenn-Verfahren by Siemens KWU) [18]
C	gasification[3]	combustion	gasification - post-combustion-process (advanced standard waste incineration) [17, 82]
D	thermolysis	gasification	thermolysis - post-gasification-process (e.g. Konversionsverfahren by NOELL [36], Thermoselect-Verfahren [94] etc.)
E	gasification	gasification	gasification - post-gasification-process (e.g. gasification and gas decomposition by LURGI [6])

[1] Includes the processes drying, degasification, gasification and combustion
[2] Includes the processes drying, degasification and pyrolysis
[3] Includes the processes drying, degasification and gasification

The flue gas purification (so-called secondary measures) of the 'classical' process in particular has been improved in recent years. Plants equipped with state-of-the-art technology meet the statutory specified limits for the emission of pollutants into the air, water and soil.

The current priority is the development, modeling and optimization of the process control (so-called primary measures) of the main thermal process. A considerable potential for development is present in the areas of grate and post-combustion in order, for example, to

- reduce the flue gas flow (flue gas purification plants, emission loads);
- improve the energy utilization;
- influence the characteristics of the residual material.

Considerable progress has been made in development with regard to the 'classical' processes with grate systems. The optimization of the design for the combustion chamber [27, 56, 77, 100], flue gas recirculation [83, 89], enrichment of the primary air with oxygen [51], water-cooled grate elements [56], further development of the control (e.g. IR-camera, [67, 81]), etc. should be noted as examples.

In addition, new, altered process controls with grate systems, such as e.g. the gasification with air on the grate, and the connected independent post-combustion of the gases produced are being tested on the pilot plant scale for future developments [17] and, recently, also on small industrial scale [8]. The desired improvements toward the above-mentioned objectives are then even more pronounced.

In the following section, general points of the process control in grate systems will be illustrated briefly. Then examples of process control practically implemented in plants and results of experiments in industrial and pilot plants as well as on the laboratory-scale will be discussed.

3.3.2 Process Control in Grate Systems

The efficiency of the thermal process for the treatment of waste is primarily determined by the process control of the main thermal process. The possibilities available for the reduction of pollutants and flue gas flows through primary measures must be exploited. The expenditure necessary in the range of secondary measures, e.g. the flue gas purification, is then adjusted accordingly. Classical waste combustion can, as mentioned at the beginning, be divided into two units. Combustion on the grate takes place in the first unit. Then, the gases and flue dust arising from the grate are combusted in the post-combustion chamber in the second unit.

When discussing the partial steps of the solid conversion and the combustion of the gases generated, the main influencing parameters mentioned in Figure 3.12 (e.g. [87]) must be considered. This is of particular value when regarding the individual reaction mechanisms (e.g. formation and degradation of pollutants). The various control possibilities along the reaction pathway are of particular importance for the optimization of the process.

Solid Conversion in Grate Systems

The solid conversion on a grate can, more or less, be divided into the following partial processes in the direction of the reaction pathway (Fig. 3.13): drying, degassing, gasification and residue burn-out.

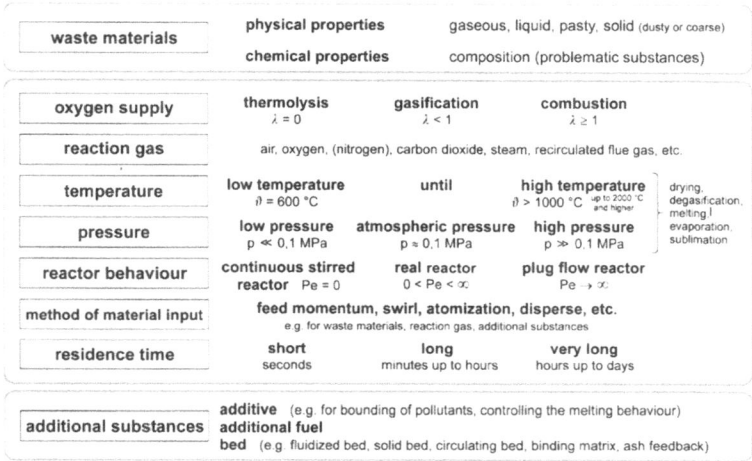

Fig. 3.12. Main influential parameters [87]

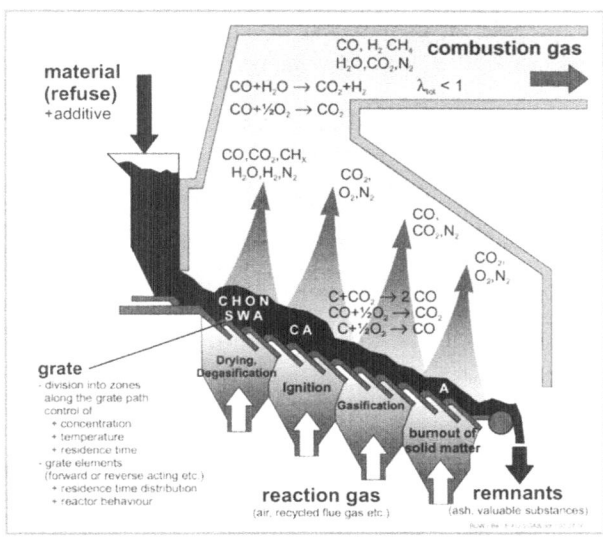

Fig. 3.13. Schematic representation of the solid conversion in grate systems during the combustion process. (CHONS: elements, W: water, A: ash).

Since the oxygen supplied to the grate (usually air) relative to the waste leads to over-stoichiometric conditions (e.g. $\lambda \approx 1.3$), it is termed "combustion on the grate."

In order to be able to discuss process-engineering options, the most influential parameters and their distribution along the reaction pathway must be considered. There are many different options for the control of the main parameters in grate systems that can selectively influence individual partial processes along the reaction pathway (Table 3.12). These control options are particularly advantageous when dealing with variations in the reaction behavior due to fluctuations in waste composition. The following representations of process-engineering options are also depicted in the overview Table 3.12 and the flowchart in Fig. 3.14.[1]

Quench reactions at the surface of the wall should be avoided by maintaining high wall temperatures using the following:

Suitable refractory lining should be guaranteed for both units (grate and post-combustion). This also means that the heat transfer must be carried out separately from the solid conversion and the post-combustion as far as possible, i.e. combustion and heat decoupling are to be connected in series as shown in Figure 3.14.

The two partial steps, solid conversion on the grate and post-combustion process, can be optimized more easily, the more decoupled they are. In order to achieve this, either a geometric separation is possible, or, as will be explained in connection with the post-combustion process, a fluid dynamic decoupling.

The division of the grate area into several grate zones, which are separate from one another with respect to the reaction gas supply must also be taken into consideration when evaluating the control options of grate systems. This makes it possible to control the amount of oxygen available and the temperature in the bed along the reaction pathway through variation of the supply of reaction gas so that the individual steps can be matched to suit the characteristics of the starting material, even with fluctuating composition.

An additional leeway with regard to distribution of oxygen and temperature is achieved when inert gas, recycled flue gas or additional oxygen is fed instead of air. A preheating of the reaction gas can also be considered for waste materials with low heating values.

By the variation in distribution, the oxygen content and the temperature of the reaction gas, the burn-out of residues from waste with varying heating values and mass flows in particular can be optimized.

---

[1] In this case, the main focus is on the process control for solid input material in grate systems. When considering the process control possibilities, the incineration of gaseous, liquid and dust-like feed material must first be discussed since generally only one unit is necessary for these. Then, using this information as a basis, one can shift to the process control for lumpy and pasty materials which generally require two units. Since the systematic representation would take up too much space, the procedure and references are mentioned here (e.g. Scholz und Beckmann [83, 87, 88]).

Table 3.12. Characterization of grate systems [85]. Process engineering options

Starting materials	
	Lumpy, also pasty when in connection with a solid or inert bed
Amount of oxygen available	
Level	Usually overstoichiometric (combustion), understoichiometric (gasification) possible; making independent post-combustion possible; oxygen exclusion (pyrolysis) not customary
Control along the reaction pathway	Very well adjustable (e.g. air/oxygen gradation, flue gas recirculation, etc.) when separated into individual zones; the partial steps of drying, degassing, burn out of the solid can be influenced in connection with temperature control
Temperature	
Level	Bed surface temperatures of up to ca. 1000 °C and higher; average bed temperature lower
Control along the reaction pathway	Very good possibilities through separation into several zones as with the control of the oxygen concentration (preheating of the air, flue gas recirculation, water/vapor cooling)
Pressure	At standard pressure, generally only a few Pascals underpressure due to the plant construction
Reactor behavior	
Solid	According to the movement of the grate elements, the individual zones can approach continuously stirred reactor characteristics (e.g. reverse-acting grate), or a plug flow reactor characteristic (e.g. traveling grate); over the entire reactor length, a plug flow reactor characteristic results
Gas	a) Oxidizing agent etc. is forced to flow through the bed and is distributed evenly over the bed surface, resulting in very good contact between gas and solid b) Flow control over the bed is possible as counter and parallel flow, gas treatment in the following process step is necessary (e.g. post-combustion)
Residence time	
Level (average residence time)	In the range of several minutes to hours; adjustable through grate speed and mass flow and can be influenced in the project design through total length and width
Control along the reaction pathway	Very good adaptation possible through separate speed regulation of the grate elements in the individual zones; control through discharge roller if necessary for additional improvement of the burn out at the end of the grate
Additives	Additives for the binding of pollutants into the solids and influence of the characteristics of the residue (ash, partly fused ash, slag); inert bed e.g. matrix for possible easily melted substances (e.g. plastic)
Functional range (examples)	
	For the solid conversion in the first step in municipal solid waste combustion plants; separation of metals from the composite at low temperatures and simultaneous understoichiometric conditions

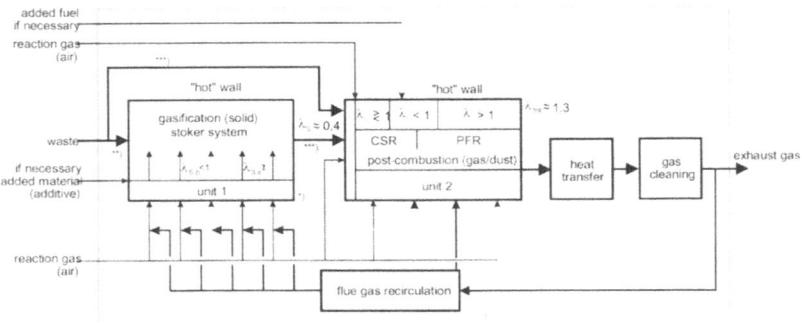

Fig. 3.14. Schematic representation of the separated process control; combustion/gasification-post-combustion process with grate and combustion chamber system [83, 88]

The gas flow over the bed can be controlled by applying parallel, center or counter flow. In parallel flow, post-combustion of the gases emitted from the beginning of the grate occurs first. In the counter flow, the hot gases from the end of the grate convey heat to the drying and degassing zone at the beginning of the grate. The post-combustion of the gases can be optimized independent of the application of parallel or counter flow regimes if separated process control of the grate (1st unit) and the post-combustion process (2nd unit) is available. This makes it possible to implement primary measures for the reduction of NO_X and at the same time to optimize the burn-out. Similarly, the temperatures in the drying and degassing stages do not depend on the flow control. The heat transfer rates can be influenced by the so-called secondary heating surface effect, i.e. by radiation exchange between the surrounding refractory-lined hot walls and the surface of the material.

An additional degree of freedom in the optimization of combustion is obtained by the individual movement of the grate elements in the different zones. This means, on the one hand, an increased construction expenditure, but on the other hand, the residence time and bed height can be controlled independently in the individual reaction zones.

By additional water-cooling of the grate elements, the reaction gas distribution, oxygen concentration and temperature can be varied over a wider range. Using the above-mentioned measures, the excess-air coefficient, the flue gas and flue dust flows can be reduced. The temperature in the burn-out stage can be increased to the point that ash sintering occurs, which leads to a marked improvement of the elution behavior of the residues [94, 105]. Furthermore, the water-cooled grate results in a lower risk of thermal damage for waste materials with high heating values, and reduced ash sifting through the grate (low thermal expansion permits little clearance between the grate elements).

The control of the combustion process on the grate using an infrared (IR) camera for the detection of the temperature at the bed surface has been mentioned. The position of the main reaction zone can be identified and controlled by redistributing the primary air. Through selective change of the primary air (reaction gas input), a fast adjustment to fluctuations in the waste composition, an avoidance of gas blow-through and streaming, an improvement in the residue burn-out and the avoidance of emission peaks can be expected.

Post-Combustion Process

The main parameters controlling the post-combustion process (Fig. 3.12) can be discussed in the same manner as for the solid conversion process. An overview is shown in Table 3.13.

The residence time is an important parameter in connection with the formation and decomposition of pollutants. In chemical engineering one differentiates between two limiting cases with respect to the residence time behavior: the ideal continuously stirred reactor (CSR) and the plug flow reactor (PFR) [58]. Figure 3.15 shows the decomposition of CO as a function of the residence time for these two reactor types, together with a real reactor. It can be seen that the PFR exhibits the best characteristics for CO oxidation. In the post-combustion process, a CSR must be connected to ensure mixing of the gas before entering the PFR (Fig. 3.14, second unit, first stage).

Mechanical structures generally cannot be applied for mixing due to high temperature, corrosion, etc. Two fluid dynamic stirring mechanisms can be used:

- so-called supercritical swirl flow with linear recirculation or recirculation of hot gases [25, 57] for plants with low capacity or
- single or multiple jets arranged above or next to each other which act as injectors to draw in and mix the surrounding gas [83, 84].

The installation of an intense mixing zone (CSR element) at the entrance of the post-combustion leads to the above-mentioned hydraulic separation as represented in Figure 3.14. A relaxation zone (PFR as suitable reactor type for complete oxidation) is situated above the mixing zone.

High intensity in the mixing zone must be achieved in order to guarantee optimum burn-out (CO, hydrocarbons, dust). Plants with greater capacities are generally equipped with injector jets.

The amount of oxygen available along the reaction pathway can be controlled by air staging, fuel staging, flue gas recirculation, oxygen supply, etc.

Sufficient options are available to optimize the process control with respect to the mechanisms of formation and decomposition of pollutants. At this point it should be noted that extensive experience and knowledge exists for the reduction of NOx and the improvement of the burn-out of gaseous, liquid and powdery fuels. This experience should be transferred to the post-combustion.

Table 3.13. Schematic representation of the separated process control; combustion/ gasification-post-combustion process with grate and combustion chamber system [85]

Starting materials	
	Gaseous, liquid, dust-like
Amount of oxygen available	
Level	Overstoichiometric to understoichiometric; variable over wide ranges; if overstoichiometric at reactor discharge: called "combustion chamber"; if understoichiometric at the end: called "gasification reactor"
Control along the reaction pathway	Very good through gradation of oxidizing agent and fuel along the reaction pathway (introduced over stirred reactor elements)
Temperature	
Level	Different combustion temperatures in the range from 1000 °C to 2000 °C or more; range is very variable
Control along the reaction pathway	In addition to the staging of oxidizing agent and fuel along the reaction pathway, intervention by flue gas recirculation, spraying of water etc. possible; indirect heat coupling and decoupling through corresponding heating or cooling systems; many possibilities
Pressure	At standard pressure, generally only a few Pascals underpressure due to the plant construction; high-pressure combustion rare; pressure gasification more common
Reactor behavior	
Dust/gas	Hydraulically, stirred reactor as well as plug flow characteristics can be approached for dust and gas
Residence time	
Level (average residence time)	In the range of seconds (longer at higher pressure); adjustable through load conditions and can be influenced in the project design through geometric dimensions
Control along the reaction pathway	Very difficult; residence time distribution can be controlled over the reactor behavior
Additives	
	Additives, in particular, introduced over the stirred reactor elements, in order to bind pollutants (e.g. sulfur dioxide, nitroxides) as well as to influence the slag characteristics and melting points of the dust
Functional range (examples)	
	Combustion of liquid residues; post-combustion of gas and dusts in the second step of the thermal treatment process; high temperature gasification of residues for the production of process gas (low and high temperatures); certain combustion processes (e.g. recirculation of chlorine as hydrochloric acid in the production cycle etc.)

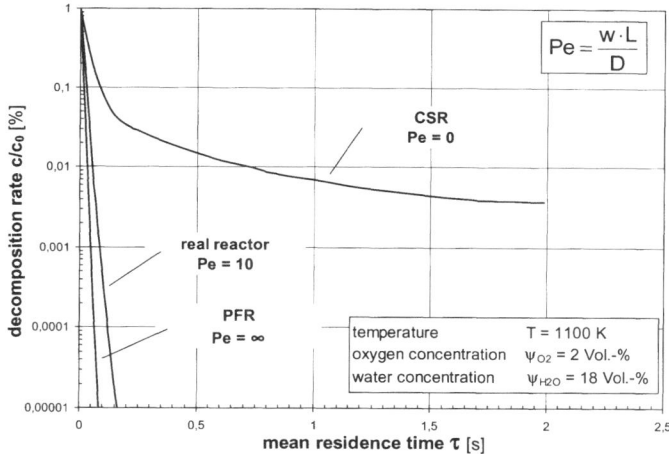

Fig. 3.15. Dependence of the CO decomposition on the average residence time and the mixing conditions in the reactor [32, 87]

3.3.3 Examples and Results

The following discussion of the effect of given process controls is important to the understanding of the processes taking place during waste incineration. However, no estimation of the total concept of the plant can be derived from the evaluation of the individual aspects.

Solid Conversion on the Grate

The main goals of the solid conversion on the grate are the achievement of high ash burn-out (low residual carbon content) and a low concentration of heavy metals and salts.

These objectives are influenced, aside from by the composition of the waste, mainly by temperature, oxygen concentration and residence time as well as reactor type with respect to residence time. A distribution of 'ignition cores' [67, 68] and stabilization of the temperature in the bed is achieved by intensive mixing using the reverse-acting grate system. Due to the early ignition of the bed in reverse-acting grate systems, the addition of primary air generally already reaches its maximum in the second primary air zone. The maximal temperature in the bed is also reached in this section. Reverse-acting grates can be approximated in the individual sections as CSR elements [12] [13]. A forward-acting grate system can be considered as a PFR element. The input of the primary air into the drying and ignition phase must be regulated carefully in order to avoid a break in the ignition front. The main air is generally added in the middle of the grate. The primary air

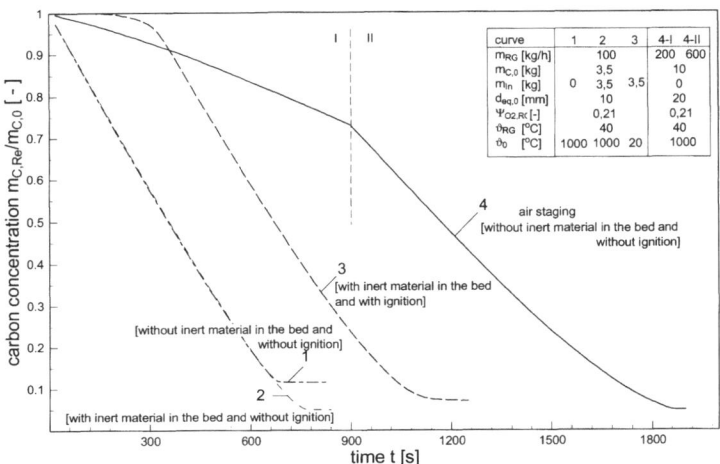

curve	1	2	3	4-I	4-II
$\dot{m}_{RG}$ [kg/h]		100		200	600
$m_{C,0}$ [kg]		3,5			10
m_{in} [kg]	0	3,5	3,5		0
$d_{eq,0}$ [mm]		10			20
$\Psi_{O2,RC}$ [-]		0,21			0,21
ϑ_{RG} [°C]		40			40
ϑ_0 [°C]	1000	1000	20		1000

4 air staging
[without inert material in the bed and without ignition]

3 [with inert material in the bed and with ignition]

[without inert material in the bed and without ignition]
1

2
[with inert material in the bed and without ignition]

time t [s]

Fig. 3.16. Calculated carbon concentration vs. time for a given bed [11]

in the burn-out zone must be adjusted carefully for high burn-out. A too high primary airflow in the burn-out zone can lead to a 'blow cold' (extinguish) of the bed and a corresponding high carbon content in the ash. The inert material acts as a heat reservoir which takes up heat at high temperatures in the main combustion zone. Due to this heat reservoir, the risk of 'blow cold' in the burn-out zone is reduced. This positive influence of the inert material should be considered in discussing pre-treatment of MSW before incineration, such as mechanical-biological processing: Reducing the amount of thermal mass in the bed might be contraproductive.

The influence of air staging and of inert material on the burn-out can be clearly shown by calculations using mathematical models for the solid conversion [11]. The calculated normalized carbon content in the bed over time for various boundary conditions is shown in Figure 3.16. A comparison of curves 1 and 2 in Figure 3.16 clearly shows the influence of the inert material as described above. A significantly better carbon conversion is achieved with inert material than without. Curve 4 shows the influence of air staging for a fuel without inert material in which the residual carbon content is as low as that of case 2 (no air staging).

To ensure an even distribution of primary air, the pressure loss in the grate element must be much higher than that in the waste bed, independent of the type of grate.

For forward-acting grate systems, a control of the independent grate speeds in several grate zones is generally possible. However, for reverse-acting grate systems, such a differentiated control of the residence time is not carried out. A so-called discharge drum for the additional control of the residence time in the burn-out zone is used more often in reverse-acting than forward-acting grates [68].

The course of combustion is influenced, in addition to by the distribution of the primary air, by the temperature and oxygen concentration of the primary air (reaction gas). The primary air is generally preheated to temperatures of $\vartheta = 140$ °C, which is particularly advantageous for the drying and burn-out phase. Higher bed temperatures are also reached in the main combustion zone by preheating the air. An increase in the oxygen concentration of up to 35 vol.-% in the reaction gas has been carried out in e.g. the "SYNCOM Process" [51]. This guarantees good burn-out [2] and causes a reduction in the flue gas flow (see below). In addition, the reduced gas flow through the bed leads to decreased entrainment of dust particles.

The temperature of the bed is influenced by the gas flow regime applied. The temperature in the drying and burn-out phases is influenced by the design of the combustion chamber geometry (secondary radiation surface). This so-called secondary heating surface is more important in the burn-out zone of a counter-flow combustion than a parallel-flow combustion.

The advantages of using water-cooled grate elements have been discussed above. In addition they can result in further improvements with respect to the ash quality (increase in temperature and reduction of the oxygen concentration), as pilot plant experiments have shown [22, 37, 105]).

The primary air supply, grate speed, waste feed, etc. are controlled in order to ensure an even course of combustion and highest possible burn-out [100]. This firing power control can be supported by appropriate temperature detectors increasingly including methods such as e.g. IR cameras [67], pyro-detectors [81], 'heating value sensors' [103].

Whereas process control for a good burn-out is state-of-the-art in the practice, the optimization of the process parameters with respect to the selective release or immobilization of heavy metals in the ash (primary measures for the improvement of the ash quality) is still a current research topic. Approximately 300 kg/Mg$_{waste}$ grate ash remain after waste combustion, from which ca. 40 kg/Mg$_{waste}$ scrap iron can be separated. Currently, ca. 3 million Mg/a grate ash accumulate in Germany, of which ca. 50 % to 60 % are used as secondary building material in road construction and the rest is deposited in landfills [104]. The aim of future developments must be the improvement of the ash quality by primary measures in order to make a high-grade utilization possible.

Depending on the combustion conditions, heavy metals, chlorine, sulfur and fluorine, etc. are released from the fuel bed and bound by the ash.

Investigations in a pilot plant using various waste compositions have shown that the evaporation of heavy metals (e.g. Cu, Zn and Pb) correlates to the bed temperature and chlorine content (Fig. 3.17, Fig. 3.18; MW = municipal solid waste, reference fuel; CSR = car shredder residue - light-weight fraction; ES = plastic scrap from electronics; PVC = polyvenil chloride) [105].

An additional parameter governing evaporation is the oxygen partial pressure. Results from experiments with radioactive tracers (^{69m}Zn) in a pilot plant [22] clearly show the differing behavior for the release of zinc under reducing and oxidizing conditions. The measured zinc evaporation for three different operation conditions is shown in Figure 3.19. The two parameters which were varied on the grate were the oxygen partial pressure and temperature. The redox conditions of

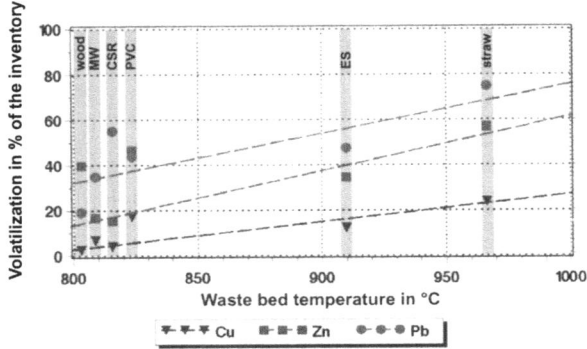

Fig. 3.17. Relationship between volatilization and waste bed temperature [105]

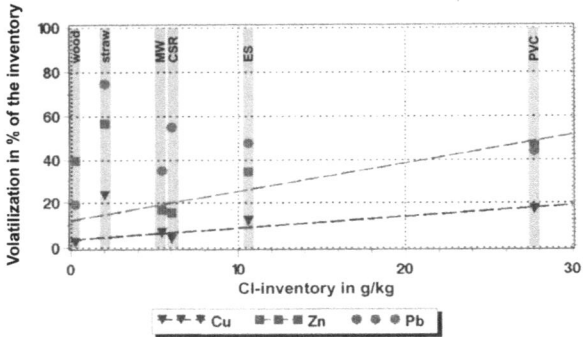

Fig. 3.18. Relationship between volatilization and chlorine (Cl) inventory (abrev. see text) [105]

the incineration process were changed from combustion to gasification operation by decreasing the flow rate of primary air to one third. It must however be noted that even in the so-called combustion operation mode, reducing conditions dominate in the solid bed in the first primary air zone. The temperature variation was achieved by adjusting the water content in the waste material. Figure 3.19 reveals that the trend of zinc evaporation is independent of the operating conditions and the amount of evaporated zinc. The evaporation takes place very fast, i.e. in a narrow area on the grate. However the location at which the evaporation takes place on the grate depends on the operating conditions. Complete evaporation occurred at locations with sufficiently high temperatures and with reducing conditions. For the experiment in which only 50 % evaporation of the radio tracer was observed (wet, oxidizing), the location of evaporation was shifted further down the grate. This was due to a delay in reaching the necessary temperature.

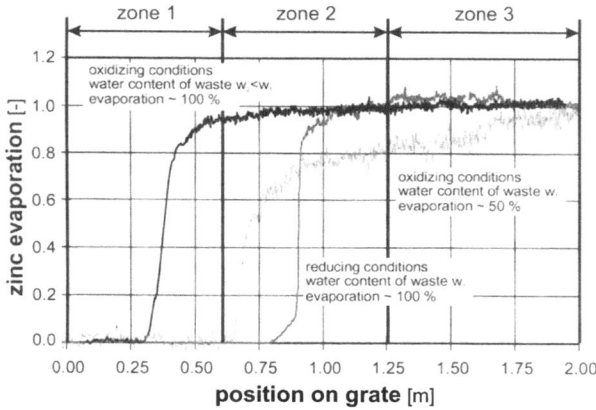

Fig. 3.19. Normalized Zn amount detected in flue gas scrubber after the pulsed addition of Zn tracer at position 0. Accumulation is monitored while tracer is transported along the grate (0.0 m to 2.0 m). oxi.: oxidizing conditions; red.: reducing conditions (gasification); w: water content of waste; evaporation: % of tracer transferred to flue gas [22]

The shift in the temperature profile brings the evaporation zone for zinc into the region where oxidizing conditions prevail (air zone 2). Increasing amounts of zinc oxide were produced which were not volatile and, therefore, complete zinc evaporation was not possible. The hypothesis that a thermal mobilization of heavy metals at high temperatures is supported by the reducing effect of the hot carbon in the fuel bed is referred to in [105].

It should also be mentioned that a marked decrease in the extractability of pollutants from ash can be achieved by sintering, and that the transition from sintering to melting achieves only relatively small further improvements [105].

Post-Combustion

As outlined in the previous section, the main objective of the post-combustion process is primarily the degradation of CO and organic trace elements as well as the burn-out of fly dust (a total high burn-out). The minimization of NO_X by primary measures is also being investigated for practical implementation and is state-of-the-art in burners for gaseous, liquid and powdery fuels [49, 65, 109, 110, 111].

The most influential parameters in the achievement of this objective are temperature, oxygen concentration, residence time and residence time behavior. These parameters are determined by a series of individual aspects of the process control and not only by the flow form, which is often discussed in connection with an initial post-combustion of the gases (e.g. parallel flow) or with the minimization of NO_X in the combustion chamber above the bed.

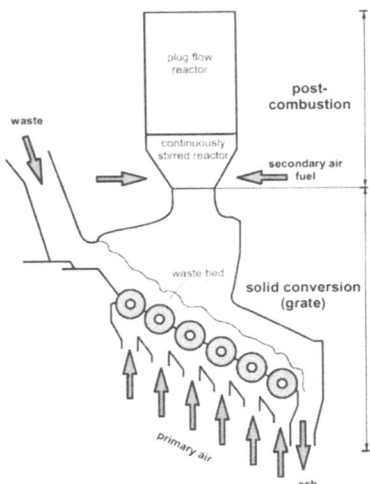

Fig. 3.20. Schematic representation of the grate and post-combustion zone [86]

The separation of the post-combustion from the grate as shown in Figure 3.12 is discussed in more detail in [87]. The separation does not have to be geometric; it can also ensue by fluid dynamic means. A fluid-dynamic separation (Fig. 3.20 and 3.21) can also be very effective when, for example, a good overlap of the cross-section, adequate suction of the residual stream to be mixed and sufficient penetrating depth of the injector jets are achieved. Optimization of these injectors generally leads to fields of jets. The positioning of the injection along a narrowed cross-section in the post-combustion zone assists the mixing (Fig. 3.21 and 3.22). This fluid-dynamic separation also leads to the formation of the important CSR element. The settling section is then located above the fields of jets (Fig. 3.21 and 3.22) (PFR zone).

The average velocity w_x of an injector stream which penetrates the combustion chamber above the secondary air jets with a speed of w_0 decreases hyperbolically with the penetration depth of the stream x. In the combustion chamber, the stream is diverted in the direction of the up-streaming gas y. The flow rate ratio of the up-streaming gases w_y to the exit velocity of the stream at the jet w_0 is proportional to the square of the penetration depth x^2. Achieving high-quality mixing in plants with large cross-sections is a problem that can be solved by the installation of a so-called prism in the combustion chamber (Fig. 3.22). In this system, first installed in the MSW-Incineration plant in Bonn, the flue gas from the grate is divided into two partial flows "A" and "B" by a membrane-wall construction in the shape of a prism. The prism is water-cooled and protected with refractory materials. Secondary air is injected into the divided flue gas streams "A" and "B" as indicated in Figure 3.22, reducing the necessary jet length correspondingly. By the installation of the prism, a much shorter and more clearly defined burn-out of the flue gases just above the prism could be achieved [77].

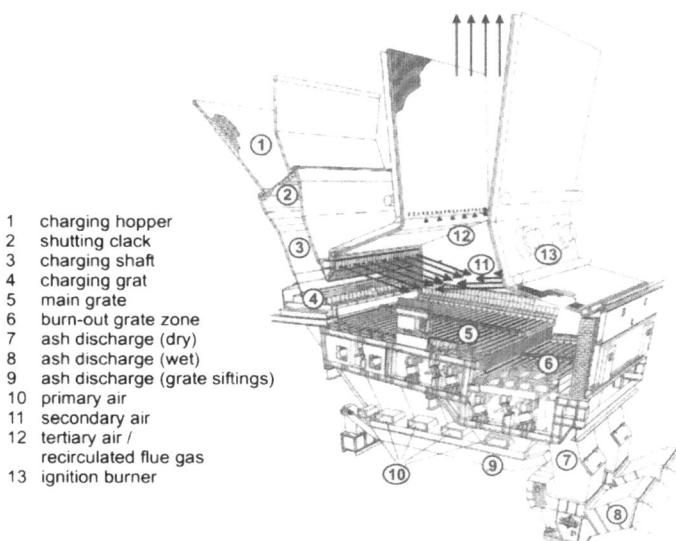

1   charging hopper
2   shutting clack
3   charging shaft
4   charging grat
5   main grate
6   burn-out grate zone
7   ash discharge (dry)
8   ash discharge (wet)
9   ash discharge (grate siftings)
10 primary air
11 secondary air
12 tertiary air /
     recirculated flue gas
13 ignition burner

Fig. 3.21. Schematic representation of the forward acting grate and separated post-combustion zone [56]

The injection of secondary air should be minimized as much as possible. The substitution of secondary air by recirculated flue gas reduces the total flue gas mass flow at the chimney, and at the same time prevents temperature peaks in the post-combustion zone if the injector jets are positioned properly.

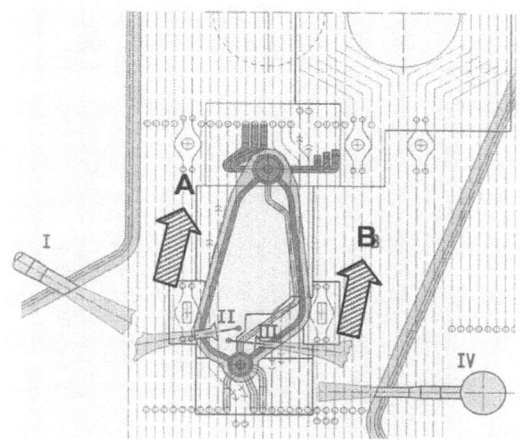

Fig. 3.22. Cross-section prism [77]

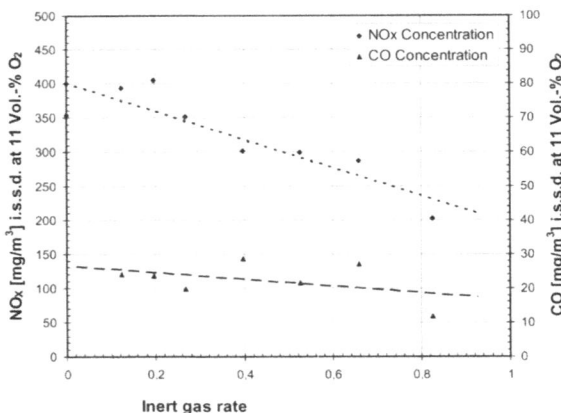

Fig. 3.23. NO$_X$ and CO gas concentration dependent upon the inert gas rate [89]

This leads to a decreased formation of thermal NO$_X$. The influence of the flue gas recirculation in the post-combustion process of a waste combustion plant is shown in Figure 3.23 [89]. In these investigations, the flue gas recirculation was simulated by adding nitrogen. The mixing of the flue gas originating at the grate was carried out at a high inert gas ratio with injector jets. Overall, a decrease in the NO$_X$ concentration was achieved. The constant low CO concentration, with regard to the inert gas ratio, shows that the CO conversion does not depend on the oxygen available in this case.

A different separation between grate and post-combustion is shown in Figure 3.24. Here, the stirred reactor element for the post-combustion of the gases emitted from the grate is situated above the first half of a drum grate. The post-combustion zone now only consists of the „calmed burn-out zone" (PFR zone).

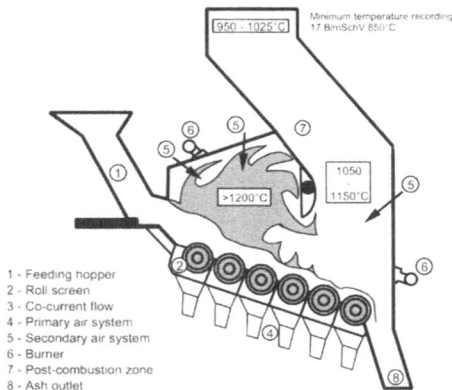

Fig. 3.24. Drum grate with 'Feuerwalze' [27]

The European Combustion Guidelines, based on the 17th BImSchV, set process-internal requirements for minimum values of the oxygen concentration, temperature and residence time (6 vol.-% O_2, 850 °C, 2 s), with the objective of guaranteeing a sufficient burn-out. This requirement forces the operator to increase the temperature in the combustion chamber by e.g. reducing the primary excess air coefficient. This intensification of the combustion requirements in connection with the increased heating values of the waste in recent years has lead to higher thermal load, more corrosion and therefore to reduced availability of the plant [75].

The most influential parameters of oxygen concentration, temperature, residence time and reactor behavior (mixing) cannot be discussed individually, especially with respect to complete oxidation of CO and organic trace elements. This was illustrated in a practical plant (MHKW Mannheim) within the framework of a research project [71]. Five experiments were carried out in the waste reactor 2 of the MHKW Mannheim in which the parameter's combustion air and amount of waste were varied in order to influence the oxygen concentration, temperature and residence time. The experimental settings were accompanied by an extensive measurement, sampling and analysis program for the flue gas flow (gas and dust) and the grate ash. The residence time range for the flue gas during the experiments at temperatures above 850 °C was between $\tau_{850} = 0.7$ s and 2.27 s. An important result of the experiments (Fig. 3.25) was the discovery that there is no direct connection between the pollutant concentration in the flue gas and the residence time in the post-combustion chamber. The experimental settings V5 (total load) and V2 (partial load) result in almost identical requirements concerning heating value, fire position, combustion chamber temperature and flue gas composition. They differ, however, with respect to the flue gas flow rates and the residence times in the post-combustion process. A shorter residence time in experiment V5, with better mixing, yields concentrations of CO, $C_{organic}$, and PCDD/F almost identical to those of the V2 experiment. The results from the practical plant support those from investigations carried out in the pilot plant [14, 15, 72].

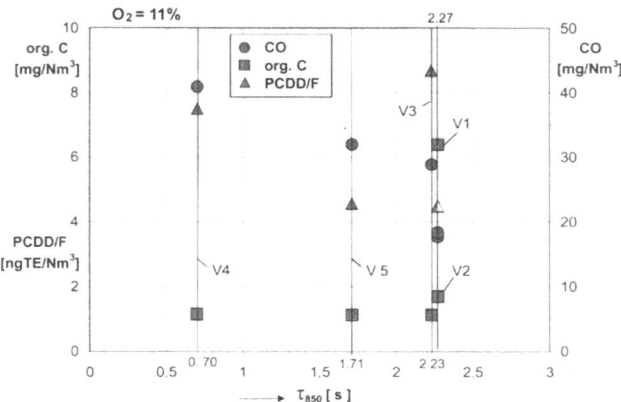

Fig. 3.25. PIC concentrations (CO, organic C, PCDD/F) of the flue gas as a function of the residence t ime of the flue gas τ_{850} above 850 °C (V1..V5: experiment number) [71].

In classical waste combustion, appropriate measures for the primary reduction of NO_X are still being investigated [90]. These measures however are far more related to reactions in the bed and directly over the bed than those applied for the oxidation of CO and organic trace elements. At the beginning of the grate, volatile nitrogen compounds (NH_i, $N_{org.}$), which can be converted to NO_X if sufficient oxygen is available, escape from the fuel bed. The conversion of NH_i radicals to HCN proceeds with increasing release of volatile organic components and the formation of CH_4 in particular. These reaction pathways are well known from batch grate and other experiments [95] and support the assumption that NO_X can be degraded to N_2 in the presence of NH_i. However, a temperature is required in the grate combustion chamber for this so-called internal "Exxon Process" which is higher than that for the classical SNCR process. In order to suppress the reaction of NH_i radicals with CH_4 leading to HCN, temperatures above 1000 °C are necessary. It can be concluded that the temperature of the gas in zone I is a highly influential parameter with respect to the release of volatile compounds and the degradation of CH_4 at the beginning of the grate. In addition, the oxygen supply or excess-air ratio λ in this zone has an increasing influence on the optimal NO_X /NH_i ratio in the first grate zone.

The effects of the most influential parameters, temperature and λ, described above were confirmed by tests in a pilot plant.

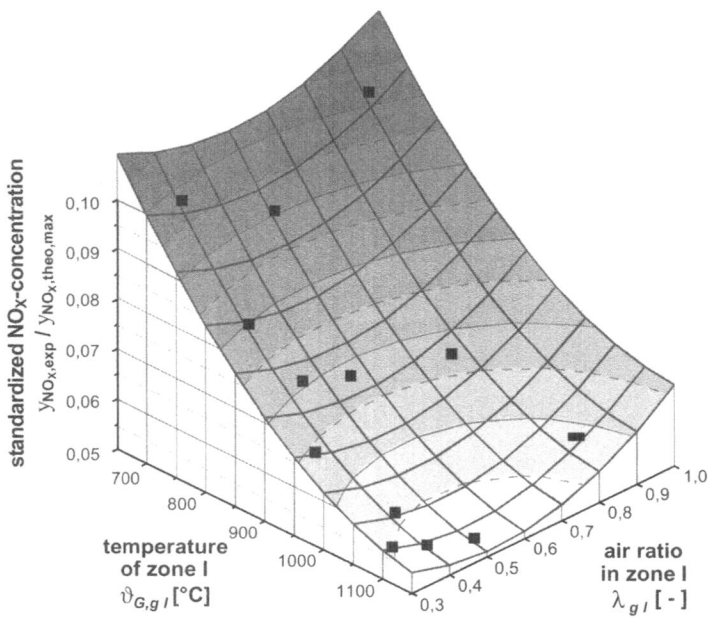

Fig. 3.26. Dependence of the NO_X concentration on the gas temperature in the ignition zone of the grate ($\vartheta_{G,gl}$) and the excess-air coefficient in this zone (λ_{gl}) during the combustion on the grate [16]

The results are shown in Figure 3.26. As expected, the NO_X values decrease with increasing temperature $\vartheta_{G,gl}$. In addition, local minima in the range of $0.4 < \lambda_{gl} < 0.6$ occur as a result of constant temperatures. Reference points supporting these results can be derived from investigations in connection with the NO_X reduction in a fixed bed gasifier with separate post-combustion [48].

As shown above, an independent optimization of the NO_X reduction, in addition to the optimization of the solid conversion, together with other advantages, is achieved when the process control of the post-combustion is clearly separated from that of the grate. Another degree of freedom for an independent post-combustion occurs as a result of the gasification on the grate.

Gasification - Post-Combustion Process

In the separated process control, as shown in Figure 3.12, the solid conversion on the grate (1st unit) can be operated under-stoichiometrically (e.g. $\lambda \approx 0.4$). Due to to the above-mentioned control options of grate systems, a complete burn-out of the grate residue (ash) can also be achieved in this mode of operation. The difference between this and the classical incineration process is the generation of a gas which can be combusted independently in the post-combustion process in the second unit. The second unit can therefore be designed as an independent gas-firing unit. This gasification-post-combustion concept, currently examined on a test-size scale, appears to be promising, as, in comparison to the conventional incineration processing in grate systems,

- the flue gas mass flows are significantly reduced (Fig. 3.27);
- combustible gases which enable an independent post-combustion process are generated;
- the post-combustion process itself can be optimized, regardless of the process on the grate, with the help of familiar primary measures for reducing the NO_X-emissions and at the same time achieving high burn-out results;
- emission loads can be reduced considerably.

These aspects are explained here in detail, with reference to first results at a pilot plant. It should be mentioned, that an industrial scale plant with a thermal power of 15 MW was also commissioned at the beginning of 2002 [8]. Initial considerations for the optimization of the process began over ten years ago [82] and the results achieved in a pilot plant in the mid 1990s [11, 17] have been confirmed by results of the recently commissioned industrial plants.

The diagram in Figure 3.27 shows that for a separated process control with a stoichiometric ratio of $\lambda \approx 0.4$ to $\lambda \approx 0.6$ for solid conversion on the grate and $\lambda \approx 1.2$ to $\lambda \approx 1.8$ for the post-combustion, total stoichiometic ratios of $\lambda \approx 1.1$ to $\lambda \approx 1.4$ are the result. The total stoichiometric conditions achieved during the separated gasification-post-combustion process are significantly lower than those for the classical gasification-post-combustion operation ($\lambda \approx 1.6$ to $\lambda \approx 2.0$) resulting in a slower flow velocity in the fuel bed (dust discharge) and a reduction in the flue gas mass flow in comparison with the classical combustion process.

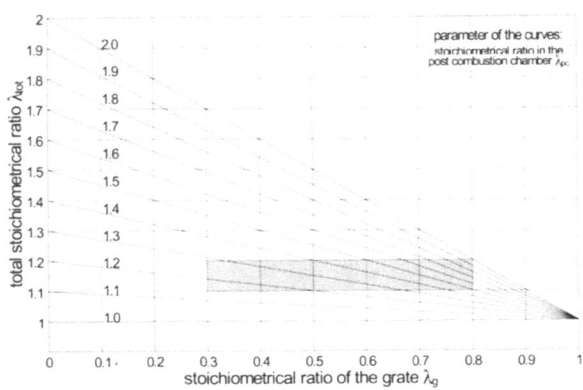

Fig. 3.27. Dependence of the total stoichiometrical ratio on stoichiometrical ratios of the grate and the post-combustion chamber [11]

The gas composition differs from calculated values [11]. The major combustible component is found to be carbon monoxide ($\psi_{CO} \approx 8..15$ vol.-%). The hydrogen content ($\psi_{H2} \approx 2..5$ vol.-%) is far below the calculated equilibrium concentration. This may be attributed to the fact that the water content in the combustible material evaporates at the beginning of the grate (Fig. 3.13).Thus, the heterogeneous water gas shift reaction between steam and the hot coke bed does not take place to the full expected extent. Furthermore, the equilibrium of the homogeneous shift reaction of CO and H_2O to form CO_2 and H_2 is not attained. Based on the assumption that the CO formation in the combustion bed of a grate essentially takes place via the heterogeneous gasification reaction of carbon with oxygen and, depending on the height of the bed, additionally via the so-called Boudouard reaction, a hot coke bed should result soon after the successful ignition of the combustible. Due to the decreasing carbon content along the length of the grate, less reaction air is required in the subsequent grate zones for gasification. This fact is confirmed by the results presented in Figure 3.29, obtained at a pilot reverse-acting grate (0.5 MW$_{thermal}$) [11, 17]. For evaluating the influence of air staging, three distinctly different air distribution settings (in each case with constant mass flows for both fuel and total air) have been tested. The main air supply is in zone 1 for the first setting and in zone 4 for the second. An even distribution over zones 1 to 4 is approached for the third setting. Figure 3.28 shows that the leveling off of hydrogen and methane concentrations ψ is more or less independent of the selected air staging settings. The wood used as a model fuel already ignites in the first stage of the grate. When shifting the main air supply from the beginning to the end of the grate, the CO concentration in the combustion gas is reduced. However, an increase in the grate bar velocity, which leads to a more intensive mixing and stoking of the combustion bed, causes an increase of the CO concentration in the example given (Fig. 3.28).

The distribution of air depends on the fuel or waste material gasified. Refuse-derived fuel (RDF) with a significant content of synthetic material requires careful degassing and ignition (Fig. 3.29). Supplying the main part majority of air at the beginning of the grate causes the degassing products to be burned immediately. This leads to high temperatures in the bed and consequently to a caking and fritting of the bed. An even flow through the bed is hindered. As with incineration, residual carbon contents of about 1% or less can be attained under gasification conditions. For example, results of combustion and gasification of wood from railway sleepers treated with coal tar are shown in Figure 3.30.

Marked differences between process conditions of the gasification and incineration mode with regard to the formation of flue dust are evident. The significantly lower mass flow of air resulting from the operation under gasification conditions causes the formation of less flue dust, as confirmed in Figure 3.30.

The gases generated by gasification in the grate process are fed to the postcombustion chamber. The net heating value is about $h_n \approx 1500 \, \text{kJ/kg}$ to $h_n \approx 2500 \, \text{kJ/kg}$ and temperatures reach levels of 750 °C to 1000 °C depending on the stoichiometric air ratio of the grate process. If the post-combustion chamber is well insulated, the post-combustion process runs independently, without additional fuel. Well-known primary measures for minimizing pollutants can then be applied (e.g. [49, 65, 109, 110, 111]). Consideration is given here to NO_X minimization with a simultaneous reduction of the CO concentration. First of all, the stoichiometric ratio of the gasification process affects the NO_X emission.

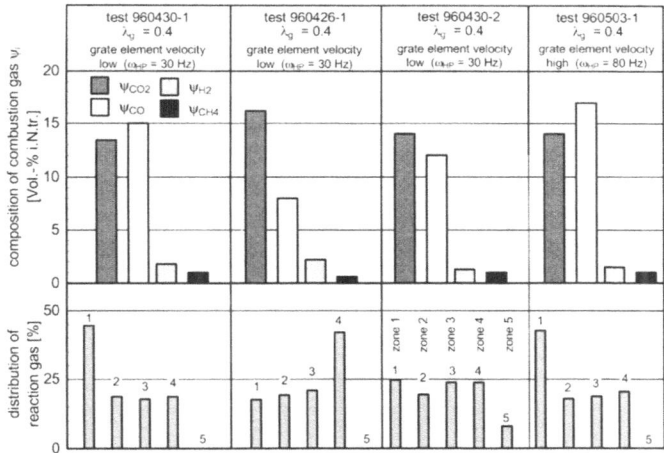

Fig. 3.28. Comparison of the composition of combustion gas with varying distribution of reaction gas along the grate path and grate element velocity (model fuel wood) [17]

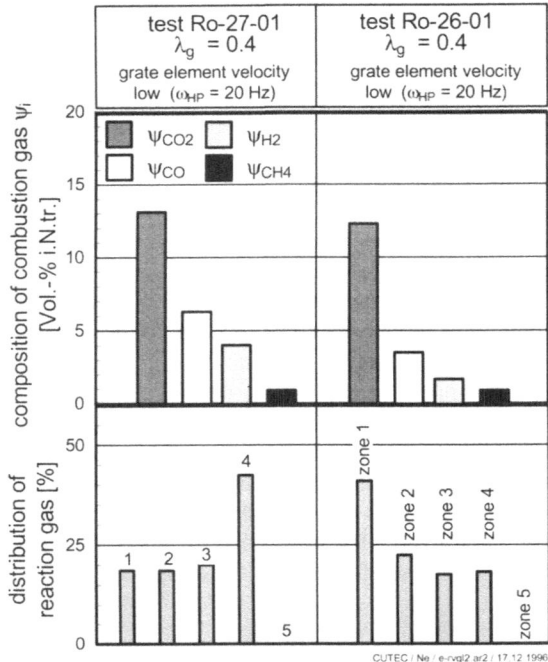

Fig. 3.29. Comparison of the composition of combustion gas with varying distribution of reaction gas along the grate path (refuse-derived fuel RDF, $\lambda_g = \lambda_{grate}$) [17].

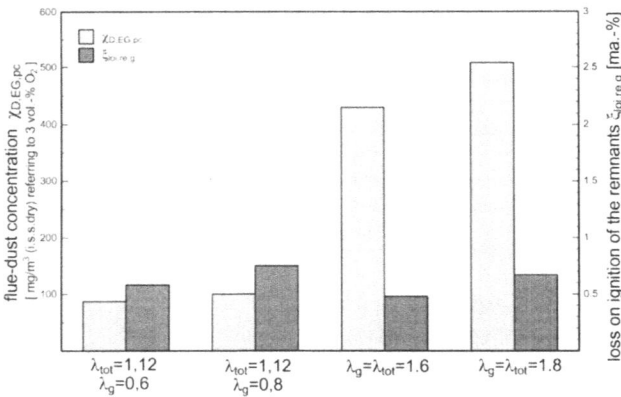

Fig. 3.30. Flue dust concentration before flue gas cleaning (χ) and loss of ignition (ξ) for different stoiciometric ratios in the stoker system ($\lambda_g = \lambda_{grate}$) [11]

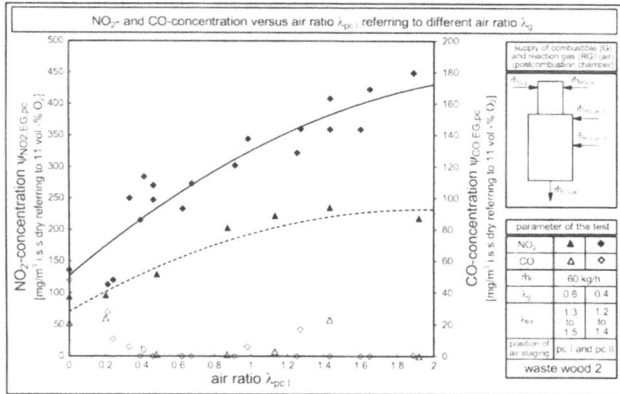

Fig. 3.31. NO$_2$ and CO concentration versus air ratio of the post combustion chamber $\lambda_{pc\ I}$ (different air ratio ($\lambda_g = \lambda_{grate}$) in the grate unit) [11]

Figure 3.31 shows the NO$_2$ and CO emissions of a staged combustion of the combustible gas generated at the grate for a primary air ratio of $\lambda_{grate} \approx 0.4$ and $\lambda_{grate} \approx 0.6$. For the higher primary air ratio, significantly lower NO$_2$ emissions are obtained. An increase in the primary air ratio is accompanied by a temperature increase from about 700 °C to about 1000 °C. Furthermore, there is an increase in the oxygen supply as a result. With these conditions, a better decomposition of volatile nitrogen components such as HCN and NH$_i$ via NO, and a reduction of already formed NO via the "NO$_X$ recycling" path is more probable. In addition to the influence of the primary air ratio, the NO$_2$ concentrations in Figure 3.31 show the typical course for the formation of NO$_2$ from the nitrogen contained in the fuel. The NO$_2$ concentrations decrease with a falling air ratio in the first stage of the combustion chamber whereby the total air ratio is kept constant at $\lambda_{tot} \approx 1.3$. For a primary air ratio of about $\lambda_{grate} \approx 0.4$, the NO$_2$ concentration drops from about 450 mg/m^3 for a single-staged post-combustion to 200 mg/m^3 for a double-staged post-combustion with $\lambda \approx 0.5$, whereby the CO concentration is far below 10 mg/m^3. In the same manner, the NO$_X$ concentration can be reduced in the case of a primary air ratio of $\lambda_{grate} \approx 0.6$, from 200 mg/m^3 to about 120 mg/m^3. For a twofold-staged post-combustion, a further NO$_X$ reduction below 100 mg/m^3 results when air is fed into the first and the third stage of the combustion chamber whereby the CO concentration remains below 10 mg/m^3 [12]. The further NO$_X$ reduction here is due to the longer residence time in combination with higher temperatures and lower oxygen concentrations in the under-stoichiometric first stage. Furthermore, if air is fed to the third stage, plug-flow like conditions prevail in the under-stochiometric part after the enlargement of the post-combustion chamber, which is preferable for NO$_X$ reduction steps [49].

3.4 Emissions from Incinerator Ash Landfills

Christian Ludwig and Jörg Wochele

3.4.1 Introduction

This section is a short introduction to the heavy metal problems associated with landfilling incineration residues. In the future, more ash residues from thermal MSW treatment will be produced. In Europe waste will be more extensively used for the production of heat and electricity (see section 5.1), and in many other countries incineration will replace common landfill practices because space is a limiting factor.

MSW incineration produces residues which contain toxic heavy metals. Landfills containing such residues are therefore potentially hazardous. However, compared to regular MSW landfills, these residues can be deposited in a much safer manner and under a higher level of control. The biological activities (see section 2.2) are much smaller due to the small amount of organic matter remaining in the residues after thermal treatment of MSW. The long-term behavior of such landfills is difficult to predict, although, relevant mechanisms determining leaching have been elucidated in the past.

The deposition of untreated filter ash is often not allowed; the upper concentration limits determined by legislation (this is generally a leaching test) can easily be exceeded. Therefore, filter ashes should be stabilized or stored in safe long-term deposits (or containments), which will never have contact with ground water (e.g. empty salt mines). Not all countries have such safe storage possibilities. Therefore, different stabilizing methods have been tested or put to practice which can substantially reduce the leachate concentrations for heavy metals, such as cementation of filter ash [23] or application of iron sulfate solutions [64]. Soluble phosphates were found to stabilize heavy metals in BA [28]. The goal of the treatment methods is to keep the concentrations below certain limits determined by the legislation. Taking the threshold value and the average discharge of a real site, the minimum durations can easily be estimated for the time it takes to totally wash out the heavy metals. Based on these calculations, it can be seen that such systems need to be under control for more than thousands of years. However, the concentrations will hopefully never exceed the limiting values.

3.4.2 Concentrations of Heavy Metals in Landfill Leachates

It must be emphasized that heavy metal concentrations in filter ashes are generally much higher than in bottom ashes (Table 2.5). It is therefore not surprising that the concentrations of heavy metals found in the leachates of filter ash landfills are higher than that of bottom ash landfills (Table 3.14).

Table 3.14. Comparison of concentrations found in the leachates of a filter ash and a bottom ash landfill. Filter ash data were obtained from a pilot landfill containing cemented filter ash, adapted from [63, 76]. Bottom ash values were adapted from [43, 45], experiments at landfill Lostorf, Switzerland. Please note that the values correspond to a given pH value/range. Threshold values of drinking water (Switzerland [5]) are given as a reference.

Heavy metal	Bottom ash [45]	Fresh filter Ash*1 [76]	Filter ash [a] after 5 a [76]	Drinking water Switzerland, limiting value
Cu [mg/l]	0.05	0.29	0.038	-
Pb [mg/l]	0.00031	5.39	0.083	0.01
Zn [mg/l]	0.002	7.85	0.98	-
Cr [mg/l]	-	4.68 [b]	0.52 [b]	0.05 [d]
Cd [mg/l]	0.0011	-	0.025 [c]	0.005
pH	9.7 - 11	ca. 12.5	ca 13.5.	-

[a] Cement-cubes of 0.5x 0.5 x 0.5 m
[b] Main fraction is Cr(VI)
[c] [60]
[d] Cr(VI)

Leachate concentrations in a landfill containing fresh ashes are higher than those found for old ones (Table 3.14). Still, leachate concentration can also change on a short term scale (days) due to rainfall (Fig. 3.32). Depending on the strength and amount of a rainfall, the concentrations in the leachate may be reduced by several factors [62, 63]. After a first washout period (a few years), concentrations approach threshold values for drinking water. Exceptions are possible, e.g. Zn (Table 3.15). However, the ashes pose a potential risk because pH may drop in the long run and the heavy metal concentration in the leachate will increase (Fig. 3.34).

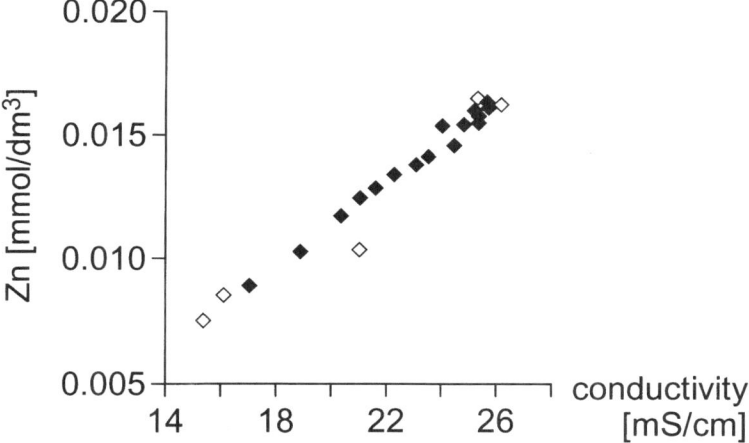

Fig. 3.32. Example of change in leachate concentration of a pilot landfill containing cemented filter ash due to a rain event. The symbols indicate the measurements before ($\diamond$) and after ($\blacklozenge$) reaching the discharge maximum. The conductivity was used as a sum parameter which correlates well with the Zn, and with many other ion concentrations [63].

3.4.3 Long-Term Predictions

Solubility Controlled Leaching of Heavy Metals

According to well known sorption phenomena of heavy metals at mineral surfaces [97], changes in pH of less than one pH unit can sometimes lead to full desorption of a heavy metal. This example indicates the hazard potential if pH drops in the long run. The characteristic pH ranges for desorption varies from element to element. The main constituents, present mineral phases, and their physical properties will also change in the long run. Due to these changes, the leaching behavior of the heavy metals will also change. This suggests what difficulties one may encounter when making long-term predictions.
Huge R&D efforts have been made in recent decades to better understand the weathering and leaching processes. A review on the fundamental chemistry of dissolution and weathering can be found in [26]. A rough summary on the mechanisms relevant for leaching from landfills is given by [45, 47, 62, 63, 70]. Figure 3.33 gives an overview of the most relevant mechanisms.

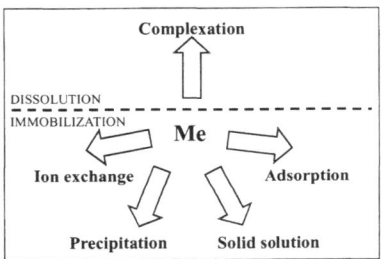

Fig. 3.33. Important effects and relevant mechanisms for the leaching of heavy metals from landfills containing ash residues from MSW incineration [adapted from 43].

Table 3.15. Saturation indices for heavy metal components. Metal hydroxides (Me(OH)$_2$), metal carbonates (MeCO$_3$), and calcium metal oxy-anions (CaMeO$_4$) are considered.

	log SI		
	CO$_3$	OH	Ca
Cu^{2+}	-4.31	1.91	
Pb^{2+}	-2.86	-3.94	
Cd^{2+}	0.33	-2.53	
Zn^{2+}	-5.19	-2.22	
Mn^{2+}	-3.34	-4.61	
MoO$_4^{2-}$			-0.18
WO$_4^{2-}$			-0.56
CrO$_4^{2-}$			-4.90

SI = [Me]$_{free}$[counter ion]$_{free}$ / K$_{S0}$, where K$_{S0}$ is the solubility constant. Experimental data cover a pH range of 9-11, bottom ash landfill Lostdorf, Switzerland [adapted from 63]

Based on [45] and [47], one can say that the mobility of heavy metals is primarily controlled by a combination of dissolution/precipitation, sorption and complexation reactions. The saturation index (SI) of a given metal with respect to a given solid phase can provide an indication of the potential role of the solid phase in controlling solubility [47] (Table 3.16). A strongly negative value signifies under-saturation and indicates that other solid phases or sorption reactions are important. Over-saturation, as found for Cu, is interpreted by the effect of organic ligands [47].

Leaching Tests

Generally, leaching tests cannot predict the leaching behavior of a landfill. Nevertheless, they can be useful for finding and assessing future leaching trends. The pH value is a highly influential factor which can change leaching concentrations by several orders of magnitudes. A decrease of pH values in a landfill is expected in the long run, and is related to its buffer capacity [43]. Cd concentrations found in leachate solutions containing different ash residues at various pH values are given in Figure 3.34 as an example.

Figures 3.34 and 3.35b show that Cd and Zn released from various ashes may increase drastically if pH in a landfill decreases in the long-term. This is especially pronounced for incineration fly ash and even more for ash residues from refuse derived fuel (RDF) utilization (Fig. 3.34). Vitrified products appear to be more resistant towards heavy metal leaching (see sections 5.3 and 5.4)

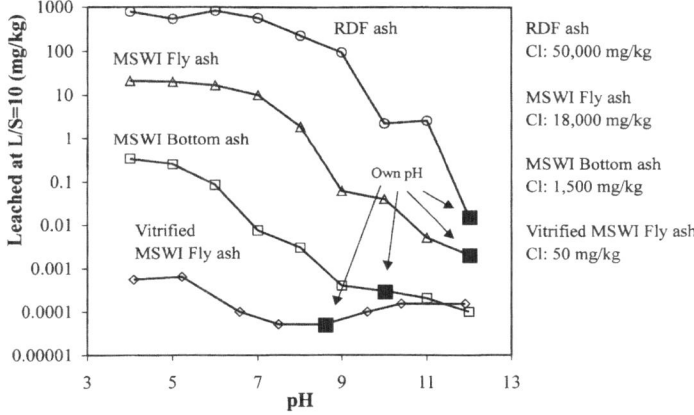

Fig. 3.34. Leaching behavior of Cd from MSW incineration residues in comparison with ash from refuse derived fuel (RDF) and vitrified MSW incineration fly ash. Cd leaching curves are largely related to the chloride content at pH>7. The amount released at pH<5 usually reflects the total content of Cd in the residue steam, as almost all Cd present is leachable. The "individual pH" of a material is the final pH of the leaching solution when the material is extracted with deionized water (marked as"own pH"). Reprinted from [102] with permission from Elsevier Science.

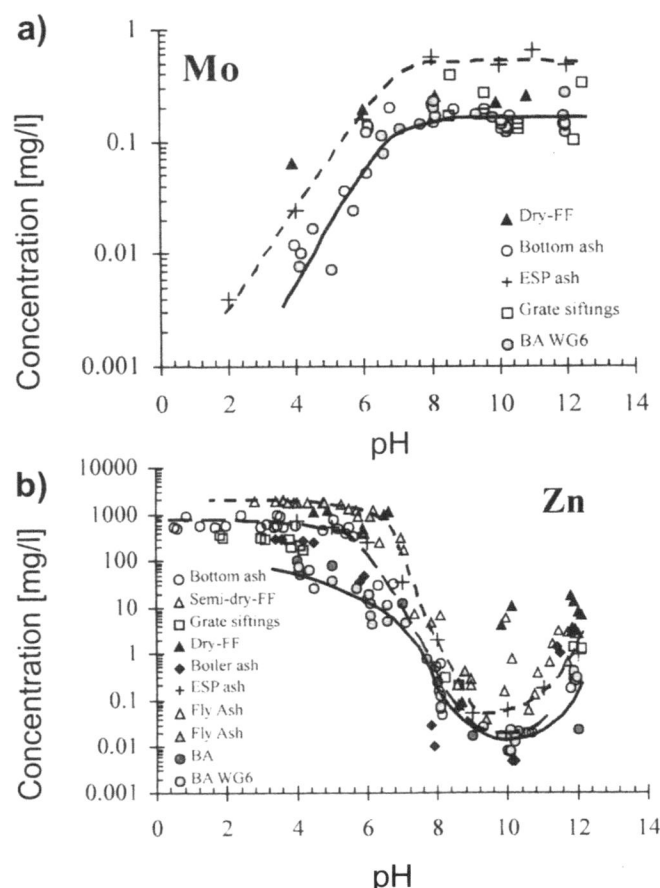

Fig. 3.35. Characteristic leaching behavior of a) Mo and b) Zn from MSW incineration residues. Reprinted from [102] with permission from Elsevier Science.

Heavy metals present as oxy-anions behave in a varied manner; whereas the solubility of cations (e.g. Zn^{2+}) increases with decreasing pH values, anions (e.g. MoO_4^{2-}) become less soluble. The pH value is a highly influential factor which can change leaching concentrations by several orders of magnitudes. A decrease of pH values in a landfill is expected in the long run, and is related to its buffer capacity [43]. Cd concentrations found in leachate solutions containing different ash residues at various pH values are given in Figure 3.34 as an example.

Figures 3.34 and 3.35b show that Cd and Zn released from various ashes may increase drastically if pH in a landfill decreases in the long-term. This is especially pronounced for incineration fly ash and even more for ash residues from refuse derived fuel (RDF) utilization (Fig. 3.34). Vitrified products appear to be more resistant towards heavy metal leaching (see sections 5.3 and 5.4)

Heavy metals present as oxy-anions behave in a varied manner; whereas the solubility of cations (e.g. Zn^{2+}) increases with decreasing pH values, anions (e.g. MoO_4^{2-}) become less soluble.

In the long run, "nature" penetrates into the landfill system. Organics from the soil on top of a landfill contain chelating agents, which are known to be able to increase the solubility [47] and, additionally, the dissolution/desorption kinetics [61] of heavy metal compounds.

It is known that the changes between dry and wet periods in a landfill may also have an influence on the alteration and weathering processes. Carbonativation reactions seem to play an important role [43] in such wet-dry cycles [80]. Alteration has been simulated in the laboratory, but little is known about these effects in a landfill.

Even, if leaching processes are not fully understood and models cannot yet exactly predict and quantify the long-term behavior for incineration residues, such residues are, however, produced every day. Therefore, pragmatic solutions are needed to test, characterize and compare different residues.

Many leaching tests have been developed and standardized by the authorities of different countries. As the behavior of the materials is so manifold depending on the processes used to treat MSW, elemental composition, different mineral phases, pH value, chelating agents, ionic strength, availability of surfaces and active surface sites, pore sizes, residence times, and other factors, slightly different leaching tests can lead to totally different results. Therefore, harmonization of these tests is necessary and substantial efforts have already been undertaken [e.g.101].

Field Investigations and Hydrology

Laboratory leaching experiments cannot account for hydrological effects in a landfill. Therefore investigations in the field are important for the elucidation of the ongoing mechanisms. The hydrology of a landfill can be investigated best by studying the composition of leachates during rain events [47, 63]. In experiments of a pilot landfill containing cemented filter ash [63], it was shown that up to about one third of the rainwater appears to pass through the landfill within days with relatively little interaction with the cemented filter ash. The greater part of the drainage discharge (>60%) of a rain event is leachate of relatively constant composition that has a much slower response to a rainfall event. This leachate is in direct contact with the cemented filter ash. A quasi equilibrium condition between the solid phase and the leachate can also be assumed because of the relatively long residence time of the leachate. In Figure 3.32 it was shown that the heavy metal concentrations are much lower after a strong rainfall; however, due to the hydro-geochemical behavior, the absolute amounts leached increase because of the increase in drainage discharge. Very similar results have been observed in a MSWI bottom ash landfill [44, 46].

A comparison of field and laboratory data, using solid samples for leaching tests from the same pilot landfill, is given in [10]. The effect of different liquid/solid (L/S) ratios (50, 100, and 200 g/L) on the leachate concentrations was tested and it was found that the major components, such as Ca, Al, Si and sulfate, can be modeled assuming equilibrium with mineral phases known to be present in cement. However, for most heavy metals the concentrations were lower in the field than in the laboratory leachates. From spectroscopical data, there is evidence that Zn is incorporated into calcium-silicate-hydrates (CSH), which play a major role in the leaching mechanism [108]. Such a mechanism could also be important for other heavy metals.

3.4.4 Lessons to be Learned for Thermal Use and Treatment of MSW

The hazard potential of different ash residues increases in the following order: vitrified MSWI ash < MSWI bottom ash < MSWI filter ash < RDF ash. A comparison of different residues from various MSW thermal treatment methods is more extensively discussed in section 5.4.

The dissolution kinetics, even of simple mineral oxides, are still not fully understood after 30 years of extensive research (mostly performed for the nuclear industry). Long-term estimates for a complex system, such as a landfill, are even more uncertain. General trends can be interpreted and for certain cases it has been shown that heavy metals are solubility-controlled; however, in general the mechanisms are only roughly understood. In most cases, models can only support possible hypotheses, but quantifications for making long-term predictions are very difficult or even impossible.

Landfills can be sealed to avoid water contact, and leaching might thus be avoided. However, it is well known that today's sealing materials last no longer than 50-100 years. It is possible to delay water contact, but the problems are then left to coming generations. Such landfills need to be controlled for centuries or even longer. Bioactivity is an unknown factor if bacteria and organic chelating agents from biological active species penetrate the mainly inorganic body of a landfill containing incineration residues in the long-term.

Better technologies for the detoxification of MSW and/or MSW residues should be developed and introduced to the market. As a first priority, residues should be cleaned from the toxic substances such as the heavy metals. Otherwise, hazardous potential remains, even if incinerator ash landfills currently pose no risk for the ground water. Only when treatment methods cannot be exerted (e.g. for ecological reasons, if the impact on the environment is higher by treatment than without treatment) should the materials be inertized. The vitrification lowers the leaching rates drastically (section 5.4), but the long-term safety, nevertheless, cannot be guaranteed.

3.5 Secondary Raw Materials from Waste

Frank Jacobs

3.5.1 Introduction

In Switzerland – with a population of approximately 7 million - 2.58 million tons of waste was burned in municipal solid waste incinerators (MSWI) in 1999. Approximately 520'000 tons of bottom ash (slag) and 60'000 tons of fly ash were produced by incineration. Hence nearly 100 kg of MSWI residues were produced per inhabitant. These materials must be disposed of, because they can not be further used. If the same amount of burned waste would be produced in Europe per capita as in Switzerland, more than 20 million tons of slag and ash should be disposed of every year. Therefore attention has been given to whether or not the slag and/or ash might be of further use. Focus was placed on the building industry and, therein, on the cement and concrete industry, which produces nearly 4 million tons of cement, 50 million tons of aggregate and approximately 30 million tons of concrete in Switzerland on a yearly basis. The slag from traditional municipal waste incinerators (MSWI) was used in Switzerland years ago in the foundation of roads. Due to environmental concerns – e.g. release of heavy metals into the environment – this use is prohibited today in Switzerland. In other European countries the use of MSWI slag is still partly allowed. Fly ash from MSWI must be disposed of all over Europe. Technologies are being developed with the goal of treating MSWI slag and ash so that they are suitable for use instead of disposal. These technologies must reduce the amount of detrimental substances in the slag and ash.

If reference is made to cement production or cement, Portland cement, not high alumina cement, is always considered. Subsequently, the European cement standards (e.g. EN 197-1) are generally discussed rather than the ASTM standards (e.g. C 150-94). To make the distinction between different types of slag and fly ash easier- for example, those from the iron industry as opposed to those from coal power plants- subsequently slag and fly ash from new MSWI-technologies are classed together under the heading "further thermally treated MSWI-slag." Mineral materials derived from further thermally treated MSWI-slag might be used as four different types of building materials:

1. Raw material: for thermal processes in, for example, brick or cement production; traditional raw materials in cement production are limestone, marl and clay;
2. Inert material: No significant reaction within the material (e.g. powdered limestone, quartz) takes place. In concrete inert materials are necessary either as aggregates (sand, gravel) having a grain size between approximately 0.1 and 32 mm or filler with a grain size of less than 0.1 mm.

3. Pozzolanic material: the material reacts chemically with the products (mainly $Ca(OH)_2$) from the reaction of cement with water (hydration). Traditional pozzolan materials are, e.g. fly ash from coal firing power plants or silica fume from the production of silicon metal and ferro-silicon alloys. Pozzolans and hydraulic material generally have a maximum grain size of less than 0.1 mm; silica fume is a very fine grained pozzolan with a maximum diameter of less than 0.001 mm.

4. Hydraulic material: the materials react in the presence of water and harden as a result. Normal cement (Portland cement CEM I according to the European cement standard EN 197-1; Portland cement is made from clinker, burned in a cement kiln, and interground additives like gypsum) is a typical hydraulic material. Materials are called latent hydraulic if the hydration reaction is slow and/or if $Ca(OH)_2$ from, e.g. the cement hydration acts as an activator in starting the reaction. If the reaction proceeds, then – in contrast to pozzolans - no further $Ca(OH)_2$ need be present.

The economic value of the materials generally increases from 1 to 4.

Figure 3.36. shows the main chemical composition of the different types of materials as mentioned before. This shows, for example, a certain similarity between Portland cement (CEM I) and granulated blast furnace slag from the iron industry.

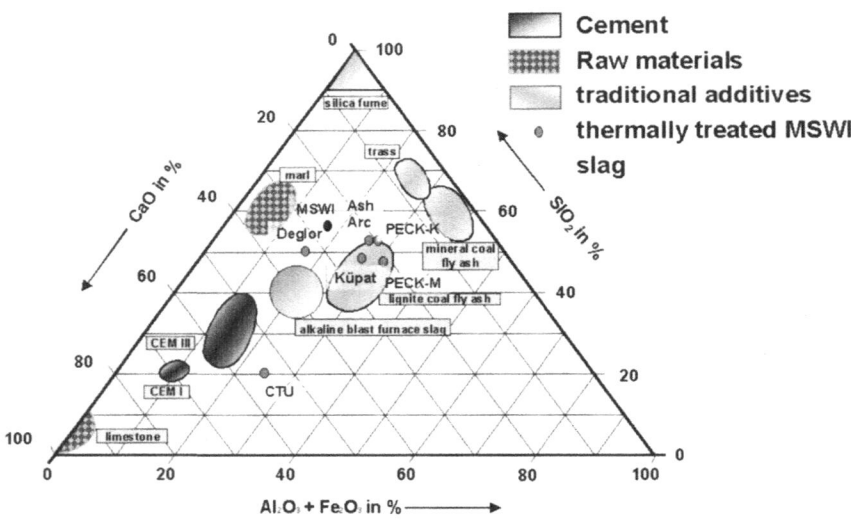

Fig. 3.36. Rankine-diagram for different types of materials: CEM I is Portland cement, CEM III a mixture of Portland cement with blast furnace slag, Trass is a natural mineral, coal fly ash and silica fume are industrial pozzolans, lignite coal fly ash is latent hydraulic, alkaline blast furnace slag is hydraulic.

Alkaline granulated blast furnace slag has an average composition of around 40 % SiO_2, 40 % CaO + MgO and 20 % Al_2O_3 + Fe_2O_3. Hydraulic materials like Portland cement have a CaO : SiO_2 ratio > 2, whereas pozzolan materials (e.g. lignite coal ash) generally have a CaO : SiO_2 ratio of less than 0.5. The composition of latent hydraulic materials (e.g. alkaline blast furnace slag) falls between these two.

3.5.2 Technical and Other Requirements for Construction Materials

Preliminary Remarks

Construction products are only allowed for public use according to Swiss or European legislation, if they fulfil several requirements for the following:

- Mechanical strength and stability
- Protection against fire
- Hygienic standards
- Health and protection of the environment
- Safety for use
- Sound protection
- Economical and rational use of energy

Subsequently, technical and environmental requirements of further thermally treated MSWI-slag are dealt with.

Technical Requirements

For commonly used building materials, two ways of specifying the technical requirements are possible:

1. In the set up of several physical and chemical requirements for each application field, the materials must be investigated intensively, to see if they fulfil the requirements or not.
2. Alternatively, one can set up a very limited set of chemical and physical requirements and define the process from which the material is derived. In this case, the process technologies of, e.g. a coal-fired power plant or of blast furnace define the basic material properties. The set up of further requirements only narrows the range of selected properties.

The second alternative is commonly used in many countries, particularly European countries. Examples can be found in European technical standards for fly ash from coal firing power plants (EN 450-1), silica fume from the production of silicon and ferro-silicon alloy (EN 13'263-1) and different types of cement according to EN 197-1. The standards are, for example, based on decades of experience in the case of cement, or for fly ash on a shorter period of experience but with very intense investigations. Around 45 million tons of this fly ash is produced in Europe in the coal power plants per year and around 2/3 of this is used.

Technical requirements comprise the fineness, influence on hardening, soundness, rheological properties, strength and durability. Decisive for durability are, for instance, the chloride and alkali content and the absence of corrosive substances for steel or concrete.

Environmental Requirements

Generally, only technical requirements are given in standards. The requirements are very strict because the field of application is generally not limited. This means that the materials in question can be used to make, for example, low quality concrete, elements to transport drinking water or high-rise buildings. The EU entrusted the incorporation of environmental properties of, for example, construction materials in the technical standards to the committee on normalisation (CEN). Up to now, this incorporation has not or has only seldom taken place. But within 5 to 10 years, the time for which the next revisions of the standards are scheduled, environmental aspects must be included. The environmental aspects to be considered will be, e.g. heavy metal content and the leaching behaviour of heavy metals. Organic substances in construction materials will probably also be included. Therefore it is necessary to know the environmental properties of further thermally treated MSWI-slag.

With the experience of several decades, for many widely used building materials (e.g. cement, aggregates, blast furnace slag) it has become evident that they cause no acute environmental risk (see e.g. [93]). The knowledge of the environmental properties of these materials can therefore serve as benchmarks for other, new materials.

In order to limit the content of toxic substances in clinker or Portland cement, standard values are listed in pertinent Swiss guidelines [24]. Among other things, the standard values relate to requirements regarding heavy metal concentrations in the secondary materials for cement production and cement clinker (Table 3.16). The levels are defined in such a way that they can be complied with through the fuels usually used, such as coal and heavy oil, as well as though natural raw materials such as limestone, marl and clay. Swiss standard values for heavy metal contents of excavated material are listed in the last two columns of Table 3.16 for comparison. Excavated material is from the rock strata lying below the root zone. This excavated material can be recycled for further use if uncontaminated, but also if below specific contamination levels (tolerable). It can be seen that the requirements for the materials used in cement production lie between the standard values for uncontaminated and tolerable excavated material. The heavy metal content of the cement can be either slightly lowered or increased through the use of secondary fuels. Overall, the standard values in Table 3.16 illustrate the high level of requirements stipulated by the Swiss authorities. Similar standard values exist in Switzerland for secondary raw materials for brickwork.

Table 3.16. [a] Standard values for heavy metals in ppm (g/t) according to [24] and further revisions for the production of CEM I cements, [b] provisional standard values for fly ash and granulated blast furnace slag and, for comparison, for uncontaminated (U) and tolerable (T) excavated material [93]

	Waste Fuels [a]	Raw meal substitute [a]	Inter-ground additives [a]	Clinker [a]	CEM I cement [a]	Fly ash, blast furnace slag [b]		Standard values for ecavated material	
						suitable	condi-tionally suitable	U	T
As	15	20	30	40		30	80	15	40
Be	5	3	3	5					
Cd	5	0.8	1	1.5	1.5	1	2	0.5	5
Co	20	30	100	50					
Cr	100	100	200	150		200	500	50	200
Cu	100	100	200	100		200	500	40	250
Hg	0.5	0.5	0.5		0.5	0.5	1	0.5	1
Ni	100	100	200	100		200	500	50	250
Pb	200	50	75	100		100	500	50	250
Sb	5	5	5	5		5	20		
Se	5	1	5	5					
Sn	10	50	30	25		30	80		
Tl	3	1	2	2		0.5	1		
Zn	400	400	400	500		400	1000	150	500

In other countries no standard values for the heavy metal content might have been published. But nevertheless, experience and data will be available which show the environmental properties of traditionally used materials. It can be assumed that the data do not deviate much from the Swiss standard values.

In addition to the heavy metal contents, levels of organic materials are also taken into consideration in examining environmental compatibility. Precise measurement of organic substances is usually very complex and in some cases still cannot provide unequivocal results. Data on the levels of organic carbon (TOC) for concrete constituents vary between 0 and approximately 5% [93].

Handling Requirements

Each type of material has it own storage and handling requirements. This depends on the condition (e.g. liquid, pasty, fine or coarse grained, soft or hard) of the materials. The condition determines the way of transporting the material to the plant and in the plant. The more difficult the transport is, e.g. for pasty materials, the more expensive is the equipment. Expensive equipment can only be amortised with large material flows and/or extra payment.

Additionally the constancy of the material properties is important in order to permit high quality production. Ideally, the material properties should also not be strongly influenced by the usual annual temperature variations.

3.5.3 Initial Experiences with Further Thermally Treated MSWI-Slags

Technical Properties

Samples from several new MSWI-processes were examined. The samples derived from small or laboratory processes and are mainly single spot samples (Table 3.17). In a first screening of the further thermally treated MSWI-slag, the investigations were based on the requirements given for the cement standard EN 197-1. In this standard, fundamental requirements are given. The following properties were determined:

- The effort required to grind the materials to a fineness similar to cement (determined property: hardness): A coarse grained cement has a fineness (Blaine-value) of approximately 2500 cm^2/g and a fine grained cement, one of about 4000 cm^2/g.
- The influence the materials have on the hardening behaviour (setting time).
- Whether the materials lead to internal damage of the concrete (soundness test).
- Whether the materials alter the rheological properties of concrete.
- Whether the materials are inert or reactive (strength development versus time).
- Whether the materials change the strength of concrete (compressive, bending strength).

Additionally the solidification of fly ash from traditional MSWI was tested by the use of further thermally treated MSWI-slag.

In Figure 3.37 results from further thermally treated MSWI-slag are shown. The cement used was always type CEM I 42.5 N (Blaine-value approx. 3000 cm^2/g), but delivered over a period of several years. Differences in the mortar strength were thereby achieved. This provides an idea of possible strength variations. The samples with further thermally treated MSWI-slag from Küpat or PECK-M showed the lowest strength. One reason for this is the gas generation in the samples. The gas pores which are generated lower the strength. A clear influence on the fineness of the further thermally treated MSWI-slag in terms of strength is visible. Increasing the Blaine-value from 2500 to 4000 cm^2/g increases the strength in general by approximately 10 to 20 %. Mortars with samples PECK-K, Deglor, AshArc and CTU-H showed a comparable strength. The contribution of further thermally treated MSWI-slag to the compressive strength of mortar was in the best case slightly better than inert quartz powder, having a Blaine-value of approx. 4000 cm^2/g (Fig. 3.37). By using 30 % of further thermally treated MSWI-slag, a reduction of the compressive strength of about 1/3 compared to cement mortar was found.

Table 3.17. Origin and nomenclature of the examined further thermally treated MSWI-slag

Process	Description of materials and process	Nomenclature*	Desity** [kg/m³]	From
VS-Küpat	Mostly partially vitrified, sharp edged, soft, porous pieces of slag (1–10 cm) originating from thermal treatment of bottom ash (slag) from MSWI	Küpat 2550 Küpat 3000 Küpat 3800	2.77	Biollaz et al. [21]
ABB Deglor	dense glassy sharp edged pieces (1–10 cm) originating from the vitrification of fly ash and bottom ash (slag) in an oxidising atmosphere	Deglor 2530 Deglor 3100 Deglor 4050	2.81	Selinger et al. [91]
ABB AshArc	dense glassy sharp edged pieces (1–10 cm) originating from the vitrification of fly ash in a reducing atmosphere	AshArc 2460 AshArc 3030 AshArc 4060	2.77	
CTU	Dense spheres with a diameter < 1 mm originating from hydrolized vitrified fly ash	CTU-H 2500 CTU-H 4270	2.98	Jakob & Mörgeli [42]
CTU	Dense spheres with a diameter < 1 mm originating from washed vitrified fly ash	CTU-W 2500 CTU-W 4830	3.08	
PECK	Vitrified bottom ash (slag) from traditional MSWI, molten and quenched sample	PECK-M 2500 PECK-M 4000	2.95	Biollaz et al. [21]
PECK	Vitrified slag from VS-Küpat process, molten and quenched sample	PECK-K 2500 PECK-K 4000	2.78	

* Numbers indicate the fineness (Blaine-value in cm²/g). The higher the number, the finer the material.
** Determined on milled material with a maximum grain size of ca. 0.1 mm

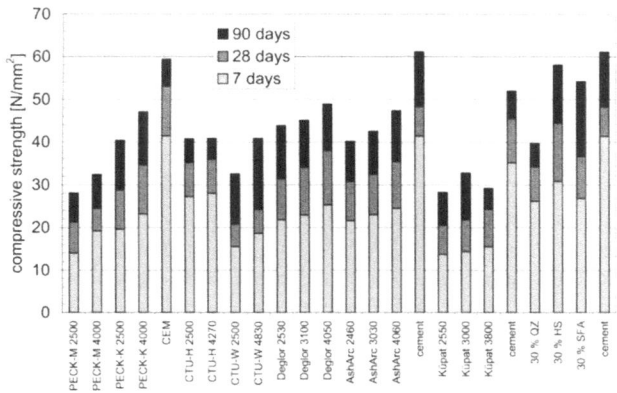

Fig. 3.37. The development of compressive strength of mortars with a binder consisting of mixtures of 70 wt.-% cement and 30 % further thermally treated MSWI-slag, quartz (QZ), fly ash (SFA), blast furnace slag (HS), and pure cement

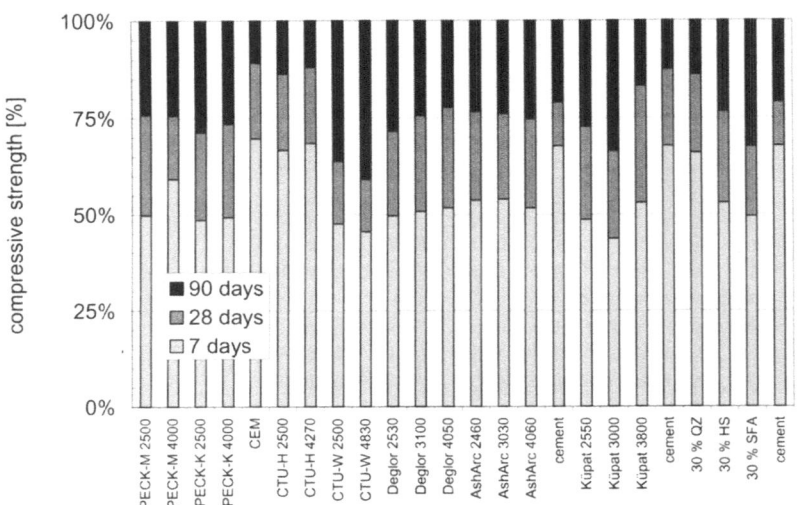

Fig. 3.38. Development of compressive strength (in percentage) for 90 day value of mortars with a binder consisting of mixtures of 70 wt.-% cement and 30 % further thermally treated MSWI-slag, quartz (QZ), fly ash (SFA), blast furnace slag (HS), and pure cement

In Figure 3.38 the strength development from the 90 day values is indicated in percentage. At 7 days, approximately 50 to 70 % of the total 90 day strength for the further thermally treated MSWI-slag samples is already reached. Between 28 and 90 days a further strength increase of about 20 to 40 % takes place. The relative strength development (over time) of the further thermally treated MSWI-slag is similar to that of fly ash from coal-fired power plants and blast furnace slag from iron industry; lower compared to quartz powder (Fig. 3.38). Samples with further thermally treated MSWI-slag CTU-H showed the lowest compressive strength and samples with further thermally treated MSWI-slag CTU-W showed the highest strength increase after 28 days. In general, for further thermally treated MSWI-slag, hardly pozzolan reactivity can be assumed.

Further experiments were carried out in order to study the use of further thermally treated MSWI-slag for stabilisation of MSWI ash. As binder for stabilisation purposes, cement and lime have been used and served as references. Further samples were made with mixtures consisting of cement, lime and further thermally treated MSWI-slag (PECK-M 2500, PECK-M 4000, PECK-K 2500, PECK-M 4000). In those mixtures, 15 % (by weight) of the cement was replaced by further thermally treated MSWI-slag. The compressive strengths of the stabilised MSWI fly ash are shown in Figure 3.39. The strengths of all samples are similar at 7 days. At 35 days the samples consisting only of cement as binder showed significantly higher strength. Nevertheless, the strength of the samples with further thermally treated MSWI-slag is sufficient.

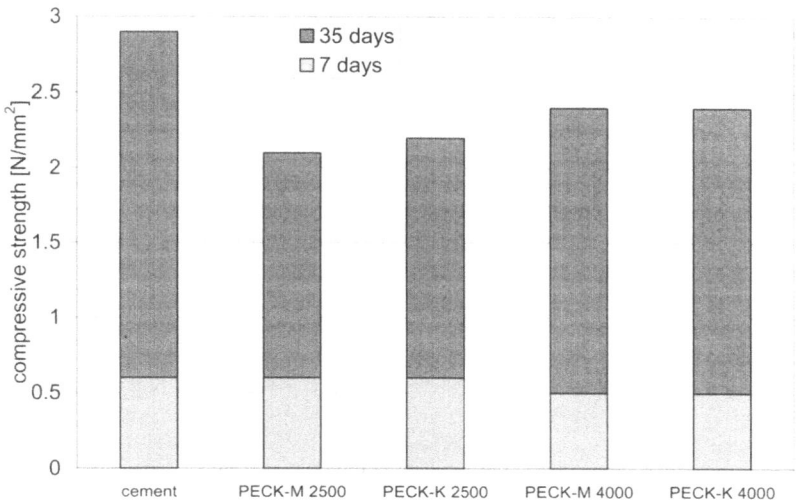

Fig. 3.39. Compressive strength at 7 and 35 days of the MSWI fly ash stabilised with cement and lime (denoted cement) or cement, lime and further thermally treated MSWI-slag replacing 15 % of cement per weight (denoted PECK)

In Table 3.18 a comparison is made between the examined further thermally treated MSWI-slag of different origin (e.g. type of input into process and process itself). It should be stated again that all samples derive from small scale processes and the properties of these single spot samples have not been confirmed by repetitions. The detailed results are given in the reports by Jacobs [39, 40].

Table 3.18 illustrates that nearly all examined materials (except Küpat slag) have both technical advantages and disadvantages. Many disadvantages might be reduced by a reduction of the quantity of material used as, for example, a cement replacement in the order of only 5 to 10 %.

Replacing cement by 30 wt-% of further thermally treated MSWI-slag significantly increased the initial and, particularly, the final setting times. Therefore for normal construction purposes, further thermally treated MSWI-slag PECK-M is generally not suitable. Further thermally treated MSWI-slag PECK-K might be suitable. The delayed setting could be caused by the heavy metal content of the further thermally treated MSWI-slag. The further thermally treated MSWI-slag PECK-K has high heavy metal contents of Zinc and Copper [33]. The further thermally treated MSWI-slag PECK-M has even higher contents (Fig. 3.40). These amounts of heavy metals, especially for the further thermally treated MSWI-slag PECK-M, are known to retard the hydration of cement [96] and hence the setting time.

The problem of swelling (soundness) might be solved by the following method: During a moist storage of several months, most of the reactions causing swelling can take place. Therefore, use after storage should cause no additional swelling.

This approach is also incorporated in the draft of [35]. This instruction leaflet regulates technical and environmental points in the use of fly ash from MSWI in road construction in Germany.

Some further thermally treated MSWI-slags cause a stiffening (increasing flow resistance: CTU), some a plasticizing (PECK) effect.

Table 3.18. Influence of cement replacement by approx. 30 % of further thermally treated MSWI-slag on mortar properties. +: positive effect, -: negative effect; 0: indifferent

Sample	Grind ability	Setting time	Soundness	Flow resistance	Remarks
PECK-M	-	-	o	+	minimal swelling of sample due to gas generation at higher ages
PECK-K	-	-	o	+	
CTU-H	+	o	o	-	
CTU-W	+	o	o	-	
Deglor	-	o	o	+	
AshArc	o	o	o	+	
Küpat	o	o	-	o	strong swelling of sample due to gas generation

Environmental Properties

The contents for selected heavy metals of further thermally treated MSWI-slag are displayed in Figure 3.40. For comparison, Swiss standard values for Portland cement clinker and for conditionally suitable fly ash from coal power plants (Table 3.16) are given. Additionally, data from a conventional MSWI slag are shown. The heavy metal contents of the further thermally treated MSWI-slag are up to approximately one order of magnitude higher than the Swiss standard values for Portland cement clinker and lay partly in the range of the Swiss standard values of conditionally suitable fly ash. Therefore, a further reduction of the heavy metal contents is desirable. If concrete is made with further thermally treated MSWI-slag as a partial replacement of cement, concrete will have higher heavy metal contents. The higher heavy metal content will usually lead, in the long term (over a period of decades or longer), to a higher heavy metal release from concrete. But the short term release due to leaching in, e.g. groundwater will not be significantly increased [93]. Aside from these environmental aspects, the reduction of heavy metal contents would help to avoid the strong retardation of the hydration reaction of cement as well.

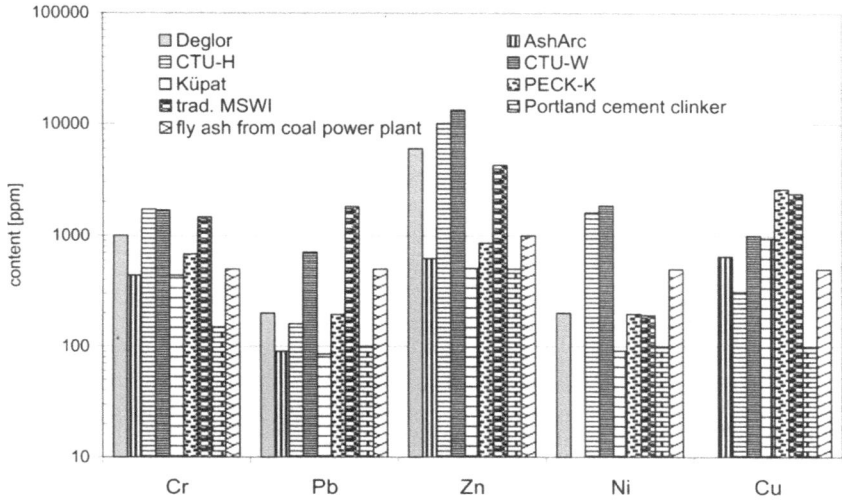

Fig. 3.40. Heavy metal contents of various types of further thermally treated and traditional MSWI slag [33, 39, 40, 99] and standard values for Portland cement clinker and fly ash (conditionally suitable) according to Table 3.16.

3.5.4 Applications of Further Thermally Treated MSWI-Slag

New materials, such as slag from MSWI, can not be used as generally accepted construction material, until very intensive investigations are carried out and bear positive results. Finally, the results must be approved by the national standardisation body. In [41] the general procedures with strict requirements for the introduction of a new product into the market are explained. Based on the preliminary results of further thermally treated MSWI-slag, Table 3.19 gives an overview of potential application areas in Switzerland and indicates relevant standards.

Thus far, it appears to be most promising to investigate the use of further thermally treated MSWI-slag in cement clinker production. [53] presented examinations for the use of MSWI-slag in cement plants and comes to the conclusion that this is a promising approach and should be further investigated. Less regulated application fields, where damages due to insufficient properties do not easily lead to severe accidents, should be pursued. Examples of these are floors or road stabilisation with no high requirements.

In different countries different standards may apply. For example, in USA the standards ASTM C 593-95 "Standard specification for fly ash and other pozzolans for use with lime" or ASTM C 595M-95a "Standard specification for blended hydraulic cements" define pozzolan as follows: "Pozzolan shall be a siliceous or siliceous and aluminous material, which in itself possesses little or no cementitious value but which will, in finely divided form and in the presence of moisture, chemically react with calcium hydroxide at ordinary temperatures to form compounds possessing cementitious properties." Further thermally treated MSWI-slag possessing the required pozzolan properties falls within the scope of these ASTM standards. Swiss or CEN-standards do not allow the use of further thermally treated MSWI-slag for, e.g. the production of standard cement.

Table 3.19. Potential application fields in Switzerland for further thermally treated MSWI-slag, the requirements and judgement of the possible suitability for using further thermally treated MSWI-slag; based on [112]; Brackets indicate that further investigations are necessary; * Legal restrictions for the use of further thermally treated MSWI-slag

Application field	Standard	Requirements for	Possibly suitable for use
Raw material for cement clinker production	no BUWAL guideline	Defined by cement plant heavy metal content	yes
Cement compound	EN 197-1	Origin	no
Concrete structures	EN 206-1	Cement, aggregates	No
Precast concrete industry	EN 13'389, EN 12'839, SIA 320	Partly for cement and aggregates	(yes)
Concrete floors	SN 640'461	Cement, not for aggregates	(no)
Seamless industrial floors coverings and cement coatings	SIA 252	None	(no)
Floating support floors	SIA V 251/1+2	Cement	(no)
Brickwork	SIA V 177	None	(no)
Natural stone masonry work	SIA V 178	Binder	(no)
Stabilization	prENV 13'282	Binder	(no)
Injections	prEN 288'006	None	(no)
Stabilization of road constructions	SN 640'500a, SN 640'506a	Binder, partly on aggregates	(no)*
No regulated areas	no	Given by builder owner / constructor	yes

An important goal should be the improvement of the properties of further thermally treated MSWI-slag, in order to use further thermally treated MSWI-slag as, e.g. cement addition or pozzolan material. High quality further thermally treated MSWI-slag and high production rates would allow for the intensification of the investigations and would open a huge market. Globally, around 1.5 billion tons of cement and approximately 0.2 million tons of industrial pozzolan materials are used in the concrete field. High deposit fees would further stimulate the investigations and process optimisations.

To increase the properties, particularly the reactivity, of further thermally treated MSWI-slag, several methods exist (based on [54] with additions):

- Increased fineness (> 5000 cm^2/g), though this is rather expensive
- Increased amount of reactive (vitreous) components in the MSWI through e.g. appropriate quenching
- Further reduction of the heavy metal content
- Forced weathering of the mineral components through, e.g. longer storage under advantageous conditions
- Devitrification (zeolitisation) of glasses
- Altered composition of the input material in MSWI through the addition of appropriate materials or omission of waste fractions; in general the higher the ratio CaO : SiO$_2$, the higher the reactivity

3.5.5 Conclusions and Outlook

Large amounts of further thermally treated MSWI-slag are produced nowadays and the tendency, at least in Europe, is toward continual increase. In the building material industry, large amounts of materials are used (exploited from nature). Therefore, at first glance it seems most rational to investigate whether further thermally treated MSWI-slag can be used in the building material sector to reduce the use of natural resources. Based on investigations thus far, further thermally treated MSWI-slag is mainly inert in cementitious materials during the investigations over 90 days. In order to create a higher level of benefit from further thermally treated MSWI-slag, a pozzolan or latent hydraulic reactivity would be desirable. The optimisation of new processes at MSWI plants with respect to the quality of the further thermally treated MSWI-slag is strongly recommended.

References

1.   Directive on Used Electric and Electronic Goods 2000/0158 COD (draft version). European Union
2.   Dritte Allgemeine Verwaltungsvorschrift zum Abfallgesetz; Technische Anleitung zur Verwertung, Behandlung und sonstiger Entsorgung von Siedlungsabfällen (TA-Siedlungsabfall), vom 14.05.1993. Deutschland
3.   German Packaging Ordinance from 1991. Germany
4.   Kreislaufwirtschafts- und Abfallgesetz von 1996 (Recycling and Waste Managing Act). Germany
5.   Swiss Federal Order: Verordung über Fremd- und Inhaltsstoffe in Lebensmitteln (FIV) 26.2.2002. Switzerland

6. Albrecht J, Loeffler J, Reimert R (1994) Restabfallvergasung mit integrierter Asche-verschlackung. GVC-Symposium Abfallwirtschaft Herausforderung und Chance, Würzburg
7. Anonymous (1993) INTECUS: Errichtung, Inbetriebnahme und Optimierung einer Pi-lotanlage mit einem Durchsatz von 5 Mg/h zur automatischen Farbglassortierung ein-schließlich der zugehörigen Konditionierung. Zwischenbericht zum Forschungsvorha-ben, Berlin
8. Anonymous (2002) Dezentrale Heizkraftwerke für den Einsatz von Sekundärbrenn-stoffen. Advertising brochure. Energos Deutschland
9. Anthony R, Mang J (1990) A Grand Tour. Recycling Today 10:115
10. Baur I, Ludwig C, A. JC (2001) The Leaching Behavior of Cement Stabilized Air Pol-lution Control Residues: A Comparison of Field and Laboratory Investigations. Envi-ron. Sci. Technol. 35:2817-2822
11. Beckmann M (1995) Mathematische Modellierung und Versuche zur Prozessführung bei der Verbrennung und Vergasung in Rostsystemen zur thermischen Rückstandsbe-handlung. CUTEC-Schriftenreihe:
12. Beckmann M, Scholz R (1995) Simplified Mathematical Model of Combustion in Stoker Systems. Vol. II. Proceedings 3rd European Conference on Industrial Furnaces and Boilers, Lisbon, pp 61-70
13. Beckmann M, Scholz R (2000) Residence Time Behaviour of Solid Material at Grate Systems. Proc. In: Reis A, al e (eds) 5th INFUB (European Conference on Industrial Furnaces and Boilers). INFUB 2000 Rio Tinto, Porto Portugal
14. Beckmann M, Zimmermann R, U, S. (1998) Gasification - Post-Combustion of Waste Materials; Influencing Parameters and On-line Monitoring of Organic Substances. Vol. II. AIChE-PTF Topical Meeting "Advanced Technologies for Particle Processing", Miami Beach, Florida, pp 661-671
15. Beckmann M, Griebel H, Scholz R, (1998) Einfluß von Temperatur, Durchmischung und Verweilzeit auf den Abbau organischer Spurenstoffe bei der thermischen Behand-lung von Abfallholz. Tagungsbericht. DGMK
16. Beckmann M, Davidovic M, Gehrmann H-J, Scholz R (1999) Prozeßoptimierung der Verbrennung und Vergasung von Abfällen in Rostsystemen. Bericht 1492. VDI-Verlag GmbH, Düsseldorf, pp 361-368
17. Beckmann M, Scholz R, Wiese C, Davidovic M, (1997) Optimization of Gasification of Waste Materials in Grate Systems. International Conference on Incineration & Thermal Treatment Technologies, San Francisco-Oakland Bay, California
18. Berwein H-J (1995) Siemens Schwel-Brenn-Verfahren - Thermische Reaktionsabläu-fe. In: Abfallwirtschaft Stoffkreisläufe, Terra Tec `95. B.G. Teubner Verlagsgesell-schaft, Leipzig
19. Bilitewski B, Härdtle G, Marek K (1996) Wastemanagement. Springer-Verlag, Berlin-Heidelberg-New York
20. Bilitewski B, Heilmann A (1995) Aufbereitung von Rohstoffen vor der Verwertung. In: Müllhandbuch KZ 2955, 3. Erich Schmidt Verlag, Berlin
21. Biollaz S, Grotefeld V, Künstler H (1999) Separating Heavy Metals by the VS-Process for Municipal Solid Waste Incinerators. In: Barrage A, Edelmann X (eds) Recovery, Recycling, Reintegration R'99 Conference. EMPA, Geneva
22. Biollaz S, Ludwig C, Beckmann M, Davidovic M, Jentsch T (2000) Volatility of Zn and Cu in Waste Incineration: Radio-Tracer Experiments on a Pilot Incinerator. In:

University of California I (ed) IT3 Conference (Incineration and Thermal Treatment Technologies). Internat. Incineration Conference, Portland Oregon

23. BUWAL / SAEFL (1995) Reststoffqualität von stabilisierten Rückständen aus der Rauchgasreinigung von Kehrichtverbrennungsanlagen. Reihe Umweltmaterialien, Switzerland:

24. BUWAL / SAEFL (1998) Entsorgung von Abfällen in Zementwerken, 1998. Richtlinie:

25. Carlowitz O, Jeschar R (1980) Entwicklung eines variablen Drallbrennkammersystems zur Erzeugung hoher Energieumsetzungsdichten. Brennstoff-Wärme-Kraft (BWK) 32

26. Casey WH, Ludwig C (1995) Silicate Mineral Dissolution as a Ligand-Exchange-Reaction. In: White AF, L. BS (eds) Chemical Weathering Rates of Silicate Minerals. Mineralogical Society of America, BookCrafters Inc., Michigan, p 87

27. Christmann A, Quitteck G (1995) Die DBA-Gleichstromfeuerung mit Walzenrost. Bericht 1192. VDI-Verlag GmbH, Düsseldorf

28. Crannell BS, Eighmy TT, Krzanowski JE, Eusden Jr JD, Shaw EL, Francis CA (2000) Heavy Metal Stabilization in Municipal Solid Waste Combustion Bottom Ash using Soluble Phosphate. Waste Management & Research 20:135-148

29. Deutschland Duales System (2000) Packaging Recycling. Cologne

30. Deutschland Duales System (2001) Consumers trust the Green Dot.

31. Deutschland Duales System (2001) Wo steht die Kreislaufwirtschaft? DS-Dokumente

32. Dryer FL, Glasman I (1973) 14th International Symposium On Combustion. Combustion Institute, Pittsburgh

33. Eggenberger U, Mäder U (2002) Charakterisierung von thermisch nachbehandelten Schlackenproben bezüglich Inhaltsstoffen und Auslaugverhalten. Report. Institute for geology of the University of Berne and technical centre for secondary raw materials, Berne

34. European Council (1994) Council Directive 94/62/EC of December 1994 on Packaging and Packaging Waste. Official Journal 1994 L 365(31/12/1994)

35. FGSV Forschungsgesellschaft für Strassen- und Verkehrswesen (2001) Merkblatt über die Verwendung von Hausmüllverbrennungsaschen im Strassenbau. Entwurf, Köln, p 17

36. Göhler P, Schingnitz M (1995) Stoff- und Wärmebilanzen bei der Abfallverwertung nach dem NOELL-Konversionsverfahren. Energie und Umwelt' 95 mit Energieverbrauchsminderung bei Gebäuden. TU Bergakademie, Freiberg

37. Hunsinger H, Merz A, Vogg H (1994) Beeinflussung der Schlackequalität bei der Rostverbrennung von Hausmüll. GVC-Symposium Abfallwirtschaft Herausforderung und Chance, Würzburg

38. INTECUS GmbH Dresden (1994-2000) Wissenschaftliche Untersuchung, Begleitung und Kostenkalkulation von fünf Modellversuchen zur getrennten Erfassung graphischer Papiere und Verpackungen aus Papier und Pappe. Studie. AGRAPA - Arbeitsgemeinschaft Graphische Papiere, Bonn

39. Jacobs F (1998) Untersuchungen von Schlacken aus dem Küpat-Verfahren. TFB-Report

40. Jacobs F (1999) Untersuchungen von Schlacken aus dem ABB-Verfahren. TFB-Report

41. Jacobs F, Timper J (1999) Anforderungen an Beton und dessen mineralische Bestandteile. TFB-Report

42. Jakob A, Mörgeli R (1999) Detoxification of Municipal Solid Waste Incinerator Fly Ash, the CT-FLURAPUR™ Process. In: Gaballah, Hager, Solozabal (eds) REWAS'99, San Sebastian, Spain

43. Johnson CA, Brandenberger S, Baccini P (1995) Acid Neutralising Capacity of Municipal Waste Incinerator Bottom Ash. Environ.Sci. and Technol. 28

44. Johnson CA, Schaap MK, Abbaspour K (2001) Modelling of Flow through a Municipal Solid Waste Incinerator Ash Landfill. J. Hydrol. 243(1-2):55-72

45. Johnson CA, Kersten M, Ziegler F, Moor HC (1996) Leaching Behaviour and Solubility - Controlling Solid Phases of Heavy Metals in Municipal Solid Waste Incinerator Ash. Waste Management & Research 16:129-134

46. Johnson CA, Richner GA, Vitvar T, Schittli N, Eberhard M (1998) Hydrological and Geochemical Factors affecting Leachate Composition in Municipal Solid Waste Incinerator Bottom Ash. Part I: The Hydrology of Landfill Lostorf, Switzerland. J. of Contaminant Hydrol. 33 (3-4):361-376

47. Johnson CA, Kaeppeli M, Brandenberger S, Ulrich A, Baumann W (1999) Hydrological and Geochemical Factors Affecting Leachate Composition in Municipal Solid Waste Incinerator Bottom Ash, Part II. The Geochemistry of Leachate From Landfill Lostorf, Switzerland. Journal of Contaminant Hydrology 40:239-259

48. Keller R (1994) Primärseitige NOX-Minderung mittels Luftstufung bei der Holzverbrennung. Brennstoff-Wärme-Kraft (BWK) 46

49. Klöppner G (1991) Zur Kinetik der NO-Bildungsmechanismen in verschiedenen Reaktortypen am Beispiel der technischen Feuerung. Technische Universität

50. Knapp D, Ward T, Liss G, Anthony R (2001) Del Norte Zero Waste Plan; Cluster Analysis. Feasibility Study and Discard Generation Study, 2000-2001

51. Knörr A (1995) Thermische Abfallbehandlung mit dem SYNCOM-Verfahren. Bericht 1192. VDI-Verlag GmbH, Düsseldorf

52. Kost T, Rotter S, Bilitewski B (2001) Chlorine and Heavy Metal Content in House-Hold Waste Fractions and its Influence on Quality Control in RDF Production Processes. In: Christensen TH, Cossu R, Stegmann R (eds) Sardinia 01 Eighth International Waste Management and Landfill Symposium, Waste Management of Municipal and Industrial Waste, Cagliari, Sardinien

53. Krähner A, 65 p. (1997) Verwertung von MVA-Schlackenfeinkorn als Rohmehlkomponente beim Zementklinkerbrand. In: Umweltplanung, Arbeits- und Umweltschutz. Hessische Landesanstalt für Umwelt, Wiesbaden, p 65

54. Kruspan P (2000) Natürliche und technische Petrogenese von Pozzolanen. PhD thesis, ETH

55. LAGA (1997) Massstäbe und Kriterien für die energetische Verwertung von Abfällen in Zementwerken. Entwurf. Länderarbeitsgemeinschaft Abfall, Deutschland

56. Lautenschlager G (1994) Moderne Rostfeuerung für die thermische Abfallbehandlung. GVC-Symposium Abfallwirtschaft Herausforderung und Chance, Würzburg

57. Leuckel W (1967) Swirl Intensities, Swirl Types and Energy Losses of Different Swirl Generating Devices. G02/a/16. IFRF, Ijmuiden

58. Levenspiel O (1972) Chemical Reaction Engineering. John Wiley and Sons, New York

59. Linne W (1992) Altglasaufbereitung und -absatz. BW 1635. VDI-Verlag GmbH, Düsseldorf

60. Ludwig C, Johnson AC (1999) Weathering of Cemented MSWI Filter Ashes in a Landfill. Recovery, Recycling, Re-integration, R'99 4th World Congress. EMPA, Geneva

61. Ludwig C, Casey WH, Rock PA (1995) Prediction of Ligand-Promoted Dissolution Rates from Reactivities of Aqueous Complexes. Nature 375:44-47

62. Ludwig C, Ziegler F, Johnson CA (1997) Heavy Metal Binding Mechanisms in Cement-Based Waste Materials. In: Goumans JJJM, Senden GJ, van der Sloot HA (eds) Waste Materials in Construction. Elsevier, Amsterdam, pp 459-468

63. Ludwig C, Johnson CA, Käppeli M, Ulrich A, Riediker S (2000) Hydrological and Geochemical Factors Controlling the Leaching Process in Cemented Air Pollution Control Residues: A Lysimeter Field Study. Journal of Contaminant Hydrology 42:253-272

64. Lundtorp K, Jensen DL, Sorensen MA, Christensen TH (2000) On-Site Treatment and Landfilling of MSWI Air Pollution Control Residues. Proccedings. WASCON 2000, Waste Materials in Construction, Harrogate England

65. Malek C, Scholz R, Jeschar R (1993) Vereinfachte Modellierung der Stickstoffoxidbildung unter gleichzeitiger Berücksichtigung des Ausbrandes bei einer Staubfeuerung. Bericht 1090. VDI-Verlag GmbH, Düsseldorf

66. Management ScfW (1986) Guidelines to Waste Management in Switzerland. Swiss Agency for the Environment, Forests and Landscape, Bern

67. Martin J, Busch M, Horn J, Rampp F (1993) Entwicklung einer kamerageführten Feuerungsregelung zur primärseitigen Schadstoffreduzierung. Bericht 1033. VDI-Verlag GmbH, Düsseldorf

68. Martin M (1988) Moderne Abfallverbrennung. Abfallwirtschafts-Journal 0/88:7-11

69. Meadows D (1974) Club of Rome, Boston

70. Meima JA, Comans RNJ (1997) Overview of Geochemical Processes controlling Leaching Characteristics of MSWI Bottom Ash. In: Goumans JJJM, Senden, GJ, van der Sloot, HA, (ed) Waste Materials in Construction. Elsevier, Amsterdam

71. Merz A, Seifert H (2000) Verweilzeit von Verbrennungsgasen. Abschlussbericht zum Forschungsprojekt. VGB Kraftwerkstechnik, Essen

72. Merz A, Seifert H, VGB-Nr. 173 (1999) Prozesskontrolle bei Abfallverbrennungsanlagen. Abschlussbericht zum Forschungsprojekt. VGB Kraftwerkstechnik, Essen

73. N.N. (2000) Trie-Inking- Vollautomatisches Verfahren zur Herstellung der Papierfabrikation De-Inking. Firma Trienekens

74. N.N. (2001) Duales System Deutschland Köln. SORTEC 3.0

75. Neukirchen B, Schirmer U, al. e (1999) Auswirkungen der 17. BimSchV auf die thermische Abfallbehandlung. VGB Technisch-wissenschaftliche Berichte "Feuerungen ". VGB Kraftwerkstechnik, Essen, p 5

76. Ochs M, Stäubli B, Wanner H (1999) Eine Versuchsdeponie für verfestigte Rückstände aus der Rauchgasreinigung von Kehrichtverbrennungsanlagen. Müll und Abfall 5:301-306

77. Périlleux M, Creten G, Kümmel J (2002) Improving Combustion and Boiler Performance of New and Existing EFW Plants with the Seghers-Ibb-Prism. 6th INFUB (European Conference on Industrial Furnaces and Boilers), Estoril Lisbon Portugal

78. Püchel H-J (1999) Die Herstellung von Substitutbrennstoffen aus Restmüll und ihre Qualitätssicherung. In: SIDAF (ed) Kostenminimierte Abfallentsorgung durch Verfahrenskombination. Kalt/Warm, Freiberg

79. Putz H-J (1998) Development of the Deinking Technology in Europe in the last Decades. Reprint. COST Conference, Gran Canaria

80. Sanchez F, Gervais C, Garrabrants AC, Barna R, Kosson DS (2002) Leaching of Inorganic Contaminants from Cement-Based Waste Materials as a Result of Carbonation during Intermittent Wetting. Waste Management & Research 22:249-260

81. Schäfers W, Limper K (1993) Fortschrittliche Feuerungsleistungsregelung durch Einbeziehung der Fuzzy-Logik und der IR Thermografie. In: Thomé-Kozmiensky KJ (ed) Reaktoren zur thermischen Abfallbehandlung. EF-Verlag für Energie- und Umwelttechnik GmbH, Berlin

82. Scholz R, Schopf N (1989) General Design Concept for Combustion Processes for Waste Fuels and Some Test Results of Pilot Plants. Incineration Conference, Knoxville, USA

83. Scholz R, Beckmann M (1991) Möglichkeiten der Verbrennungsführung bei Restmüll in Rostfeuerungen. Bericht 895. VDI-Verlag GmbH, Düsseldorf

84. Scholz R, Jeschar R, Carlowitz O (1984) Zur Thermodynamik von Freistrahlen. Gas-Wärme-International 33

85. Scholz R, Beckmann M, Schulenburg F (1995) Waste Incineration Systems; Current Technology and Future Developments in Germany. In: Leuckel Wea (ed) 3rd INFUB (European Conference on Industrial Furnaces and Boilers). INFUB Rio Tinto, Lisbon, Portugal

86. Scholz R, Beckmann M, Schulenburg F (1996) Entwicklungsmöglichkeiten der Prozessführung bei Rostsystemen zur thermischen Abfallbehandlung. In: Proceedings: Die thermische Abfallverwertung der Zukunft. FDBR 23 Februar 1996, Rostock

87. Scholz R, Beckmann M, Schulenburg F (2001) Abfallbehandlung in thermischen Verfahren - Verbrennung, Vergasung, Pyrolyse, Verfahrens- und Anlagenkonzepte. Teubner, BG, Stuttgart, Leipzig, Wiesbaden

88. Scholz R, Jeschar R, Schopf N, Klöppner G (1990) Prozessführung und Verfahrenstechnik zur schadstoffarmen Verbrennung von Abfällen. Chemie-Ingenieur-Technik 62:877-887

89. Scholz R, Beckmann M, Horn J, Busch M (1992) Thermische Behandlung von stückigen Rückständen - Möglichkeiten der Prozeßführung im Hinblick auf Entsorgung oder Wertstoffrückgewinnung. Brennstoff-Wärme-Kraft (BWK)/TÜ/Umwelt-Special 44

90. Seifert H (1999-2002) Primärseitige Stickoxidminderung als Beispiel für die Optimierung des Verbrennungsvorgangs in Abfallverbrennungsanlagen. HGF-Strategiefond-Projekt. HGF-NOx

91. Selinger A, Schmidt C, Wieckert C, Steiner C, Rüegg H, (eds.): Conference proceedings (1999) Advanced Treatment of Waste Incinerator Ashes. In: Barrage A, Edelmann X (eds) Recovery, Recycling, Reintegration R'99 Conference. EMPA, Geneva

92. Sher B (1989) California Integrated Waste Management Act. Report

93. SIA (2000) Environmental Aspects of Concrete. SIA documentation, Zurich, p 59

94. Stahlberg R, Feuerrigel U (1995) Thermoselect - Energie und Rohstoffgewinnung. In: Abfallwirtschaft Stoffkreisläufe, Terra Tec `95. B.G. Teubner Verlagsgesellschaft, Leipzig

95. Starley GP, Bradshaw FW, Carrel CS, Pershing DW, Martin GB (1985) The Influence of Bed-Region Stoichiometry on Nitric Oxide Formation in Fixed-Bed Coal Combustion. Combustion and Flame 59:197-211

96. Stephan D, Knöfel D, Härdtl R (2001) Examination of the Influence of Cr, Ni and Zn on the Manufacture and Use of Cement. ZKG International 6:335-346

97. Stumm W (1992) Chemistry of the Solid-Water Interface. John Wiley & Sons, Inc New York, p 428

98. Thomé-Kozmiensky K-J (1992) Stellung der Aufbereitung in integrierten Abfallwirtschaftssystemen. In: Materialrecycling durch Abfallaufbereitung. EF-Verlag, Berlin, pp 7-264

99. Traber D (2000) Petrology, Geochemistry and Leaching Behaviour of Glassy Residues of Municipal Solid Waste Incineration and their Use as Secondary Raw Material. PhD Thesis, University of Berne

100. UBA (1994) Emissionsminderung bei Müllverbrennungsanlagen. Endbericht eines Verbundvorhabens zwischen Firma MARTIN GmbH, München, NOELL GmbH, Würzburg, L.&C. Steinmüller GmbH, Gummersbach, UBA, Berlin

101. van der Sloot HA, Heasman L, Quevauviller P (1997) Harmonization of Leaching/Extraction Tests. Elsevier, Amsterdam, p 281

102. van der Sloot HA, Kosson DS, Hjelmar O (2001) Characteristics, Treatment and Utilization of Residues from Municipal Waste Incineration. Waste Management & Research 21:753-765

103. Van Kessel LBM, Leskens M, Brem G (2001) On-Line Calorific Value Sensor and Validation of Dynamic Models Applied to Municipal Solid Waste Combustion. Contribution. 3rd Int. Symp. on Incineration and Flue Gas Treatment Technologies, Brussels

104. Vehlow J (1996) Verwertung von Reststoffen aus der Abfallverbrennung. Brennstoff-Wärme-Kraft (BWK)/TÜ/Umwelt-Special 10:10-13

105. Vehlow J, Hunsinger H (2000) Einfluss verschiedener Abfallmenüs auf die Metallfreisetzung bei der Verbrennung auf dem Rost. BAT- und preisorientierte Dioxin/Gesamtemissionsminimierungstechniken 2000. VDI Bildungswerk Seminar, München

106. Vesilind A, Worrell W, Reinhart D (2001) Solid Waste Engineering. Brookcole Publishing Co, Monterey

107. World Watch Institute (1983) State of the World Report 1983. Yearly Reports. World Watch Institute,

108. Ziegler F, Scheidegger AM, Johnson CA, Dähn R, Wieland E (2001) Sorption Mechanism of Zinc to Calcium Silicate Hydrate: X-ray Absorption fine Structure (XAFS) Investigation. Environ. Sci. Technol. 35:1550-1555

109. Kolb T, Leuckel W (1988) NOX-Minderung durch 3-stufige Verbrennung - Einfluß von Stöchiometrie und Mischung in der Reaktionszone. 2. TECFLAM- Seminar, Stuttgart

110. Kolb T, Sybon G, Leuckel W (1990) Reduzierung der NOX-Bildung aus brennstoffgebundenem Stickstoff durch gestufte Verbrennungsführung. 4. TECFLAM- Seminar, Heidelberg

111. Kremer H, Schulz W (1985) Reduzierung der NO_X-Emissionen von Kohlenstaubflammen durch Stufenverbrennung. Berichte 574. VDI-Verlag GmbH, Düsseldorf

112. Mäder U, Traber D, Jacobs F, Eggenberger U (2000) Technical and ecological requirements. Concrete Plant and Precast Technology 11:76-84

4 Biological and Bio-Mechanical Processes

Contributions by Konrad Soyez, Sebastian Plickert, Konrad Schleiss, Saburo Matsui, and Helmut Brandl

Robert F. Curl, Nobel Prize winning chemist, stated in 1996 that *"this was the century of physics and chemistry, but it is clear that the next century will be the century of biology."* Great advances are expected from a better understanding and control of biological processes, and raise the question of whether biotechnology will be the preferred technology for solving environmental problems in the future.

The trick to applying biological processes in waste treatment is to exploit the large potential of microorganisms for efficiently transforming organic and inorganic constituents in a way that avoids harmful emissions to the environment. Environmental biotechnology includes air pollution control, waste-water treatment, soil remediation, and groundwater clean-up as well as waste treatment. In the course of a recent conference it was argued that *"process-integrated biotechnology plays an important role in reducing environmental damage as well as cost reduction."* Process-integrated biotechnology is characterized by the application of biocatalysts (enzymes, microorganisms) in an industrial process and the substitution of existing processes or the development of entirely novel processes. These concepts can also be applied for novel municipal solid waste management systems: Mechanical-biological waste pretreatment is an example of such a technique and is currently applied in Germany. The pretreatment combines mechanical treatment (shredding, sieving, sorting) and aerobic composting as well as anaerobic fermentation. As a result, MSW is stabilized and the hazardous potential of the waste is reduced. Natural biological processes occurring in any waste dump (e.g. the degradation of organic matter) are "domesticated" and integrated into an industrial process. Usually, these techniques do not apply genetically modified microorganisms, but rather stimulate naturally-occurring microbial populations in the respective ecosystems. In addition to the biological treatment of MSW fractions rich in organic matter, the metal-containing MSW fractions can also be subjected to a biological treatment. This is particularly the case for fly and bottom ashes resulting from MSW incineration. Ideally, metals can be recovered and re-used in metal-manufacturing industries. Biological processes can also be integrated into landfill management. Novel approaches of landfill techniques mainly include odor control measures and enhanced landfill stabilization by the selective inhibition or stimulation of the present microflora. These methods have the potential to improve landfill techniques in developing countries as well.

Because they are based on the natural biogeochemical elemental cycles, biological processes have a great potential to contribute to the development of future sustainable MSW treatment technologies. It is also a fact that biological techniques are highly accepted by the public, suggesting a natural method of solving environmental problems.

Biological techniques have some advantages in comparison to chemical or physical methods applied in MSW treatment: i) as previously mentioned, microorganisms act as natural biocatalysts; ii) biologically-based techniques are characterized by low hazardous emissions; and iii) are generally considered as "low-tech" and therefore "low-cost" systems. However, biological treatment technologies reveal certain disadvantages as well: i) biological processes are susceptible to changes in process conditions such as differences in available oxygen, pH, or temperature and therefore need balanced process control; ii) biological systems have only a limited tolerance towards toxic compounds; and iii) most important, often require very long residence times and, therefore, large reaction volumes.

Waste treatment is not a process which creates large added value, and the economic boundary conditions thus limit the feasibility of many new technical solutions. Clever combinations of physico-chemical and biological processes exploiting the specific advantages of both types of technology seem to be most promising for the time being. Biological science and technology are expected to advance dramatically in the coming century, as the above citation and many others suggest, and it is very likely that these developments will shape the environmental technology of the 21st century.

4.1 Mechanical-Biological Treatment of Waste (MBP)

Konrad Soyez and Sebastian Plickert

Mechanical-biological treatment of municipal solid waste (MBP) is defined as the processing or conversion of waste from human settlements (households, etc.), which include biologically degradable components, by a combination of mechanical processes (e. g. crushing, sorting, screening) and biological processes (aerobic "rotting", anaerobic fermentation) [1].

At the beginning of the development of MBP, it was applied as a pre-treatment technology for residual waste before landfill (hence the common abbreviation "MBP"). It aimed primarily at the reduction of the mass, volume, toxicity and biological reactivity of the waste, in order to minimise environmental impacts from waste deposition such as landfill gas and leachate emissions as well as settlements of the landfill body. Concerning these points, MBP competed with waste incineration. The recovery of reusable waste components such as metals and plastics, then, was only an incidental to the minimization of the waste amounts. In recent years, the recovery of waste components for industrial re-use has become an integral part in the development of MBP, especially concerning the production of refuse derived fuels (RDF). Thus, MBP is now an integrated technology for the material flow management of MSW, where almost half of the input flow is recovered for industrial re-use, and only one third remains for deposition (Fig. 4.1). A further 20% are process losses in the biological stage, converted into biogas in the case of an anaerobic process.

The mechanical-biological treatment of municipal solid waste has been applied for approximately ten years, specifically in Germany, Austria and Switzerland on technical scale, but also in several developing and emerging countries on a pilot plant scale. In Germany about 1.8 million tons of MSW are treated in 29 mechanical-biological pre-treatment plants [2], compared to twelve million tons treated by incineration in 51 facilities.

Since March 1st 2001, the application of MBP in Germany has been ruled by the German Ordinance on Environmentally Compatible Storage of Waste from Human Settlements (Abfallablagerungsverordnung, AbfAblV) [1]. It defines quality limits for the pre-treated waste, e.g. limit values for heavy metals, AOX and the reactive organics. Standards for the process emissions are defined by the 30th Ordinance on Execution of the Federal Immission Control Act: Ordinance on Facilities for Biological Treatment of Waste (30. BImSchV). For material flow management, required limit values are defined for TOC and the upper thermal value of the output material for deposition (18% resp. 6000 kJ/kg). To meet these requirements, all MBP facilities in Germany have to separate out a considerable fraction of high calorific waste components, which is then predominantly utilized as RDF. Thus, the new ordinances support the development of MBP from a waste disposal technology towards a technology for material flow management.

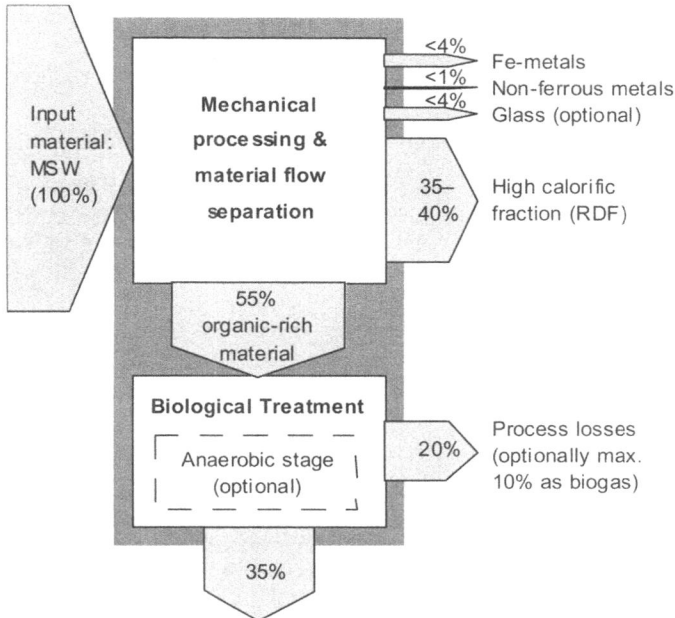

Fig. 4.1. Principal material flow diagram of a MBP plant

For this aim, MBP comprises several mechanical and biological process steps and combinations thereof. The mechanical step is important for the pre-processing of the material before the biological step, but its main task is the separation of the material streams. The biological step mostly determines the residual organic content and thus the landfill behaviour, but may also influence the separation behaviour of the material. Hence, there is not necessarily a strict sequence between the mechanical and the biological stage. Part of the mechanical processing may take place after or even within the biological step, e.g. the separation of metals from the output material or the removal of heavy, mineral-like substances at the bottom of a biogas reactor.

4.1.1 Separation of Organic and RDF Fractions by Mechanical Processing

In general, the mechanical processing inside an MBP plant has the following functions:

1. Removal of contaminants and components which impede the mechanical or biological processes;
2. Adjustment of the particle size distribution for the subsequent processes;

3. Recovery of waste components, such as ferrous and non-ferrous metals, optionally glass and plastics, for recycling;
4. Recovery and processing of a high calorific fraction destined for energy recovery as refuse derived fuel (RDF);
5. Pre-processing of the remaining material for biological treatment, e.g. homogenisation and adjustment of the material's water content.

To fulfil these tasks, a combination of various mechanical processing devices is applied in MBP plants, mainly crushing and screening units, which are also used in traditional waste processing [53]. There is still substantial potential for the optimisation of mechanical units used for MBP.

As MSW is a very inhomogeneous and complex mixture of both organic-rich, high calorific, metal and mineral-like components, the separation of material fractions is generally not suited to compromise concerning decisions between maximum output flows and high product qualities. Thus, an optimisation has to be made for each mechanical step, according to the individual waste composition and the quality demands for the products recovered.

The separation of high-calorific or organic-rich fractions is even more complicated, because neither the calorific value nor the organic content is a suitable property for material separation (in contrary, e.g. to magnetism, which is an unique property in the separation of ferrous metals). Thus, only secondary material properties such as particle size and density or a varied crushing behaviour can be used for physical separation, but this causes a low selectivity. In addition to this, the separation effect of technical screening units is significantly worse than in laboratory, especially when the throughput is close to the machine's capacity [22].

Another compromise situation can be seen in the fact that the organic-rich components also contribute to the waste's calorific value. Hence a maximum output of the energy content, on one hand, and a maximum organic load in the remaining material on the other, are also conflicting goals. As the purpose of the biological treatment is to reduce the input of reactive organics into the landfill, the organic load of this fraction must not necessarily be maximized. However, the portion of total organic load in the remaining fraction also has to be taken into consideration to ensure an optimal material flow separation.

4.1.2 Case Study: RDF Separation by a Two-Step Mechanical Process

Figure 4.2 exemplifies a compromise situation, like the one described above, for a simple two-step mechanical separation of RDF from the input material for the biological step in the MBP plant in Quarzbichl, Germany [22]. In this case, a compromise has to be found for the conflicting goals of a) a maximum transfer of the waste's energy content into the coarse-grained fraction for energy recovery, and b) a maximum calorific value of this fraction.

In Figure 4.2, the assumed quality demand for RDF, due to market reasons, has a calorific value of 15'000 kJ/kg (see vertical broken line; the minimum for energy recovery according to German law is 11'000). As no screen fraction of the non-crushed material reaches this value (see the graph on the left), crushing the waste is indispensable to the production of RDF. The highest calorific values are reached by the hammer mill or the roll crusher, when screened at 150 mm (points 'A' and 'B'). In order to choose an optimally suitable crushing device, the portion of total energy in the RDF output has to be considered: At the points 'A' and 'B', only 7% resp. 16% of the waste's total energy is recovered. When crushed by a revolving composting drum, the portion of total energy in the RDF output rises to 31% at 80 mm screen overflow (point 'C'). The 40 mm screen overflow (point 'D') contains 48%, while its calorific value is only slightly below 15'000 kJ/kg.

Concerning the organic load in the remaining fraction, the combination of a revolving composting drum and a 40 mm screening device is equally acceptable, as it almost doubles the organic load in the 40 mm screen underflow compared to non-crushed MSW (determined as organic dry matter of biological origin). Only the hammermill crusher reaches higher accumulation of reactive organics in the underflow, but the difference in the 40 mm screen underflow is negligible compared to the sizable difference in the RDF output (see above).

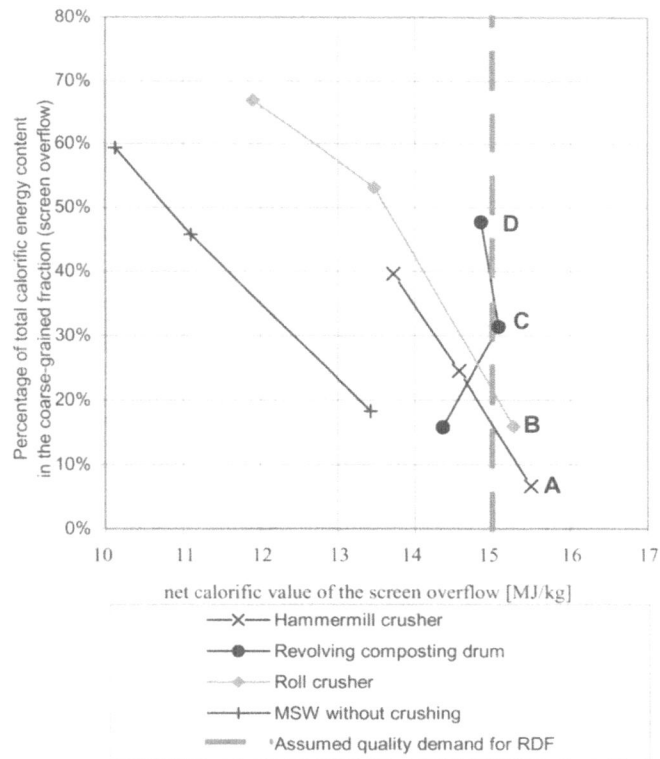

Fig. 4.2. Effects of different crushing devices and screening sizes on the calorific value of the screen overflow and on the total calorific energy recovered as RDF (the 3 data points on each graph represent – from top to bottom – the screen overflows at 40, 80 and 150 mm)

Under real conditions, the demands on both product and separation quality are more critical, thus requiring a more sophisticated system for mechanical processing and separation. Naturally, the optimisation of such a complex system is much more demanding than the example given here, especially as the process expenditures also have to be considered, both economically and ecologically.

4.1.3 Improved Material Flows by the 'Dry Stabilate' Process

Another approach to material flow management by mechanical-biological processes is the 'dry stabilate' process developed by the Herhof Company [48], applied in four facilities in Germany with a capacity of 75'000 to 180,000 tons each. The goal of this process is to make a maximum percentage of waste available for industrial re-use, thus minimizing the amounts for deposition.

To achieve this, the entire waste is brought into the biological stage, where it is dried using physical and biological processes for 7 days under high aeration rates. Within this process, the organic content is reduced only slightly, and the separation behaviour is improved significantly. This is followed by a separation into a heavy and a light fraction. The light fraction is used as RDF after a further separation of metals. The heavy fraction (about 15 %) is separated into metals, glasses, batteries, and mineral components.

Table 4.1 shows that in this process more than two thirds of the input flow are recovered for industrial re-use, while only 4% remains for waste disposal, i.e. waste incineration. The rest is lost in the process and stripped out with the process air, which is purified by a sophisticated regenerative incineration technology.

Table 4.1. Output flows of the 'dry stabilate' process [48]

Output fractions	% of total input
Fractions for industrial re-use:	
RDF (calorific value: 15–18 MJ/kg)	53
Ferrous metals	4
Non-ferrous metals	1
Batteries	0,05
White glass	3
Brown glass	0,5
Green glass	0,5
Minerals	4
Total	66,05
Others:	
Fine grain and dust.	4

4.1.4 Biological Processing and Effects on Landfill Characters

Technical Performance

The aim of the biological stage of a regular MBP plant is to reduce the content of organic material to form a low-reactive "stabilised" product for deposition. The techniques applied are aerobic rotting, anaerobic fermentation, or combinations thereof. Aerobic systems are in widespread use. They include windrows with or without aeration, containers or boxes, drums, and tunnels. Aerobic treatment shows wide variation in both process intensities and duration. Low-tech windrow systems operated directly on the landfill site require only a minimal technical outlay. They can be aerated through drainage pipes which act as vents; the required process time varies from 5 to 15 months (Fig. 4.3).

The majority of contemporary plants include an encapsulated, controlled, intensive biological treatment of 4 to 8 weeks, followed by an extensive maturation process in an open windrow without forced aeration. Actual aeration rates are between 3000 and 8000 m^3/ton waste, while the oxygen needed for the biological process itself is only 2000 m^3/ton in the case of a typical residual MSW and a 60% degradation of the reactive organics.

The new parameters AT_4 and GB_{21} have been introduced as indicators for waste stabilisation in the case of biological waste treatment. They are applied as a substitute to the ignition loss, which is used as an integral criterien for mineralisation by incineration. The GB_{21} indicates the experimental gas formation within 21 days. Alternatively, the respiration coefficient AT_4 is used, and is

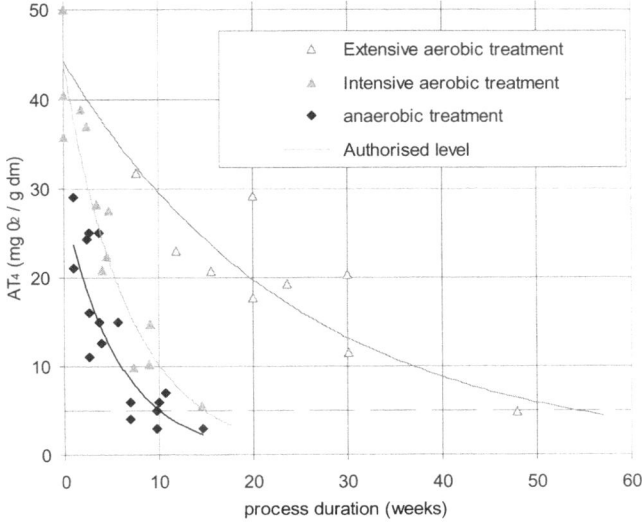

Fig. 4.3. Stability parameter AT_4 as a function of the treatment duration [52]

defined as the amount of oxygen consumed in 96 hours in microbial processes for a specified experimental device [4]. AT_4 is most easily determined: it is preferred for practical process control. For untreated material, typical values of AT_4 are in the range of 40 to 50 mg O_2 per g of dry matter. A sufficient biological treatment results in an AT_4 value of less than 5 mg O_2 per g of dry matter, and in a $GB_{21} < 20$ l/kg. Figure 4.3 illustrates the process duration required to reach a given AT_4-value dependant upon the intensity of the biological process applied.

Deposition Behavior of MBP Output

The level of contaminants in both leachate and gas emissions of the treated residual waste is reduced by more than 98% in comparison to untreated waste. One kilogram of treated waste potentially releases a total load of 1-3 g COD, 0.5-1.5 g TOC, and 0.1-0.2 g NH_4-N into the leachates. The real numbers clearly depend on the intensity resp. the duration of the pre-treatment (see Table 4.2 for leachate concentrations)

Table 4.2. Leachate parameters as functions of the process duration in windrows [13]

Parameter	Process duration [months]			
	3	6	9	13
COD [mg/l]	701	425	320	215
BOD_5 [mg/l]	46	4.6	4.3	3
TOC [mg/l]	328	174	128	80
NH_4-N [mg/l]	20	10	3.7	1

The waste treated can be compacted in the landfill to a density of 1.5 tons/m^3 (wet). This leads to a better use of landfill capacity. The hydraulic conductivity (k_f-value) is approximately $k_f = 10^{-8}$ m/s or even lower. Water flow through the landfill is therefore partly prevented, and the leachate volume decreases considerably.

To achieve these effects, the landfill construction is to be optimised according to the conditions determined by the waste, the shape of the landfill surface, the area open for the material input, etc. A special aspect is the very low level of remaining gas production, especially methane, which at a maximum is less than 1 l CH_4/m^2*h [18]. On this level, no active landfill gas collection is possible. To reduce the remaining emissions to an environmentally acceptable minimum, a passive oxidation is applied by a bio-active oxidising landfill cover.

MBP Emissions and Cleaning Technologies

The emissions from the biological pre-treatment process are the following:
1. Carbon dioxide and methane produced by aerobic resp. anaerobic biological activities;
2. Organic compounds metabolised or generated by biological reactions;
3. Volatile substances stripped out from the original waste;

4. Heavy metals and heavily volatile substances which mostly remain in the residues;

5. Germs, as bacteria and moulds, emitted from the system as a result of the bioprocess.

The total organic compounds emitted amount to 600 g per ton of original material, measured as non-methane volatile organic compounds (NMVOC). Methane may amount to approximately 100 g per ton of original waste, or even more in the case of difficulties in the process. Ammonia amounts to about 500 g per ton of original waste. In the bio-filter it can be transformed into N_2O, which is a climate relevant gas [15].

To avoid health and climate risks from the emissions, the waste gas is collected and cleaned during this period. Scrubbers and bio-filters are traditionally used. They achieve a mean reduction by 50%, so that about 300 g TOC per ton of waste remain [15], including difficult to degrade chlorinated carbohydrates. Further detoxification by an extra thermal process has to be applied in order to reach the exact TOC limit value of 55 g/Mg, which is analogous to the limit in the case of waste incineration.

Evaluation of the Technology, Future Trends and Research Needs

By producting components for industrial re-use, MBP contributes directly to an improved resource management. The future development of the MBP technology will strongly depend upon successes in the marketing of the products. The most important fraction is the RDF, which can be used as a substitute for fossil fuels, but also as a raw material for synthesis gases in the chemical industry or even for hydrogen as an energy source. To broaden the applicability of RDF, a standardisation of the product quality and a stable high quality production is necessary. In Germany, a quality certificate (RAL-GZ 724) was introduced in 2001 [21].

Future improvements for high standard MBP can be seen in an integrated design of the MBP plants with particular respect to the waste composition, infrastructure and ecological situation. Potential combinations with existing waste management facilities, such as incineration plants and landfills, have to be considered. The design process must be adapted to market demands concerning the amount and the quality of recycling products.

With respect to the realisation of the technology itself, air flow management inside the plant is of major interest. The air flows have to be reduced by closing loops and implementing a specific treatment of waste air streams according to their individual qualities. However, for the waste gas of the intensive biological step, a thermal treatment is imperative. The mechanical separation processes must be improved by defining clear separation goals, according to which an appropriate machinery has to be developed.

With respect to waste deposition, MBP technology has some benefits in comparison to other treatment options. In the case of poor management of landfills or a low availability of suitable grounds, MBP can contribute to a rapid improvement of the waste management situation, with respect to landfill gas production, leachate emissions, and settlements. One advantage is a higher flexibility in comparison to incineration. It can be easily adapted to changes in waste volume or composition. Thus, it seems well suited to the situations in many developing and emerging countries as a first step toward an effective waste management. Under higher developed waste management conditions, MBP has to compete with other options of waste management according to ecological and economical preferences. Ecologically, MBP has advantages with respect to risks to human health, but disadvantages in climate effects. Typical costs of a high-tech MBP are in the range of 80 to 100 € per ton, which corresponds to the lower range of incineration. The minimum capacity for an economical operation of MBP facilities is between 50.000 and 80.000 tons p.a.

4.2 Composting and Anaerobic Digestion

Konrad Schleiss

Composting is the oldest form of recycling and has long been used to treat waste. In the last century, municipal solid waste (MSW) was composted in many places. Over the last decades, some countries have shifted to separately collected green waste, so as to ensure a marketable quality of the produce. This tendency was particularly marked in Germany, Holland, Austria and Switzerland, where the agricultural use of compost from MSW is forbidden since 1986 [11]. In many countries, composting is still considered a treatment process for municipal waste prior to landfilling.

The processes described here do not apply to neighbourhood- or home-composting initiatives, but above all to facilities treating more than 1'000 tons annually. For marketing reasons, the author considers that only separately treated green waste has a future. Biological treatment before landfilling has other types of requirements than the composting discussed here.

Composting techniques evolved from open-air composting, often plagued by odour problems, to covered composting to anaerobic digestion. Anaerobic digestion, which proceeds in the absence of air, has shown to be particularly suited to water-rich, easily degradable wastes. It therefore complements aerobic composting, where problems were often encountered to aerate poorly structured materials. Anaerobic digestion is a globally isothermic biological process, so requires a good heating for the thermophilic variant, while aerobic composting is highly exothermic, releasing large amounts of heat and is therefore a self-heating process.

The types of waste most adapted to anaerobic digestion or composting are schematically presented in Figure 4.4. This classification is only valid for the dominant waste type of a mixture. For the minor components it is not relevant.

Composting (aerobic)   Anaerobic digestion

Tree and bush cuttings
Agricultural green waste
Urban green waste
Kitchen and food waste
Restaurant waste
Slaughterhouse waste

Increasing water content

Increasing structural material

Fig. 4.4. Suitability for aerobic or anaerobic processes [29]

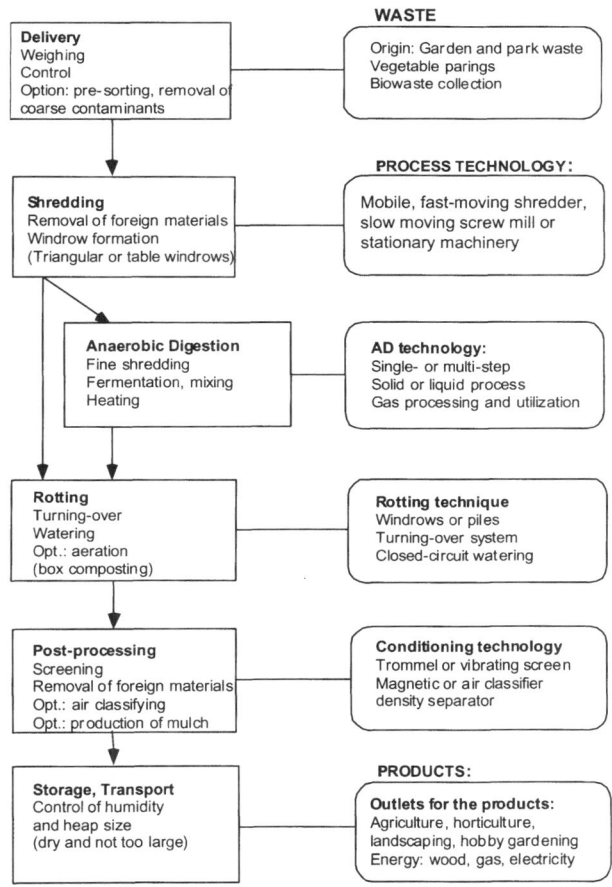

Delivery
Weighing
Control
Option: pre-sorting, removal of coarse contaminants

WASTE
Origin: Garden and park waste
Vegetable parings
Biowaste collection

Shredding
Removal of foreign materials
Windrow formation
(Triangular or table windrows)

PROCESS TECHNOLOGY:
Mobile, fast-moving shredder, slow moving screw mill or stationary machinery

Anaerobic Digestion
Fine shredding
Fermentation, mixing
Heating

AD technology:
Single- or multi-step
Solid or liquid process
Gas processing and utilization

Rotting
Turning-over
Watering
Opt.: aeration
(box composting)

Rotting technique
Windrows or piles
Turning-over system
Closed-circuit watering

Post-processing
Screening
Removal of foreign materials
Opt.: air classifying
Opt.: production of mulch

Conditioning technology
Trommel or vibrating screen
Magnetic or air classifier
density separator

Storage, Transport
Control of humidity
and heap size
(dry and not too large)

PRODUCTS:
Outlets for the products:
Agriculture, horticulture,
landscaping, hobby gardening
Energy: wood, gas, electricity

Fig. 4.5: Sequence of operations in a composting or anaerobic digestion facility [23]

Green waste processing has three main aims: to favour the biological processes, to improve the quality of the produce and to ensure operational efficiency. This also includes a regular control of the sanitary harmlessness of the produce (no germinating weed seeds and disease supressivity).

Figure 4.5 shows the sequence of steps required for composting or anaerobic digestion. When processing green waste in centralised facilities, a general requirement is the setting up of corresponding efficient collection logistics. The single treatment steps are described in the following paragraphs.

During *delivery and weighing* the compostability of the waste received is checked. Quality compost requires that the larger contaminants be removed. Weighing the waste is essential for waste management planning and correct billing of the amounts received. *Shredding* is the preparatory step for biological degradation. It must be adapted to the structure of the material, so as to allow for adequate oxygenation: neither too fine (powder), nor too coarse (sticks will favour air bypass). Fine shredding is a prerequisite for efficient degradation in anaerobic digestion. *Anaerobic digestion* (AD) is best for easily degradable materials, that may produce offensive odours if they are composted. AD is also a source of renewable energy. In general the material is shredded a second time in a rotary shredder and the substrate is made into a pulp. As AD is mainly a biological degradation process, a post-maturation phase is necessary to produce compost. The *rotting process* (including turning-over, aeration and watering) is the next step. The turning-over influences the biological process most. Its finality is the mixing, aeration, and watering of the substrate, and these must be carefully balanced one against the other. Turning-over may be partially, but not completely, replaced by forced aeration, as the turning-over mixes the dryer and wetter parts, thus optimising the water balance. *Post-processing and conditioning* consists mostly of culling out the foreign materials and calibrating the product to the required particle size. This is generally carried out after the rotting phase is over. Sometimes mature compost is also stored as is, and only conditioned shortly before sale. In general screens are used to remove foreign materials.

Separately collected biowaste can be divided into garden and park waste and household biowaste (vegetable and kitchen waste). The amounts collected vary greatly depending on the type of settlement and the collection logistics, ranging between 50 and 150 kg per inhabitant per year. In Switzerland, nearly 100 kg per inhabitant of separately collected biowaste are composted or digested each year. Of this, over half comes from municipal collections. The rest stem from gardening and landscaping (Fig. 4.6). Obviously, these materials cannot be reused on-site and are therefore handed over for another use to composting or digestion facilities.

4.2.1 Overview of the Various Processes

Two-thirds of the biowaste is composted in traditional open-air windrows. This technique is characterised by relatively low investment costs and the generally high quality of the product. Its drawbacks are the emission of odours, when too much unsuitable materials are processed (kitchen wastes, etc.), or when the facility

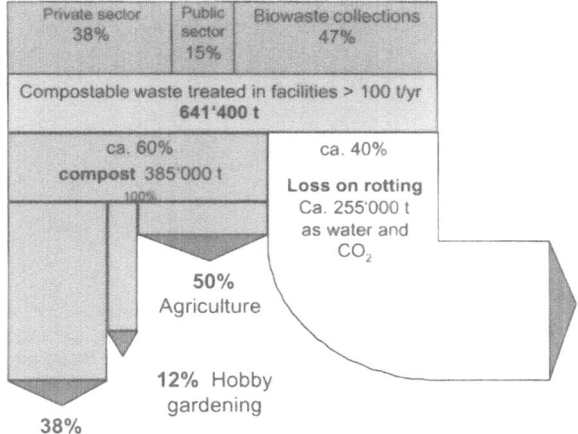

Fig. 4.6. Origin of biowaste and destination of compost in Switzerland in 2000 [12]

exceeds a certain size (5'000 to 10'000 t/yr). Field-edge composting counts as windrow composting, since piles are made along fields, though the amounts treated rarely exceed 1'000 t/yr.

Hall and box composting can be grouped under the heading of covered systems. Covering the facility avoids waterlogging by rainfall. By completely housing a plant, it becomes possible to treat the exhaust air through a biofilter. In Switzerland, only a few plants are completely covered. In Germany, Holland and Austria, this type of plant is more common for large facilities (> 20'000 t/yr). The energy consumption of fully housed facilities, for the forced aeration of the windrows and the air cleansing system, is very high.

In Germany and Switzerland, some 10 to 12 % of the total biowaste is processed by anaerobic digestion (Fig. 4.7). In the canton of Zurich, a pioneer in the field, this percentage rises to over 25 %. With such a ratio, the energy balance of the total collection and processing is positive: biogas production are about double the consumption of diesel fuel by the composting facilities.

Windrow composting	403'400 t	62%
Box composting	41'700 t	7%
Field-edge composting	62'300 t	10%
Hall composting	55'400 t	9%
Anaerobic digestion	78'600 t	12%
TOTAL	**641'400 t**	**100%**

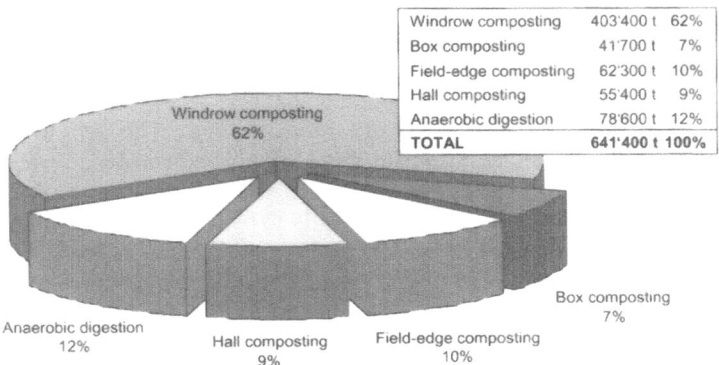

Fig. 4.7. Types of processes used in Switzerland for the year 2000

Table 4.3. Net energy balance (energy production minus energy consumption) of the various biowaste treatment processes

Useful residual energy [kWh/ton]	Electricity	Heat	Fuel	**Total**
AD (100 % electricity production)	130	* 320	0	450
AD (100 % fuel production)	-120	-100	600	380
AD combined heat and power (50/50)	5	* 110	300	415
Wood chips heating	0	1000	-25	975
Composting, open system	0	0	-30	-30
Composting, housed system	-100	0	-20	-120

* Up till now the heat could practically not be used, since most industrial buildings are already equipped with an individual heating system. Therefore, the values for electricity and fuel are most significant.

AD produces on average 100 m^3 of biogas, with a methane content of about 65 %, 34 % carbon dioxide and 1 % trace gases. The energy content amounts to about 6 kWh per m^3 biogas. This gas can be transformed into about 130 kWh electricity by a block heating and generation plant (see Table 4.3). If all the biogas is used as biofuel, 120 kWh electricity are consumed, as well as some heat for the fermentation tank. Optimization is possible with combined heat and power generation.

4.2.2 Prospects: What Should be Improved for a Sustainable Development?

Techniques adapted to every kind of exploitation exist. But the market prices are not adapted to sustainable processing. For example, an important argument in favour of AD is the production of energy. However the proceeds from the energy cover only at most 10 % of the operation costs, since comparable oil and electricity prices are much lower.

Advocates of a purely energetic exploitation of biowaste claim that composting is no more necessary. The woody fraction could be burnt in wood chips heating systems and the remainder anaerobically digested. This point of view overlooks the fact that without aerobic composting there is no compost to speak of since compost is defined as the product of aerobic degradation. Therefore, after the digestion process, a post-composting phase is mandatory. The sense of anaerobically processing agricultural biowaste with a high percentage of wood that cannot be easily separated out is still in dispute. It is however certain that, in the future, digestate will have to be matured aerobically to be marketable.

Recycling only makes sense when a market exists for its products [51] This means that green waste processing plants must produce marketable goods. Up till now, it is only after an increase in production that an outlet with minimum costs is sought out. If a technology is to acquire a practical value, there must be a demand for its products. There is a general need for market investigations *before* any production is initiated. As long as the waste producers continue to pay for treatment without any investigation being done on the product marketing side, inappropriate technical investments will continue to be made. The existing technology is good, but there is much to be gained in better defining to what end it is applied. For example, there is a latent demand for growth substrates with suppressive capacity towards seedling diseases, etc. Global knowledge of these techniques exist, but specific recipes have never been elaborated. Similarly, when biogas is used for electricity production, some 50 % of the energy is lost as waste heat. However, AD plants are practically never built on sites were this energy could be beneficially exploited.

These examples show that if the market prices also covered the production costs, it would be possible to approach zero waste levels, but this is not the case. The difference between the production costs of marketable goods and the market price for such goods must then be paid by waste taxes. However, if their treatment led to the production of marketable goods, these waste materials could then lose their status as waste.

4.3 Active Landfill Control and Stabilization of MSW

Saburo Matsui

Municipal solid waste (MSW) is worldwide mostly managed by landfilling without incineration. Only a limited number of countries and cities practice incineration followed by landfilling of the ash. Incineration of MSW has the advantage of reducing the bulk of MSW by oxidation of organic materials, whereas direct MSW landfilling shows very slow oxidation and reduction of organic materials followed by a very long time period of stabilization. Generally, if space for slow oxidation and reduction processes of MSW is available, all municipalities select this method. If space is limited, incineration of MSW is inevitable and advanced incineration technology and subsequent pollution control technology is required.

Developing countries generally have a low technology standard in the application of MSW incineration. In addition, MSW itself is not suited for incineration, due a low caloric value (energy content). There is, therefore, a need to develop improved methods of MSW landfilling in contrast to today's conventional "sanitary landfills". However, most developing countries practice so called "open dumping", resulting in the generation of many pollution problems such as gas emissions, odor, waste water formation, or ground water pollution. Sanitary landfills can provide better solutions than open dumping reducing many of the problems, yet, there is still a potential for improvement.

A novel approach to solve the problems mentioned is called "Active Control Landfill and Stabilization Method (ACLSM)". This introduces technologies such as a) odor control during landfill processes; b) methane collection after closure of the landfill site; c) enhanced landfill stabilization by sulfate-reducing bacteria which suppresses methane formation; d) ensuring the decrease of toxic pollutants in leachates from the landfill site. Methane collection as well as suppression of methane formation during the various stages of landfilling are important objectives regarding issues of global warming, i.e. the reduction of greenhouse gases [47]. Methane is a much stronger warming agent than carbon dioxide. Therefore, developing countries can contribute to the efforts to stop global warming by introducing ACLSM.

4.3.1 How to Proceed in ACLSM?

There are five stages of MSW landfill and stabilization processes, namely 1) landfilling process; 2) landfill closure; 3) methane collection; 4) stabilization enhancement by stimulating bacteria from the sulfur cycle; 5) completion of landfill process (Fig. 4.8).

Stage 1 - Landfilling process: What are major problems in stage 1 when the landfilling begins? Odorous gas is immediately generated and leachate treatment becomes necessary to reduce high BOD values and ammonia concentrations in the waste water. Odorous gas consists mainly of hydrogen sulfide. During decomposition of organic matter, sulfate is easily reduced to sulfide by sulfate-reducing bacteria. Sulfide formation is successfully inhibited by adding nitrate to the landfill (e.g. irrigation with nitrate solutions) and bacterial denitrification is easily stimulated. Denitrification activity (i.e. the formation of nitrogen gas) usually outcompetes sulfate reduction so that sulfide is not formed. As a consequence, landfill leachates must be aerobically treated by ammonia-oxidizing

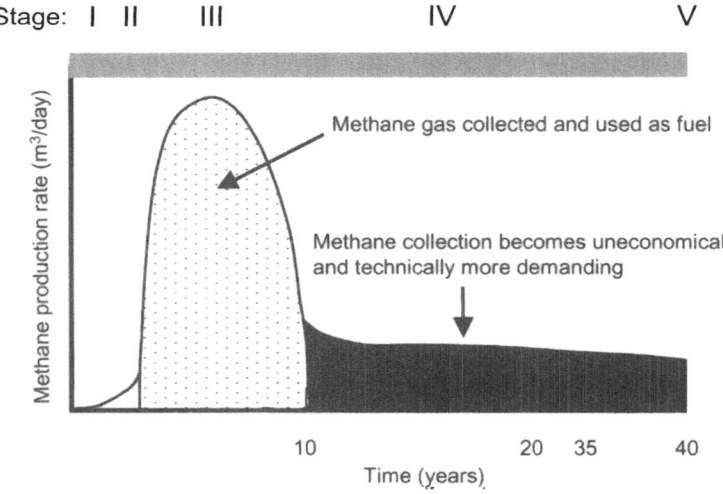

Fig. 4.8. Five stages of active control landfill and stabilization processes

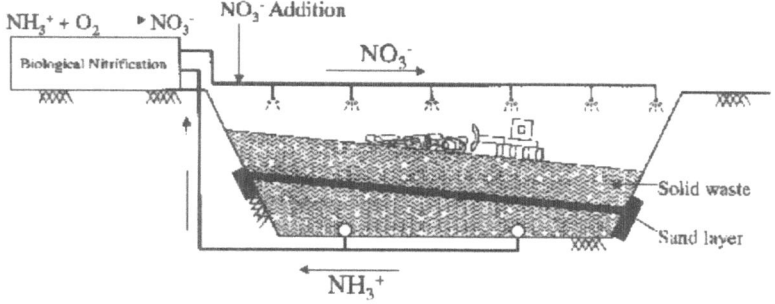

Fig. 4.9. Odor control by denitrification

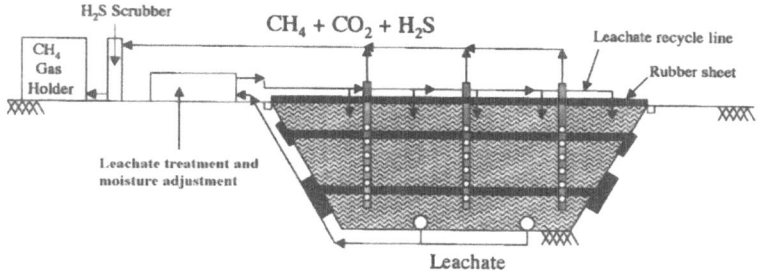

Fig. 4.10. Closure of the landfill site and collection of gaseous emissions

microorganisms. The leachate can be recycled and added back to the landfill where odorous gas control is accomplished (Fig. 4.9).

Stage 2 - Landfill closure: At stage 2, a landfill is sealed by covering the landfill site with a rubber or plastic sheet to prevent rain water seepage which is an important technique in the control of pollution problems. Additionally, gas can be collected from the site. Odorous gas (hydrogen sulfide) has to be eliminated by gas scrubbers and methane gas must be stored in a gas holders. Leachate treatment is one of the important control methods that require additional energy and financial expenditure [17]. If rain water seepage is controlled, leachate treatment costs are highly reduced. Treated leachates can be recycled to the site (Fig. 4.10).

Stage 3 - Methane collection: An important aspect is the collection of methane as biogas (fuel gas recovery for households). As an additional consequence, the proper management of MSW can also contribute to the reduction of greenhouse gases which contribute to global warming processes. This is of particular importance for developing countries.

Stage 4 - Stabilization enhancement by stimulating sulfate-reducing microorganisms: This stage lasts more than 40 or 50 years after the landfill closure, depending on the local climatic conditions of the site. It is very important to control methane emission during this stage. After a peak of high methane concentration during the first five years of stage 3 (Fig. 4.8), emission of low methane concentrations occurs for a very long time period. Due to the low concentrations, methane collection it is no longer economically beneficial. However, methane release (even at low concentrations) over an extremely long period has nonetheless an important impact on global warming. The solution to this problem is the introduction and stimulation of the microbial sulfur cycle in the active control landfill system: When leachate is recycled after aerobic treatment, the level of the sulfate concentration can be adjusted to approximately 1g per liter. Sulfate is the terminal electron acceptor for sulfate-reducing bacteria. Sulfate reducers can outcompete methane-forming bacteria, so that organic matter is degraded to organic acids while methane production is suppressed. As a result, hydrogen sulfide is formed. Organic acids are treated by aerobic bacteria during the leachate treatment process, which oxidize sulfide to sulfate. Sulfate originating from this process can be recycled and utilized when the landfill surface is irrigated with treated effluent. Hydrogen sulfide gas from deep landfill sites has to be oxidized by the aerobic soil filter covering the top of the landfill site. The top soil cover is approximately 1.5m deep and represents the habitat for sulfur-oxidizing bacteria (Fig. 4.11). Hydrogen sulfide rises from the bottom and is exposed to oxygen provided by air pumps at a depth of 1.5 m. Consequently, sulfide is biologically oxidized to sulfate or chemically oxidized to elemental sulfur within the soil filter. Basically, two main sulfur cycles have been established in the system, one being trough the landfill site and the leachate treatment, which is the major sulfur reduction and oxidation cycle, the other between the bottom of landfill and the landfill surface which might be a minor cycle. During the two sulfur flows being cycled, there is a slow build up of elemental sulfur in a landfill. The rubber or plastic sheet is placed over the biological soil filter in order to control the penetration of rain water and moisture level inside the landfill.

Stage 5 - Completion of landfill process: One of the major questions (especially for environmental engineers) is how to terminate the landfill process and how to re-use the landfill site for further purposes, such as agricultural or recreational use.

For this reason, the decomposition of degradable organic matter has to be complete. Although this is difficult to evaluate, there are some possible indicators: The occurrence of humic substances as well as some organic acids with trace phenol compounds which are basic elements of lignin polymers in leachate can indicate that the remaining organic matter is basically lignin and that any other degradable organic matter is present in only minor amounts [45, 47].

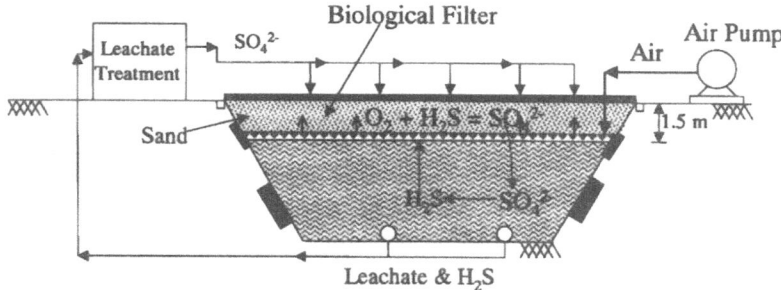

Fig. 4.11. Stabilization enhancement by stimulating the microbial sulfur cycle

Lignin is a natural polymer that cannot be degraded within a time span of several decades [30-32]. In addition, it is known that many synthetic chemicals are degraded by sulfate-reducing bacteria. Furthermore, most heavy metals, such as lead, cadmium, chromium, copper, zinc, aluminum, or iron, which are present in the various soluble forms, are finally stabilized in the form of metal sulfides.

4.3.2 Conclusions

Current landfill practices need improvement in many ways. ACLSM is a new approach for MSW management. Developing countries could introduce this approach where energy recovery as biogas can be extremely valuable and can contribute to the control of global warming efforts. There are many disadvantages in direct MSW landfilling practices: One typical case worth noting is the new landfill practice in Hanoi City, where the city introduced a standard MSW landfill method that had been developed based on Japanese experiences. As a major fatal mistake, the huge differences in MSW quality had not been considered. Japanese MSW is mostly incinerated, reducing the organic content and, consequently, decreasing methane and sulfide formation. The Japanese standard landfill technology does not provide much gas control measures because incineration of MSW is a premise. In contrast, Hanoi City MSW contained fractions rich in organic matter. This resulted in immediate odor problems as soon as landfilling activity started. Residents around the landfill site demonstrated against the activity and picketted in front of the landfill gate preventing the city from transporting MSW to the landfill site for three days. This issue was temporally settled by controlling sulfide emissions through chemical methods. The city is now considering changing the landfilling practice according to the MSW quality. Moreover, the leachate treatment facilities have not been established yet. Further problems are high BOD, COD, and elevated nitrogen as well as color formation in the leachate. The new technique has to provide odor control measures, leachate treatment, and toxic chemical management. Therefore, the introduction of active control landfill and stabilization processes in developing countries is strongly recommended.

4.4 Biotechnology for the Treatment of Inorganic Wastes

Helmut Brandl

4.4.1 Biogeochemical Element Cycles

It is an important prerequisite for future sustainable waste treatment technologies to integrate industrial civilization-dependent processes into natural element cycling. These technologies have to be oriented according to the natural biogeochemical cycles of elements found in the ecosphere, e.g. carbon, sulfur, or copper cycle. In nature, microorganisms are the driving force of the biogeochemical cycles. They show among all living creatures the highest biological diversity and represent the foundation of the biosphere. However, it is estimated that the vast majority of microorganisms is still not isolated, cultivated, and characterized [5]. As comparison, approximately 1.4 millions organismic species have been described until today, (Table 4.4): insects represent the group with the highest number of known species, whereas only a very small percentage of bacteria is known.

Microorganisms are of vital importance for the cycling of elements in the ecosphere, mainly either as primary producers (photosynthesis) or destruents (mineralization of organic matter). Figure 4.12 summarizes a general biogeochemical cycle driven by microorganisms. In a first step, carbon (as carbon dioxide) is utilized by primary producers (autotrophic microorganisms, plants) and biomass is formed by photosynthesis. The energy to drive this process is provided by the sun (light). Biomass is utilized by first order consumers (herbivores) which are themselves consumed by second order consumers (carnivores). Saprophytes (destruents) are responsible for the complete mineralization of organic matter (biomass, waste) formed by primary producers, herbivores or carnivores. Mineral

Table 4.4. Numbers of known organismic species in comparison to total estimated numbers

Organismic group	number of described species	estimated total number of species	percentage of known species
insects	950'000	8'000'000 - 100'000'000	1 - 12
plants	250'000	300'000 - 500'000	50 - 85
spiders	75'000	750'000 - 1'000'000	7.5 - 10
mollusks	70'000	200'000	35
fungi	70'000	1'000'000 - 1'500'000	5 - 7
vertebrates	45'000	50'000	90
protozoae	40'000	100'000 - 200'000	20 - 40
crustaceae	40'000	150'000	25
algae	40'000	200'000 – 10'000'000	0.4 - 20
viruses	5'000	500'000	1
bacteria	4'000	400'000 – 3'000'000	0.1 - 1

compounds are liberated and made available for primary producers. With this process the cycle is closed. In summary, the fundamental importance of microbial processes in the biogeochemical cycling is based on the ability of microorganisms to mobilize and convert elements [34].

Approximately 60 elements are involved in the turnover of the biosphere [34]. In completely closed biogeochemical cycles, elements are temporary immobilized in the biomass and subsequently recycled in relatively short time periods, whereas in incomplete cycles certain elements are immobilized in the hydro- or lithosphere for geological time spans.

Besides their fundamental importance in the mineralization and complete degradation of organic matter, microbes are naturally involved in the weathering of rocks, in the mobilization of metals from sulfidic minerals, in metal reduction and oxidation, in metal precipitation and deposition, and in isotope fractionation. These microbiological principles and processes might have the potential to be adapted for technical waste treatment applications. Besides the huge microbial potential to degrade a wide spectrum of organic substances (as known in composting or in the bioremediation of polluted soil), microbial abilities can also be used to cycle inorganic compounds (e.g. metals). Most of these capabilities are related to geological processes [19].

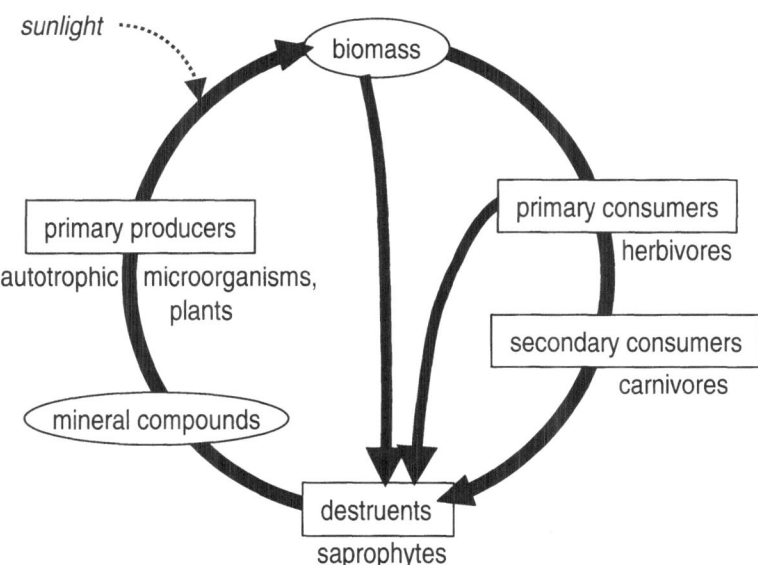

Fig. 4.12. General biogeochemical cycle driven by microorganisms both as primary producers for the formation of biomass and as destruents for the mineralization of biomass.

4.4.2 Organic Aspects: Mechanical-Biological Waste Pretreatment

Prior to final treatment (incineration, landfilling), municipal solid waste can be pretreated by mechanical-biological techniques. Microorganisms are involved in the transformation and degradation of the organic fraction of the waste. As alternative to thermal pretreatment (incineration in waste treatment plants) municipal solid waste is subjected to a combined mechanical and biological treatment to improve certain properties (e.g. stabilization) of waste fractions and to reduce the hazardous potential of the waste [58].

The first step of the treatment consist of a mechanical process. Municipal solid waste can be sorted, sieved, shredded, magnetically separated, and homogenized. As result, waste is classified in several fractions, namely reusable materials, a fraction of high calorific value, a heavy mineral fraction, and a fraction rich in organics, which is readily biodegraded [28]. The main goals of the mechanical treatment are the recovery of valuable and reusable components, the conditioning (volume reduction, particle size reduction, concentration of certain compounds) of the waste for an optimal subsequent biological or thermal treatment [35].

In a second step, a biological treatment under either oxic or anoxic conditions follows mechanical treatment. Under oxic conditions, the fraction rich in organic matter is composted in drums or bins as well as in tunnels or windrows requiring periodical agitation (turning) [28]. Most of the organic waste materials (agricultural and food industrial wastes, sewage sludge, etc) are composted in simple piles or prisms. These can be mechanically turned or undisturbed (static) prism, which are equipped with a built in aeration system. Fermentation under anoxic conditions requires a closed system where resulting gases (mainly methane) can be collected. The gas can be utilized for heating or as energy source (e.g. for cars). Consequently, the net energy gain from biogas formation is a major advantage of this technology [28]. Finally, residual fractions originating from the mechanical-biological pretreatment can be incinerated for volume reduction and energy recovery or disposed in landfills.

Landfilling of mechanical-biologically pretreated waste is mainly characterized by reduced volume, reduced gas and leachate emissions of environmentally hazardous compounds as well as a reduced settling behavior of the landfill [58]. Several goals are pursued with a mechanical-biological waste pretreatment [28]:

1. Reduction of landfill volume. As a result from separation and recycling of reusable waste fractions as well as from biological degradation of organic matter, the landfill volume is reduced by 30% after mechanical treatment and by 60% by a combined mechanical-biological treatment [28].
2. Reduction of the emission potential of landfill gas and leachate. In comparison to untreated municipal solid waste gas formation can be reduced by 75 to 90% resulting in gas emission rates of 20 to 40 m^3 per ton of dry matter [52]. During pretreatment, the main fraction of readily degradable organic compounds is decomposed under oxic conditions by microorganisms whereas recalcitrant substances remain in the landfill. Organic compounds include cellulose, non-

cellulose carbohydrates, proteins, lipids, and lignin. Cellulose is the compound which is easily degraded, lignin is the most-recalcitrant substance [52]. Long-term microbiological (and chemical) processes in the landfill are shortened to a few months by the mechanical-biological treatment [28]. Additionally, the organic load of landfill leachate is also reduced by up to 80% [28]. The acidic phase in the landfill is virtually eliminated because the organic material has been degraded prior to landfilling [52].

3. Reduction of deposit formation (incrustinations) in the leachate collection system [28].
4. Increase of the quality of the residual waste due to the removal and separation of contaminants and unwanted components.
5. Improving landfill operation due to dust reduction and odor emission [28].
6. Reduction of landfill settling. Mechanical-biological waste pretreatment also influences the physical behavior of the landfill. After separation of waste fraction with a high calorific value and a biological treatment for several months, landfill settlement is in the range of only 1% [52].

An ecological evaluation compared mechanical-biological waste treatment technologies with waste incineration regarding the effect on several environmentally relevant parameters [52]: It was demonstrated that both waste incineration and mechanical-biological pretreatment reduced the demand for fossil fuels only to a small extent. However, according to a recent study [52], waste incineration contributed to a decrease of the formation of summer smog whereas mechanical-biological waste pretreatment resulted in an increase. In contrast, mechanical-biological pretreatment was responsible for an enhanced ozone degradation. Contributions of both waste incineration and mechanical-biological pretreatment regarding global warming (greenhouse effect) was only of minor importance [52]. The study conducted in Germany concluded that the results obtained strongly depend on the local basic conditions regarding existing infrastructure and economical considerations. A generalization favoring either waste incineration or mechanical-biological waste pretreatment can hardly be deducted.

4.4.3 Inorganic Aspects: Microbe-Metal-Interactions

In addition to biotechnological processes regarding the transformation and degradation of organic compounds in municipal solid waste, microorganisms can also play an important role with respect to the portion of inorganic compounds (metals) in the waste: Metal-containing wastes can be biotechnologically treated to either recycle metal values from the waste or to remove unwanted metal-compounds resulting in an improvement of the waste quality.

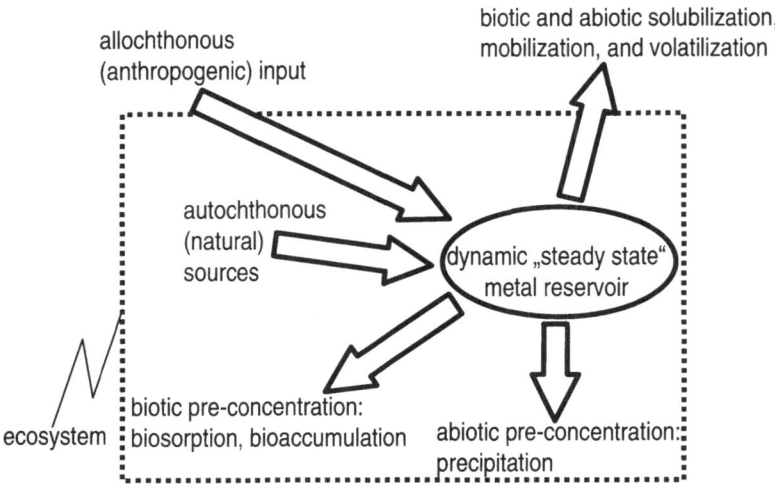

Fig. 4.13. Sources and sinks of metals in an ecosystem

Generally, it is assumed that a large fraction (>90%) of microbial populations is attached to particulate phases in environments where solid surfaces (e.g. mineral particles, biomass) are present. Particle-associated microbial populations are responsible for transformations of organic and inorganic substances. It is generally accepted that cells are not in direct contact with the solid surface, but they are surrounded by a layer of extracellular polymeric substances (EPS) which represent together with other compounds a complex matrix and are part of natural biofilms. This EPS layer can be several micrometers thick and fills the distance between the solid surface and the microbial cell.

Regarding the presence of mineral and metallic compounds in an ecosystem, several metabolic reactions (metal transformations) can occur when microbes are in contact with solid metals or metals in solution [6]. Metals are present in an ecosystem in a dynamic "steady state" reservoir (Fig. 4.13). This metal reservoir (pool) is supplied from either allochtonous or autochthonous sources. Allochthonous input includes anthropogenic activities such as industry, mining, or agriculture, whereas geogenic sources such as rock or soil are autochthonous (natural) metal sources. Simultaneously, metals are removed from this pool by biotic and abiotic processes (sorption, precipitation), are immobilized and remain in the ecosystem. By biotic or abiotic mobilization processes metals can also be remobilized and leave the ecosystem (Fig. 4.13). In natural environments, microbial cells are constantly exposed to stress conditions (presence of metals) and have, therefore, developed several mechanisms to overcome this pressure. Besides metal detoxification, microorganisms can also gain energy from some metal transformations (e.g. oxidation, reduction).

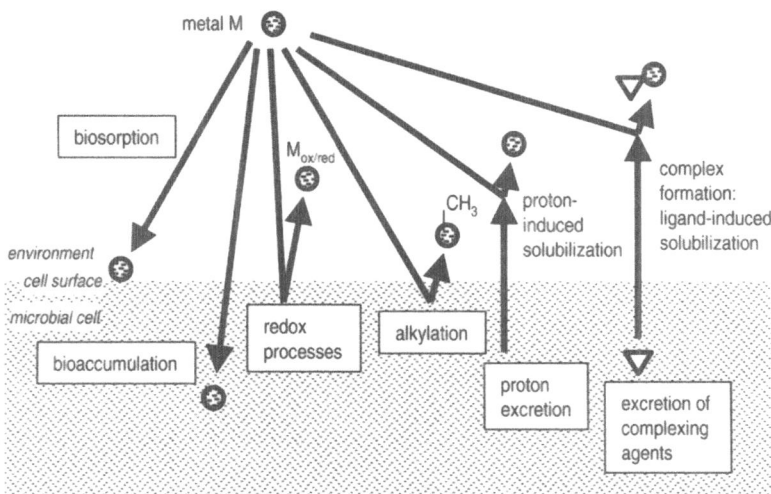

Fig. 4.14. Schematic presentation of possible mechanisms of microbe-metal-interactions

Figure 4.14 gives a schematic overview on possible reactions and mechanisms:

1. Metals can be concentrated by dead or living microbial biomass. Metal compounds are extracellularly bound onto cell surfaces and removed from an aqueous solution, a meachanism described as "biosorption". Biosorption is defined as a passive process of metal sequestering and concentration by chemical sites naturally present and functional in biomass [57]. As a possibility, EPS layers act as sorbing site. EPS usually contain functional groups such as carboxyl, sulfonate, phosphate, hydroxyl, amino, or imino residues, which can interact with metal ions by their electrical charge. Generally, more than one functional group of the biomass is involved in biosorption, depending on the environmental conditions (e.g. pH, salinity), type of biomass, and the type of metal present [16]. Metal sorption can be more or less selective [57]. As example, it has been demonstrated that freeze-dried cells of cyanobacteria (*Synechococcus* sp.) can remove more than 98% of copper and lead, respectively, from an aqueous solution [26].

2. Soluble metals can be transported through the cell membrane by ion pumps, ion channels, or other carrier compounds and are accumulated within the cells. This process is termed "bioaccumulation". In general, bacteria can sequester metals by an energy-consuming process involving the transport to the interior of the cell. As example, iron can be removed from the environment and be accumulated by specialized bacteria which have the ability to orient themselves in the earth magnetic field, a phenomenon called magnetotaxis. The metal is transported through the cell membrane and accumulated as solid particles (magnetosomes) in the cell plasma [55]. These magnetosomes are made of ferromagnetic mineral magnetite, but can contain also iron sulfides such as greigite, pyrrhotite, and pyrite. By the sequestering high concentrations of iron intracellularly as magnetite or iron sulfide, magnetotactic bacteria have an

important impact on their environment [55]. Another examples is the intracellular accumulation of silver crystals. *Pseudomonas stutzeri* is capable of forming nanocrystals of silver when growing in the presence of growth medium containing dissolved silver salts [33]. Most frequently, however, living cells store toxic metals in immobile and ineffective forms bound to high molecular-weight proteins.

3. During oxidation-reduction processes (also termed redox processes), metals can be microbially oxidized or reduced. From both reactions energy is provided for microbial growth. As result from the transformation, metal mobility is either decreased or increased depending on the type of metal and its oxidation state. In addition, certain metals such as cobalt, copper, iron, magnesium, manganese, molybdenum, or zinc are essential for the support of microbiological functions, e.g. as cofactors in metal-containing enzymes (metalloproteins).

A variety of metals can be enzymatically reduced in metabolic processes which are not related to metal assimilation, a process described as dissimilatory metal reduction [40]. During the degradation and mineralization of organic matter, metals serve as terminal electron acceptor when alternatives (e.g. oxygen, nitrate) are absent. Iron(III) and manganese(IV) reduction is widespread in the microbial world. A series of microorganisms from mostly anoxic soils, sediments, or aquifers are known to couple metal oxidation of simple organic acids, alcohols, aromatic compounds, or hydrogen with the reduction of oxidized metal species [40]. Reduction of uranium, selenium, chromium, and mercury can also result in a metabolic energy gain, whereas the reduction of technetium, vanadium, molybdenum, gold, silver, and copper are investigated to a lesser extent and leave, therefore, some open questions on biochemical reactions.

Contrasting reduction, metals can also microbially be oxidized to provide energy for growth and other vital metabolic processes [43]. Iron (II) oxidation is probably the best investigated case: A series of microorganisms are known to form ferric iron from metal-containing solids such as pyrite or arsenopyrite. These organisms, which are important for the biogeochemical cycling of iron, are mostly found in acidic environments (at pH values of 2 to 3) related to ore resources or mining sites [7]. Iron oxidizers are responsible for the generation of often extremely acidic effluents from ore mines (acidic mine drainage) which can pose severe environmental problems [42]. Microorganisms capable of oxidizing manganese (II) can be detected in metal-rich freshwater sediments [54]. It has been demonstrated that these organisms contribute significantly to the formation of certain types of ore.

4. Certain metals can be alkylated resulting in an increased metal mobility [56]. This process is described as "metal volatilization". Metals can be emitted from ecosystems as gaseous compounds. Methylation is the best known alkylation process. A methyl group is enzymatically transferred to the metal and covalently bound. A series of metals and metalloids such as antimony, arsenic, cadmium, chromium, gold, lead, palladium, platinum, selenium, tellurium, thallium, and tin can be methylated [37]. However, some methylated metals,

e.g. lead and cadmium, show only a week stability in natural ecosystem, especially in the presence of light, oxygen, and water.

5. The microbial secretion of protons can result in changes of the metal mobility. Usually, metals are more mobile under acidic condition. The process is called "acidolysis" or proton-promoted metal solubilization. Under these conditions protons are bound to the surface resulting in the weekening of critical bonds as well as in the replacement of metal ions leaving the solid surface [24]. This is especially the case in the presence of metal oxides.

6. The microbial formation of complexing or chelating agents can lead to an increase of metal mobility, a mechanism termed "complexolysis" or ligand-promoted metal solubilization. On metal surfaces, complexes are formed by ligand exchange polarizing critical bonds and facilitating the detachment of metals species from the surface. Organic ligands (bi- or multidentate ligands) which are able to form surface chelates are particularly effective in solubilizing metals [24]. Oxalate and citrate – both very common microbial products formed by bacteria or fungi – belong to this group. A kinetic model of the coordination chemistry of mineral dissolution has been developed which describes both the dissolution of oxides by the protonation of the mineral surface as well as the surface concentration of suitable complex-forming ligands such as oxalate, malonate, citrate, and succinate [24]. Proton-promoted and ligand-promoted mineral solubilization occurs simultaneously in the presence of ligands under acidic conditions. It has been shown that dissolution rates coefficients increase with increasing numbers of ligand functional groups and can be predicted from the reactivities of aqueous ligand-metal complexes [41].

Conversely, soluble metal species can also be immobilized (precipitated) by suitable complexing agents. Under anoxic conditions and in the presence of sulfate, sulfate-reducing bacteria form hydrogen sulfide as a final product which can react with soluble metals resulting in the formation of highly insoluble metal sulfides [8]. Metal sulfides show very low solubility products, so that metals are efficiently precipitated even at low sulfide concentrations [25]. Additionally, the activities of sulfate-reducing bacteria can result in a reduction of the acidity in an environment leading to the precipitation of metals as hydroxides, e.g. as copper and aluminum hydroxide [25].

In summary, mineralytic effects of bacteria and fungi on minerals are based mainly on three principles, namely acidolysis, complexolysis, and redoxolysis. As described earlier, microorganisms are able to mobilize metals by the formation of organic or inorganic acids (protons), by oxidation and reduction reactions; and by the excretion of complexing agents. Sulfuric acid is the main inorganic acid found in leaching environments. It is formed by sulfur-oxidizing microorganisms such as *Acidithiobacillus* species. A series of organic acids are formed by bacterial (as well as fungal) metabolism resulting in organic acidolysis, complex and chelate formation [3]. It has been suggested that a combination of all three mechanisms might be responsible for metal solubilization, termed "bioleaching" in the case of industrial applications. These microbial activities can be applied in the industry for the recovery of metals from solid materials. The technology has successfully found practical application in copper and gold mining (also termed "biomining") where low-grade ores are biologically treated to obtain metal values which are not accessible by conventional (mechanical or thermal) treatments [9]. Metals can be obtained from leachates by suited techniques (e.g. precipitation, cementation, or electrowinning). Besides industrial interests, the application of bioleaching technologies must also be seen in the context of a sustainable future in which industrial technologies have to be increasingly in harmony with the global material cycles of the biosphere.

4.4.4 Biological Treatment of Heavy Metal Rich Wastes

In general, bioleaching is a process described as being "the dissolution of metals from their mineral source by certain naturally occurring microorganisms" or "the use of microorganisms to transform elements so that the elements can be extracted from a material when water is filtered trough it" (Atlas and Bartha 1997, Parker 1992). Additionally, the term "biooxidation" is also used [27]. Usually, "bioleaching" is referring to the conversion of solid metal values into their water soluble forms using microorganisms. In the case of copper, copper sulfide is microbially oxidized to copper sulfate and metal values are present in the aqueous phase. Remaining solids are discarded. "Biooxidation" describes the microbiological oxidation of host minerals which contain metal compounds of interest. As a result, metal values remain in the solid residues in a more concentrated form.

One of the first reports where leaching might have been involved in the mobilization of metals is given by the Roman writer Gaius Plinius Secundus (23 - 79 A.D.). In his work on natural sciences, Plinius describes how copper minerals are obtained using a leaching process. The Rio Tinto mines in South-Western Spain are usually considered as the cradle of biohydrometallurgy. These mines have been exploited since pre-Roman times for their copper, gold, and silver values. However, with respect to commercial bioleaching operations on an industrial scale, biohydrometallurgical techniques have been introduced to the Tharsis mine in Spain ten years earlier [49]. As a consequence to the ban of open air ore roasting and its resulting atmospheric sulfur emissions in 1878 in Portugal, hydrometallurgical metal extraction has been taken into consideration in other countries more intensely. In addition to the ban, cost savings were another incentive for the development: Heap leaching techniques were assumed to reduce transportation costs and to allow the employment of locomotives and wagons for other services.

Efforts to establish bioleaching to the Rio Tinto mines have been undertaken in the beginning of the 1890ies. Heaps (10 m in height) of low-grade ore (containing 0.75% Cu) were built and left for one to three years for "natural" decomposition [49]. 20 to 25% of the copper left in the heaps were recovered annually. It was calculated that approximately 200'000 tons of raw ore could be treated in 1896. Although industrial leaching operations were conducted at Rio

Tinto mines for several decades, the contribution of bacteria to metal solubilization was confirmed only in 1961, when *Thiobacillus ferrooxidans* (reclassified as *Acidithiobacillus)* was identified in the leachates. Earlier, in 1947, *Thiobacillus ferrooxidans* has already been identified as part of the microbial community found in acid mine drainage [14]. A first patent was granted in 1958 [59]. The patent describes a cyclic process where a ferric sulfate sulfuric acid lixiviant solution is used for metal extraction, regenerated by aeration (ferrous iron oxidation by iron oxidizing organisms), and re-used in a next leaching stage.

In addition to mining operations, biohydrometallurgy is also a promising technology useful either to recover valuable metals from industrial waste materials (e.g. slag, galvanic sludge, filter dust, fly ash) or to detoxify them for a less hazardous deposition. Biohydrometallurgical processing of solid waste allows the cycling of metals similar to biogeochemical metal cycles and diminish the demand for resources such as ores, energy, or landfill space.

As a case study, fly ash from municipal waste incineration (MSWI) represent a concentrate of a wide variety if toxic heavy metals (e.g. Cd, Cr, Cu, Ni). The low acute and chronic toxicity of fly ash for a variety of microorganisms and a low mutagenic effect of fly ash from MSWI has been demonstrated [38]. Nevertheless, the deposition of heavy metal containing material bears severe risks of spontaneous leaching of heavy metals due to natural weathering processes and due to uncontrolled bacterial activities [39]. In the light of Agenda 21 established at the Earth Summit in Rio in 1992 there is a strong demand to support sustainable development which include also the ecological treatment of waste and their safe disposal.

A biological metal leaching of fly ash is a very important step in this direction. To optimize the process, a semi-continuous laboratory-scale leaching plant (LSLP) was constructed in order to achieve high leaching efficiencies resulting in an increased overall load of elements in the effluent. In addition, treatment times were as compared to batch cultures [10]. A mixture of *Acidithiobacillus thiooxidans*, which forms high concentration of sulfuric acid due to bacterial energy metabolism, and *Acidithiobacillus ferrooxidans*, which is able to oxidize reduced metal compounds resulting in an increased solubility of these metals, is used to perform the leaching. However, biohydrometallurgical processing of fly ash poses severe problems especially at higher pulp densities, because of the high content of toxic metals and the saline and strongly alkaline (pH >10) environment [10]. By employing a semi-continuous process, higher pulp densities can be applied.

In a recent study, *Acidithiobacillus* species*, Pseudomonas putida, Bacillus megaterium,* and *Aspergillus niger* were used as test organisms and incubated with fly ash obtained from municipal waste incineration [36]. Elements such as cadmium, copper, or zinc were mobilized by >80% whereas others (e.g. Pb) were solubilized only by a small percentage (Fig. 4.5). The fungus *Aspergillus niger* proved especially effective for the leaching of Pb. Results show the potential of *Acidithiobacilli* together with different microorganisms to leach substantial quantities of toxic metals from fly ash.

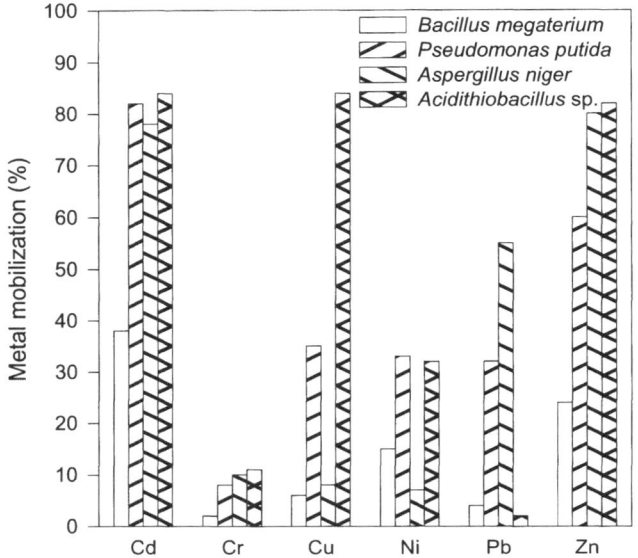

Fig. 4.15. Metal mobilization in a suspension of municipal waste incineration fly ash (20 g/l) by different microorganisms such as bacteria (*Bacillus megaterium, Pseudomonas putida, Acidithiobacillus* sp.) and a fungus (*Aspergillus niger*)

Table 4.5. Metal-containing solid wastes treated by biohydrometallurgical processes

material	source	metal content [g/kg]							
		Al	Cd	Cr	Cu	Ni	Pb	Sn	Zn
bottom ash	MSW incineration	46	tr	0.4	1.5	0.1	0.7	0.3	2.5
fly ash	MSW incineration	70	0.5	0.6	1	0.1	8	8	31
dust	electronic scrap	240	0.3	0.7	80	15	20	20	25
sludge	galvanic industries	nd	nd	26	43	105	nd	nd	166
soil	mining (Zn/Pb-mine)	nd	0.1	0.2	9	0.1	25	1.6	24
ore[a]	rock	300	0.1	300	2	7	4	10	10
soil[b]	earth crust	72	0.0003	0.1	0.03	0.04	0.02	0.01	0.05

nd not determined, *tr* traces
[a] metal concentration which makes a recovery economically interesting
[b] average metal concentration

Depending on the point of view, the mobilization or bioleaching of these metals could be either a hazard for the environment (leachates from landfills or ore deposits) or a chance to reduce toxic elements and to recover valuable metals by a low cost and low energy level technology compared with thermal treatment (e.g. vitrification or evaporation) [10]. The experimental installation described seems to be a first promising step on the way to a pilot plant with high capacities to detoxify fly ash (for a re-use of these materials for construction purposes) and for an economical recovery of valuable metals (e.g. zinc).

Slag and ash from municipal waste incineration represent concentrates of a wide variety of elements. Some metals (e.g. Al, Zn) are present in concentrations that allow an economical metal recovery, whereas certain elements (e.g. Ag, Ni, Zr) show relatively low concentrations (comparable to low-grade ores) what makes a conventional technical recovery difficult. Table 4.5 includes also other metal-containing waste materials having the potential to be treated microbiologically. Especially in these cases, where wastes contain only low amounts of metals, microbial processes are the technique of choice and basically the only possibility to obtain metal values from these materials.

A second case study reports the microbiological treatment of fine-grained residues originating from the mechanical recycling of used electric and electronic equipment [7]: Relatively short lifetimes of electrical and electronic equipment (EEE) result in the production of increased amounts of waste materials. In Switzerland, approx. 110,000 tons of electrical appliances have to be disposed yearly, while Germany these quantities are ten times bigger, reaching 1.5 million tons. Specialized companies are responsible for recycling and disposal of EEE which is dismantled and sorted manually. The resulting material is subjected to a mechanical separation process. Dust-like material is generated by shredding and other separation steps during mechanical recycling of electronic wastes: approx. 4% of the 2400 t of scrap treated yearly by a specialized company is collected as fine-grained powdered material. Whereas most of the electronic scrap can be recycled (e.g. in metal manufacturing industries), the dust residues have to be disposed in landfills or incinerated. However, these residues can contain metals in concentrations which might be of economical value. Provided that a suitable treatment and a recovery process is applied, this material might serve as a secondary metal resource. Results indicate that it is possible to mobilize metals from electrical and electronic waste materials by the use of microorganisms such as bacteria (*Acidithiobacillus thiooxidans, Acidithiobacillus ferrooxidans*) and fungi (*Aspergillus niger, Penicillium simplicissimum*). After a prolonged adaptation time, fungi as well as bacteria grew also at concentrations of 100 g/l [7]. Both fungal strains were able to mobilize Cu and Sn by 65%, and Al, Ni, Pb, and Zn by more than 95%. At scrap concentrations of 5 to 10 g/l *Thiobacilli* were able to leach more than 90% of the available Cu, Zn, Ni, and Al. Pb precipitated as lead sulfate while Sn precipitated probably as tin oxide. For a more efficient metal mobilization a two-step leaching process is proposed where biomass growth is separated from metal leaching. Metals were precipitated and recovered as hydroxides by the stepwise addition of sodium hydroxide.

4.4.5 Conclusions

Overall, it can be concluded that microbial metal mobilization can be applied for the recovery of metals from metal-containing solid wastes. It is possible to recycle the leached and recovered metals and to re-use them as raw materials by metal-manufacturing industries. Bioleaching represents a "clean technology" process on a low cost and low energy level compared with some conventional mining and waste treatment techniques. Government regulations and research policies that favor "green" technologies are a key incentive for developing such processes. These processes find a wide acceptance in public and in politics and represent innovative technologies with a proved market gap. However, the development of a process which is technically feasible and economically as well as ecologically justifiable is an important prerequisite.

References

1. Ordinance on Environmentally Compatible Storage of Waste from Human Settlements and on Biological Waste-Treatment Facilities, Federal Minister for Environment, Nature Conservation and Nuclear Safety (BMU); 2001. Germany
2. Thermal, Mechanical-Biological Treatment Plants, and Landfills for Residual Waste in Federal Germany (in German). German Federal Environmental Agency (UBA); 2001. Germany
3. Berthelin J (1983) Microbial Weathering Processes. In: Krumbein WE (ed) Microbial geochemistry. Blackwell, Oxford, pp 223-262
4. Bockreis A, Brockmann C, Jager J (2000) Testing Methods for the Evaluation of the Landfill Suitability of MBP-Treated Waste (in German). In: Die Zukunft der mechanisch-biologischen Abfallbehandlung. In: Soyez K, Hermann T, Koller M, Thrän D (eds) Potsdamer Abfalltage, Potsdam, Germany
5. Brandl H (1998) Vom Nutzen der Mikroben: Bakterien als Helfer bei der Produktion von industriellen Werkstoffen. Vierteljahresschrift der Naturforschenden Gesellschaft in Zurich 143:157-164
6. Brandl H (1999) Bioleaching: Mobilization and Recovery of Metals from Solid Waste Materials. In: Selenska-Pobell S, Nitsche H (eds) Bacterial-Metal/Radionuclide Interaction: Basic Research and Bioremediation. Wiss.-Tech. Berichte TZR-252. Forschungszentrum Rossendorf, Rossendorf /Dresden, pp 20-22
7. Brandl H (2001) Microbial Leaching of Metals. In: Rehm HJ, Reed G (eds) Biotechnology. Vol. 10. Special processes. Wiley-VCH, Weinheim, pp 191-224
8. Brierley CL, Brierley JA (1997) Microbiology for the Metal Mining Industry. In: Hurst CJ, Knudsen GR, McInerney MJ, Stezenbach LD, Walter MV (eds) Manual of Environmental Microbiology. ASM Press, Washington, D.C., pp 830-841
9. Brombacher C, Bachofen R, Brandl H (1997) Biohydrometallurgical Processing of Solids: A Patent Review. Appl. Microbiol. Biotechnol. 48:577-587
10. Brombacher C, Bachofen R, Brandl H (1998) Development of a Laboratory-Scale Leaching Plant for Metal Extraction from Fly Ash by Thiobacillus Strains. Applied and Environmental Microbiology 64:1237-1241

11. BUWAL SAftEFaLS (1986) Guidelines on Swiss Waste Management. Schriftenreihe Umweltschutz 51

12. BUWAL SAftEFaLS (2001) Umweltmaterialien. Abfallstatistik (Swiss Waste Statistics):

13. Collins H-J, Maak D (1997) MBP Plant Meisenheim (in German). In: Bilitewski, B.; Stegmann, R. (ed.) Mechanical-Biological Processes for a Substance-Specific Waste Disposal (in German). Müll und Abfall 33:123-127

14. Colmer AR, Hinkle ME (1947) The Role of Microoganisms in Acid Mine Drainage. Science 106:253-256

15. Doedens H, Cuhls C, Mönkeberg F, et al (1999) Balancing Environmentally Relevant Chemicals in the Biological Pre-Treatment of Residual Waste - Phase 2: Emissions, Pollutant Balances and Waste Gas Treatment (in German). In: Final Report for the German Federal Research Project on mechanical-biological treatment of waste before landfill. University, Hannover

16. Eccles H (1999) Treatment of Metal-Contaminated Wastes: Why Select a Biological Process? Trends Biotechnol. 17:462-465

17. Echigo S, Yamamoto K, Yamada H, Shishida K (1997) Some Aspects on the Application of Ozone/Vacuum Ultraviolet Process to Leachate from Solid Waste Landfill Site. In: Association IO (ed) 13th Ozone World Congress. International Ozone Association, Pan American Group, Kyoto, Japan

18. Ehrig H-J, Höring K, Helfer A (1998) Demands on Biological Pre-Treatment of Waste before Landfill and Evaluation of its Effects (in German). Final Report for the German Federal Research Project on mechanical-biological treatment of waste before landfill. University, Wuppertal

19. Ehrlich HL (1998) Geomicrobiology: Its Significance for Geology. Earth-Science Reviews 45:45-60

20. Escobar B, Huerta G, Rubio J (1997) Influence of Lipopolysaccharides on the Attachment of Thiobacillus Ferrooxidans to Minerals. World Journal of Microbiology & Biotechnology 13:593-594

21. Flamme S, Gallenkemper B (2001) Inhaltsstoffe von Sekundärbrennstoffen, Ableitung der Qualitätssicherung der Bundesgütegemeinschaft Sekundärbrennstoffe e.V. (Ingredients of RDF, quality management by the German quality association for RDF). Müll und Abfall 12:699-704

22. Fricke K, Müller W (1999) Stabilisation of Residual Waste by Mechanical-Biological Treatment and Consequences for Landfills (in German). In: Final Report for the German Federal Research Project on Mechanical-Biological Treatment of Waste before Landfill. IGW, Witzenhausen, Germany

23. Fuchs J, Galli U, Schleiss K (2000) VKS-Grundkurs für Mitarbeiter von Vergärungs- und Kompostieranlagen, Kursordner. VKS-Geschäftsstelle, Schönbühl-Urtenen

24. Furrer G, Stumm W (1986) The Coordination Chemistry of Weathering: I. Dissolution Kinetics of δ-Al_2O_3 and BeO. Geochim. Cosmochim. Acta 50:1847-1860

25. Gadd GM (2001) Accumulation and Transformation of Metals by Microorganisms. In: Rehm HJ, Reed G, Pühler A, Stadler P (eds) Biotechnology. Vol. 10. Special processes. Wiley-VCH, Weinheim, pp 225-264

26. Gardea-Torresdey JL, Arenas JL, Francisco NMC, Tiemann KJ, Webb R (1998) Ability of Immobilized Cyanobacteria to remove Metal Ions from Solution and Demonstration of the Presence of Metallothionein Genes in Various Strains. Journal of Hazardous Substance Research 1:2/1-2/21

27. Hansford GS, Miller DM (1993) Biooxidation of a Gold-Bearing Pyrite-Arsenopyrite Concentrate. FEMS Microbiology Reviews 11:175-182
28. Heerenklage J, Stegmann R (1995) Overview on Mechanical-Biological Pretreatment of Residual MSW. In: Christensen TH, Cossu R, Stegmann R (eds) SARDINIA 95, Fifth International Landfill Symposium. CISA, Environmental Sanitary Engineering Centre, Caglari, S.Margherita di Pula, Italy, 2-6 October 1995
29. Kern M (1994) Verfahren zur aeroben und anaeroben Behandlung von Bioabfällen. In: 5. Hohenheimer Seminar. Böhm, R, DVG Giessen, ISBN 3-924851-98-0
30. Kim SK, Matsui S, Shimizu Y, Matsuda T (1997) Biodegradation of Recalcitrant Organic Matter under Sulfate Reducing and Methanogenic Conditions in the Landfill Column Reactors. Water Sci. Technol 36:91-98
31. Kim SK, Matsui S, Pareek S, Shimizu Y (1998) Biodegradation Characteristics of Recalcitrant Organic Matter under Sulfate Reducing and Methanogenic Conditions in Landfills - Batch Experiment. Proceedings (7). Japanese Solid Waste Society 9, pp 310-317
32. Kim SK, Matsui S, Pareek S, Shimizu Y (1998) Biodegradation Characteristics of Recalcitrant Organic Matter under Sulfate Reducing and Methanogenic Conditions in Landfills - Continuous Experiment. Proceedings (2/3). Japanese Solid Waste Society 9, pp 79-86
33. Klaus-Joerger T, Joerger R, Olsson E, Granquist CG (2001) Bacteria as Workers in the Living Factory: Metal Accumulating Bacteria and their Potential for Materials Science. Trends in Biotechnology 19:15-20
34. Köhler M, Völsgen F (1998) Geomikrobiologie. Wiley-VCH, Weinheim
35. Komilis DP, Ham RK, Stegmann R (1999) The Effect of Municipal Solid Waste pretreatment on Landfill Behavior: a Literature Review. Waste Management & Research 17:10-19
36. Krebs W, Brombacher C, Bosshard PP, Bachofen R, Brandl H (1997) Microbial Recovery of Metals from Solids. FEMS Microbiol. Rev. 20:605-617
37. Krishnamurthy S (1992) Biomethylation and Environmental Transport of Metals. Journal of Chemical Education 69:347-350
38. Lahl U, Struth R (1993) Verwertung von Müllverbrennungsschlacken aus der Sicht des Grundwasserschutzes. Vom Wasser 80:341-355
39. Ledin M, Pedersen K (1996) The Environmental Impact of Mine Wastes - Roles of Microorganisms and their Significance in Treatment of Mine Wastes. Earth Science Reviews 41:67-108
40. Lovley DR (1993) Dissimilatory Metal Reduction. Annual Reviews in Microbiology 47:263-290
41. Ludwig C, Casey WH, Rock PA (1995) Prediction of Ligand-Promoted Dissolution Rates from Reactivities of Aqueous Complexes. Nature 375:44-47
42. Nordstrom DK, Alpers CN, Ptacek CJ, Blowes DW (2000) Negative pH and Extremely Acidic Mine Waters from Iron Mountain, California. Environmental Science & Technology 34:254-258
43. Norris PR (1989) Mineral-Oxidizing Bacteria: Metal-Organism Interactions. In: Poole RK, Gadd GM (eds) Metal-Microbe Interactions. Oxford University Press, Oxford, pp 99-117
44. Ostrowski M, Sklodowska A (1993) Bacterial and Chemical Leaching Pattern on Copper Ores of Sandstone and Limestone Type. World Journal of Microbiology & Biotechnology 9:328-331

45. Pareek S, Matsui S, Shimizu Y (1998) Biodegradation of Ligno-Cellulosic Material under Sulfidogenic and Sethanogenic Conditions in the Landfill Column Reactors. Environmental Technology 19:253-261

46. Pareek S, Matsui S, Shimizu Y (1998) Hydrolysis of (LIGNO)Cellulosic Materials under Sulfidogenic and Methanogenic Conditions. Water Sci. Technol 38:193-200

47. Pareek S, Matsui S, Kim SK, Shimizu Y (1999) Mathematical Modeling and Simulation of Methane Gas Production in Simulated Landfill Column Reactors under Sulfidogenic and Methanogenic Conditions. Water Sci. Technol. 39:235-242

48. Puchelt A (2000) Dry Stabilization of Residual Waste - Exemplary Plant in Rennerod / Westerwaldkreis, Germany (in German). In: Die Zukunft der mechanisch-biologischen Abfallbehandlung. In: Soyez K, Hermann T, Koller M, Thrän D (eds) Potsdamer Abfalltage, Potsdam, Germany

49. Salkield LU (1987) A Technical History of the Rio Tinto Mines: Some Notes on Exploitation from Pre-Phoenician Times to the 1950s. Institution of Mining and Metallurgy, London

50. Sand W, Gerke T, Hallmann R, Schippers A (1995) Sulfur Chemistry, Biofilm, and the (In)Direct Attack Mechanism - a Critical Evaluation of Bacterial Leaching. Appl. Microbiol. Biotechnol. 43:961-966

51. Schleiss K (2002) Success Factors for the Marketing of Compost. R'02 6th World Congress on Integrated Resources Management, Geneva

52. Soyez K (2001) Mechanisch-biologische Abfallbehandlung. Erich Schmidt, Berlin

53. Soyez K, Thrän D, Hermann T, Koller M, Plickert S (2001) Results of the BMBF-Joint Research Program on Mechanical Biological Waste Treatment (in German). In: Hösel, Bilitewski, Schenkel, Schnurer (eds) Müllhandbuch, No. 5613. Erich-Schmidt-Verlag, Berlin

54. Stein LY, La Duc MT, Grundl TJ, Nealson KH (2001) Bacterial and Archaeal Populations associated with Freshwater Ferromanganous Micronodules and Sediments. Environmental Microbiology 3:10-18

55. Stolz JF (1993) Magnetosomes. J. Gen. Microbiol. 139:1663-1670

56. Thayer JS, Brinckman FE (1982) The Biological Methylation of Metals and Metalloids. Adv. Organometal. Chem. 20:312-356

57. Volesky B (2001) Detoxification of Metal-Bearing Effluents: Biosorption for the Next Century. Hydrometallurgy 59:203-216

58. Zach A, Binner E, Latif M (2000) Improvement of Municipal Solid Waste Quality for Landfilling by Means of Mechanical-Biological Pretreatment. Waste Management & Research 18:25-32

59. Cyclic Leaching Process Employing Iron Oxidizing Bacteria (1958). US Patent 2,829,964

5 Advanced Thermal Treatment Processes

Contributions by Kai Sipilä, Marcel A.J. van Berlo, Jörn Wandschneider, Michael Beckmann, Reinhard Scholz, Martin Horeni, Fréderic Vogel, Jörg Wochele, Christian Ludwig, Samuel Stucki, Harald Lutz, Rudolf P.W.J. Struis, Serge Biollaz, Rainer Bunge, Shin-ichi Sakai, Didier Perret, Kaarina Schenk, Marc Chardonnens, Peter Stille, Urs Mäder, and Charles Keller

Waste incineration is a technology which is more than 100 years old. The technology was first introduced at the end of the 19th century to reduce the volume and mass of MSW in order to save landfill space. In a time when smoking stacks were synonymous with progress, nobody cared about the emissions from incinerators into the air. Only when air pollution became a high priority issue in the early 1980s was it found that MSW incinerators were in fact contributing substantially to the overall emission of air pollutants, notably of HCl, dioxins and various heavy metals. Once this drawback to MSW incineration was acknowledged, legislation intervened and forced the environmental technology industry to develop suitable air pollution control (APC) technology, which in turn required that operators of incinerators invest large sums to limit the emission of acid gases, particulates and toxic trace compounds. The era of environmental technology from 1980 to 2000 has in fact brought down the emissions from incineration to levels that have made MSWI a minor contributor to air pollution for all emission categories [48].

Most of the pollutants that used to be emitted to the atmosphere before the installation of scrubbers and filters are now turned into APC residues in state-of-the art incinerator plants. To more effectively and efficiently handle these products, R&D has been carried out with the objective of destroying those pollutants that are in fact destructible (such as dioxins, other toxic organics and

NOx) and to either immobilize or separate and concentrate those pollutants which cannot be destroyed (notably the heavy metals). Optimizing the incineration process and complementing it with additional ash treatments with the objective of creating fractions that contain the heavy metals in highly concentrated form has been the challenge for a number of research projects which are presented in section 5.2. The alternative concept of forming leach-proof glassy residues by melting the residues at high temperatures has been introduced to the market in Japan (section 5.3), but has not yet been successfully implemented in Europe. Although a long-term (centuries!) prognosis of the leaching stability of vitrified residues is extremely difficult, section 5.4 supplies ample evidence that such materials can be considered safe for final deposition.

Melting of ashes at high temperatures requires energy inputs that are preferably supplied by the incineration process itself. However, melting processes will always lead to losses in the energy that can't be entirely recovered in the form of power or process heat. LCA studies have shown (chapter 6) that energy consumption or equivalent losses in energy production have a strong impact on the overall ecological performance of a process. There is a tradeoff to be made between energy conservation and the optimization of the material output with respect to recyclability or long-term stability. There is considerable potential for improved recovery of high value energy (electricity or high temperature process heat) from MSW (section 5.1) which is by no means exploited in current installations. Thermal processes have a high potential for the optimization of both material and energy recovery (see e.g. section 5.2.3).

Incinerators are 'at the end of the pipe' and therefore must be designed to accommodate any material that can be fitted into a MSW collection bin. The development of material and energy optimized processes can, along with traditional emissions control, to some extent guarantee that the solid products leaving the installation meet environmental standards, irrespective of the input treated. This raises the question as to whether advanced incinerators have reached standards with respect to materials processing which make the separate collections of e.g. electronic scrap or spent batteries redundant. Any materials separating process, however, will work more efficiently if operated using input streams with high concentrations of the material to be recovered. Therefore, it does indeed make sense to separate problematic waste fractions from MSW at the source and to treat them in dedicated, efficient plants (section 5.5).

5.1. Energy Recovery from Waste

There are currently more than 300 waste incineration plants in operation in Europe with a total capacity exceeding 50 million tons/a. The primary function of incineration has been volume reduction and mineralization of waste. The recovery of energy for heat and power production has been a secondary, but increasingly important goal. The role of waste to energy recovery is dependent on many national and local conditions- and in particular on the national waste management strategy and landfilling policy. In industrialized countries the composition of waste has changed over the past 50 years due to changing consumption habits. The increasing contents of dry organic matter (mainly plastics and paper) in MSW have led to a continual increase in the specific heating value of MSW. In Switzerland, the lower heating value of MSW delivered to incineration reached 12.6 MJ/kg in the year 2000. In fact, one of the most valuable resources wasted or lost with MSW is a mix of renewable (biomass) and fossil (plastics) energy.

Mixed waste massburning or incineration is currently the dominating technology. Most of today's incinerators operate with a poor energy recovery rate. Electric efficiency is typically in the order of 15%. Heat recovery depends on the availability of sufficient capacity for relatively low process heat. District heating requires expensive infrastructure and in most cases the heat demand is subject to seasonal variations. To some extent a better match of the heat output of an MSWI with local demand might be achieved by, for example, using seasonal storage of stabilized MSW or high heating value fractions of MSW. MSWI are, however, optimized for continuous, constant load operation and, therefore, applying heat demand driven load management of an MSWI is limited.

Because local heat demand only in rare cases matches the output of continuously running installations, there are good incentives for increasing the efficiency of MSWI boilers and steam cycles for power production. The case study (5.1.2) of the AVR plant in Netherlands is an example of how to improve the potential of power production from waste. Efficiency gains are possible through higher temperatures in the boilers. Though higher boiler temperatures can lead to corrosion problems; countermeasures are available but costly. Boosting the energy recovery rate from waste has a price.

By separating high heating value fractions from MSW using mechanical or biological mechanical pre-processing, a relatively well characterized fuel (Refuse Derived Fuel, RDF) can be made (section 3.1). Advanced power plant technology and high quality fuels are in an active development phase; some examples are fluidized bed combustion and gasification technology. RDF co-firing in existing utility boilers and power plants will be an economically attractive option in co-operation with waste management and energy companies. RDF can also be used as a substitute fuel in thermal processing of materials such as cement production, metallurgy, etc. (section 5.1.3). The fuel properties are crucial for the design of such processes running on conventional fossil fuels and admixtures of RDF. Whether a mechanical pre-sorting for producing RDF pays from the point of view of overall energy efficiency depends on the specific requirements of producing the substitute fuel.

All advanced processes for utilizing the energy in waste efficiently will in some way or other have to rely on suitable mechanical pre-treatment of the waste in order to obtain a sufficiently homogeneous and well characterized fuel which is a prerequisite for designing a high efficiency energy conversion process. This holds true for all kinds of RDF applications, as well as for more futuristic processes, such as the hydrothermal route outlined at the end of this chapter.

5.1.1 Waste to Energy: New Integrated Concepts

Kai Sipilä

Introduction and Trends in Waste to Energy

In Europe there were 304 waste incineration plants in operation at the end of 2000 [15], and in the EU 15 member states there were 269 facilities. The total capacity of these units is 50.2 million tons of MSW and related waste. Energy recovery was 49.6 TWh/a, 70 % was used for heat generation and 30 % for electricity generation. The size and electricity output of the incineration plants is increasing, electricity generation from 100 GWh/a in 1980`s to typically to 400 GWh/a in 1990`s, processing from 125 000 t/a to an average of 225 000 t/a. The largest plant is AVR in the Netherlands processing 1.5 million tons and generating 500 MW electricity [15].

The waste management practices are locally and nationally reflecting the national waste management strategy and landfill policy, which is indicated the Figure 5.1.

The megatrends which will be affecting the future waste to energy practices in Europe are presented in Table 5.1. EU directives and national policies will decrease landfilling and call for growing capacities for material recycling and energy recovery. Especially higher efficiencies in electricity and combined heat and power production with advanced emission control will be the key trends.

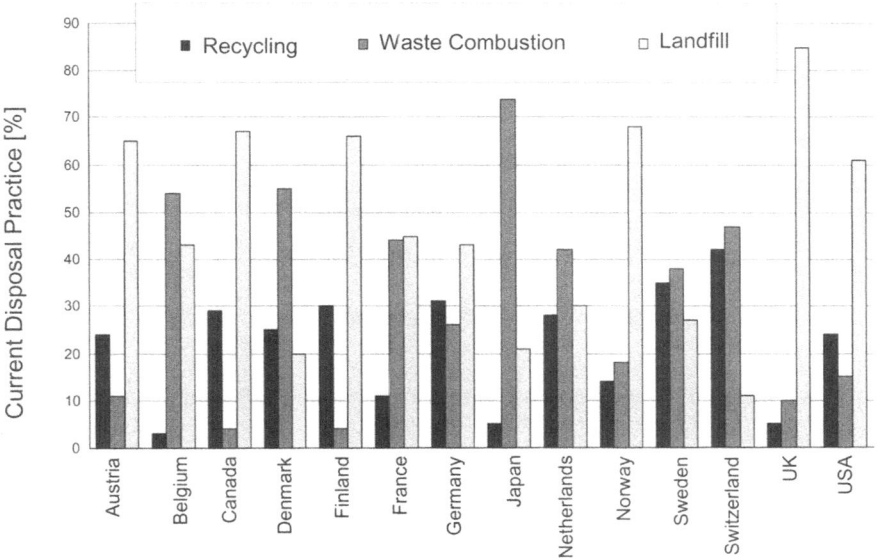

Fig. 5.1. Waste management practices in various countries [91]

Table 5.1. Megatrends of future waste management and MSW to energy policies

WASTE TO ENERGY – MEGATRENDS	
1.	KYOTO 2010 process
	- landfill gas recovery and volume reduction
	- EU White Paper increasing renewable energy sources from 6 to 12%
	- EU-RES-E directive boosting power from bioenergy including waste
2.	EU Landfill directive
	- MSW volume reduction, MSW pretreatment, and new WtE capacity needed
	- methane recovery and energy application or flaring
	- pre-treatment of MSW to be landfilled
3.	EU Waste Incineration directive and LCP directive
	- reduction of emissions, stand alone and co-incineration plants
4.	Higher efficiencies in waste to energy, low emissions
	- high steam values, new superheater materials, RDF quality and standards
5.	Balance of materials and energy recovery, integrated concepts
6.	Waste policies, taxes and WtE acceptance by the public opinion

Traditionally MSW is collected as mixed waste and incinerated in a grate boiler for steam and electricity generation with low efficiencies, typically 15-22 % of the heating value of MSW. The new EU waste incineration directive gives the guidelines and limits for emission control.

In Figure 5.2 the share of renewable energy sources, RES, especially bioenergy including waste, is presented. The EU level target in the White Paper, is to double the RES contribution from 6 % to 12 % by 2010 as an instrument in the Kyoto process.

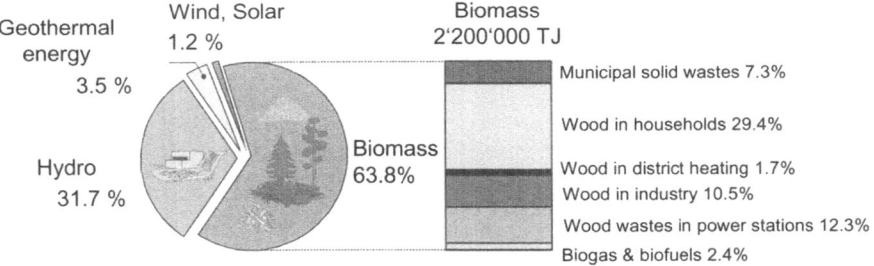

Total RES 82.1 Mtoe (= 5.8% of gross inland consumption)

Fig. 5.2. The share of renewable energy sources, bioenergy and MSW in 1997 [20]

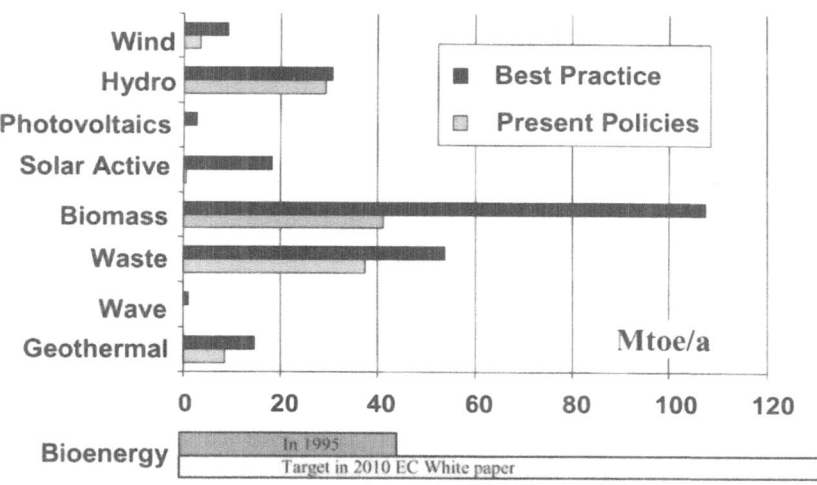

Fig. 5.3. The renewable energy penetration, high potential of waste to energy [149]. The EU target in the White Paper is to increase bioenergy from 45 to 135 (Mtoe/a) in the period 1995–2010.

Waste to Energy Technologies, WtE

The traditional MSW incineration plants are grate boilers generating hot water or steam for electricity production. Steam values are quite low, typically 380 – 400 C and pressure 40 bar, and the efficiency is low 15 – 20 % to power production, compared to efficiency of a solid fuel fired, utility size power plants, 35 – 42 % (based on lower heating vales of the fuel). There also still many MSW or RDF plants, which are not generating power, only hot water for district heating or low-pressure steam for industrial applications. In Northern European countries the heat market is economically attractive with a typical 3500 – 5000 h/a peak running time and high 75 – 80 % efficiency to heat in 10 – 50 MW size class. The RDF annual input is 10 – 80 000 t/a. The trend is to increase to power production efficiencies with the following measures and R&D&D objectives:

- higher steam values in traditional grate boilers + process integration;
- higher steam values up to 480 C, 65 bar in fluidized bed boilers;
- co-firing of good quality RDF in power/utility boilers;
- integration of RDF boiler steam cycle with existing power/utility boiler;

- co-firing in coal boilers with the RDF gasification and advanced fuel gas cleaning , co-firing the producer gas in gas turbines or gas engines;
- co-firing of RDF based liquid fuels in existing utility boilers or engines;
- other advanced concepts and technologies.

The driving force boosting the higher power efficiency is also the new EU RES-E directive, which will increase the economic value as green electricity. The biomass and biodegradable fraction of MSW based power production is classified as renewable energy source. Traditionally the efficiency is 15-20 % and in fluidized bed boilers 30 –35 % with high quality RDF fuels. In co-firing option the cleaned producers gas could be fired in coal based utility boiler with 40 – 44 % efficiency, so the overall efficiency can be as high as 30 – 40 % depending on the gasification plan process parameters. This will have a significant effect on the power production capacity:

- mixed waste massburning, 15 % efficiency, power yield 450 kWh/t MSW;
- RDF in advanced fluidized bed boilers, 33 %,   1 150 kWh/t RDF;
- RDF gasification at coal PC boiler,   1 400 kWh/t RDF.

In modern high efficiency boilers the RDF quality is essential, if it cannot be controlled, the low steam values and lower technical and economic risks will be chosen. Key elements are heating value and chemical compositions of impurities like chlorine, heavy and alkali metal content, aluminum and sulfur level etc. Also the physical properties, like particle size and bulk density, are specified in the work of CEN Recovered fuel standardisation activity [79] and in national standards [8].

Emission control is a key element. The EU Directive on waste incineration [7] was adopted in December 2000 and will be applied for new units in 2002 and existing ones in 2005. This gives emission limits for existing and new waste to energy plants. There are also principles given how emission limits and operational procedures for co-firing at existing power plants should be applied. Emission limits and some co-firing examples are given in Table 5.2.

A description of grate fired incinerators, including the downstream process for gas cleaning is shown in section 3.3. An overview of the emissions and material balances from various waste incineration plants, heavy metals, dioxins, furans etc. is reported by Vehlow [245]. Bottom ash, fly ash and residues from air pollution control (APC) contain potentially harmful material like heavy metals and traces of micro-pollutants. Ferrous metals can be magnetically separated from the bottom ash for recycling. (Composition and WtE residue practices are discussed also in chapter 3.2. in this book.)

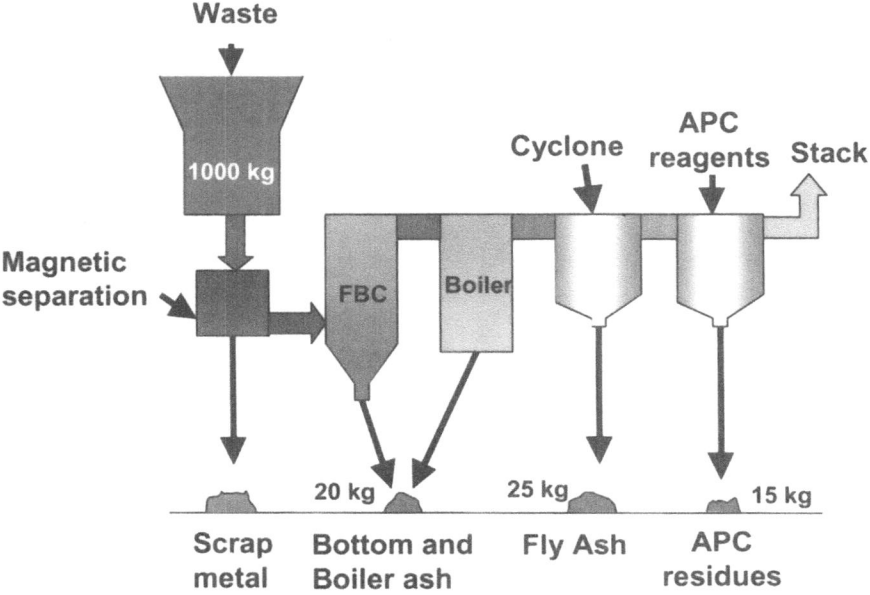

Fig. 5.4. Residues from RDF fluidized bed combustion plants [14]

Fluidized Bed Combustion and Gasification of RDF

Fluidized bed combustion, in bubbling BFB and in circulating CFB mode, has been applied for solid fuels combustion successfully with coal, lignite, peat and biomass. It is technically and economically competitive option in size classes typically 20 – 300 MWth. The benefits are high efficiency combustion of inhomogeneous fuels due to heavy mixing in the sand bed, low NOx emissions due low combustion temperature compared to pulverized coal combustion etc. There are many RDF fired fluidized bed combustion units in operation - an overview of their operational characteristics is presented in a report of the IEA Bioenergy task 23, Energy from Thermal Conversion of MSW and RDF [90]. Six fluidized bed MSW incinerators from five boiler manufacturers were reported by Granatstein [90]; power production ranged from 8 MW to 50 MW. Process overview, MSW and RDF properties, operational experiences and emission data is presented, also investment and operational costs are indicated. In Table 5.3 the fluidized bed units and the statistics are presented. The largest unit is the Robbins Resource Recovery Facility (RRRF), located in Robbins Chicago, Illinois, US handling 1 450 t/day MSW and generating 50 MWe power. The overview and experiences of the plant is presented by Granatstein and Hesseling [91]. Emissions, bottom and fly ash analyses are given.

Table 5.2. Emission limits in EU waste incineration directive [7] and some co-firing examples.

Emission	Waste Incineration	Solid fuel		Coal-fired boiler + 30% REF from flue gas	Biomass	Biomass boiler + 30% REF from flue gas
	red. 11% O_2	red. 6% O_2		red. 6% O_2	red. 6% O_2	red. 6% O_2
	[mg/Nm^3]	[mg/Nm^3]		[mg/Nm^3]	[mg/Nm^3]	[mg/Nm^3]
		100-300 MW	> 300 MW	200 MW	100-300 MW	200 MW
SO_x	50	850-200	200	390	200	162
NO_x (<6 t/h)	400	300	200	300	300	300
NO_x (>6 t/h)	200					
Dust	10	30	30	25.5	30	25.5
CO	50			*		*
NH_3				10		10
TOC	10			10		10
HCl	10			10		10
HF	1			1		1
Hg	0.05			0.05		0.05
Cd+Tl	0.05			0.05		0.05
As+Pb+Mn +Ni+Cr+Cu +Co+V+Sb	0.5			0.5		0.5
PCDD/F, 1-TEO2	10^{-7}			10^{-7}		10^{-7}

Continuous measurements	CO, TOC, total dust, HCl, Hf, SO_2, NO_x
Continuous measurements	Temperature, O_2, pressure, moisture content of exhaust gas
Heavy metals (12 components)	PCDD/F twice a year, one measurement every three months for the first 12 months of operation*
Annual measuring costs	Typically euro 60 000 - 120 000, for RDF euro 2-5/MWh

*If the waste consists only of certain sorted combustible fractions, national quality criteria are available and the emissions are always under 50% of the limit values, the reduction of frequency may be authorised by competent authority.

Table 5.3. Fluidized bed combustion plant statistics, reported by Granatstein [90]

Plant	FBC type	Driver	Input	Output (max.)	Efficiency [%]	Capital [US$M]	O & M [US$/t MSW]
Robbins	2 x FW CFB	Community benefits	1450 t/d MSW = 1090 t/d RDF	50 MWe	23 (net)	301	29
Lidkoping	2 x Kvaerner BFB	District heat supply	50'000 t/a (MSW/ISW) + 20'000 t/a wood waste	34 MWth −2 MWe [1]	88 (gross)	-	61
Toshima	2 x IHI BFB	Landfill use reduction	400 t/d MSW (source separated)	7.8 MWe + 4.3 MWth	30.5 (gross)	140	127
TirMadrid	3 x Rowitec TIF BFB	Landfill use reduction	440'000 t/a MSW = 260'000 t/a RDF	29 MWe	19.4 (net)	125	31
Dundee	2 x Kvaerner BFB	Meeting emissions limits	120'000 t/a MSW (source separated coarse RDF)	10.5 Mwe	20.9 (net)	56	31
Valene	3 x TMC BFB	Insufficient landfill capacity	100'000 t/a unsorted MSW	8 MWe (net)	21.5 (net)	42	70

[1] Purchased externally

Residue production and treatment from the RDF combustion and MSW incinerator plants are in a joint IEA – ISWA report [14]. In the traditional mass burning incineration plant residue amount is about 30 wt-% of input, bottom ash residue is dominating about 225 kg/t MSW. In RDF fluidized bed combustion typical residue streams are presented in Figure 5.4. Bottom ash, fly ash and APC residues are about equal amounts, totally 60 kg/t RDF input excluding the fuel preparation rejects [14]. The volume and the content of the residues depend significantly on degree of local source separation and the quality of the RDF streams and the power plant technology applied. In the IEA report some detailed chemical data was available on the residues from fluidized bed combustion.

There are several companies manufacturing fluidized bed boilers and gasifiers. Some modifications of fluidized bed boiler plants for good quality RDF operation are: feeding line is modified in order to receive and feed low bulk density material; in the boiler coarse material and the bottom ash must be removed during normal operation; superheater can be placed in the bed material recycling loop; high resistance superheating materials are used; flue gas treatment will be installed including fabric filters with lime and sorbent injection and an optional scrubber based on local conditions and emission limits. RDF properties must be better compared to conventional grate firing and low steam temperatures. Especially level of impurities should be low. The good quality recovered fuels can be produced typically from source separated, dry commercial packaging waste. Also co-firing in fluidized bed boilers with coal, lignite and/or biomass is possible with high steam values if the level of impurities is low (chlorine, metallic aluminum, alkali metals etc.). For superheater corrosion the sulfur/chlorine ratio in the fuel is a critical factor. When the ratio is higher than 4, rate of corrosion is low. However often in household based RDF the ratio is lower than 1, and life-time of the traditional superheater materials can be as low as one year [126]. In order to meet the EU waste incineration directive requirements, 850 C and 2 second residence time conditions can be reached.

In Figure 5.5 a typical recovered fuel production and MSW source separation scheme is presented. This is an optimal solution for source separation of biowaste, paper, metal and glass, and finally dry waste for RDF processing. The dry waste flow will be crushed, screened and ferrous and non-ferrous metals will be separated. Finally, the product is a recovered fuel with specified characteristics for the energy market. Based on this principle typical fuel properties are analyzed, Table 5.4. In Finland RDF produced from source separated, quality controlled MSW, is called recovered fuel, REF. The fuel specifications are given in a national standard [8].

In fluidized bed boilers the fuel quality requirements (for physical properties) are demanding - typical particle dimension is less than 50 mm. For corrosion and slagging aspects the chlorine, sulfur, alkali metal and metallic aluminum content are important. Chlorine content is the most crucial one, reflecting the PVC and sodium chloride content in the RDF. In household waste it is often higher than

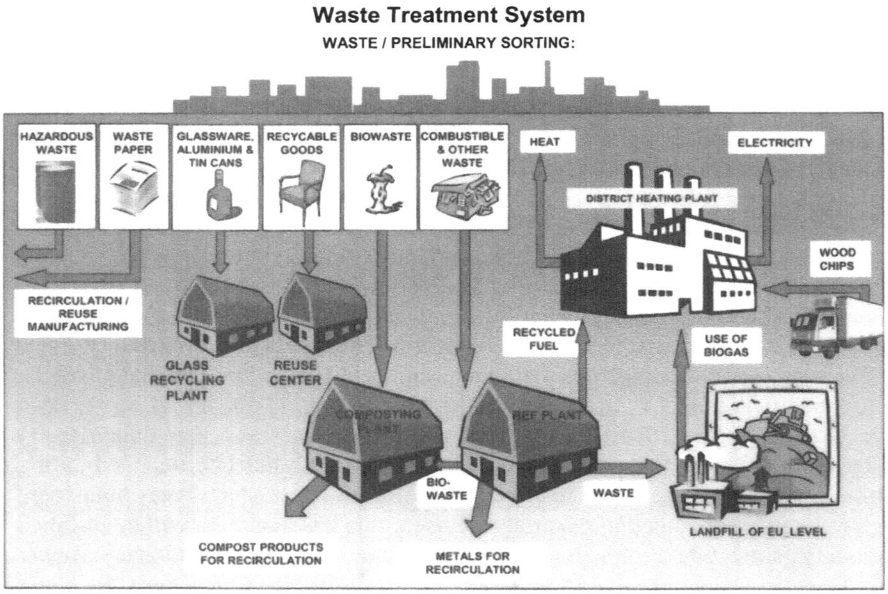

Fig. 5.5. Source separation of MSW and recovered fuel production, LoimiHämeen Jätehuolto Oy, Finland

0.7 %, however in good quality commercial waste it can be less than 0.3 %. Source separation and selection of packaging material have an important role in decrease the chlorine content. At the RDF production site some PVC separation can still take place but, so far, long-term experiences are not available. For the future additional R&D work is needed for RDF quality improvements and quality control. Another aspect is the high reject amount at RDF stations, including mainly biowaste, metals, glass etc. which is landfilled. In the future the target will be reduction of reject flows to landfill and additional material recycling of non-combustible material.

RDF Gasification Cases in Lahti and Helsinki, Finland

In the city of Lahti, Finland, the local waste management and energy company, Lahden Lämpövoima Oy started a biomass and RDF fired CFB fluidized bed gasifier in 1998 connected to an existing coal fired 350 MWth boiler [133]. The unit is one of the demonstration plants of this type in Europe. Fuel effect is 40 MWth when feeded with high moisture (> 50 %) content forest residues and up to 80 MWth when waste plastic is the dominating fuel component. This corresponds to about 40 000 t/a REF I quality [8] feed rate. The unit has been successfully in operation with high availability, Figure 5.6. This RDF gasification plant replaces up to 20 % of coal providing significant CO_2 emission savings. NOx, SOx and particulate emissions are decreased and no changes were measured for dioxins, heavy metals and CO emissions [163]. Due to high steam values in the coal boilers, high RDF based power generation will be reached. In Europe there are more than 300 existing pulverised coal fired power plants, where this type of co-gasification with biomass and RDF may be introduced economically.

Based on the encouraging experiences from Lahti power plant and similar cases in Austria and in Italy, significant RDF gasification R&D&D has been carried out in various countries and companies in Europe. For example a 40 MWth gasifier has been successfully started in 2000 in Varkaus, Finland by Corenso Ltd. Post consumer liquid packages (capacity 70 000 t/a) are recycled by defibering and the waste plastic including aluminum folio is feeded to the BFB gasifier. Metallic aluminum is recovered (2 000 t/a) from the hot cyclone and polyethylene based product gas is used in a steam boiler.

Additional hot gas filtering can be introduced to a RDF gasification process and significant emission control improvement can be achieved. RDF ash can be collected separately from coal ash for end use alternatives, due to hot gas filtering. In Figure 5.6 test results by Kurkela and Nieminen [133] at VTT is given on RDF fluidized bed gasification with fabric filter at 400 C temperature. Chlorine reduction up to 90 % and non-evaporating heavy metals up to 99 % can be reached according to pilot test results at VTT. Mercury is passing the hot temperature filter as expected without any sorbent injection.

Table 5.4. Average RDF properties from various sources in Finland [147]

		Construction wood REF I	Commercial waste REF I	Household waste REF III
Moisture	wt-%	14.2	9.1	28.5
Density (bulk)	kg/m³	200	180	210
Calorific values	MJ/kg			
gross for dry matter		19.8	24.7	22.9
net for dry matter		18.5	23.1	21.5
net as received		15.5	20.8	14.6
Ash content	wt-%	3.0	5.9	9.5
Volatile content	"	78.8	81.0	76.4
Carbon - C	"	49.4	56.0	52.9
Hydrogen - H	"	6.0	7.4	7.3
Nitrogen - N	"	0.7	0.63	0.71
Chlorine - Cl	"	0.06	0.19	0.71
Sulfur - S	"	0.07	0.16	0.13
Sodium - Na	mg/kg	1400	1360	3800
Potassium - K	"	1330	960	2100
Calcium - Ca	"	4470	9530	13700
Cadmium - Cd	"	0.40	0.36	2.75
Thallium - Tl	"	<0.5	<0.5	-
Mercury - Hg	"	<0.1	<0.1	0.18
Antimony - Sb	"	7.8	24.9	-
Arsenic - As	"	132	21.3	5.3
Lead - Pb	"	103	37.3	84
Chromium - Cr	"	129	40.7	67
Cobalt - Co	"	<5	6.4	-
Copper- Cu	"	71	61.7	215
Manganese - Mn	"	92	54	135
Nickel - Ni	"	6.4	16.4	12.3
Vanadium - V	"	1.6	4.0	-
Tin - Sn	"	<10	<10	-
Aluminum - Al	"	1700	4400	10700
Zinc - Zn	"	340	380	200

BIOMASS GASIFICATION - COAL BOILER - LAHTI PROJECT

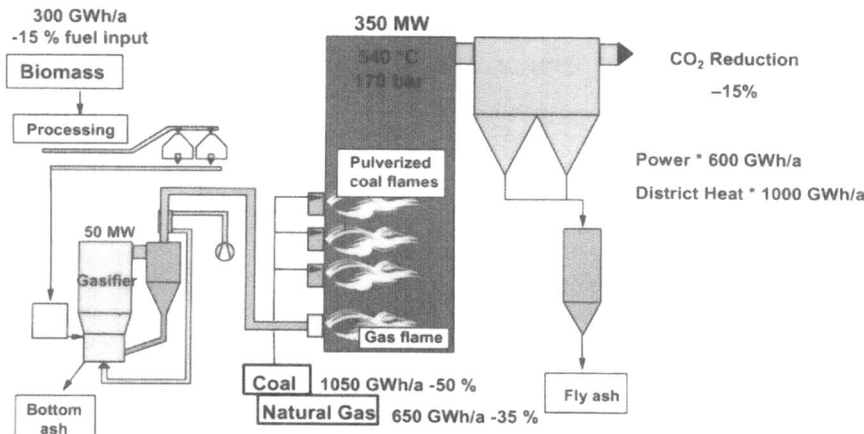

Fig. 5.6. RDF fluidized bed gasification plants in Lahti, Finland [163]

For the Helsinki metropolitan area in Finland, with a population of one million people and a RDF volume of 280 000 t/a combustible material, a RDF novel gasification concept is planned. Based on these gasification tests, a gasification process is proposed with three 80 MWth gasifiers, á 90 000 t/a, connected to existing coal fired power plants. All RDF produced from commercial and household waste could be utilized with high efficiency replacing coal in combined heat and power production. Thermal efficiency of the gasification line is 90-95 % so coal can be replaced effectively and CO_2 reduction is significant. A gasification process including hot gas filtering is presented in Figure 5.7 by Foster Wheeler. No final investment decisions exist.

The gate fee of the proposed concept is lower compared to a scenario where condensed mode power production or additional heat generation capacity would come to local free power market, Figure 5.8. It will be difficult in the future find suitable heat loads for new waste to energy installations in CHP or heat generation mode in countries like in Finland where the CHP capacity has already been built during the last decades. Separate power production without heat benefit is possible, however the low power price in the grid, typically 2 – 2.5 €c/kWh, will require the highest gate fee.

In Figure 5.9 the lowest gate fee is for the 80 MWth RDF gasification concept where the hot filtered fuel gas will be co-combusted in the existing pulverised coal fired CHP plant. Coal can be replaced up to 20 – 30 % without major modification of the existing boiler. The value of the produced power was 23 euro/MWh and district heat was 13 euro/MWh in the calculation.

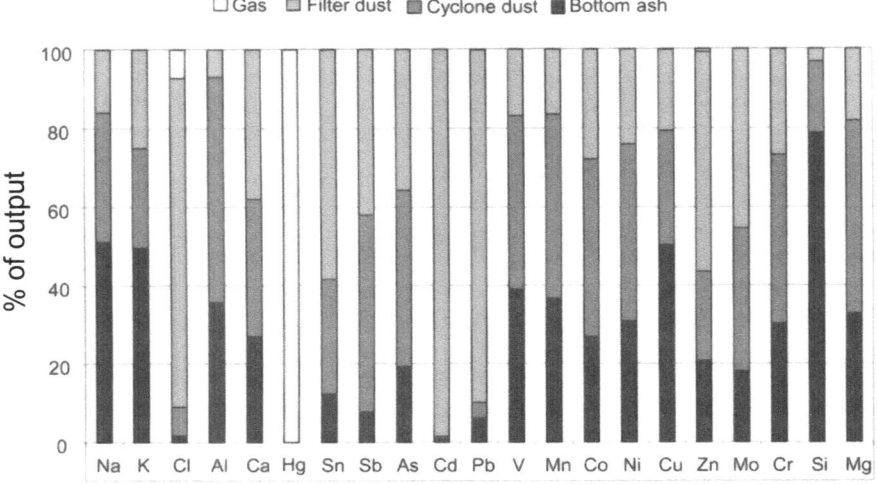

Fig. 5.7. RDF based PDU-gasification and product gas filtering test by Kurkela and Nieminen, VTT [133]

RECYCLED FUEL GASIFICATION PLANT

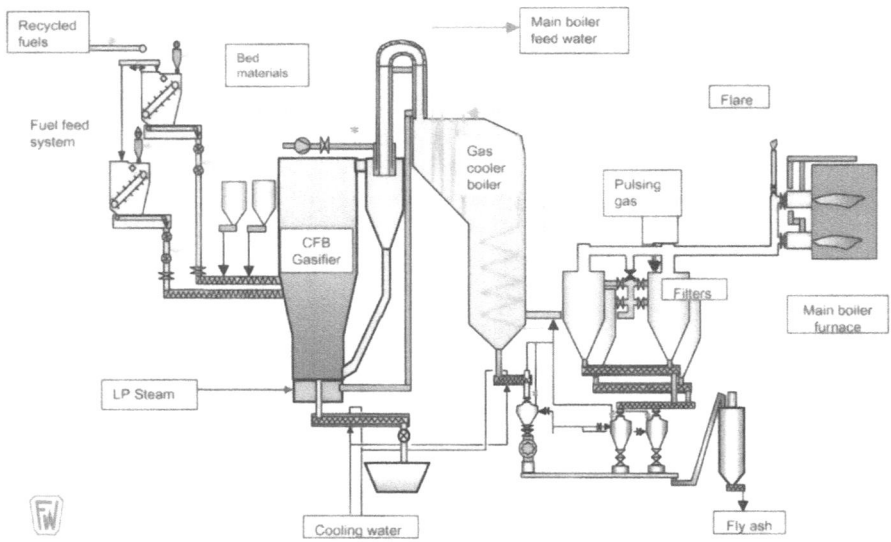

Fig. 5.8. RDF CFB-gasifier with hot gas filtering, Helsinki case (Foster Wheeler)

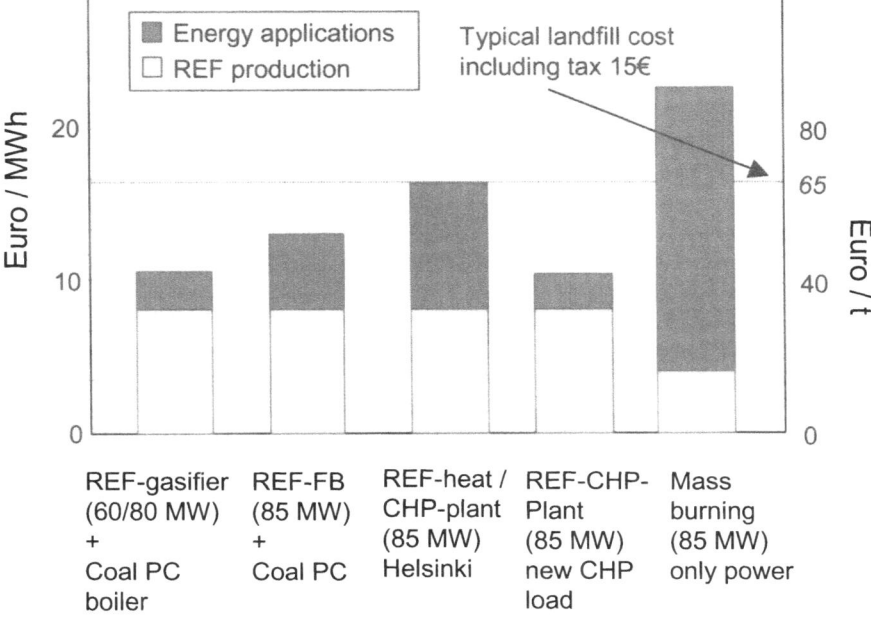

Fig. 5.9. Estimated gate fees in various waste to energy concepts [147]

In Figure 5.10 the high MSW recovery rates, > 70 %, are presented as a target for the year 2005; MSW recovery rate is today 46 %. The Finnish Waste action plan is setting more than 70% recovery rate target for MSW management strategy until 2005. With three gasifiers with 80 MWth effect, totally 280 000 RDF/a, in co-firing at existing power plants, this target could be reached. In this scenario biowaste from household etc. would be composted and dry MSW would be used for energy replacing coal - this scenario would stop landfilling of any combustible material.

WASTE RECYCLING VOLUMES IN HELSINKI
- present values and vision to 2005

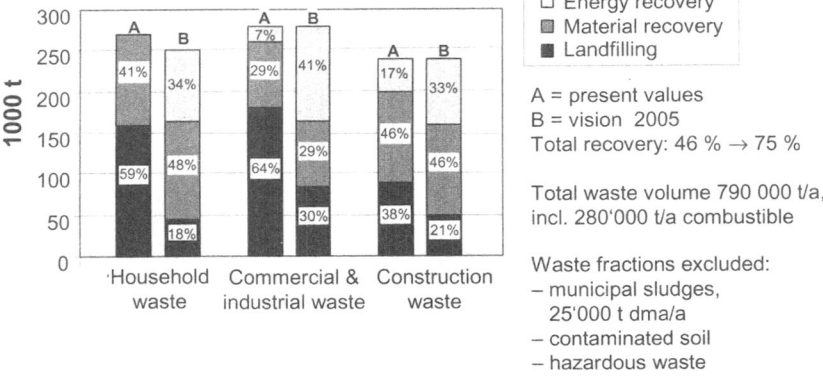

Fig. 5.10. The present MSW recovery rate and scenario for 2005 in Helsinki area, Finland [147]

Integrated MSW Concept with Paper Recycling and Energy Recovery

Metso Ltd has been developing an integrated concept, Urban Mill, where especially from commercial waste the high quality paper and board can be recycled and low grade fibers will be used for energy applications. In the Helsinki region the composition of various recovered fuels favors advanced paper recycling concepts. In some measurements the paper and board composition has been quite high, Figure 5.11 [147]. Based on this background a challenge of higher material recovery can be raised instead of RDF energy applications if economically attractive.

Metso Ltd [180] has been testing at the pilot scale a pulper concept with more than 90 % fiber recovery from high quality RDF. The reject has good fuel properties with high heating value due to plastics and low impurities due to water pulping [180]. In Figure 5.12 a vision is presented on integration of the Urban Mill concept at an existing landfill, power plant or metropolitan type city paper mill. The plastic and wood containing reject can be processed to high value products instead of co-gasification and energy options depending on local markets of the products. The clean or cleaned waste wood can be used for particle-board etc. production or as clean biomass fuel. The waste wood, classified under the EU waste incineration directive including heavy metals or halogenated organic substances, should be utilised in boilers fulfilling the requirements presented in the directive. Waste plastics could be processed to cleaned solid, gaseous or liquid fuels or recycled as material, typically as extruded products for construction application.

RDF FRACTIONS IN HELSINKI
Challenges for additional material vs. energy recovery ?

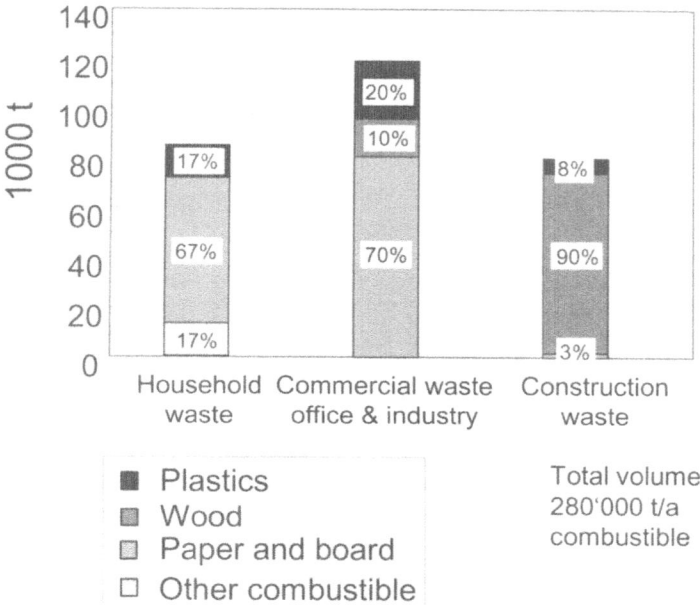

Fig. 5.11. Composition of the combustible material in RDF fractions [147]

In Figure 5.12 a vision is presented how this concept could be utilised in Helsinki area. The commercial waste based REF has typically low biological activity, so REF could be used as feed material to a pulper process. The produced pulp quality is expected to be high enough to be mixed at the present paper mills to existing and new paper products. Typically newsprint or packaging products can be produced, the color of the recycle fiber is dependent on the raw materials, number of the pulper lines and a possible de-inking step. In this scenario household waste is processed with a mechanical biological process in order to stabilize the reject from RDF production for landfilling. RDF fraction could be used for replacing coal in existing CHP power plants via fluidized bed gasification and advanced hot gas cleaning. From RDF and recycle pulp production the non-combustible reject like metals, aluminum, glass etc. should be recycled instead of landfilling. By this concept it would be possible to stop landfilling of organic material and significant benefits can be reached in total material and energy recovery from MSW and reducing all negative effects of the landfilling to water and air emission. This additional fiber recovery is not included in Figure 5.10, the Urban Mill concept would mainly increase the material recovery instead of energy recovery.

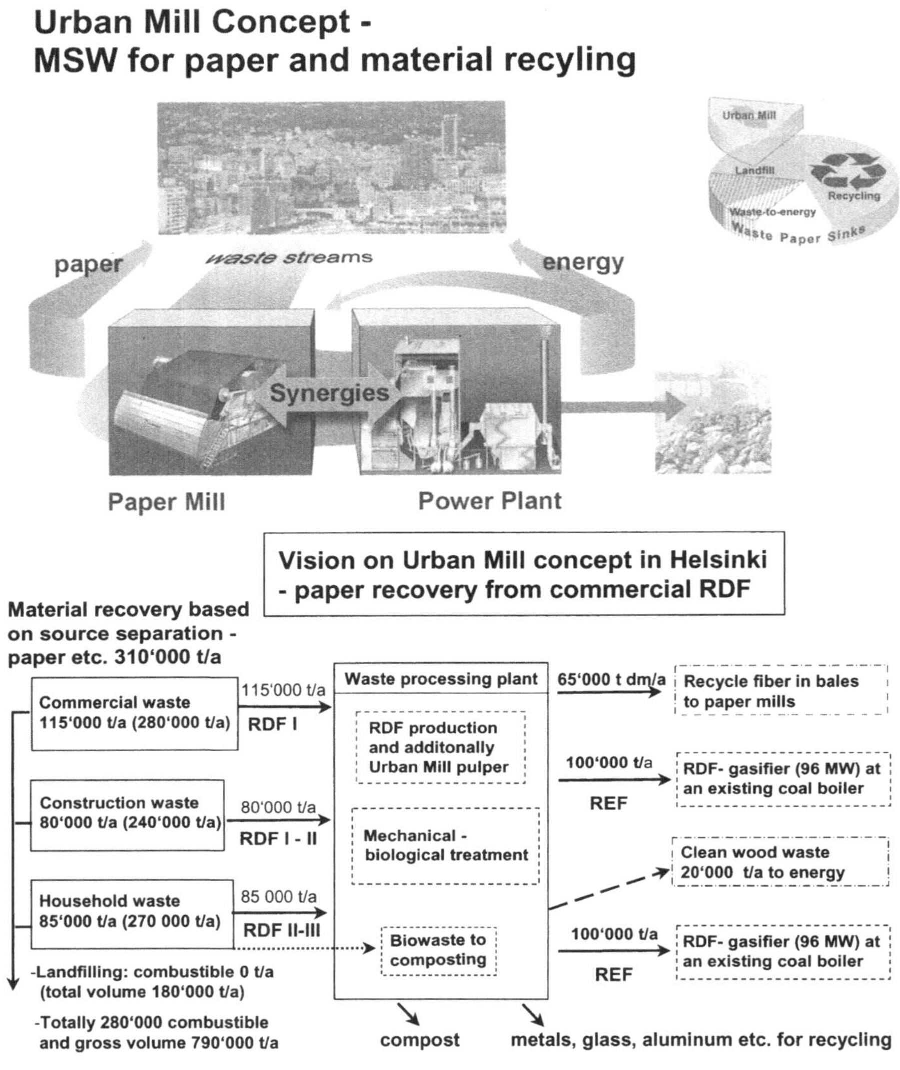

Urban Mill Concept -
MSW for paper and material recyling

Fig. 5.12. Integrated paper recycling and energy recovery, Urban Mill concept and the distribution of the European waste paper end use, Ristola [180]

Waste to Energy and Green House Gas Emissions: The Helsinki Case

In Europe one of the key drivers is the GHG (Green House Gas) and Kyoto policy when reducing the landfilling of organic material and increasing material recycling and energy recovery. In the Helsinki case GHG emissions are calculated

in various scenarios with traditional and advanced waste to energy scenarios
[242]. In Figure 5.13 four scenarios are presented. In the first column traditional
landfilling is presented with methane recovery and landfill gas utilization in a gas
engine with 33 % power efficiency. The second column is traditional RDF
incineration with 22 % efficiency in separate power generation. In the RDF
production ferrous and non-ferrous metals and glass are recycled as material.
Third column is the RDF co-gasification case where coal can be replaced with 90
% energy efficiency in a CHP-boiler. Additional benefit will be based on metal
and glass recycling from RDF production plant with 90 % recovery. Fourth
column presents on top of the previous case the Urban Mill concept with 90 %
paper recovery from commercial waste, commercial waste based RDF combustion
is reduced by 60 wt-% to the level of 200 000 t/a of RDF. The amount of RDF
production is 280 000 t/a, the specific $CO_{2\ eq}$ reduction is in the 3rd RDF
gasification scenario about 890 kg $CO_{2\ eq}$ / ton RDF. If one assume that RDF yield
is 80 % from MSW based source separated dry waste, then the specific $CO_{2\ eq}$
reduction would be 670 kg $CO_{2\ eq}$ / ton MSW.

Significant GHG reduction can be seen, the cost of reduced CO_2 tons in modern
WtE are minimal compared to other GHG reduction options. Additional costs can
be even negative, if the gate fee will be lower than local landfilling cost incl. taxes
like in the REF gasification and Urban Mill concepts.

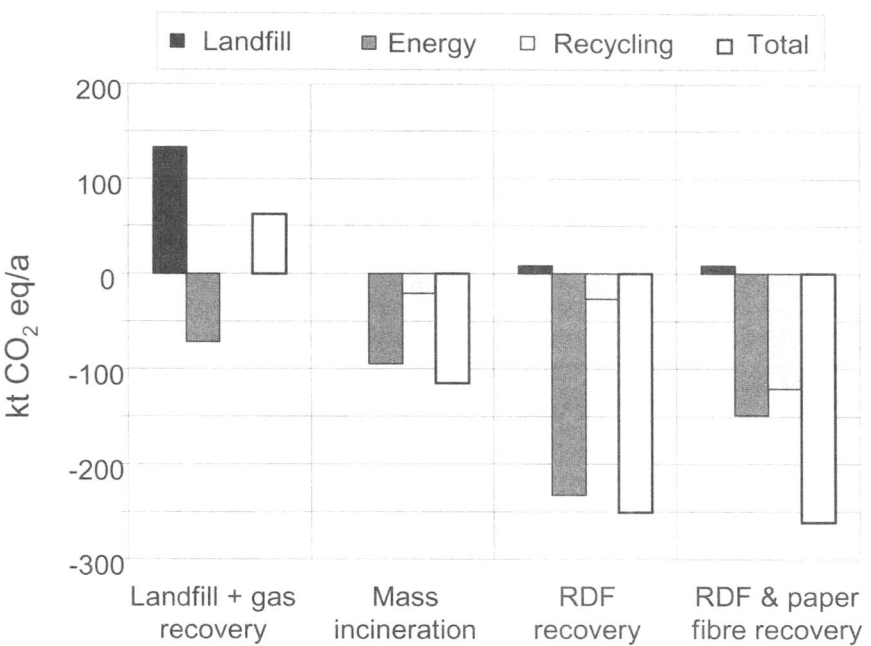

Fig. 5.13. Green house gas emissions in various waste management strategies [242]

The role of waste management options in Europe has been recently estimated in a report from AEA Technology [216]. Various present cases and future scenarios are presented. The landfilling policy is the worst, waste recycling, mechanical-biological treatment, composting and waste to energy have better reduction potential. Concerning waste to energy, the most important economical and GHG aspect is the net energy production and efficiency in heat and power generation. The selected energy technology and the local market will have also a clear role to play.

Conclusions

The need to reduce significantly the landfilling of organic waste in Europe, will increase new investments for waste volume reduction, additional material recycling and energy recovery. The source separation, high quality RDF production, advanced energy technology and novel integrated recycling and energy concepts will be the key development elements in the near future. Co-firing will be a low capital option for bioenergy including RDF reflecting the EU waste, energy and climate change policy. In Europe additional demand of waste to energy capacity is estimated to be 20 Mt within the five years. Parallel to large scale waste incineration plant there will be additional demand for decentralized, co-operation concepts integrating the traditional waste management companies to new partners like energy, paper and packaging companies. In European metropolitan areas higher material and energy recovery can be reached with zero landfilling of organic waste in the future.

5.1.2 High Efficiency Waste-to-Energy Concept

Marcel A.J. van Berlo and Jörn Wandschneider

Introduction

This article describes the concept for a High Efficiency Waste-to-Energy (HE-WTE) plant for waste incineration developed by AVI Amsterdam, the incineration division of the Amsterdam Municipal Waste Processing Department (GDA / AVI Amsterdam). We shall first explain the context in which the idea for the plant was developed.

AVI Amsterdam has been working on this extension of its waste incineration facility since late 1998. The planned extension consists of two new incineration lines with a total capacity of 500,000 Mg/a. After a detailed evaluation of all options, it was decided that a grate incineration should be used for the extension; primarily because it is most highly suited for untreated domestic and non-industrial commercial waste.

A new process setup has been developed to achieve an overall net energy efficiency rate of over 30%, which is a major step in comparison to the current 17-22% rate of efficiency in conventional waste-to-energy plants.

National Outline

In order to reduce CO_2 emissions, the Netherlands must derive 10 per cent of its total energy production from sustainable sources by 2020. Biomass is one such renewable source. Waste is recognised as consisting of 50% biomass (on energy content). Moreover, the other 50% could even be considered a renewable source. Part of the political objective is based on the decision that combustible waste must, as far as possible, be processed in the most energy efficient way. This is outlined in a moratorium that effectively rules out conventional grate combustion as a potential technology for the future, since it allegedly offers a lower efficiency than other so-called modern technologies.

Since August 1, 1999 the MWTEs in the Netherlands have received somewhat higher rates for the electricity they generate, which is paid for from energy taxes. In recognition of the acknowledged 50 per cent proportion of biomass in waste, remuneration for half of the total electricity production is increased by € 0,0077 per kWh in the form of a tax rebate.

At the same time, a covenant was agreed upon between the MWTE operators and the government. In compliance with the agreement, the operators have undertaken to increase their energy production by 23 per cent between 1999 and 2003, at which time the covenant expires. Specific measures which each operator can select in accordance with the characteristics of its own plant are partially covered by individual subsidy arrangements.

Current AVI Amsterdam Plant

The existing plant consists of four incineration lines with a nominal capacity of 25 Mg/h each and a total capacity of 800,000 Mg/a. Featuring a conventional furnace with air-cooled grates, the unit works very satisfactorily from a technical and commercial viewpoint, with an availability rate of over 92 per cent. The energy efficiency achieved is 22.2%. As there is insufficient incineration capacity in the Netherlands, an extension of the installation has been planned. The planned extension will consist of two new incineration lines with a total capacity of 500,000 Mg/a.

Preliminary Research

In developing a master plan, a feasibility study concerning an extension of the existing plant site was carried out. Alternatives to classic grate combustion were also examined – in particular fluidised bed incineration and gasification. For both planning and economic reasons, however, grate combustion was eventually chosen despite the moratorium on grate installations in the Netherlands. One of the main reasons for this decision was the fact that only grate combustion can fully incinerate domestic and commercial waste without intensive pre-treatment. Pre-treatment (industrial separation of waste into different streams) has been evaluated as having no contribution to the overall efficiency of the waste-processing chain, generating substantial streams that have to be landfilled and pose a major risk to the hygiene of those working there as well as the surroundings.

In July 1999 the development of an alternative concept for a so-called high efficiency boiler was considered as a supplement to the master plan. Various aspects of this were extensively examined in an additional study that was completed in December 1999.

The high efficiency boiler should provide a net efficiency of approximately 30 per cent in order to compete with the potential performance levels of so-called modern technologies, in particular the fluidised bed. What is meant by net efficiency is the ratio of actual net electricity output (= production minus in-plant consumption) to energy input (= the energy contained in the incoming waste). This definition of net efficiency is based purely upon electricity production and does not include any heat generated, for example, for district heating.

The principle of the high efficiency boiler involves an innovative combination of the latest grate combustion technology with a steam boiler, the maximum possible yield being extracted using a wide range of individual techniques. For the time being, the principle of reducing and optimising operating costs has been sidelined so as to be able to assess feasibility and potential energy yield. It goes without saying that the high efficiency boiler must have a sufficiently high rate of availability, because only then can the processing of the amounts of waste agreed upon by contract be guaranteed. The availability of the entire plant has been set at greater than 90 per cent, compared to the 92 per cent achieved by the current conventional unit.

The following individual measures were identified as that which will enable the high efficiency boiler to achieve a net efficiency of 30.5 per cent:

- enhanced steam parameters of 480°C/125 bar;
- cladding of the entire boiler with Inconel (empty draughts and steam superheaters);
- intermediate re-heating of the steam using saturated steam;
- boiler outlet temperature of 180°C, independent of fouling;
- a second economiser after the cloth filter for condensate preheating;
- a third economiser after the washers for condensate preheating;
- flue gas recirculation for primary and secondary air;
- water-cooled grate with heat recovery;
- various other measures during flue gas scrubbing and in the auxiliary equipment.

Furnace

As far as the furnace is concerned, it can, in principle, be stated that the opportunities for improving energy yields from the grate combustion of waste by taking structural measures are relatively limited. Only the surplus combustion air can influence the yield of the boiler, but even this effect is limited by the possibilities and freedoms of incineration technology.

Because future EU directives will no longer put a limitation on a minimal O_2 content in flue gases, only the minimum air surplus necessary for operation of the high efficiency boiler will be used as a basis.

It is absolutely necessary to work with an air surplus in order to guarantee the complete combustion of the flue gases under all operating conditions – so as to prevent, for example, boiler corrosion and other incineration irregularities in the event of a sudden change in the combustion behaviour of the waste resulting in increased emissions. On the other hand, the high combustion temperatures in the furnace, which result from a low air surplus, are a limitation. Taking into account these contradictory tendencies, an air surplus of 1.4 (corresponding with 6 volume per cent O_2 dry) has been taken as the basis for the high efficiency boiler. Increases in the combustion temperature are limited by a very substantial recirculation of flue gases. The chosen O_2 content is a compromise between the innovative demands of the high efficiency concept and sufficient operating reliability.

Clearly, a water-cooled grate is needed and has been planned for the extension. Entirely in accordance with the high efficiency concept, the heat from the grate cooling system is thus not released unused into the environment, but is used for the final stage in preheating the condensate. Experience shows that the heat from the grate cooling system represents approximately 1.5-2.0 per cent of the gross incineration energy, and so is available in considerable quantities for preheating.

As already described, the amount of primary air must be minimised; virtually stoichiometric combustion is aimed for. The oxygen required for complete flue gas incineration is supplied via the recirculated flue gases and secondary air.

The primary air can be preheated to 160°C and regulated independently for zone 1, 2 and 3 so as not to have to heat any more than is strictly necessary. The last zone is supplied with cold air only. The first zone will be supplied with recirculated flue gases from after the cloth filter. The objective is to dry the waste at high temperature (>200°C), with incineration in this zone kept to a minimum thanks to the low oxygen levels. The result of this is a well-defined fire on the 2nd and 3rd zones, improved burnout of the slag, and sharply reduced steam consumption for preheating. However, these technological benefits require exceptional measures to combat flue gas leakage and corrosion in the recirculation system.

Concerning the supply of secondary and tertiary air, extensive theoretical research has been conducted with the aid of computational fluid dynamics (CFD) simulations. It has been decided that recirculated gasses be used to create thorough overall mixing (front- to backside over the grate) without producing an O_2-excess, because with the hot spots in the flames, this leads to NO_x formation. The tertiary air is then normal air and provides for the required O_2-excess at a place with lower and evenly distributed temperatures.

It should be pointed out that the recirculation of flue gas is not intended to achieve a reduction in flue gas volume flow, but is technically necessary in the process so as to keep the temperature of the flue gas at a manageable level. Flue gas recirculation, with all its known problems, is therefore essential for a high efficiency boiler with extreme steam parameters. The amount of recirculated flue gas totals 20 to 30 per cent.

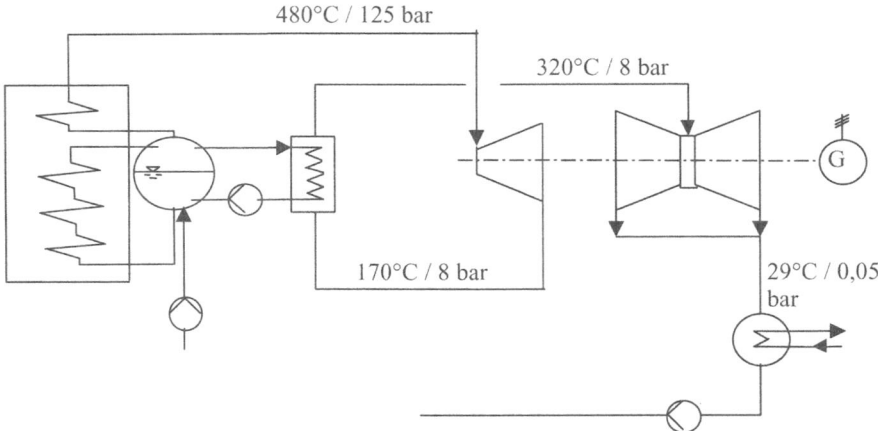

480°C / 125 bar

320°C / 8 bar

170°C / 8 bar

29°C / 0,05 bar

Fig. 5.14. Sketch of indirect reheating with saturated steam from the drum

Steam Parameters

One crucial aspect was the choice of steam parameters. An extensive study of steam parameters in German waste incineration plants revealed that, of all the boilers examined, those with high steam parameters were of an older type and so were in unfavorable condition regarding flow and temperature. Much of the widely known corrosion is hence ascribable to the design of the furnace and the boiler. Previous experience with high steam parameters will therefore be used to design a furnace and boiler for the high efficiency system which should, in turn, set a new technological standard.

Initially, steam parameters of 520°C/120 bar were discussed. These were necessary to restrict the humidity levels in the last stage of the turbine at the high yield desired. However, the heat calculations in respect to the boiler indicated that the desired flue gas temperature of less than 630°C upon its introduction into the convection element of the boiler made these parameters entirely unachievable. Therefore, intermediate indirect reheating using saturated steam from the steam drum was introduced. Under the principle of reheating by saturated steam, the steam in the high pressure section of the turbine is superheated using saturated steam from the steam drum. This made it possible to achieve clear efficiency improvements despite reduced steam parameters. The principle of this system is illustrated in Figure 5.14.

In broad terms, intermediate heating using saturated steam achieves the following changes.

- A new degree of freedom in setting dimensions for the boiler is established. By choosing the pressure of the steam for reheating, the heat load can be freely partitioned over the radiation and convection section of the boiler.

- The additional input of energy during intermediate heating increases the enthalpy differential available for the production of power, thus higher yield and efficiency can be achieved.
- The steam in the low pressure section of the turbine is considerably drier. This reduces the burden on the low pressure section of the turbine and reduces turbine wear.
- There is a shift towards increased vaporization surface, rather than superheaters. In other words, there is relatively more corrosion resistant surface in the boiler. This has advantages for its availability.
- The required flue gas temperature before the superheaters can be substantially lower, which strongly decreases boiler fouling if it is below 630°C.

Compared to intermediate *direct* reheating using a flue gas heat exchanger, as is common in conventional power stations, this process has the advantage of the required additional vaporization surface (membrane-wall) being much less vulnerable to corrosion than additional surface for steam superheaters. This is of particular advantage for waste incinerator boilers.

With the aid of a computer program which calculates an equilibrium in the water-steam circuit in accordance with the process diagram (see Figure 5.14 and 5.18), the variants with and without intermediate heating can be compared, with a view to achievable generator power. The results of the calculations are shown in Figure 5.15 below as a function of the temperature of the fresh steam.

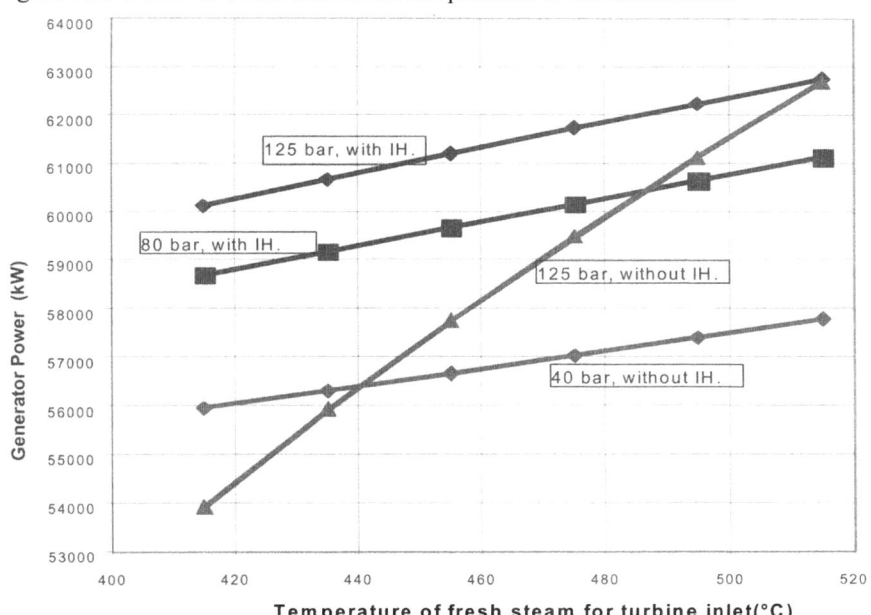

Fig. 5.15. Generator power with and without reheating as a function of the turbine inlet temperature

Figure 5.15 shows the following.

- At 520°C turbine inlet temperature, intermediate heating results in practically no improvement in power. On the other hand, this has the advantage of a reduction in the fresh steam temperature to below the maximum value of 520°C being linked to relatively limited losses of power. This offers greater flexibility in the boiler design and also makes it possible to keep the flue gas temperature lower than 630°C when it enters the convection element with the superheater. Without intermediate heating, the steeper curve means that those losses are significantly greater.
- Intermediate heating using saturated steam necessitates a higher fresh steam pressure. Reduction of the pressure to 80 bar leads to a significant decline in turbine power. The value of 125 bar chosen for the high efficiency boiler is a justifiable maximum. A higher pressure would worsen the ratio of heat of evaporation to superheating, and so necessitate a larger superheater. Moreover, under even greater pressure the limit of natural circulation in the boiler would be reached.
- Compared with the standard parameters (40 bar/400°C), intermediate reheating enables a significant increase in turbine power – in other words, a higher yield and a higher energy efficiency.
- Figure 5.15 clearly shows that intermediate heating using saturated steam enables an operationally desirable reduction in the temperature of the fresh steam, from 520°C to 480°C, without significantly reducing yield. There is also the possibility of further reducing the temperature of the fresh steam without losing much power. This would be useful in the case of unexpected problems with the superheaters.

Intermediate indirect reheating offers flexibility in the boiler design because the relationship between the superheater and vaporizer surfaces shifts towards the less sensitive vaporisers. The steam parameters can thus be set to levels which have a favorable effect upon the boiler's downtime and are likely to improve its availability in the long term without resulting in significant energy loss.

Boiler Design

The design of the boiler is shown in Figure 5.16. One essential requirement is the horizontal construction with knockers on the pipe clusters. Clearly visible are the extremely high first boiler draught and the very long horizontal section needed to place all the clusters in the convection section. The high first draught is necessary in order to cool the flue gases properly. This is also the section where the temperature window for the injection of ammonia water from the SNCR lies. Special types of high- temperature-corrosion resisting steel (nickel-chromium alloy cladding) must be used in most parts of the membrane wall and the first steam superheater. The greater height of the first draught simplifies the manufacturing of these components. The second and third boiler draughts are optimized for flue gas mixing and fly ash removal.

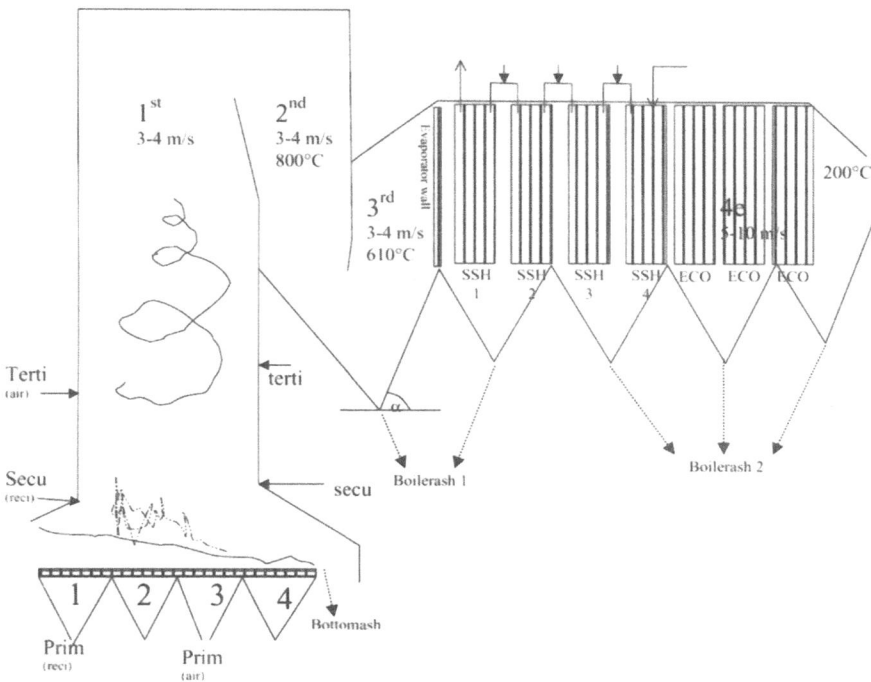

Fig. 5.16. High efficiency WTE boiler design

The arrangement of the pipe clusters and the distances between the pipes have been deliberately chosen so as to facilitate easy access and thus simplify repair work. The critical superheater clusters should be able to be replaced as complete units or rows of pipes. The evaporator group placed before the first superheater acts primarily as a flow equalizer.

Figures 5.17 and 5.18 show the optimizations planned thus far in the form of corrosion diagrams. It can clearly be seen that, thanks to these optimizations, parts of the first superheater already lie in the low-corrosion zone.

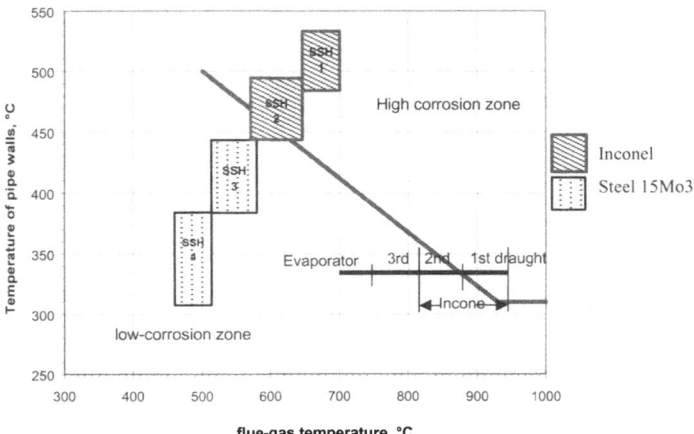

Fig. 5.17. Corrosion diagram for the steam parameters 520°C, 80 bar as required for high efficiency without reheating

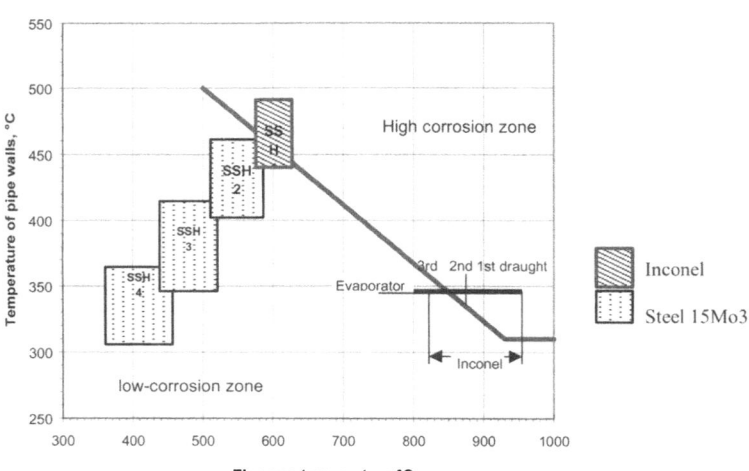

Fig. 5.18. Corrosion diagram for the steam parameter 480°C, 125 bar as a result of the use of reheating

Second and Third Economiser

The aim of the second economizer is to regain heat from the flue gas behind a dust cleaning section (bag house or ESP) which is placed after the boiler-economizer. The purpose of this second heat exchanger is that it is designed to allow acids from the flue gas to condense on the tubes of the economizer. In waste incineration, this leads to a very aggressive cocktail of acids. Therefore, the tubes have to be highly corrosion resistant.

The third economizer is placed after the wet scrubbers in order to regain additional amounts of energy from the flue gas. The flue gas here is essentially clean, about 60°C and saturated with water vapor. Here, condensation will be the heat source. This leads to the production of water with low salt content which can be used elsewhere.

The history of the ECO 2 and 3 began as part of the feasibility research into improving the efficiency of the AVI Amsterdam plant. Once its feasibility and rough cost had been established in an initial design and subsidies had been secured, tenders were requested and the project began. Parallel to the tender process, extensive tests were conducted, during which various materials were exposed to the flue gases. These tests were designed to replicate actual operating conditions concerning the flue gas as closely as possible.

Twenty materials and combinations of materials were tested. It soon became clear that, of those materials regarded as partially suitable, only graphite, ceramic and enameled pipes resisted the corrosive flue gases. All high-alloy steels and coatings corroded very quickly. For this reason, the heat exchangers now built in are double-enameled pipes. The sidewalls of the heat exchanger housing have a PTFE coating. In the existing installation the flue gas is cooled from 135°C to 100°C; in the new one a wider range will be used.

The water-steam cycle is shown in the schematic diagram (Figure 5.19). The pre-heating of the condensate using ECO 2 and 3 can be clearly seen. This is not

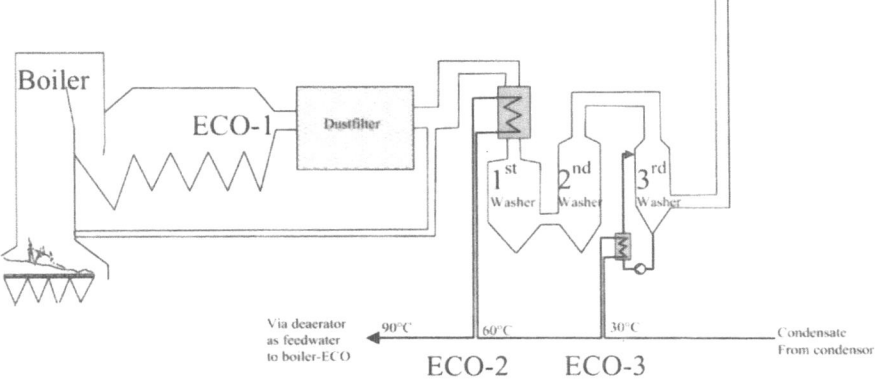

Fig. 5.19. Overview of the water-steam cycle with ECO 2 and 3 in the existing installation (no reheating!)

only incorporated in the planned high efficiency boiler, but is also being built into the existing system as part of an energy saving program. The flue gas heat exchangers have already been added to the flue gas channel and are currently being connected to the water-steam system. The ECO 2 is scheduled to enter service in the beginning of 2002.

The condensate is first heated in ECO 3 from 30°C to 60°C using plate heat exchangers. These are integrated into the washing water cycle of the flue gas washers. Plate heat exchangers similar to these new ones have been used to cool flue gases at AVI Amsterdam for the past ten years. The titanium plates have proven highly effective and have not yet had to be replaced. The installation of new plate heat exchangers is therefore not regarded as an operational risk. The condensate is then heated further, from 60°C to 90°C, in the flue gas heat exchangers of ECO 2.

Other Measures

Naturally, measures to improve energy efficiency have also been incorporated in the flue gas scrubbers and peripheral installations surrounding the high efficiency boiler. But these are limited by the established requirements. The flue gas scrubbers, for example, must comply with the very strict Dutch emission limits prescribed in the BLA (Air Emissions from Waste Incineration Order), which make the use of wet scrubbers unavoidable. Nevertheless, several of these measures are making a crucial contribution to the success of the project.

The SNCR process was chosen to reduce nitrogen oxides. The replacement of 4 bar steam with high pressure water as the atomization medium for the ammonia water also contributes to improving energy efficiency. Even now, the AVI Amsterdam plant has only one injection level for ammonia water in order to reduce steam consumption. This has no negative effect upon the efficiency of nitrogen oxide removal, with emission values of less than 50 mg/Nm^3 (normalized at 11% O_2) being achieved without any problem.

The dimensions of the boiler will keep flue gas speeds very low, 3-4 m/s, in the first draught so as to reduce dust, the amount of fly ash, fouling, erosion and corrosion. In this respect very much attention is being paid to the design of the boiler and superheater details.

For the primary air supply, a separate ventilator is used in each grate zone. This means that there are 12 ventilators in all. In this way, the air supply to each grate zone can be controlled directly by adjusting the ventilator speeds without valves causing throttling loss.

Conclusion

The use of the measures described is shown to achieve a net total energy yield of over 30.5 per cent. The objective of designing an energy-optimized high efficiency boiler is thus achieved.

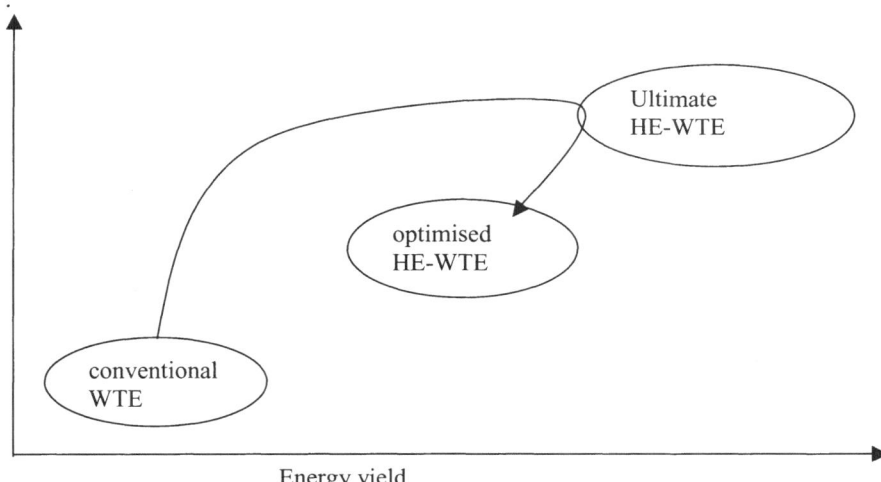

Fig. 5.20. Scheme of the design process for the optimization of energy efficiency and economics

Despite all the preventive measures taken, it is indisputable that were this boiler to be built in this way, we would have to reckon with significant operational risks For this reason a fallback scenario has been devised, under which this boiler could also be operated using reduced steam parameters.

In the current design phase, the high efficiency boiler concept is now being further developed and crystallized. Based on the resulting design, an economically justifiable case for construction, which considers all operational aspects, availability and investment costs, will have to be established.

Figure 5.20 shows that the variant with the greatest chance of actually being built falls short of the extreme variant described here. As costs are evaluated, a compromise has to be made; ultimately, it is a political decision. Initial estimates indicate that the high efficiency boiler will require a significantly higher investment and incur higher maintenance and repair costs. These, however, will largely be offset by the income from the additional energy generated. The economic position of the high efficiency boiler is therefore comparable to that of a conventional waste incineration plant. And, of course, every improvement upon the conservative assumptions made will have a direct and positive impact upon the operating results.

However, political decisions will be necessary to make it possible to actually achieve this 30.5 percent efficiency level. This particularly applies to facilities for selling the electricity produced at market prices for sustainable energy, for the 50% of the energy originating from biomass. As the energy from the *extra 8 to 10 percent* efficiency is really emission-free, it could even be considered extra-green energy, and at least comparable with wind energy or hydro power!

After a period of over 20 years in which emission control was the main focus, now optimisation of the reuse of emissions will be the name of the game, including fixed materials (ashes, iron, non-ferro, etc.) as well as the energy in the waste.

Innovative thinking characterises the construction of this high efficiency boiler, and with it a whole new generation of waste incineration plants. Recognition of this thinking has already come in the form of a EU subsidy for the boiler.

5.1.3 Using Substitute Fuels in the Basic Materials Industry

Michael Beckmann, Reinhard Scholz, and Martin Horeni

Various approaches to improving energy recovery from MSW have been discussed. The one described here involves separating fractions with high heating value, known as substitute fuels e.g. RDF, (see 4.1, 5.1) for efficient conversion. The material and energetic utilization of such fractions is being applied to an increasing degree in the production of basic materials (cement, lime, steel industry, etc.).

A series of investigations to find ways of optimizing process control using substitute fuels rather than fossil fuels have been carried out in high-temperature production processes, such as the clinker calcining process, and in the field of energy conversion in power plants. From these investigations fuel technology criteria appropriate for the individual processes have been derived. According to fuel technology, a fuel is characterized by its chemical, mechanical, caloric and reaction technological properties. Based on these properties, a corresponding classification for the substitute fuel can be made, similar to the classification of conventional primary fuels, such as fuel gases, heating oils and coal.

The differences between the ignition and burn-out behaviors of substitute fuels and conventional fuels with a substitution rate of e.g. 1 to 5 % generally have no significant effects on the process. The following statements concerning the requirements for substitute fuels therefore refer to a significant substitution rate in the range of 20 % and more. Since there is little empirical data on the fuel technology properties of substitute fuels, e.g. they have not been systematically characterized, these must be inferred from practical experience with various conventional fuels. Moreover, the process control may also have to be adapted when using substitute fuels. When substitute fuels are used, the boundary conditions change, and the following topics must be addressed:

- process optimization;
- energy recovery by internal and external energy combinations;
- examination of the process chain with cumulative material, mass and energy balances in the sense of the total analysis.

Substitute fuels are, as the name suggests, used to substitute conventional fuel materials (fossil fuels) in power plant and process combustion. Therefore, the total process chain (including expenditure on the production of substitute fuels from

MSW), must present energetic advantages over the reference case of using conventional fuels in the individual processes, and incinerating unsorted waste in a municipal solid waste incineration plant. Approaches to producing and using substitute fuels selected from the energetic point of view must then be assessed with respect to emissions and profitability. Fuel properties alone do not suffice for the evaluation of a substitute fuel.

After a comprehensive description of the fuel technology properties of substitute fuels, findings about their significance from an energetic point of view will, therefore, be presented. In this case, the so-called energy exchange ratio, i.e. specific energy consumption and efficiency, has to be considered. When evaluating the use of a substitute fuel, the process chains, consisting of the substitute fuel production process (e.g. mechanical-biological pre-treatment, MBP), the process for the utilization of the substitute fuel (e.g. production process) and the process for the treatment of the remaining residual fraction from the substitute fuel production (e.g. thermal waste treatment, dumping), must be examined [35, 204, 207].

Fuel Properties

An evaluation of the chemical, mechanical and reaction kinetic properties of substitute fuels depends upon empirical results (e.g. [95, 103, 117, 150, 192]). As described in [35], the interaction between the main influencing parameters can be demonstrated with the help of simplified, but effective, practical mathematical models of the caloric properties.

The **chemical properties** of conventional fuels are divided into noncombustible (ash, water) and combustible substances in the fuel. This classification can be transferred directly to substitute fuels. The combustible substances can then be divided into plastics and other organic components, which is useful for a comparison of the processes. Furthermore, the element and trace element composition, the proportion of the organically bound and volatile components and the ash-melting temperature can be viewed as chemical properties. These properties can, of course, also be used in evaluating substitute fuels.

The so-called trace element contents (heavy metals, chlorine, phosphorus, sulfur, etc.) also belong to the category of chemical properties. The trace elements are particularly important in the evaluation of emissions and product quality.

The **mechanical properties** of conventional and substitute fuels consist of: the density of the combustible and noncombustible substances, flow properties, grindability, grain-size distribution and general handling with respect to the storage (mechanical) and feeding of the fuel to the plant (burner, etc.).

The **caloric properties,** such as heating and calorific value, specific minimum air requirement, specific minimum amount of flue gas, adiabatic combustion temperature, heat capacity, thermal conductivity, etc. can, as with conventional fuels, also be used with substitute fuels.

Assessing the **reaction technology properties** of waste materials is more difficult than for conventional fuels. Reaction technology properties are basically dependent upon the chemical, mechanical and caloric characteristics of the fuels. Since the early 1930s, investigations of the ignition and burn-out rate in dependence upon e.g. the quantity of volatile components, the grain size or the heat transferability, thermal conductivity, etc. have been carried out for conventional fuels. A series of mathematical models and corresponding data concerning the reaction coefficients, pore radius distribution, activation energy, diffusion coefficients, etc. are available. For substitute fuels, which have a more heterogeneous and variable composition than conventional fuels, it is possible to use a simplified model to describe the burn-out with cumulative kinetic parameters determined with a special experimental apparatus (e.g. batch reactor, thermobalance). An additional criterion, closely related to the reaction technology properties, concerns the storage life, which should be examined from more than just a purely mechanical standpoint.

When assessing the above-mentioned criteria it is important to consider the technical process that the fuel is to be applied to. Questions concerning the process optimization and energy recovery by internal and external energetic combinations for the altered boundary conditions with fuel substitution must be answered. The various energy exchange ratios will be discussed in the following section.

Energy Exchange Ratio in the Partial Steps of the Substitution

If a certain amount of the primary fuel PF (e.g. hard coal) is replaced by a substitute fuel in a partial process (e.g. main combustion in the cement clinker calcining process, melting process of iron scrap), then the process as a whole, with respect to the production goal (e.g. clinker brick production and quality, melting capacity), should remain unchanged. This means that the solid temperature ϑ_s and the furnace capacity, e.g. the transmitted heat flow $\dot{Q}$, must, from the point of view of heat-engineering, remain constant. The energy exchange ratio E:

$$E = \frac{\dot{m}_{SF} \cdot h_{n,SF}}{\dot{m}_{PF} \cdot h_{n,PF}}$$

expresses the energetic significance of the substitute fuel (SF) relative to the primary (conventional) fuel (PF). With the help of simplified mathematical models of the balancing and heat transfer in the furnace chamber (e.g. [35], the energy exchange ratio, e.g. as a function of the corresponding gas temperature, is determined for given boundary conditions (in particular, heating values for the primary and substitute fuel, preheating of air and fuel, gas temperature, solid temperature) as shown in Figure 5.21.

The energy exchange ratio E (Figure 5.21) increases, the higher the temperature to be reached and the lower the heating value of the substitute fuel in comparison with the primary fuel. Where the heating value of the substitute fuel is greater than that of the primary fuel, the energy exchange ratio assumes values less than one. This happens, for example, when hard coal is replaced with high-calorific plastic waste during the primary combustion in the clinker calcining process.

Energy Exchange Ratio for the Total Process

The influence of air preheating can be seen from the curves for the energy exchange ratio E of the partial processes Figure 5.21. When we compare two processes in which a substitute fuel with $h_{n,SF\ I} = 20$ MJ/kg (SF I) and no air preheating is combusted in the first process and a substitute fuel with only $h_{n,SF\ II} = 15$ MJ/kg (SF II) but with air preheating in the second, then, at process temperatures of $\vartheta_G > 1600$ °C, the substitute fuel SF II has "more value" than the higher-calorific SF I. This shows that the energy exchange ratio is clearly not only dependent upon the caloric characteristics of the fuel, but is also influenced significantly by the process control and optimization, e.g. through improved heat recovery.

During the preheating of the air and fuel, the fact that the efficiency of the heat recovery is determined substantially by the minimum air requirement l_{min} and the corresponding excess-air ratio λ [e.g. 220] must be taken into account.

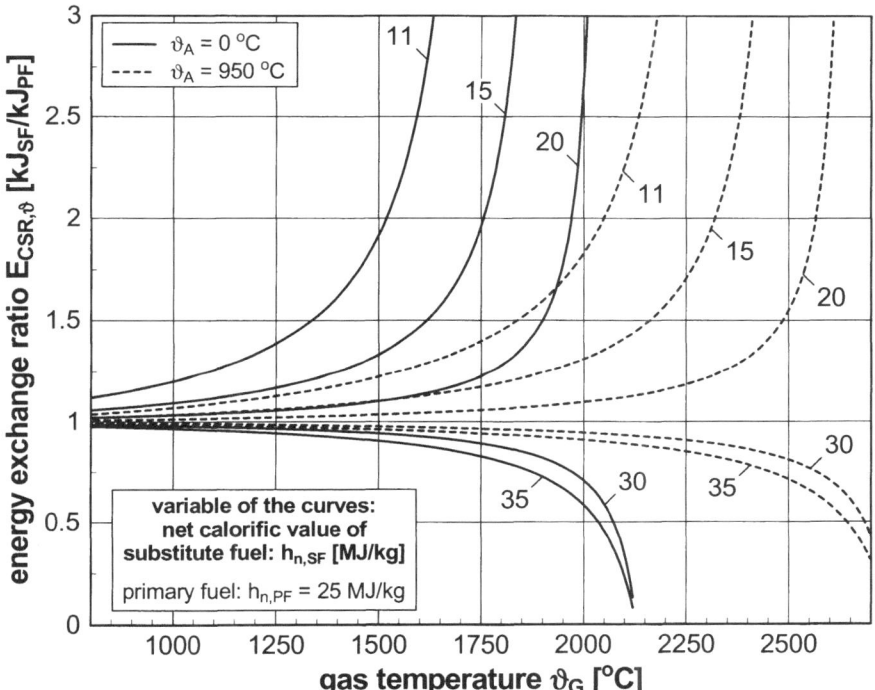

Fig. 5.21. Energy exchange ratio for statistical evaluation [35]

Another characteristic feature which must be considered with regard to the energy exchange ratio of the total process is that the minimum air requirement l_{min} and the minimum amount of flue gas v_{min} do not decrease at the same rate as the heating values $h_{n,SF}$ / $h_{n,PF}$. For solid fuels, the minimum amount of flue gas can be approximated by statistical combustion calculation [42]. If a constant specific energy requirement for the process is assumed, then the ratio of the minimum flue gas concentration from the substitute fuel to that of the primary fuel $v_{min,SF}$ / $v_{min,PF}$ can be approximated dependent upon the heating value ratio $h_{n,SF}$ / $h_{n,PF}$. The specific energy requirement of the total process increases primarily with the flue gas flow rate. In this case, a reduction of the specific energy requirement is necessary e.g. through heat recovery (preheating of the air or fuel, utilization of the heat lost for other processes, etc.).

Another possibility is for the heat recovery to be made process-internal, e.g. by adjusting the process control by changing the ratios of heat capacity flows or by fuel staging. Fuel staging is particularly effective for processes that are clearly divided into several partial steps, each with different temperatures (e.g. preheating, de-carbonisation, combustion or melting, overheating of the melt). If a relatively high energy exchange ratio results, e.g. from the requirements for the combustion process with a corresponding high temperature for a partial process (i.e. the energy input when using the substitute fuel must be higher than when using the primary fuel), then a portion of the additional energy input can be "recovered" in the subsequent partial steps dependent upon the temperature in the partial processes (i.e. the energy input in the following partial step, e.g. de-carbonisation of raw material, can then be minimized). For the total process, markedly reduced energy exchange ratios with values only a little above unity can result. The usable enthalpy flow in the subsequent partial process is, as mentioned above, dependent upon the temperature of the partial step, which is determined independent of the input of substitute fuel.

At this point it should be mentioned that fuel staging, such as in the cement clinker calcination process, is possible.

As shown in the plant scheme in Figure 5.22, the fuel is added at three feed points: the main firing (furnace discharge, I in Figure 5.22), the second firing (calcinator, II) and the supplementary firing at the furnace entrance (III). Theoretically, substitute fuel can be added at all of the three feed points if certain specific boundary conditions are considered. Figure 5.23 shows the results of a simplified but effective mathematical model of the energy exchange ratio $E_{CSR1,\vartheta}$ as a function of $h_{n,SF}$ for the substitution of a primary fuel in the main firing chamber (I) with $h_{n,PF} = 25$ MJ/kg, a balance temperature of $\vartheta_{G1} = 1.900$ °C and with air preheating to $\vartheta_A = 950$ °C. Furthermore, the curve for $E_{CSR2,\vartheta}$ for the second firing (II) is plotted for a certain balance temperature, which, due to the saving of primary energy for the substitution presented here, reaches values less than one. $E_{CSR12,\vartheta}$ is generally greater than one, but as a resulti of the staged fuel input, significantly lower than $E_{CSR1,\vartheta}$.

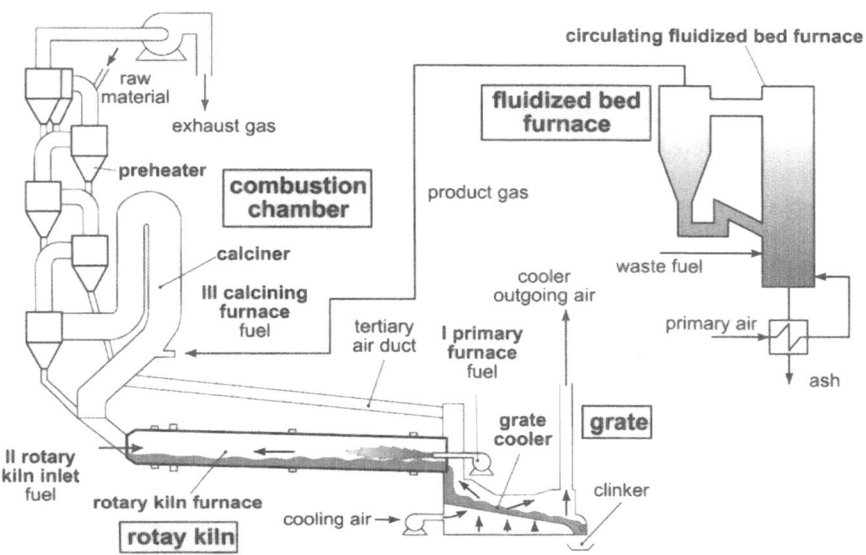

Fig. 5.22. Schematic representation of the clinker brick calcination after the drying process with the calcinator and external gasification in a circulating fluidized bed (CFB)

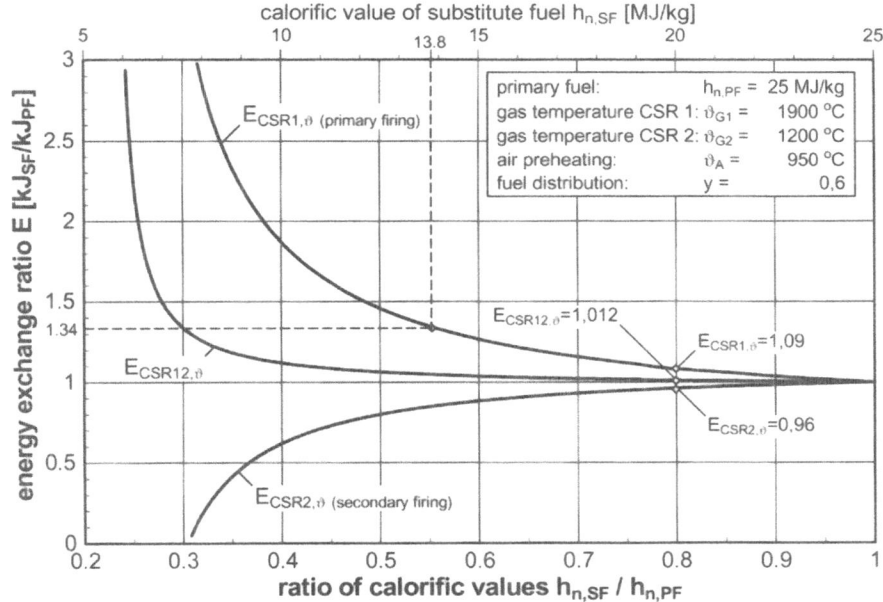

Fig. 5.23. Energy exchange ratio E dependent upon the heating value of the substitute fuel [35] (explained in the text)

Without additional measures for heat recovery, an increase in the specific energy expenditure is generally to be expected when substituting primary fuel with low heating value fuels. In [105], the practical experience from the substitution of heating oil ($h_n \approx 40$ MJ/kg) by lignite coal dust ($h_n \approx 20$ MJ/kg) is reported. This shift leads to an increased flame length, an elongated sinter zone in the direction of the furnace entrance and to higher temperatures at the furnace entrance and after the preheater. Of course, the different burn-out behavior of lignite coal dust and heating oil also plays a role. All in all there was an increase in the specific energy expenditure of approx. 170 kJ/kg$_{Cl}$. Half of the additional expenditure is due to the increase in the flue gas volume flow and fluctuations in the grain size, and the other half to the effects of the changes in the temperature, which can also be partially expressed through a corresponding energy exchange ratio with the above-mentioned relationship.

Energy Utilization Ratio from the Comparison of the Process Chains

The increase in the specific energy consumption of a basic materials industrial process on substituting primary fuels must not necessarily reflect negatively on using substitute fuels. In the same way, the simple fact that a certain amount of primary fuel can be saved by using fuel substitution may not be sufficient argument for doing so. All processes have to be considered. First, the substitute fuel must be generated, which requires energy (electrical energy, thermal energy, primary fuel). Furthermore, those waste fractions not used to produce the substitute fuel must also be treated (waste incineration, landfill). The requirements for a cumulative analysis that considers the mass, material and energy balances have been discussed in detail elsewhere (e.g. [152, 204, 207, 243]). The procedure will be outlined here briefly. The importance of defining the balance circles is illustrated in the schematic representation in Fig. 5.24. This shows an exemplary comparison between a conventional system (ConvS), consisting of the separate individual processes, with:

- the use of residual waste from household waste in a classical waste incineration plant (AT+T1),
- a high-temperature process for the production of e.g. cement clinker (simplified representation, labeled here as process line (G1), calcination process),
- and a combined system (CombS) consisting of:
- a pretreatment (A2, A3...),
- the use of a residual waste fraction with a high heating value in the high-temperature process (G1),
- the treatment of the residual waste fraction with a low heating value in a waste incineration plant (T1) and the (virtual) conversion of the primary fuel energy saved into electricity in a conventional power plant (K1). This allows the energetic comparison of the two systems on the basis of electrical energy under the same input conditions (input equality).

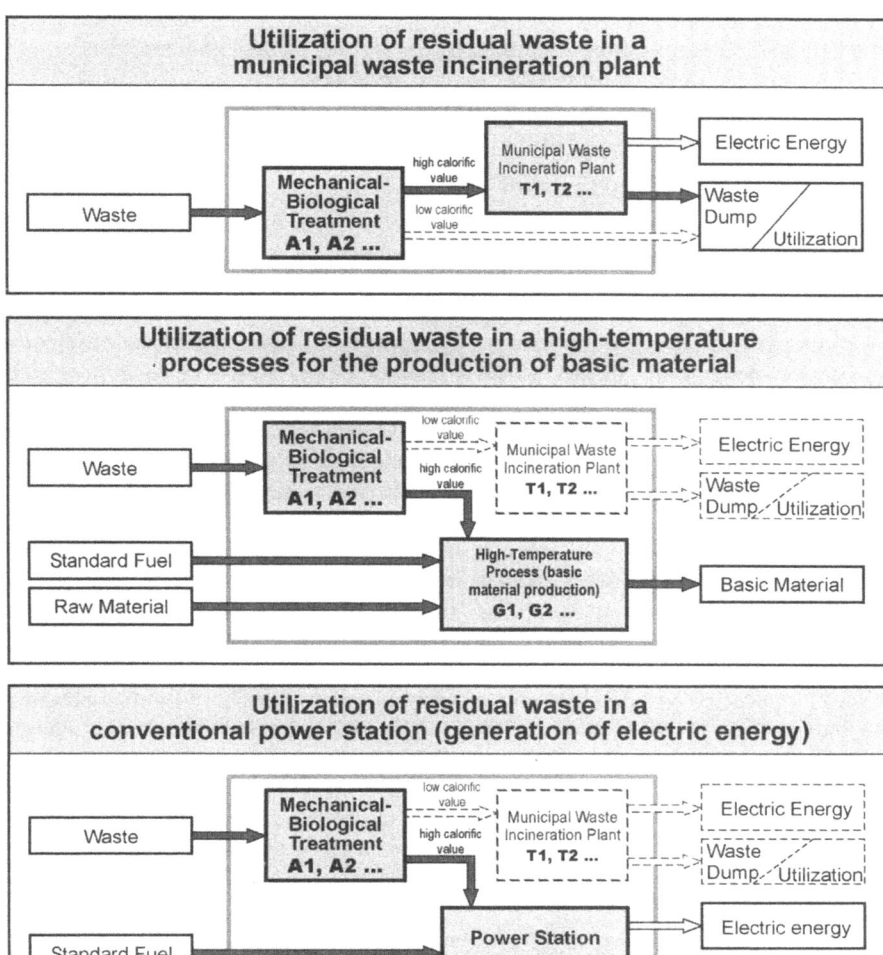

Fig. 5.24. Comparison of a conventional and a combined waste treatment system. In the conventional system the waste was treated in a waste incineration plant. Basic materials production process with an input of primary fuel. In the combined system the waste was pretreated and the residual waste fraction with a high heating value was subsequently used in a basic materials production process and the remaining waste fraction with a low heating value could,potentially, be used of in a waste incineration plant [204].

The Effective Energy Ratio z is defined by:

$$z = \frac{energy_{combined\,system}}{energy_{conventional\,system}}$$

Based on a cumulative consideration, the use of a substitute fuel can be assessed with the help of the energy utilization ratio z. Figure 5.24 shows clearly that, even for a higher specific energy input for the materials handling process, substitute fuel implementation can prove appropriate due to the better efficiency for power generation in the (virtual) power plant compared with a waste incineration plant. The expenditure during the generation of the substitute fuel, the intended quality of the substitute fuel and the substitution ratio in the material treatment process are decisive [36]. The discussion of the energy transfer ratio clearly shows that there is no need for a linear correlation between the effective energy ratio and the fuel substitution ratio. The low heating values of the substitute fuel mean that the heating values of the fuel mixture decrease with increasing substitution ratio. However, the energy exchange ratio increases more strongly with lower heating value, i.e. more substitute fuel is then needed for a given product and the same saving of primary energy (or the generation of electricity in the imaginary power plant). This causes an increase in the residual waste flow necessary for generating the substitute fuel. For the individual process, the increase in the amount of residual waste means an increase in the energy generated. These simple considerations show that the effective energy ratio must change continuously with the substitution ratio. It should be mentioned at this point that the analyses, which assume a linear correlation between the effective energy ratio and the substitution ratio from the start, do not sufficiently reflect the energy process-engineering requirements of the fuel substitution.

5.1.4 Hydrothermal Processes

Frédéric Vogel

Hydrothermal Processing of MSW

In supercritical or near-critical fluids have emerged over the last decades as promising alternatives to gas-phase or liquid-phase processes [67, 194, 202, 214]. This section discusses the application of near-critical and supercritical water processes to the treatment of MSW for energy production and destruction of hazardous compounds. Three routes will be presented: supercritical water oxidation (SCWO), hydrothermal gasification, and hydrothermal liquefaction.

Oxidation in supercritical water (SCWO) shows substantial promise for clean and efficient decontamination of many aqueous organic wastes [32, 83, 156, 170,

217, 233]. SCWO is oxidation in an aqueous medium above the critical temperature and pressure of pure water, *i.e.*, 374°C and 221 bar. Supercritical water (SCW) at typical process conditions of 500 to 600°C and 250 bar, has appreciably lower density (< 100 kg/m^3), dielectric constant (< 2), and dissociation constant ($K_w = $ [H$^+$][OH$^-$] $< 10^{-22}$ (mol/kg)^2) than water at ambient conditions. Consequently, SCW behaves as a low-polarity, largely unassociated solvent in which molecular oxygen and many organic compounds are totally miscible. SCWO can rapidly and efficiently destroy diverse organic substances, even when very dilute, in a self-contained process with very clean by-product streams (little or no NO$_x$, SO$_x$, soot). In the SCWO process, organic carbon is oxidized to CO$_2$, hydrogen to water, and nitrogen is converted to N$_2$, small amounts of N$_2$O and ammonia. S, P, and Cl containing organics form the corresponding inorganic sulfates, phosphates, and chlorides. The elevated temperature and the absence of mass transfer resistances across a phase boundary lead to almost complete mineralization in short residence times (*i.e.*, 99 to 99.99% at 500 to 650°C in 1 to 100 s). This performance has been demonstrated for many compounds and real wastes in bench-scale reactors as well as in pilot plants [83, 156, 233]. Modell *et al.* [157] provide an exemplary demonstration of the SCWO of biological sludge from a pharmaceutical plant in a continuous pilot plant. Organic carbon destruction efficiencies of 99.94% were reached at a reactor peak temperature of 590°C and a feed rate of 60 kg/h. Commercial processes for SCWO of liquid wastes containing solids are available from Chematur Engineering AB (Karlskoga, Sweden) and HydroProcessing (Austin/TX, USA; [92]).

Gasification in a hydrothermal environment (sub- or supercritical water) has been proposed particularly for converting wet biomass into a fuel gas with a high heating value [26, 203, 209, 256]. Depending on the catalyst and the process conditions, either a methane-rich or a hydrogen-rich fuel gas can be produced from a variety of biomass sources and organic wastes. The residual organics content in the effluent is usually very low (reduction of organic carbon $> 99\%$). The TEES (Thermochemical Environmental Energy System) process developed at Pacific Northwest National Laboratory in the USA has been demonstrated for a variety of organic wastes in a pilot scale reactor. It operates around 350°C and 200 bar and uses nickel or ruthenium based catalysts [209]. Supercritical gasification at pressures around 280 bar and temperatures up to 650°C using a carbon catalyst to yield mainly hydrogen has been the subject of extensive research at the University of Hawaii [26, 256]. Biomass gasification in the absence of a heterogeneous catalyst has been studied at the Forschungszentrum Karlsruhe. Experiments with straw conducted in an autoclave yielded a fuel gas containing 34-42% H$_2$, 0.3 % CO, 40-45% CO$_2$, and 16% CH$_4$ at 500°C, 315-350 bar and 2 hours residence time [203]. Hydrothermal catalytic gasification of wood at feed concentrations up to 30 wt% was studied at the Paul Scherrer Institut in a batch reactor using a nickel catalyst [249]. At 400°C and 270-320 bar, a colorless aqueous phase containing less than 1 wt% of the initial feed carbon was obtained. The highest methane yield was 0.24 g CH$_4$/g wood, corresponding to 67% of the theoretical maximum.

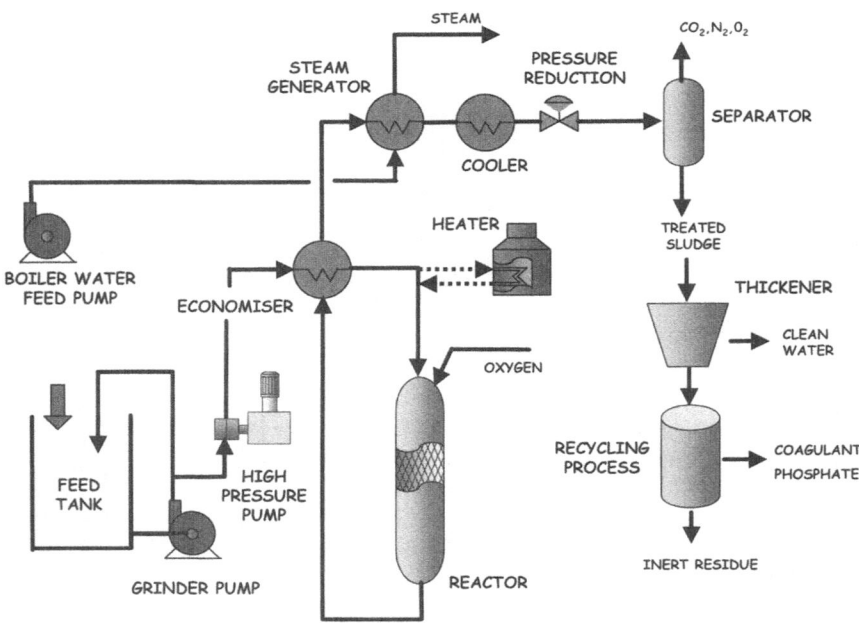

Fig. 5.25. *Above*: Schematic process flowsheet for the supercritical water oxidation of sewage sludge (Aqua Reci™); *Below*: Sewage sludge before and after treatment. Reproduced with the permission of Chematur Engineering AB and Feralco AB, Sweden

Hydrothermal liquefaction of lignocellulosic biomass has been proposed by Goudriaan and Peferoen [86]. Wood chips are treated at 300-350°C for 5-15 min. without a catalyst, releasing CO_2 to produce a "biocrude" with a lower heating value of 30-35 MJ/kg. Further hydrodeoxygenation and thermal separation yields gasoline, kerosene and gas oil. The overall thermal process efficiency is 67% (based on the wood's lower heating value). Recent advances in process design allow cost projections for the biocrude of 4.6 USD/GJ for zero-cost feedstocks [87].

Swiss MSW consists on average of 28% paper and cardboard, 23% vegetable matter, 14% plastics, 11% composite materials, 6% mineral materials, 5% wood, 3% metals, 3% glass, 3% textiles, 5% fines (see section 2.1.2, Tab. 2.4). Based on results from hydrothermal biomass and waste conversion including sewage sludge [157], plastics [257], tire waste [164], rubber [72], nylon [218], and de-inking solid residue [73], conversion of MSW using hydrothermal technology such as SCWO, hydrothermal gasification, or hydrothermal liquefaction is conceivable and would have several advantages over the state-of-the-art incineration technology. In SCWO, about 84% of the MSW would be converted to CO_2, H_2O, N_2, and inorganic salts. Glass, ceramics, and other mineral matter would leave the SCWO process mostly chemically unchanged, but trapped as a slurry. Metals would be converted to their insoluble oxides or carbonates [73]. The aqueous effluent would have a very low organic carbon content, but it would contain chloride and other halides, sulfate, ammonium, and bicarbonate [157]. Heavy metals from sewage sludge were found to be transferred up to 99% into the solid residue [157]. If required, an additional polishing step (e.g. ion exchange) for the aqueous effluent would yield nearly clean water.

SCWO is able to destroy even the most hazardous substances in a self-contained process. In late 1996, the U.S. Army recommended SCWO for the treatment of the chemical neutralization product of the nerve gas VX [213]. For this particular feed, very high destruction efficiencies of 99.9999% had to be achieved [97]. Formation of dioxins from precursors such as phenol has been observed at relatively low temperatures (300-420°C, [237]). At typical reaction severities (> 550°C) proposed for commercial applications, however, dioxins can be effectively destroyed [191, 211, 230, 258].

A promising SCWO reactor concept for handling corrosive and salt-producing wastes is the transpiring wall reactor (TWR). Foster Wheeler Corp. has commissioned a 1 ton-per-day installation treating hazardous materials from the U.S. Navy. The operating conditions are 600-820°C and 241 bar. Air is used as the oxidant. Liquid effluent TOC levels were below 3.5 ppm; NO_x and CO were below 25 and 100 ppm, respectively. Processing chlorine and fluorine-containing organics did not lead to corrosion of the reactor [60].

At high feed concentrations and appropriate reactor conditions, hydrothermal flames will occur during SCWO. Ebara Corp. investigated hydrothermal flames during SCWO of isopropyl alcohol using air as the oxidant [211]. Flame temperatures higher than 1100°C were already achieved at moderate feed concentrations. At an air equivalence ratio of 1.1 and a TOC concentration in the feed of 24 g/L, no soot, CO or NO_x could be observed in the effluent stream. A test conducted with a mixture of different dioxin isomers achieved greater than 99.9% destruction efficiency.

The advantages of MSW incineration by SCWO can be summarized as follows:

- Very clean effluent streams: little or no NO_x, SO_x, soot, dust, dioxins in flue gas, water stream with very low organic content, inert ash containing metal oxides and carbonates and other insoluble compounds such as phosphates, sulfates and aluminosilicates;

- Simple exhaust gas cleaning: Most compounds such as HCl remain in the liquid phase; neither soot nor dust is generated;

- The CO_2 can be separated from the pressurized flue gas by liquefaction for sequestration or use in industry;

- Self-contained system (no flue gas) if oxygen is used as oxidant and CO_2 is liquefied;

- Small footprint of the plant due to small equipment size and simple exhaust gas cleaning;

- Well suited to wet organic waste and hazardous organic waste;

- Heavy metals are converted to insoluble oxides or carbonates.

Hydrothermal gasification can produce either a methane-rich or a hydrogen-rich fuel gas, depending on the temperature and the use of a catalyst. Equilibrium calculations teach us that methane is formed preferably at lower temperatures whereas the hydrogen concentration increases at higher temperatures. A calculation performed for a mixture consisting of 30 wt% MSW (dry matter; composition see section 2.1.2, Tab. 2.5) and 70 wt% water at 420°C and 300 bar yields a fuel gas containing 57% CH_4, 40% CO_2, 2% H_2, and 1% other components. For the same mixture at 600°C and 300 bar, the gas composition is 46% CH_4, 37% CO_2, 15% H_2, 1% CO, and 1% other components (dry basis). Gas cleaning is expected to be much simpler than in conventional gasification due to the absence of tars, soot, and low concentrations of CO. Heat for the gasification can be supplied by partial oxidation in the reactor (autothermal operation) or by using an externally fired heater (allothermal operation). General Atomics [114] reports on experimental results and the design of a pilot plant for the supercritical partial oxidation of organic slurries to hydrogen. Feeds with up to 40 wt% solids can be pumped to high pressure. Bench-scale experiments at 650°C and 235 bar with sewage sludge and MSW yielded 98% carbon gasification.

For the economic production of electricity from MSW, the following routes are proposed:

- SCWO of the MSW with steam generation and subsequent electricity production in a steam turbine (large-scale application);

- Hydrothermal gasification of the MSW to a clean fuel gas (methane- or hydrogen-rich, free of CO and tars) and subsequent production of electricity in a fuel cell (small-scale application) or in a gas engine, gas turbine, or combined cycle (large-scale application).

Combustion of MSW via SCWO to produce steam for a steam turbine is an option if high feed concentrations are used and the heat from the reactor can be removed and transferred efficiently to the steam generator. The highest efficiencies for electricity production will be reached when using a supercritical steam cycle.

Hydrothermal gasification and subsequent conversion of the clean fuel gas into electricity in a fuel cell or using a combined cycle (gas turbine plus steam turbine) would exhibit a high efficiency for power production. Lin *et al.* [136] have proposed a hydrothermal route to convert carbonaceous feedstocks to a clean hydrogen-rich fuel gas. They converted coal and water at 700°C and 100 bar to a fuel gas containing 80% H_2, 19.5% CH_4 and 0.5% N_2. CO_2 was absorbed *in-situ* by adding CaO that was regenerated by separating the $CaCO_3$ and the char formed in the gasification reactor and burning the char. A cold-gas efficiency exceeding 90% was calculated from a mass and energy balance.

Technological challenges for a successful demonstration of energy generation from MSW using hydrothermal technology are:

- Reactor/injector design allowing for high solids loading without plugging;

- Efficient heat management for processing wastes at a high feed concentration;

- Development of stable catalysts for the hydrothermal gasification;

- Feed preparation (size reduction, equipment for high-pressure feeding);

- Preheating (scaling, corrosion);

- Depressurization (erosion of control valves).

Feed preparation is crucial for the smooth operation of a continuous hydrothermal system. Reduction of the feed particle size must take into account the diversity of the waste constituents (brittle, fibrous, tough, hard, elastic, etc.) encountered in MSW. The size of the largest particles must be adapted to the size of the smallest restriction in a hydrothermal system. This is typically found in pump check valves and control valves. Feed preparation could also include a digestion in hot water of around 230°C and 180 bar, as proposed by Goudriaan *et al.* [87] for the liquefaction of biomass. During this treatment, lignocellulosic material is transformed into a paste-like substance that can be pumped to high pressure easily.

Hydrothermal Processing of Electronic Scrap

Hydrothermal technologies, as described in section 5.1.4, can also be used to treat electronic scrap. Hirth *et al.* [102] oxidized shredded printed circuit boards in supercritical water and found that the brominated epoxy resins could be converted to CO_2, water and bromide. Glass fibers and residual metals were not changed in the oxidation process. During studies on the SCWO of the flame retardant tetrabromo bisphenol A, no formation of bromine, hydrogen bromide and dioxins could be observed. Heck and Schweppe [98] report on continuous experiments with printed circuit boards. Before SCWO the printed circuit boards underwent mechanical treatment to separate most of the metals. The remaining solids had a particle size smaller than 100 μm and contained 4.6 wt% bromine. SCWO at temperatures of 500-600°C in a tubular reactor yielded destruction efficiencies for the organic matter of 99.3% and higher. In contrast to Hirth et al.'s [102] findings, Heck and Schweppe recovered about 60-80% of the bromine as bromine gas, Br_2. Chien et al. [53] oxidized printed circuit board wastes in supercritical water with and without added NaOH. At 420°C, 250 atm and 10 min. residence time they obtained a carbon conversion of 87% without and 99% with NaOH. With the added base, 63% of the bromine was recovered in the liquid phase, compared to 40% without base. No bromine was found in the solid residue for both cases. DaimlerChrysler has studied several options for the treatment of electronic scrap. Their evaluation resulted in the development of an SCWO process [173] exhibiting several advantages over incineration, pyrolysis and gasification.

Hydrothermal technology offers a range of potential advantages over state-of-the-art MSW incineration. These include simple gas cleaning, destruction of hazardous compounds, clean effluent streams, inert solid residue, small plant footprint, production of liquid and gaseous fuels with a high efficiency. Whether these advantages will also lead to better process economics and better public acceptance cannot be answered at this time. Hydrothermal technologies have, in the past, focused on liquid wastes with little or no solids. Hydrothermal treatment of MSW requires new processes able to handle high solids concentrations and exhibiting a high reliability to compete successfully with the established incineration processes. Once the technological and economic barriers are overcome, hydrothermal processes will have a high potential to contribute significantly towards a sustainable use of MSW and other wastes.

5.2 Optimizing Incineration for Heavy Metal Recovery

According to Swiss Waste Regulations [16], bottom ash (BA) has to be stored in expensive, sealed reactor deposits. Filter ash (FA) is classified as hazardous wastes and needs to be cemented to stabilize heavy metals or it must be stored underground in abandoned salt mines. BA is by far the largest solid residual fraction from MSW incinerators. About 210 to 350 kg BA and 25-35 kg FA is produced from 1 ton of solid waste. Landfills containing these residues exhibit a potential risk for polluting ground water (see section 3.4).

A simple optimization of conventional incineration (grate furnace) could improve the quality of the residues. The heavy metal contents in FA are much higher than in BA (see section 2.1), due to the fact that most heavy metals or their compounds are volatile at incineration conditions. Therefore, the aim for optimizing the incineration is to reduce the heavy metal contents in BA by choosing the best operation conditions during the incineration process for the volatilization of the heavy metals. Thus, the heavy metals are concentrated in the FA. In a subsequent step the FA can be treated for heavy metal recovery. Some of the heavy metals are toxic substances (Cd, Hg), others are valuable goods (Fe, Mn,) and some belong to both classes (Zn, Pb, Cr, Ni, Cu). Detoxified BA and FA can be further used as secondary raw materials (see section 3.5). To improve the incineration process it is necessary to make large-scale experiments (see section 3.3). However, such investigations are very costly and time consuming. Section 5.2.1 shows how the evaporation process on a grate furnace can be simulated in simple laboratory experiments. In conjunction with thermodynamic modeling, the volatilization processes of heavy metals can be understood and recommendations for improving the incineration process can be given. To further separate the heavy metals from FA, new processes such as the Fluapur process, have been developed (section 5.2.2). However, the separation of copper does not seem feasible following the strategy of volatilizing, and the operation with hydrochloric acid in the Fluapur process puts high demands on the facilities. Today, Automobile Shredder Residues (ASR), which are mainly the organic fraction from the car shredder process, are still mixed and treated together with MSW. New solutions are being developed [198, 210]. Switzerland has decided to separately treat MSW and Automotive Shredder Residues (ASR) in the future. The Reshment technology, presented in section 5.5.1 is currently being planned for construction in Switzerland. An interesting side effect of this technology is that FA can be treated simultaneously. The FA get treated such that a inert glass and Zn/Pb concentrate are produced, which can further be used in the metallurgy industry. However, Reshment only partly solves the problems of MSW incineration. Therefore, an integral solution has still to be found, which will allow for the separation of the most abundant (Zn, Pb, Cu) and toxic (Cd, Hg) heavy metals. A new technology named PECK, using only conventional and established units and facilities, is proposed in section 5.2.3. PECK also treats the copper present in MSW using a mechanical post treatment of the solid residues.

Section 5.2 illustrates the possibilities for heavy metal recovery by optimizing today's incineration process and by combining present technologies. The financial scope for operating new or adapted technologies is very small, i.e. additional costs have to be covered by the benefit of avoiding landfill costs.

5.2.1 Heavy Metal Volatilization During MSW Incineraton

Jörg Wochele, Christian Ludwig, and Samuel Stucki

Transferring Heavy Metals into the Flue Gases

In section 3.3 the functionality of a grate furnace is discussed in detail. MSW incinerators are optimized for volume reduction, sanitizing and inertisation of waste, and for energy production (see section 5.1.1). During incineration, the organics are separated from the inorganics and the gases from solids. The volatilization characteristics of heavy metals differ depending on various parameters, which are discussed below. This section focuses solely on the possibilities for the optimization of the volatility of heavy metals during incineration.

The aim is to transfer the heavy metals entering the MSW incinerator into the flue gases. The result of such an optimization is the reduction of heavy metals in BA. Real or pilot scale experiments are very difficult and costly. An example of a field campaign in which the volatility of Zn was studied using radioactive Zn tracers is presented in section 3.3. Here, a new method is presented which allows for the simulation of the heavy metal behavior during incineration using laboratory equipment.

Production of Inert Materials with Low Heavy Metal Conent

The transfer of heavy metals from modern grid furnaces into the flue gas (including filters, washers and the stack) or to BA are shown in Figure 5.26. The concentration of heavy metals in BA depend on the input composition of MSW and on the applied thermal treatment technology. In Table 5.5, the heavy metal concentration ranges achieved by different alternative plant designs are compared with earth crust and "inert material" of the Swiss classification system. Unlike other countries, in Switzerland residues are not only classified by leaching tests, but also by maximum concentration contents for different elements. If the values are below a certain threshold (Table 5.5, "TVA.inert" [16]) they classify as inert materials, and can therefore be deposited of at low costs. Today, no incineration technologies achieve the levels classified as "inert material" or for earth crust; this is due to excesively high heavy metal content. More than 90% of the world MSWI plants are of the conventional grid type (see section 3.3). Therefore, this section focuses mainly on the optimization of this technology.

Sustainable thermal treatment for reusable products aims to optimize the separation ability by internal measures on the grid and to enhance the evaporation of the heavy metals from the BA. Such measures, if successful, will accumulate the heavy metals in the FA, which has concentrations in the range of ores (see chapter 2.1). The recovery of the heavy metals from FA is discussed in the next section (5.2.2). The mechanisms which control the evaporation of heavy metals on grid furnaces, leading to cleaner BA, have only been poorly investigated until now. In the near future a rapid replacement of conventional incineration by new technologies is unlikely and optimization of the conventional technologies is much more feasible than establishing new systems. New melting technologies are currently under development and some plants are already in operation, though they have not yet entered the market, except for in Japan (see section 5.3). Reasons for this might be high costs and energy consumption.

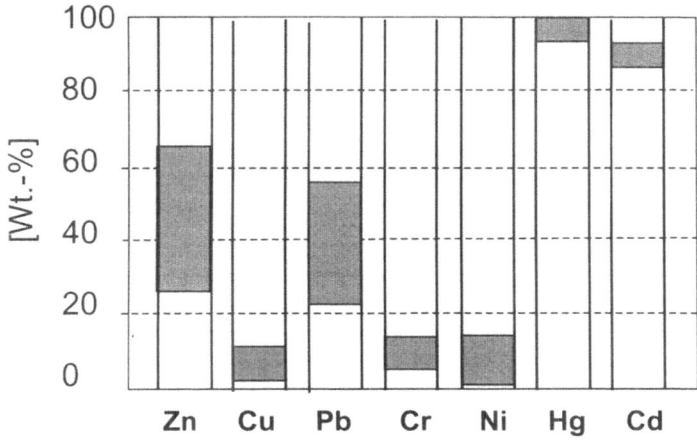

Fig. 5.26. Heavy metal distribution in flue gas above grid (volatility [wt-%]) in conventional MSWI. The remainder is in BA. Range of values of 10 different European (D, A, CH, NL) plants from literature 1988-1998

Table 5.5. Heavy metal concentration in BA of conventional MSWI and in melted granulates of 'new' or alternative technologies compared with earth crust and TVA 'inert' [16].

	Earth crust	TVA inert	conventional MSW bottom ash	Thermoselect melted granulates	Siemens-KWU melted granulates	VS Process melted granulates
Pb	0.013	0.5	0.6–5.2	<0.1	0.8	0.12-0.148
Cd	0.0002	0.01	0.0001–0.082	?	<0.004	0.001-
Cu	0.055	0.5	0.2–0.7	0.4-1.0	1.9	0.0043
Ni	0.075	0.5	0.04–0.76	<0.1	0.64	0.4-1.8
Hg	0.00008	0.002	0.0001–0.02	<0.00005	<0.0005	0.043-0.16
Zn	0.07	1	0.2–0.7	0.1-0.55	2	<0.0005
						0.84-6.8
Ref	[178]	[16]	[178]	[222]	[19]	[132]

Tools for Predicting Volatility of Heavy Metals During Incineration

Simple investigation tools for finding the principal parameters influencing the volatility, and hence the separation capability of heavy metals in grid furnaces, are thermodynamic equilibrium calculations and laboratory scale experiments considering similarity theory.

Table 5.6. Heavy metal concentration in FA of conventional grid MSWI compared with concentrations in ores

	Ore	Filter ash
Pb	50	6-12
Cd	4.4	0.2-0.6
Cu	10	1-5
Ni	4-55	0.2-0.3
Hg	4-55	0.002-0.014
Zn	40	13-39
Reference	[85]	

Estimation of Heavy Metal Volatility: Thermodynamic Calculations

Thermodynamic calculations were carried out [252] using a composition of average Swiss municipal solid waste (Table 5.7). Calculations were performed for reducing atmosphere ($\lambda = 0.5$, gasification), which is typical for conditions during early stages of the incineration process, and for oxygen rich atmosphere with an air ratio of $\lambda = 1.5$ (combustion), which is representative for the conditions found in the second halve of a grid furnace.

Table 5.7. Elemental analysis of average Swiss municipal solid waste [37, 38] and composition of synthetic waste

Element	Swiss MSW [g/kg(wet)]	Synthetic waste [g/kg(wet)]	Element	Swiss MSW [g/kg(wet)]	Synthetic waste [g/kg(wet)]
C	370	323			
H	40	45	Zn	1.40	1.40
O	200	178	Pb	0.70	0.70
N	5	0.4	Cu	0.43	0.43
S	1.3	n.v.	Cr	0.06	n.v.
Cl	7.1	5.0	Ni	0.04	n.v.
P	0.6	n.v.	Cd	0.02	n.v.
F	0.19	n.v.	Hg	0.003	n.v.
Si	37.0	37.0			
Fe	28.0	16.0			
Ca	27.0	24.0			
Al	10.0	14.3			
Na	5.7	n.v.			
Mg	4.2	n.v.			
K	2.4	n.v.			
Moisture	250	250			
Ash	146	199			
Lower Heating Value [MJ/kg(wet)]	12 (ca.)	14.2			

n.v. = no value

Thermodynamic equilibrium calculations indicate the chemical reactions possible under given conditions (temperature, pressure, chemical composition). They describe the stable products (species) which will theoretically be formed, without a detailed knowledge of the complicated chemistry of the multi-component system of waste. Such calculations are in agreement with the laboratory findings and the observation in practice [253]. The calculation results have turned out to be plausible and are in agreement with the evaporation behavior of heavy metals found in MSW incinerators (compare Fig. 5.26 and Fig. 5.27). The calculations indicate a potential for removing (evaporating) the most problematic toxic heavy metals by choosing the appropriate set of parameters (Figure 5.28).

For example: Zn should evaporate under reduced conditions (λ=0.5) in the incineration temperature range of 850-1000°C; Oxidizing conditions and chlorine seem to be favorable for Cu evaporation; Pb should evaporate under reducing and oxidizing conditions. The fact that Pb belong to the intermediate group is explained by the condition that Pb probably melts during heating and flows down the waste into cooler zones where it probably also binds with ashes.

For Cd and Hg easy evaporating is indicated. Cr and Ni as non-volatile metals are hard to remove from BA.

Simulation of Incineration Process in the Laboratory

The design and interpretation of *laboratory experiments* were based on similarity considerations. *Similarity laws* allow one to carry out small scale experiments and to transfer the results to real plants. The model laws for chemical and thermal similarity between full-scale incinerators and a model system were used considering fluid dynamics (Navier-Stokes equations), heat transfer (heat conduction, convection, heat radiation and diffusion) and chemical processes (heterogeneous reactions, thermodynamic equilibria) [254].

The incineration process in a MSW grate furnace was simulated in a quartz-glass tube (QRR) of 70 mm diameter (Fig. 5.28). Samples of synthetic waste in an alumina crucible were subjected to the same temperature and air supply profile as in the solid waste bed of a real MSW incinerator. A standard combustion procedure of 22 min. was established in order to simulate the 100 min. residence time in a MSW plant [253]. Model experiments consisted of 3 min. drying at 120 °C, 3 min. pyrolysis from 300 to 900 °C, 10 min. combustion at 900°C, and 6 min. of afterburning from 900 to 400 °C (Fig. 5.28). The synthetic waste was made up of 54% sawdust, 26% lava, 19% polyethylene and 1% PVC and approximated the average elemental compositions of reference MSW (Table 5.7). The concentrations of Zn, Cu, and Pb as metals and oxides corresponded to average values of MSW. The solid residues of each incineration experiment were analyzed for their heavy metal content by HNO_3/HF pressure digestion and ICP-OES (Inductively Coupled Plasma Optical Emission Spectrometer). The results were used to calculate the transfer coefficient to raw gas (volatility coefficient) for the relevant elements.

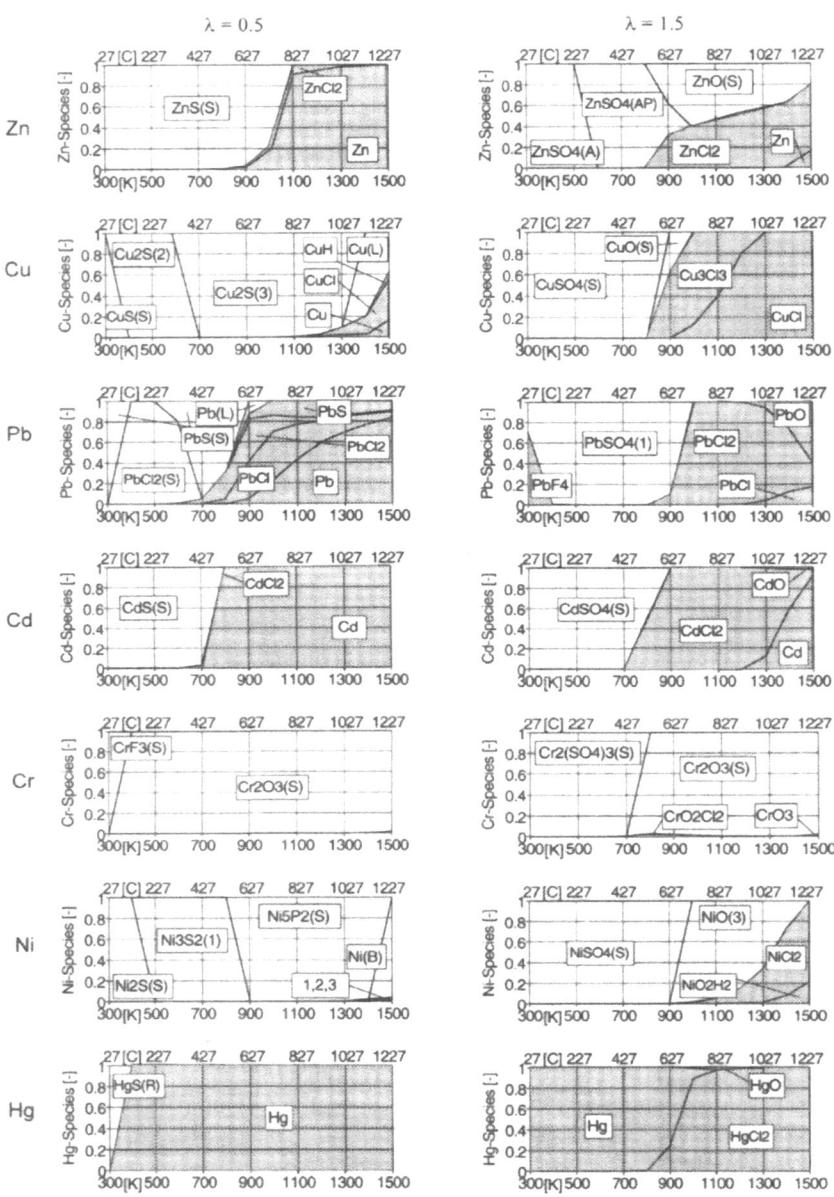

Fig. 5.27. Thermodynamic calculations for typical Swiss municipal waste for gasification (λ =0.5) and combustion (λ=1.5)

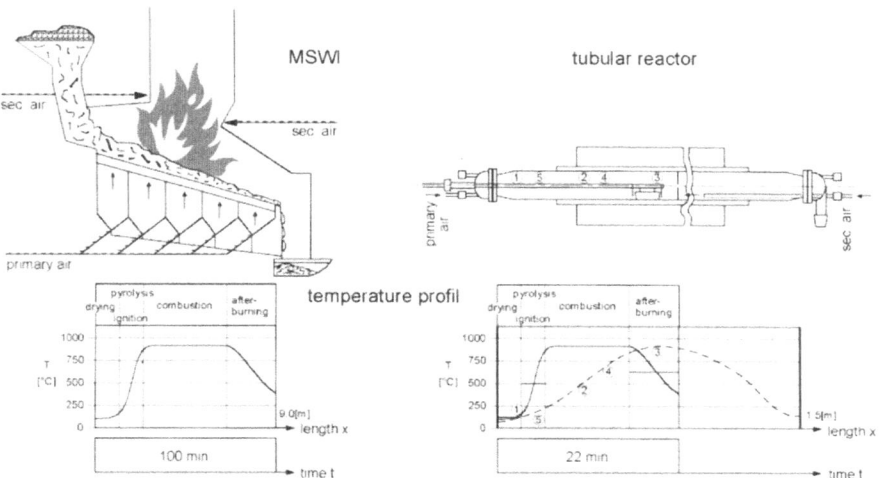

Fig. 5.28. Comparison of MSWI grid furnace and quartz-glass tubular reactor model. In the model, a crucible takes the place of the grate. It is placed at position 1-.5 for an adapted time interval in order to simulate the incinerators temperature profile during an experiment. Reproduced from [254] with permission from Wiley-VCH

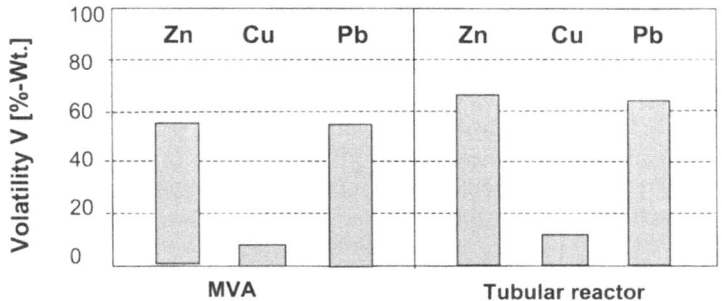

Fig. 5.29. Comparison of volatility of a MSW incinerator [37] with model experiments [254]

Figure 5.29 compares the experimental determined volatility of Zn, Cu, Pb with data of a material flux analyses from a Swiss MSW incinerator plant [141]. The typical behavior of the elements is clearly simulated and offers confidence that such experiments can be used to model, to a certain extent, the evaporation behavior of metals in waste incineration processes [253].

Parameters Influencing the Volatility of Heavy Metals

The toxic heavy metals can be divided into three groups: volatile heavy metals (Hg, Cd), intermediate group (Zn, Pb), and non volatile metals (Ni, Cr, Cu). These groups are clearly seen in Figure 5.26 (real plants), thermodynamic calculations (Fig. 5.27) and laboratory experiments (Fig 5.29). The following parameters have been found to influence the volatility:

Influence of temperature level: Conventional MSWI operates between 750°C and 1000°C. Infrared cameras allow for the measurement of temperatures above the grid (see section 3.3), however, detailed analyses of temperatures in the bed are not available. A comparison of model experiments between a combustion temperature of 830°C and 930°C clear shows an improved evaporation of heavy metals for higher treating temperatures. For Zn results in this temperature range, a better volatilization of 1-2% per 10°C higher temperature, for Pb ca. 1%/10°C and for Cu 0.5%/10°C, was achieved [255].

Influence of temperature profile: The operation of MSWI depends on manufacturer and operating team. Some run it the "soft burning" way, i.e. with an equal temperature level during whole burning zone, others do "hard burning", i.e. with a marked temperature peak at the beginning of the combustion zone, using a lot of primary air and a deep end temperature at the end of the combustion. The volatility of heavy metals decreases by "hard burning" because of the short reductive zone and lower mean temperatures [255].

Influence of residence time: In MSWI the residence time is optimized for good ignition loss. An overlong residence time (e.g. double) has a positive influence on heavy metal separation (clearly for Zn and Pb, less pronounced for Cu). But this operating mode reduces the throughput of the plant and the heavy metals can nevertheless not be totally removed. The long reductive phase probably makes the improved evaporation possible.

Influence of waste composition: MSWI must cope with different wastes. Synthetic waste with less plastic, hence less carbon (C=31.7%) but more ash (25.9%), provides lower Zn and Cu and equal Pb evaporation in comparison to standard MSWI. The results are attributed to an improved binding of heavy metals in the ash under these richer oxygen conditions.

Table 5.8. Compilation of the parameters influencing the heavy metal separation during MSW incineration. Arrow up denotes an increase in volatility. Turbulence and pressure are important parameters, which cannot be simulated in the presented equipment.

Factor		Influence on volatility		
		Zn	Cu	Pb
T Temperature	↗↗	↗↗	↗↗	↗↗
T Time	↗	↗	↗	↗
T Turbulence		n.i.	n.i.	n.i.
p Pressure		n.i.	n.i.	n.i.
c Concentrations				
O$_2$ reducing atmosphere		↑	(→)	↑
Cl	↗↗	↑	↑	↑
Heavy metal content	↗↗	↘↘	↘↘	↘↘
Si, Al, Fe (Ash content)	↗			

n.i. not investigated; Arrow up means increase, raise.

Influence of reducing atmosphere: Conventional MSWI operate with excess air (λ=1.6-2) in order to ensure complete oxidation of the organic fraction. For the experiments in reducing atmosphere, a nitrogen atmosphere was maintained during the first half of the residence time and then air was added for total oxidation. Reducing conditions enhance the evaporation of Zn and Pb significantly (Fig. 5.30, B), but have practically no influence for Cu, which is in agreement with thermodynamic calculations.

Influence of chlorine: Chlorine is known to assist the evaporation of metals. PVC and NaCl in waste contain chlorine. Experiments with excess chlorine in various forms, such as HCl, NaCl, KCl, PVC, were carried out [255]. In Figure 5.30, data are shown with HCl in the primary air.

The most promising measures for heavy metal separation are by reducing atmosphere and chlorine addition (Table 5.8). A comparison of the results of model combustion experiments showing the potentially achievable heavy metal separation is compared with the targets provided by Swiss law and the quality of earth crust materials. The standard case (A) compares well with the reference of average bottom ash from Swiss MSW incineration (a). For Zn and Pb, the limits of TVA inert could be reached by reducing atmosphere (B); and by HCl addition, even earth crust quality seems possible. With heavy doses of HCl (ca. 10 %), nearly all of the heavy metals can be transferred to the flue gas (Fig. 5.30, C).

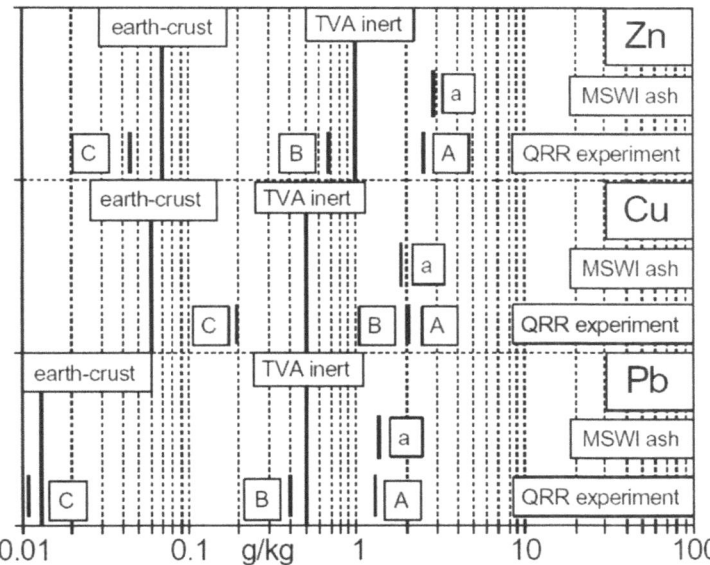

Fig. 5.30. Concentration [g/kg dry ash] of heavy metals in MSWI bottom ash and lab experiments compared with limits of earth crust and inert material [16]. MSWI ash: a Swiss reference. QRR experiments: A) standard case, B) reducing atmosphere, and C) excess HCl in primary air

Conclusions

Thermal treatment is a proved method for treating MSW. Today's incinerators could be improved for a more environmentally sound process by the optimization of heavy metal separation. However, not all heavy metals are volatile enough under incineration conditions to be volatilized. E.g. Cu cannot be separated except with high levels of chlorine in the primary air. But this does not seem feasible, because of problems with corrosion in a real plant. Separation of Cu should be performed by a mechanical (section 5.2.3) rather than thermal process. However, Cu can also be separated from BA using melting technologies [260]. Simulation and modeling allow one to investigate the volatility behavior of metals in thermal processes. The method is simple and the effects of different parameters on the volatility behavior can be tested much more easily, more quickly and more cost-effectively in comparison to real scale experiments. Still, the method has to be verified; here, e.g. shown in Figure 5.29 or 5.30, thermodynamic equilibrium calculations and laboratory experiments based on similarity proved to be useful tools for validating the heavy metal separation behavior. Simulation experiments and thermodynamic equilibrium calculations, however, cannot replace full scale testing.

The heavy metal separation of MSW incinerators can be improved by maintaining a reducing atmosphere or with HCl. HCl is a rather aggressive substance and corrodes firebricks and boiler tubes, cutting down their life span. Nevertheless, MSWI plants have always to cope with HCl, which comes from the waste input. Using internally recycled HCl from flue gas scrubbing systems allows for enhanced heavy metal recovery without changing the chlorine balance in the environment. The possible higher dioxin formation is regulated in modern plants by the already existing catalytic converters or carbon boxes.

Cu as a non-volatile element is not easy to remove from bottom ash. Only massive HCl-application (ca. 10%) can evaporate Cu species in a significant manner. A more promising solution here is a mechanical after treatment (sieving) of the bottom ash [44], as discussed in section 5.2.3.

Zn and Pb can clearly be volatilized by reducing atmosphere and HCl even enhances the volatility of Cu. It can be concluded that the Swiss inert specification should be attainable by optimizing current incineration technology [140]. In section 5.2.3, an example of one such adapted plant is presented.

5.2.2 Detoxification of Filter Ash

Christian Ludwig, Harald Lutz, Rudolf P.W.J. Struis, and Jörg Wochele

Introduction

During MSWI the organic fraction is separated from the inorganics quite thoroughly and the gas cleaning system properly removes toxic substances, but heavy metals still cause problems when deposited in landfills (section 3.4). They may be removed from the residues by optimizing incineration (see section 5.2.1) and applying an additional treatment to the FA, where the heavy metals are enriched.

Even without further enrichment the concentrations of the heavy metals in FA, their concentrations are much higher than in BA (see section 2.1). For reasons that were discussed in section 3.4, the separation of heavy metals is preferred to their inertization. Acid washing is possible, but is by no means complete and dioxines present in FA are thus not destroyed. Therefore, in the past different processes were developed for separation and/or the inertization of the heavy metals in FA (and BA), i.e. for increasing leaching resistance of the heavy metals. These processes, such as those developed by ABB (Deglor [236], AshArc) or RedMelt [74], are all based on high temperature melting technologies. However, radically new technologies have a difficult position in the market of large-scale plants. Only in Japan were the melting technologies able to break into a new market (see section 5.3).

It is therefore proposed that future development should focus on processes which are either based on existing equipment, or which can be integrated into existing plants as additional plants. Thermal treatment is also recommended for destroying toxic dioxins present in FA. But, it was found, that for the separation by evaporation of the heavy metals, melting of the ashes is not preferable and that best results were found at high temperatures just below the melting point of the ashes [107, 110].

Studying the Evaporation of Heavy Metals from Filter Ashes

A new tool was developed [139, 142, 143, 146, 240] to analyze volatilized heavy metals entrained in hot carrier gases on-line. The challenge was to ensure a reproducible transfer of these heavy metals from a furnace, where they are evaporated from the ash sample, to a heavy metal detector, an ICP-OES (Inductively Coupled Plasma Optical Emission Spectrometer). an interface was built to establish the connection. The so-called "Condensation Interface" quenches the hot carrier gas to room temperature, which leads to a controlled condensation of the heavy metal vapors. During the condensation, the heavy metals are transferred into aerosols, which can further be injected to the ICP-OES. Figure 5.31 shows the principle of interface, the so-called "Condensation Interface."

Using this tool, the thermal behavior of materials can easily be tested by performing Thermo-Desorption-Spectroscopy (TDS), i.e. heavy metal evaporation into a carrier gas is measured during temperature and constantly increased to about 950°C.

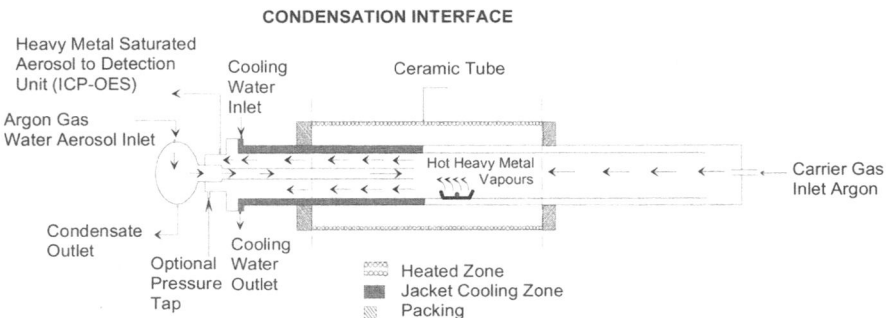

Fig. 5.31. Design of "Condensation Interface": The controlling design is such that the hot gas consisting of the supplied carrier gas (e.g. argon) and the evaporated chemical elements is cooled rapidly in a quench zone by two means. Not only is the outside of the quartz tube in this zone cooled with the aid of a water cooling jacket; the gas flow leaving the hot center of the oven is additionally mixed with a second, cool one. The latter can contain additives which promote condensation (e.g. water aerosol droplets).

Evaporation Behavior of Heavy Metals

Filter Ash Samples. An extensive study has been performed for studying the evaporation of heavy metals from three different filter ashes: BCR, FLU, and FNU. "BCR176", henceforth called BCR, is an official reference material and was certified by the European Community Bureau of Reference Materials in the 1980s [265]. FNU is the dry scrubber residue of the waste incinerator of Niederurnen, Switzerland. FNU samples were collected in 1999 and contain high amounts of calcium (Ca), chlorine (Cl), and sulfur (S). Due to the addition of $Ca(OH)_2$ the other elements were diluted. FLU was also collected in 1999, but at the incinerator in Lucerne. The gas cleaning system in Lucerne is comprised of an electrostatic filter and a wet flue gas scrubber. Due to the existence of a wet gas scrubbing system, the chlorides and sulfates passing through the electrostatic filter are not incorporated in the filter ash FLU. Only subsequently are they removed by a spray tower. Those salts, once extracted from the spraying liquid, are not mixed with the filter ash. A more detailed description of the ashes used is found in [144]. Figure 5.32 shows the elemental composition of the three ashes investigated.

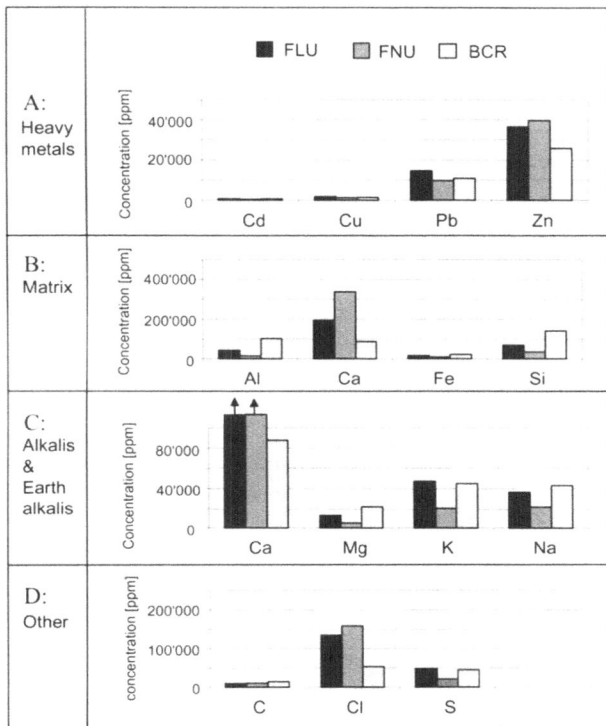

Fig. 5.32. Elemental composition of the different filter ashes FLU, FNU, and BCR [144]

Additives. Heat treatment in an inert gas atmosphere (in the laboratory: argon) is deemed insufficient for the projected heavy metal removal efficiencies (>90 % of Zn, Cu, Pb, Cd) from both the filter ash of a wet flue gas treatment system (FLU, BCR) and a dry scrubber residue (FNU) as first experiments showed.

Evaporation Curves. Several additives were tested for the improvement of heavy metal evaporation. Gaseous additives such as H_2 and O_2 were mixed to the inert carrier gas. Solid additives such as $CaCl_2$, NaCl, $CaSO_4$, or Sewage Sludge (SS) were mixed with the filter ash. Because the thermal treatment of the samples is done by applying a constant heating rate, the temporally resolved peak patterns provide valuable information on rates of evaporation (in µmol/min), temperature range of evaporation (300 °C to 950 °C) and evaporation yield (in $µmol_{TOTAL}$). Figure 5.33 shows the results of such TDS experiments for the element Zn. The change in the pattern is a result of the change in chemistry. Figure 5.34 shows that this change in chemistry is also reflected in the total amounts evaporated during an experiment.

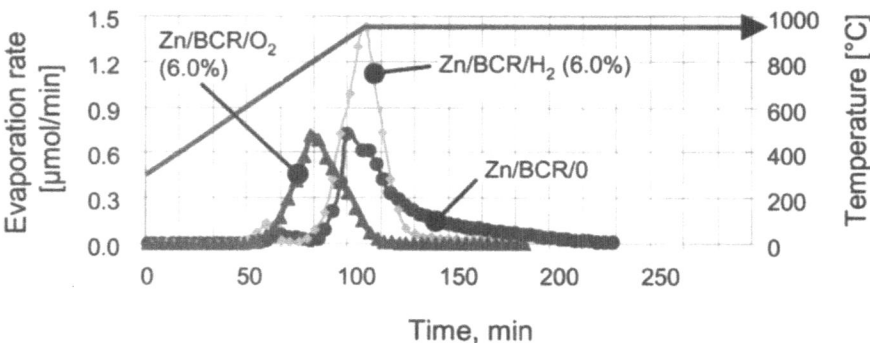

Fig. 5.33. Rates of Zn-evaporation for the ash BCR under reducing and oxidizing conditions. H_2 and O_2 were mixed into the inert carrier gas argon. Zn/BCR/0 is the reference case, i.e. only argon was used as carrier gas. The steepness of the straight the line corresponds to the heating rate.

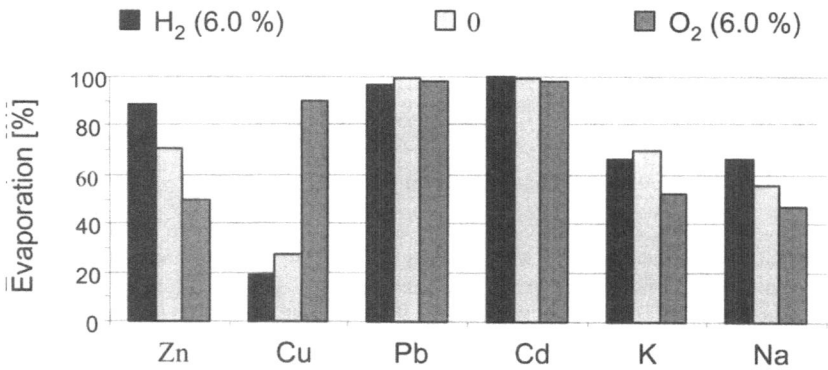

Fig. 5.34. Evaporation of heavy metals from BCR in H_2 enriched, O_2 enriched, and inert atmospheres.

For comparison, the effects on other elements are also shown. It was found that the effect of H_2 and O_2 is reverse for Zn and Cu, and that there is little effect on the total amount Cd and Pb evaporated due to redox conditions. However, it has to be stated that optimum evaporation temperatures may shift for a given element and that the composition of the ash has a strong influence on the evaporation characteristics.

EXAFS (Extended X-ray Absorption Fine Structure Spectroscopy) allowed for the assigning of the structural changes of Zn in the BCR filter ash (Fig. 5.35), [228] during TDS experiments. It was therefore possible to understand the chemical mechanisms responsible for the observed behavior of the Zn during evaporation [228]. A detailed discussion does not fall within the scope of this section, but

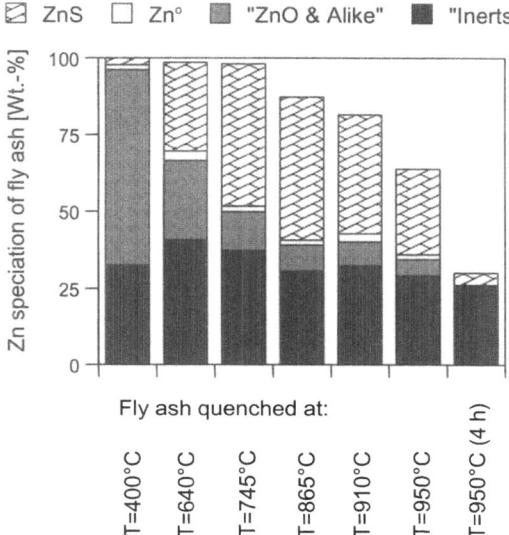

Fig. 5.35. Formation of Zn-compounds in BCR during a TDS experiments. Samples have been quenched after reaching a certain temperature. EXAFS Spectroscopy allowed for the obtainment of information on the structural composition [228]

it should be mentioned that the interplay and competition of sulfur, oxygen and chlorine for Zn is of major importance due to the various evaporation characteristics of the forming Zn compounds.

In Figure 5.36 representative experiments are summarily selected for all three ashes (FLU, FNU, BCR) and all four heavy metals (Zn, Cu, Pb, Cd) in order to describe variations both in percentage and in temperature of maximum evaporation as a function of additive type (gaseous O_2, H_2, and solids $CaCl_2$, NaCl, $CaSO_4$, Sewage Sludge).

Substantial variations regarding the percentage of evaporation can be seen both for Zn and Cu, but are largely absent for Pb and Cd (Fig. 5.36, y-axis). The latter fact has to be viewed in light of the fact that Pb- and Cd-evaporation are generally close to complete, so only decreasing values would be possible for the percentages. Additives mostly improve the Zn-evaporation (with the exception of the ash BCR under oxid176ng conditions), while for Cu extreme differences can be seen.

The additives shifted the peaks of the TDS curves mostly to lower temperature for Zn and Cu, and, by contrast, to higher temperatures for Pb and Cd (Fig. 5.36, x-axis). Gaseous reducing additives are preferable to solid reducing additives because the former may lower the treatment temperature (as found for Zn in FLU) or not reduce the Cu-evaporation as drastically as the latter.

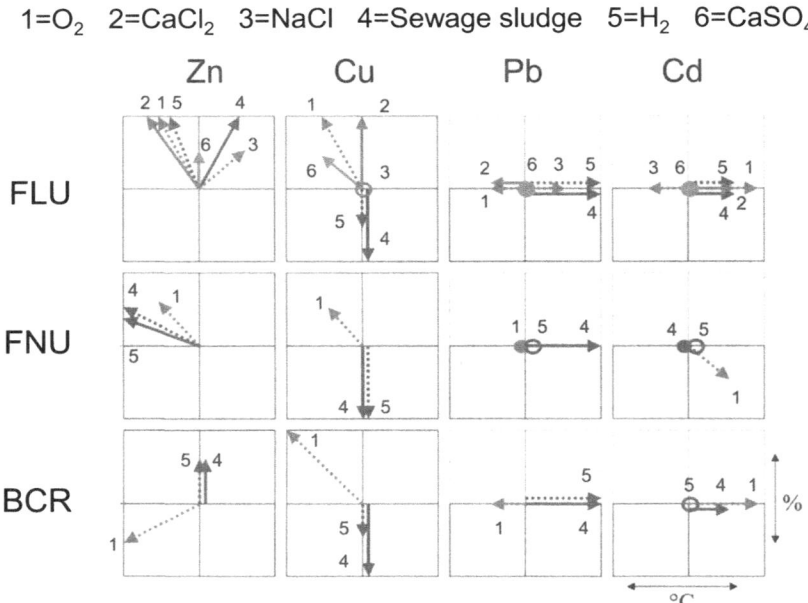

Fig. 5.36. The center of each box represents the case of ash treatment under inert (reference) conditions. Variations along the x-axis denote that the evaporation peak in a TDS experiment was shifted to lower or higher temperatures. Variations along the y-axis denote a lower or higher percentage of evaporation. For an industrial process, the upper left corner is desirable, since more of a given element can be removed at lower temperatures. The lower right is to be avoided because the capacity for removal shrinks and the treatment temperature is higher. The arrows indicate a three steps scale: insignificant changes- less than 5% (circles), significant changes (short arrow), and substantial changes of more than 20% (long arrow).

Unexpectedly, heat treatment in the presence of oxygen is advantageous for both Cu and Zn, except for Zn in the BCR ash. I.e., for Zn the removal might not be sufficiently large depending on the ash type [145].

As was found, a crucial factor is the amount and type of Cl-donor compounds and additional elements like sulfur (S when present as sulfides) already present in the ashes because they compete for the heavy metals and lead to species of different volatility. Cl-addition aids the evaporation of Zn and Cu if a comparatively strong Cl-donor like $CaCl_2$ (instead of NaCl) is taken. Sulfides often prevent or at least retard the heavy metal evaporation, while an addition of $CaSO_4$ is found to both slightly enhance Zn- and Cu-evaporation. Thus the role of sulfur has to be viewed with consideration of its speciation in the sample. The effects observed from the extensive studies on all three ashes have been summarized in Table 5.9.

Table 5.9. Summary of effects on heavy metal evaporation

Actor	Positive effect	Negative effect
O_2	Suppresses sulfide influence Stimulates Cl-donor salts to transfer their chlorine to heavy metals	Produces non-volatile species in case of Cl-deficiency
Cl	Produces volatile heavy metal chlorides	Melting might cause retardation of evaporation
S as sulfide	Leads to volatile Zn despite Cl-deficiency	Inhibits or at least retards the chlorination of heavy metals
S as sulfate[1]	Leads to slight improvements for Zn and Cu upon decomposition by functioning as a weak oxidizing agent	?
H_2/C	Compete with chlorine for heavy metal oxides and sulfides (Zn for FNU)	Compete with chlorine for heavy metal species (Cu in general)

[1] Due to the scarcity of data, the assumptions require further testing [chapter 13.2.].

Case Study: The FLUAPUR-Process

In cooperation with the Paul Scherrer Institut, CT Environment developed the Fluapur concept [254]. Figure 5.37 shows the invented CT-Fluapur process for the thermal treatment of MSW incineration residues. The residues have to be conditioned before thermal treatment, i.e. filter ash has to be granulated. As an option, bottom ash can be treated as well after breaking, iron removal, and granulation. The thermal detoxification of these residues is performed in a fluidized bed reactor at 900 °C. A part of the incineration gas which is enriched with about 10% HCl is used as a reaction gas and simultaneously as a thermal source for the reactor. A natural gas burner can supply additional heat. HCl for the process is obtained from the wet flue gas scrubber.

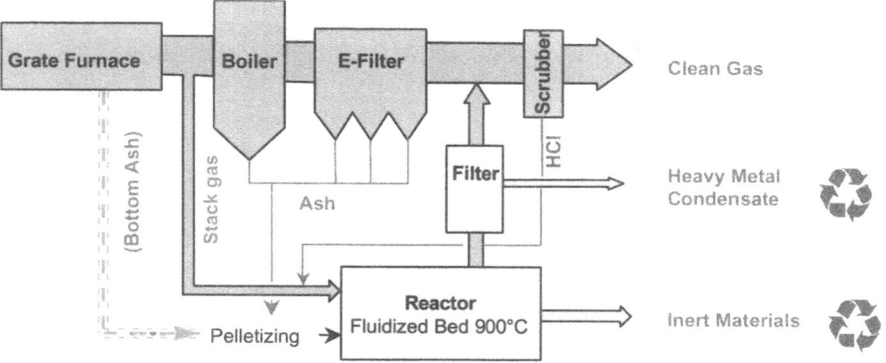

Fig. 5.37. Scheme of Fluapur-process [108]

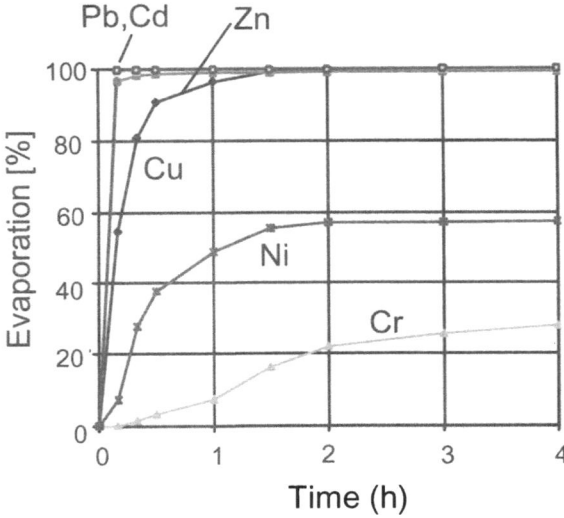

Fig. 5.38. Evaporation of Cd, Cr, Cu, Pb, Zn, and Ni from a mixture containing MSW incineration filter ash and 10% coal. The temperature was kept constant at 900°C and the atmosphere was enriched with 10% HCl. Adapted from [44].

The residues leave the reactor as mineral products poor in heavy metals. The heavy metals that are volatilized in the reactor are collected, after quenching, as pulverized condensate. The off-gas is fed back into the incinerator's flue gas treatment system. The most important requirements for the success of a filter ash treatment process, i.e. the possibility of achieving good product quality at reasonable costs, have been demonstrated.

Figure 5.38 shows the evaporation in % of the total amount of originally present heavy metals as a function of time for a laboratory pilot reactor. Under the given conditions Zn, Pb, Cd, Cu can reasonably be well separated.

The FLUAPUR-Process was optimized for a high efficiency removal of heavy metals from FA. This is necessary if FA is to be purified in a post incinerator process in order to obtain pure materials which might classify as "inert materials."

The FLUAPUR-Process requires special equipment and technological know-how for the handling of HCl. However, once correctly installed into an existing incineration plant, the heat produced during incineration can be used and the size of the reactor can easily be adjusted to allow for the treatment of the FA of several incinerators. The process is therefore expected to be much cheaper than a melting process and much more efficient with regard to heavy metal separation than a conventional acid washing procedure.

5.2.3 PECK Incineration Technology

Serge Biollaz and Rainer Bunge

PECK technology is based on a modular combination of technologies with thermal and mechanical separation steps for the recovery of heavy metals (Fig. 5.39). The remaining systems for a compete incineration plant, such as waste handling, boiler, flue gas purification, power generation, waste water treatment, etc., are comparable to conventional incinerators and not shown in Figure 5.39. The block diagram of a whole PECK incinerator is presented in Figure 5.40 and the material flux analyses in Figure 5.44.

In the PECK process, MSW and sewage sludge are treated such that minerals and metal concentrates (hydroxide sludge, ferrous and copper scrap) are the main solid secondary materials. No hazardous fly ash remains. In this chapter the key elements of the PECK process are described.

To reduce the fixed costs of the fly ash and bottom ash treatment these modules can be designed for a larger throughput than what is generally processed by a single MSWI. In such a case several conventional incinerators would deliver their fly ash and bottom ash to the PECK plant, providing regional optimisation of waste management.

While the PECK process is a combination of the three processes, it is possible to build the bottom ash treatment at a different location than the incinerator and to treat bottom ash from various incinerators at one plant.

The PECK technology is the result of the combined research and development efforts of a research institute and three Swiss companies active in the field of waste treatment. The knowledge from each individual partner was combined and consolidated, and finally resulted in the PECK process. PECK stands for: Paul Scherrer Institut - Eberhard Recycling - CT Umwelttechnik - Küpat.

- The research institute "Paul Scherrer Institut" initiated the co-operation and has assisted the process development scientifically.
- The company "Eberhard Recycling AG" has developed mechanical bottom ash treatment technologies and has operational experience of several years in the treatment of conventional MSWI bottom ash.
- The company "CT Umwelttechnik AG", a member of the Babcock Borsig Power group, has developed a fly ash treatment process "Fluapur" for the detoxification of fly ash.
- The company "Küpat AG" has developed and demonstrated in full scale the MSW incineration process "VS-process."

Until now, PECK technology has been a concept with industrially proven equipment and sub systems. The entire process has not yet been built and put into operation. The next step is to build an MSW incinerator based on PECK technology.

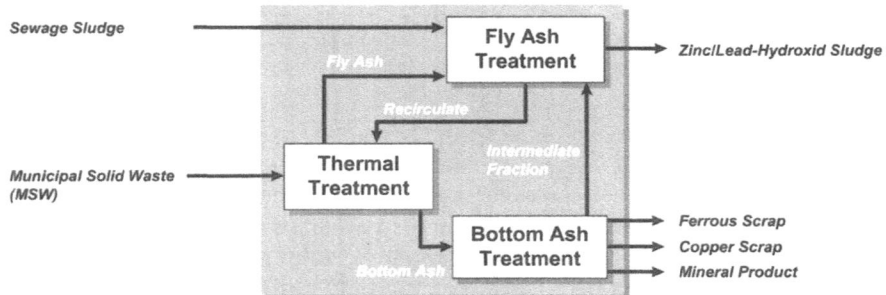

Fig. 5.39. Block diagram of the PECK process consisting of three key processes: thermal treatment, fly ash treatment and bottom ash treatment.

Comparison of Conventional MSWI and the PECK Process

Generally, the objective of a conventional MSWI is to reduce the volume of waste and to produce heat and power without polluting air and water. These two objectives can be reached using the PECK process. At the same time the PECK process produces higher quality residues with the potential of achieving sustainable material management. There are some significant differences in terms of the quality of the solid residues between a conventional MSWI and the PECK. The most important are:

- The bottom ash of conventional MSWI is typically partly sintered and not stable against leaching. The PECK bottom ash is treated at temperatures up to 1300 °C. Detailed investigations have shown that the bottom ash is essentially vitreous and is stable against leaching [238, 239]. Therefore the mineral product can at least be landfilled without any hazard or can be used as building material.
- The vitrification of the minerals is integrated in the thermal process of the PECK process without significantly compromising the useful energy output of the plant. No valuable auxiliary energy, such as oil or electricity, is used for the vitrification.
- The PECK process includes the treatment of the toxic fly ash, which contains most of the volatile heavy metals and dioxins. Depleted fly ash, so-called 'recirculate', is recycled from the fly ash treatment to the thermal treatment and becomes part of the bottom ash. Therefore no fly ash remains and potential long-term environmental hazards are avoided.
- With the PECK process, valuable metals from MSW are recovered and recycled. The PECK process as a whole, the fly ash treatment and the bottom ash treatment, is operated such that the metal concentrates can be recycled to the metal refining industry.
- Due to air staging within the PECK process, the nitrogen oxide emissions in the flue gas are at $150\,mg/m_n^3$, 11 Vol% O_2. On the other hand the staged combustion also leads to a reduced volume of excess air in the flue gas. This

increases the thermal efficiency of the incineration plant and reduces the size of the flue gas purification unit.

To achieve the above-mentioned quality improvement, only modest changes in the operation of an incinerator are needed. In conventional incinerators the objective of the grate is to incinerate the MSW and to avoid unburned carbon in the bottom ash. In the PECK process the objective of the grate is to gasify MSW. This operation mode is comparable to the first stage in "Low-NO$_x$" MSW grate incinerators. The subsequent vitrification of the bottom ash takes place in a rotary kiln which is directly combined with the grate gasifier. The gasification products from the grate are burned with additional combustion air at slightly over stoichometric ratio. This part of the PECK process is comparable to rotary kiln incinerators as used for hazardous waste treatment. Within the vitrification process, any organic material is destroyed in the bottom ash. The staged combustion leads, at the same time, to increased evaporation efficiency of volatile heavy metals. Both the reducing atmosphere on the grate, and the high temperature in the rotary kiln have been found to assist evaporation of heavy metals such as Zn, Pb and Cd (Fig. 5 46).

The evaporated volatile heavy metals are concentrated in the fly ash and recovered as a recyclable metal concentrate in the fly ash treatment. The remaining mineral material which is depleted in heavy metals is recycled to the end of the grate and melted together with the bottom ash in the rotary kiln. Conventional MSWI produce a hazardous fly ash which must be deposited in special dumps, e.g. former salt caverns or roofed dumps in order to avoid water incursion.

The molten bottom ash is quenched in water to form glass-like phases. The resulting bottom ash can be treated later on, as an option, in the bottom ash treatment. In the first step the bottom ash is comminuted. In the second step non-volatile metals, such as copper, nickel and iron, are mechanically separated from the mineral matter. This treatment of bottom ash is quite comparable to the primary ore treatment applied in the metallurgical industry. Some conventional MSWI remove ferrous scrap by magnetic separation, some remove non-ferrous metals by eddy current separation without comminution of the bottom ash before the separation. Fine iron particles are therefore not removed and copper remains in the treated bottom ash. These differences in metal separation efficiency between a conventional incinerator and the PECK process permit the treatment of copper-rich wastes such as ASR or electronic wastes with much less hazards to the environment.

In all other respects there is no significant difference between conventional MSWI and the PECK process: Steam for power is produced in a conventional boiler. The overall electrical efficiency of the PECK process is comparable to that of conventional MSWI including the internal electricity consumption. The PECK process is preferably equipped with a conventional wet flue gas purification unit. For the treatment of the fly ash, HCl is needed, which can be recovered from the washing water. The required infrastructure, such as land and construction volume, is comparable. All these facts result in capital investments for PECK not being much higher than those of conventional MSWI.

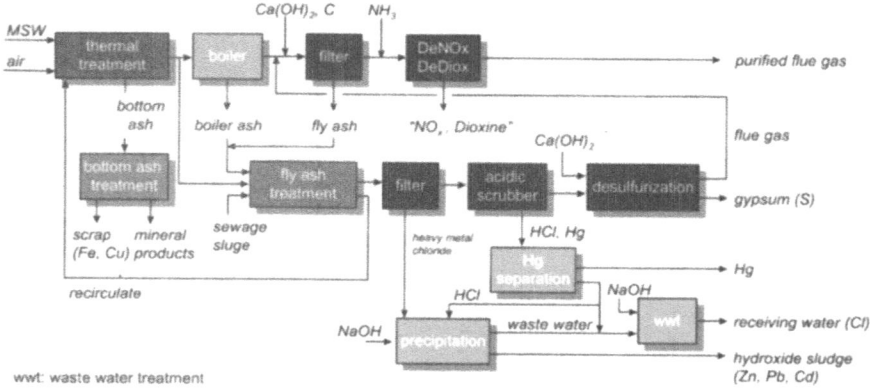

Fig. 5.40. Detailed block diagram of thermo-mechanical treatment of MSW with three key sub processes, which are different form conventional MSW incinerator: thermal treatment, fly ash and bottom ash treatment

The complete block diagram of the PECK process is shown in Figure 5.40 along with the wanted sink for the elements S, Cl, Hg, Zn, Pb, Cd, Cu and Fe. The substantial technical differences in conventional MSWI are the thermal treatment, the fly ash treatment and the bottom ash treatment. In the next chapter these three key processes are described in more detail and the process engineering of the technologies is explained.

Technical Description of Key Elements of PECK

Objective and approach of thermal treatment of municipal solid waste. The thermal treatment of MSW is performed with the VS-process of Küpat AG, Zurich. The objective of the VS-process is to improve bottom ash quality with respect to the heavy metal content and its leachability [41, 131]. The conversion takes place in a so-called VS-combi-reactor, a process combination of a grate and a rotary kiln (Fig. 5.41). This is the original and old Volund configuration, but with a different flue gas flow and different design specifications. In addition, the heavy metal depleted fly ash from the fly ash treatment is recycled to the end of the grate (see section fly ash treatment). With this measure all minerals originally contained in the MSW leave the VS-process as bottom ash. Therefore, looking at the PECK process as a whole, no hazardous fly ash is produced and the volatile heavy metals such as Zn, Pb an Cd are concentrated in the hydroxide sludge (Fig. 5.40).

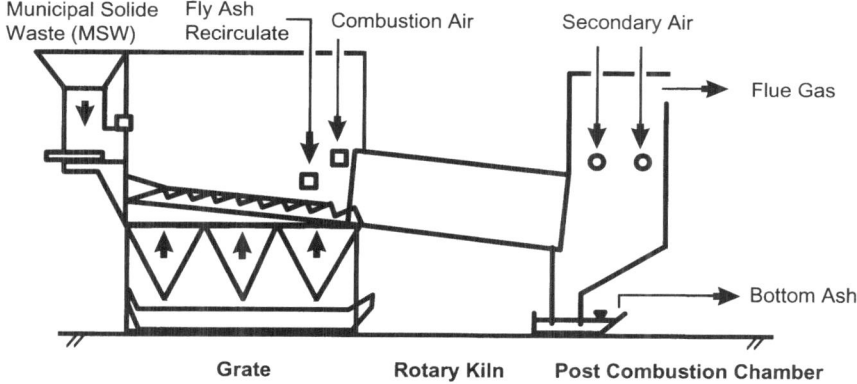

Fig. 5.41. Thermal treatment of municipal solid waste. VS-combi-reactor is a process combination of a grate and a rotary kiln. The most important solid secondary material is vitrified bottom ash

Process engineering of thermal waste treatment. First, MSW is gasified with primary air on a grate. In order to evaporate volatile heavy metals at reducing conditions, a bed temperature of at least 950°C is maintained. From the fly ash treatment, the depleted fly ash is recycled in pelletized form to the end of the grate. This measure increases the chloride content in the bed and improves the evaporation of heavy metals.

Then the products from the grate section, i.e. low BTU gas and coke, are burned in a rotary kiln with additional combustion air. With an excess air ratio from 1.1 to 1.3 and with typical heating values of waste around 12 MJ/kg, gas temperatures of up to 1400°C have been reached without using auxiliary fuels or oxygen enrichment [131]. With such a high gas temperature, a bed temperature of 1300°C is reached. This temperature is sufficient to melt the mineral materials. The remaining finely distributed heavy metals form glasses and stable phases, such as zinc silicates. In contact with water, such phases show very favourable leaching behaviour. Larger metal particles, i.e. non-volatile metals such as copper, nickel and iron, are enclosed in the molten bottom ash matrix [238, 239]. The bed temperature is low enough to prevent iron from melting and forming, for example, unwanted alloys with copper. At the outlet of the rotary kiln, the molten bottom ash is quenched in water to form glass-like phases. The produced bottom ash can be treated later, as an option in the bottom ash treatment, as described later.

The exhaust gases leaving the rotary kiln are mixed with secondary air in a post combustion chamber. This ensures the complete destruction of organic material in the flue gas. The flue gas is treated afterwards in a conventional flue gas purification unit.

The feasibility of the VS-process has been proven in numerous tests with waste throughputs of 8 to 10 t/h. The investigations were performed in the MSWI in Basel (Switzerland), which, until 1995, had an old Volund configuration as an incinerator line [131].

Objective and approach of fly ash treatment. The treatment of the fly ash is performed with the Fluapur process of CT Umwelttechnik, Winterthur. The main objective of the fly ash treatment within the PECK process is to significantly lower the contents of volatile heavy metals as well as to lower the content of dioxins and furans in the fly ash [109]. The motivation for this is twofold. One reason is the reduction of the long-term environmental hazards that may arise from landfills. The other reason is the recovery of valuable metals for substituting natural resources. As a by-product, the fly ash treatment produces a metal concentrate. This concentrate is compatible with established metallurgical processes and can therefore be recycled by the metal refining industry.

The depleted fly ash reaches heavy metal concentrations equal to the respective concentrations in bottom ash. Therefore the treated fly ash, so-called 'recirculate', can be recycled to the VS-process. The mineral material originally contained in the fly ash becomes part of the bottom ash.

Process engineering of fly ash treatment. In Figure 5.42, the flow chart of the fly ash treatment is shown. The fly ash from the boiler and the electrostatic precipitator is mixed with dewatered sewage sludge and fed to a pelletizer. The resulting dry pellets are treated in a fluidised bed reactor, where chlorination and evaporation of the metals take place at 900°C. The evaporation reactor is heated by flue gas from the post combustion chamber. This gas stream has a temperature sufficiently high to operate the reactor. Only a small portion of the total flue gas is used, i.e. approximately 2%.

The evaporated metals leave the fluidised bed reactor together with the flue gas. By a partial quench, the heavy metals are condensed and filtered afterwards. After that filter, surplus hydrochloric acid is separated from the flue gas in the scrubber. The flue gas is then fed back to the scrubber of the incineration plant for final cleaning and leaves the plant via the stack. Hydrochloric acid is produced in an internal loop from the washing water of the scrubber of the fly ash treatment by a distillation system. Hydrochloric acid make-up comes from the scrubber of the incineration plant. The depleted washing water is used as quenching water.

The depleted fly ash, the so-called 'recirculate', is removed from the evaporation reactor and fed back through a buffer silo to the VS-combi-reactor. The filtered heavy metal concentrate is collected in Big Bags and transported to the zinc and lead refining industry.

The system for the fly ash treatment consists predominantly of standard components. Both pelletizer and dryers are established components from the chemical industry. The gas cleanup is based on conventional components which are used in similar systems. The distillation system for hydrochloric acid has been a standard process in the chemical industry for years.

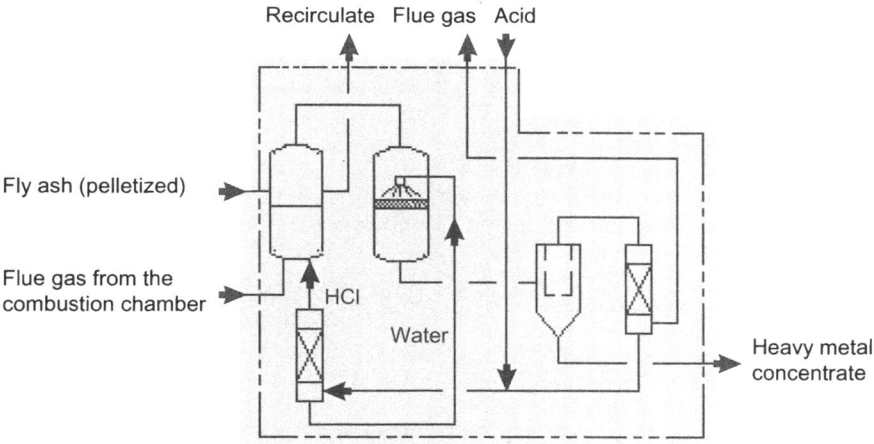

Fig. 5.42. Flow chart of fly ash treatment Fluapur. The fly ash is detoxified at 900°C in a fluidised bed reactor

Objective and approach of bottom ash treatment. The treatment process for the bottom ash is based on technology developed and operated by Eberhard Recycling, Kloten. The objective of the bottom ash treatment is the separation and recovery of metallic species of non-volatile metals from the bottom ash. The motivation for this is twofold. One reason is a reduction of the long-term environmental hazards that may arise from heavy metals in bottom ash landfills. The other reason is the recovery of valuable metals for the conservation of natural resources. The treated bottom ash can either be used as building material, or deposited without hazard.

The process for the extraction of the metals from the bottom ash consists of various stages of comminution and separation. One important sub-process is preferential breakage. The fact that the glass-like matrix of bottom ash is brittle, and the metal particles therein ductile, can be exploited. The matrix is fragmented upon mechanical stress, while the ductile metals are merely deformed. When such a product is sieved, the oversize will essentially consist of metal particles while the undersize will be fine-grained mineral matter. The quality of the recovered heavy metal concentrate is of smelter grade. Consequently, the recovered metals can be recycled.

Process engineering of bottom ash treatment. In Figure 5.43 the flow chart of the bottom ash treatment is shown. Directly after the high temperature process, the bottom ash is quenched in a water bath and forms glass-like phases. After the separation of scrap metal the bottom ash is comminuted in a hammer crusher. The crusher has a discharge grate of 16 mm gap and a discharge for oversize ferrous scrap. This scrap is separated in a subsequent step into valuable material fractions. Afterwards, further breakage is achieved in a ball mill, which is followed by an air-classifier with a separation cut of approximately 0.5 mm. The milling and

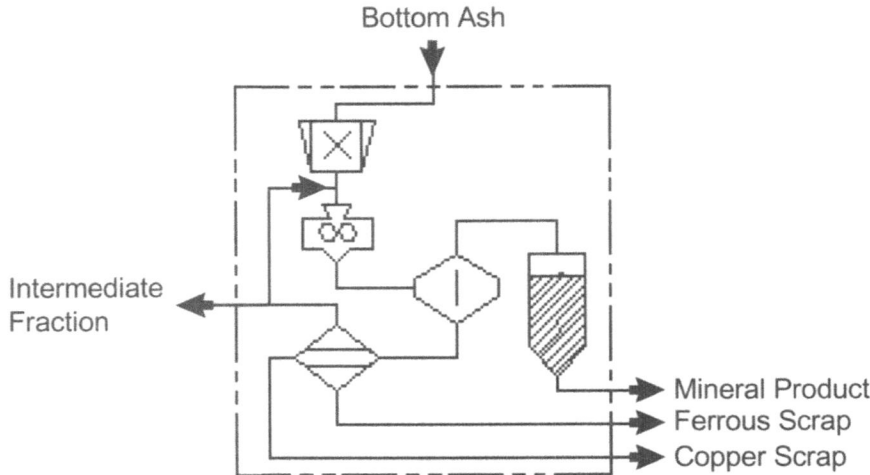

Fig. 5.43. Material flow presentation for the most relevant fluxes within the PECK process. The flows are normalised to the input stream "MSW", which corresponds to 100%

separation unit also works as a dryer which is operated with surplus heat from the thermal process. The over-size from the separator is fed to a magnetic separator and an eddy current unit, separating the ferrous metals from the non-ferrous metals, such as copper.

Since the efficiency of the eddy current separation of non-ferrous metal drops sharply at particle sizes below approximately 4 mm, the metal-fines accumulate in the grinding circuit. Therefore, a small amount of the circulating load is bled off into the fly ash treatment process. This product fraction, the so-called intermediate fraction, consists of approximately 1% of the entire bottom ash that is being processed. The air-classifier undersize is recovered by cyclones and electrostatic filters and stored in silos.

The system for bottom ash treatment consists of standard mineral processing equipment. Several comparable bottom ash treatment systems are in operation. If a further separation of heavy metals is wanted, wet separation technologies have to be taken instead of dry separation technology.

Metal Partitioning

Typical mass flows and partitioning of heavy metals. In Figure 5.44, the most relevant mass flows within the PECK process are shown. The flows are normalised to the input stream "MSW" which corresponds to 100%. The most relevant material flows produced by the MSWI are the flue gas and mineral product.

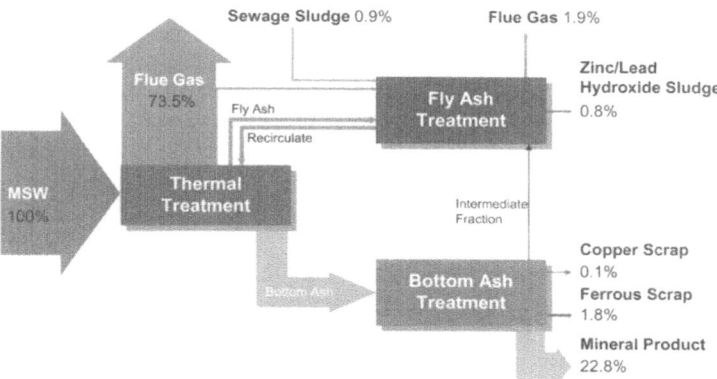

Fig. 5.44. Material flow presentation for the most relevant fluxes within the PECK process. The flows are normalised to the input stream "MSW" which corresponds to 100%

The flue gas primarily contains the combustion products from the organic material of MSW and water. The bottom ash, i.e. the mineral product, consists primarily of the minerals, such as silica and calcium, originally contained in MSW and sewage sludge. The heavy metals are concentrated in the output flows hydroxide sludge, ferrous and copper scrap. The heavy metals flows via the mineral product and the purified flue gas are negligible.

Figure 5.45 shows the particle size distribution of copper and magnetic iron in bottom ashes from conventional MSWI and PECK processes. Metal separation efficiency depends on the mechanical separation technology and reaches 70 % for copper rsp. 85 % for iron with dry mechanical separation technology. As can be seen, there is no significant difference in particle size distribution for the different bottom ashes. From Figure 5.45 one can also conclude that an efficient mechanical separation of non-volatile metallic metals, such as Cu, Fe, Ni and Cr, can only be successful if MSW is mineralised. Otherwise MSW has to be comminuted to particle sizes smaller than 1 mm.

Partitioning of the heavy metals zinc, lead, cadmium and copper between the different output flows is shown in Figure 5.46 and compared to conventional incinerators. The shown metal partitioning is based on measurements and simplified calculations [109, 131]. In particular, the metal partitioning of the thermal treatment was determined by measuring the fluxes of the different output streams bottom ash, fly ash and flue gas. It is clearly visible that the PECK process leads to significantly less heavy metals in the mineral residues than conventional incinerators. Therefore the objective concerning metal-depleted mineral residues is clearly reached. At the same time, the majority of the heavy metals are concentrated in recyclable metal concentrates. With the PECK process, an important step towards more sustainable waste management is made.

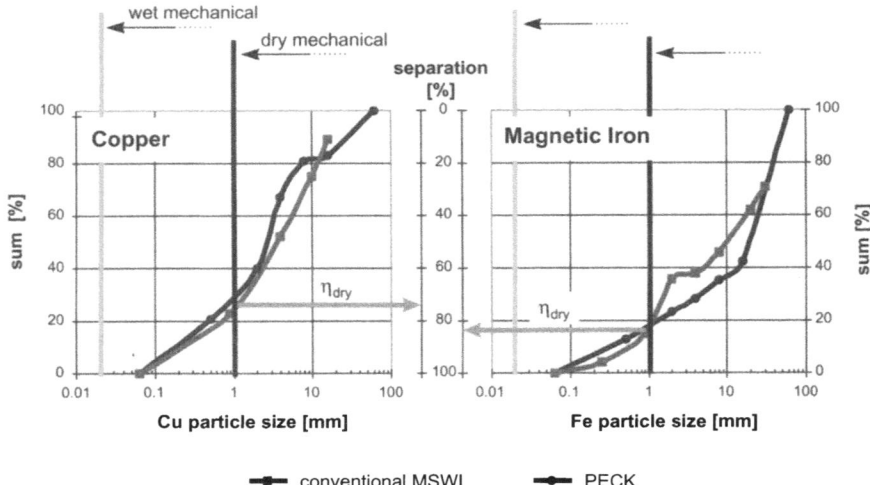

Fig. 5.45. Particle size distribution of copper and magnetic iron in bottom ash from conventional MSWI and PECK processes. Metal separation efficiency depends on the mechanical separation technology and reaches 70 % for copper rsp. 85 % for iron with dry mechanical separation technology

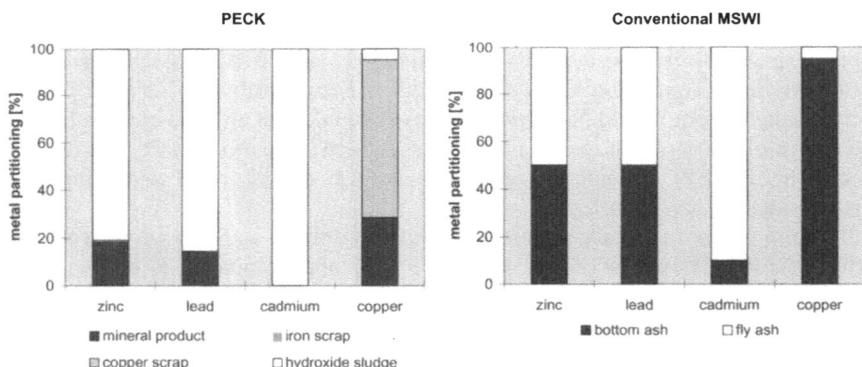

Fig. 5.46. Comparison of metal partitioning in PECK technology vs. conventional MSWI. For the PECK process, the most important secondary materials are: Minerals, ferrous scrap, copper scrap and hydroxide sludge. The most important secondary wastes of a conventional incinerator are bottom ash and fly ash

Economic Considerations of PECK

For a rough economic assessment of the PECK technology a comparative investment and cost calculation was performed for a conventional MSWI and the PECK technology. It must be mentioned that the absolute figures are highly dependant on local conditions and the chosen plant configuration.

For the MSW treatment plants, a throughput of 150'000 tons MSW/a and a heating value for the MSW of 12 MJ/kg were assumed. The PECK technology uses an additional 1300 tons sewage sludge/a. For both plants, MSW is incinerated in two separate lines which can be operated independently. The throughput per line is 8 to 10 tons MSW/h. The significant difference between a conventional MSWI and the PECK technology are the thermal treatment, the fly ash treatment and the bottom ash treatment. All remaining processes, such as flue gas treatment, are comparable to conventional plant design.

For the estimation of the total investment costs of the PECK process, a detailed analysis of all sub-processes was performed. Investment cost of the individual sub-processes are based on estimations or known costs of comparable projects. Some further explanation for the most relevant cost positions of the PECK technology are as follows:

- The combination of grate and a rotary kiln leads to an increase in investment costs.
- Due to improved combustion air control, i.e. air staging, the quantity of flue gas per ton of MSW can be reduced to a minimal level: a level some modern MSWI reach today. The boiler and flue gas purification unit could therefore be designed for a smaller throughput than typical conventional MSWI. This has a positive effect on the cost. For the estimation of total investment costs, only the cost bonus resulting from the flue gas purification is used.
- The NO_X emission in the raw gas is significantly below 200 mg/m^3_n. Nevertheless, a SCR DeNO$_x$ system is assumed for the PECK process. With the SCR DeNO$_x$ system a NO$_X$-emission limit of 80 mg/m^3_n is reachable.
- The fly ash treatment and bottom ash treatment are additional process steps compared to conventional MSWI and therefore the total investment cost is increased.

The calculations show that the total investment for a PECK process is higher than a conventional MSWI by 8 to 9°%.

The total treatment costs are composed of capital, operation, maintenance and disposal costs as well as employment costs. In Table 5.10, mass flows for the 150'000 ton/a conventional MSWI and PECK process are summarised. The amount of fly ash corresponds to 3 % of the MSW. Some modern MSWI produce less fly ash, i.e. as low as 1.5 % of the MSW. The estimated total disposal costs are based on the following specific disposal cost for Switzerland:

- 70 Euro/ton bottom ash and 300 Euro/ton fly ash for the conventional MSWI
- 14 Euro/ton mineral product and 250 Euro/ton hydroxide sludge for the PECK process

Table 5.10. Comparison of mass flows [t/a] for conventional MSWI and the PECK process for a 150'000 ton MSW/a installation

Mass flow (dry substance)	MSWI	PECK
Production from thermal waste treatment of 150'000 t/a MSW		
Fly ash	**4'500**	4'500
Bottom ash	**33'000**	37'391
Fly ash treatment		
Fly ash	0	4'500
Sewage sludge	0	1'300
Intermediate fraction	0	374
recirculate	0	4'391
Zinc/lead hydroxid sludge	0	1'133
Bottom ash treatment		
Bottom ash	0	37'391
Iron scap	0	2'625
Cupper scrap	0	158
Mineral product	0	**34'234**

In Figure 5.47 the estimated treatment costs per ton of MSW are shown for the conventional MSWI and the PECK process. The capital, operation and maintenance costs are slightly higher for the PECK process. This is due to the increased investment cost. On the other hand the disposal costs are significantly lower for the PECK process. This results in an overall reduced treatment cost per ton of MSW. The revenues for power sales are comparable for the conventional MSWI and the PECK technology. The resulting gate fee is therefore lower for the PECK process.

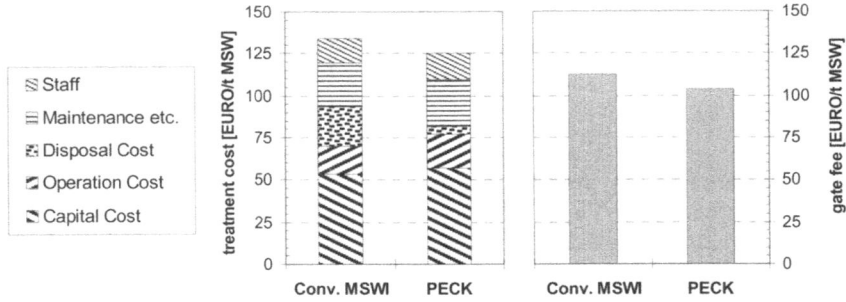

Fig. 5.47. Estimated treatment costs per ton MSW for conventional MSWI and the PECK process. The capital, operation and maintenance costs are slightly higher for the PECK process. The disposal costs are significantly lower for the PECK process and the revenue for selling power are comparable for both technologies. The resulting gate fee is therefore lower

5.3 High-Temperature Melting of Municipal Solid Waste

Shin-ichi Sakai

5.3.1 Development of High-Temperature Melting Technologies

Melting Technologies

The melting of municipal solid waste (MSW) is carried out by using either electricity or the combustion of fuel to achieve a temperature of approximately 1400 °C or more in a high-temperature furnace. Through this method, the mass and volume of the residues are greatly reduced, and the altered physical and chemical state, brought about by melting the residues at such high temperatures, produces a highly stable, high-density slag. However, this technology needs to be improved in certain areas: e.g., reducing the rate at which refractory materials need to be repaired and improving the control technology to ensure a stable operation of the high-temperature melting process. The melted–solidified slag can be used as a construction material, in roads for example, and is also a useful material in land reclamation, since the bulk of the material is reduced to between half and one-third of that of the original incinerator ash. Another advantage of melting MSW relates to the fact that incinerator fly ash contains hazardous substances such as heavy metals, which can cause problems when they leach out into waterways. Through this process of melting and solidification, metallic compounds are stabilized in the molecular structure of the waste product, and are thereby prevented from leaching out and diffusing into the surrounding environment.

At present, a variety of furnace melting systems have been developed and are being put into use. These systems can be divided roughly into two categories: those that burn fuel as the energy source and those that use electricity. The systems can be further classified as follows:

(1) Fuel-burning melting systems
- Surface melting furnaces
- Swirling-flow melting furnaces
- Direct coke-bed melting furnaces
- Rotary kiln melting furnaces
- Internal melting furnaces

(2) Electric melting systems
- Electric-arc melting furnaces
- Electric resistance melting furnaces
- Plasma melting furnaces
- Induction melting furnace (high frequency, low frequency)

Some of the fuel-burning melting systems, e.g., coke-bed melting and rotary kiln melting, can also directly melt MSW. Another type of melting system is the gasification/melting system.

In systems using a gasification furnace, MSW is first heated under non-oxygen conditions to about 450 °C whereby the combustibles are gasified by thermal decomposition. Secondly, the gases are combusted and the remaining incombustibles are melted at high temperatures of approximately 1400 °C or more. Owing to the high temperatures, fewer PCDDs/DFs are formed and the waste heat can be efficiently used to generate electricity. Moreover, since the molten slag that remains at the end of the process can be mixed with base materials and recycled, such gasification/melting systems rarely produce wastes to be disposed of in landfill sites.

Present Status of Development of the Melting Process [100, 186]

In Japan, the sewage sludge melting process was developed in the 1980s and is in practical operation at approximately 10 full-scale plants [188, 232]. MSW fly ash and bottom ash can also be melted. The first melting plants used thermal surface melting furnaces, electric arc-type furnaces, and coke-bed type melting furnaces. Recently, new melting technologies, such as plasma melting furnaces, electric-resistance melting furnaces, and low-frequency induction furnaces, have been developed and are being put into use. The 47 MSW treatment facilities in Japan that presently use (or are scheduled to use) a melting system are listed in Table 5.11. Some of these systems are still in the trial stage of operation. Each company is, however, making efforts to proceed in their research and development and to bring their technology to the marketplace.

The moisture content of municipal solid wastes in Japan is 50% on average, and the lower calorific value is 8,500 - 9,000 kJ/kg on the wet basis. Under these conditions, the necessary operating and management costs in melting are 11 – 14 thousands yen/ton-MSW for MSW melting. Fusion or vitrification of MSW incinerator residues is not practiced in Europe or North America, [51] but detoxification of thermal filter ash is under development [101].

Table 5.11. Full-scale municipal solid waste melting plants in Japan

	Municipalities	Completion	Capacity [t/d]	process type	Slag cooled by	Manufacturer
1	Iwaki City, South	04/2000	40	Plasma	Air	Mitsubishi
2	Utsunomiya City, Mohara	01/2001	40	Electric-Arc	Water	Daido Steel
3	Okutone Amenity Park	04/1998	3/16	Electric	Air	IHI
4	Saitama City, West	04/1993	75	Electric-Arc	Water	Daido Steel
5	East Saitama, Plant 1	10/1995	160 (80x 2)	Electric-Arc	Water	Daido Steel
6	Koyamakawa Clean Center	03/2000	30	Plasma	Water	Kawasaki Heavy Industries
7	Chiba City, Kitayazu	01/1998	24	Plasma	Water	Kawasaki Steel
8	Ohta Clean Plant 2	04/1991	500 (250x 2)	Electric-Arc	Water	Daido Steel
9	Hachioji City, Tobuki	03/1998	36 (18x 2)	Electric	Air	NKK
10	Tama River, Clean Plant	04/1998	50 (25x 2)	Electric-Arc	Water	Daido Steel
11	Yokohama City, Kanazawa	03/2001	60	Electric	Air	NKK
12	Sasayuri Clean Park	04/1999	60 (30x 2)	Plasma	Water	Hitachi Zosen
13	Komakilwakura Center	04/1996	9.6 (4.8x 2)	Electric	Water	ABB
14	Kyoto City, Northeast	03/2001	24	Plasma	Water	Kawasaki Heavy Industries
15	Clean Center Mima	03/1997	5	Plasma	Water	Kobelco
16	Matsuyama City, South	03/1994	52	Plasma	Water	Ebara
17	Sagae Area Clean Center	11/2000	14	Swirling-flow	Water	Kubota
18	Clean Plaza Ryu	07/1997	24 (12x 2)	Swirling-flow	Water	Kubota
19	Itako-Ushihori Union of Environmental Hygiene	11/1999	12	Rotary kiln	Water	SHI
20	Meltec, Oyama East	04/1998	120	Coke-bed	Air	ShinMaywa
21	Sayama City, Center 1	04/1991	15	Swirling-flow	Water	Kubota
	Sayama City, Center 2	12/1996	7	Radiant Surface	Water	Takuma
22	Sakado City, West	07/1995	9.6	Radiant Surface	Water	Takuma
23	East Saitama, Plant 2	04/1985	28.8	Radiant Surface	Water	Takuma
24	Yachiyo City Clean Center	04/2001	10.68	Surface	Water	
25	Abiko City Clean Center	07/1994	15	Radiant Surface	Water	Hitachi Zosen
26	Togane City and 3 towns Clean Center	04/1998	26	Radiant Surface	Water	Takuma
27	Shirane Green Tower	11/1994	7	Swirling-flow	Water	Kubota
28	Sado South Clean Center	04/2000	1	Rotary kiln	Water	Chugai Steel
29	Tokai City	10/1995	30 (15x 2)	Coke-bed	Water	Shin Nippon Steel
30	Clean Center Koromoura	10/1995	30 (15x 2)	Internal	Water	IHI
31	Kohoku Crystal Plaza	03/1999	11	Surface	Water	MHI
32	Minami Kawachi Cleaning Union	04/2000	76 (38x 2)	Radiant Surface	Water	Takuma
33	Anan Clean Center	11/1999	9.6 (4.8x 2)	Radiant Surface	Water	Takuma
34	Isehaya Cïty	12/1999	24	Swirling-flow	Water	Kubota
35	Nanko South Cleaning Union	04/2000	14	Radiant Surface	Water	Hitachi Zosen
36	Hioki Area Clean Plant	11/1999	16 (8x 2)	Surface	Water	Tanabe
37	Chubu Kamikita Broader-based Union	10/2000	60	Fluidizing bed	Water	Kobelco
38	Kamaishi City, Plant	04/1979	100 (50x 2)	Shaft furnace	Water	Shin Nippon Steel
39	Joetsu Sludge Recycle Park	04/2000	15	Fluidizing bed	Water	Ebara
40	Owari East Center	04/1998	24	Shaft furnace	Air	Kawasaki Giken
41	Kameyama City	03/2000	200 (100x 2)	Shaft furnace	Water	Shin Nippon Steel
42	Ibaraki City, Plant 1	04/1996	300 (150x 2)	Shaft fumace	Water	Shin Nippon Steel
	Ibaraki City, Plant 2	04/1999	150	Shaft furnace	Water	Shin Nippon Steel
43	Iryuu Clean Center	04/1997	130 (65x 2)	Shaft furnace	Water	Shin Nippon Steel
44	Eastern Melting Clean Center near Kagawa Pref.	04/1997	130 (65x 2)	Shaft furnace	Water	Shin Nippon Steel
45	Iizuka City	04/1998	180 (90x 2)	Shaft furnace	Water	Shin Nippon Steel
46	Fire&Welfare Union (Ito island)	03/2000	200 (100x 2)	Shaft furnace	Water	Shin Nippon Steel
47	Yame West Clean Center	03/2000	220 (110x 2)	Kiln	Water	Mitsui Zosen

5.3.2 Material and Energy Balances of High-Temperature Melting

Direct Coke-Bed Melting [11]

The basic flowchart of a direct coke-bed melting system is shown in Figure 5.48. In this system, the addition of small amounts of coke and limestone flux to the melting furnace makes it possible to directly melt waste, including incombustibles, at high temperatures, allowing all slag and metal byproducts to be effectively utilized as resources. The gas generated from thermal decomposition in the melting furnace is completely combusted in a downstream combustion chamber and, after that, a heat recovery system, such as a waste heat boiler, effectively recovers the energy. The waste gas is then discharged from a stack after rapid cooling by a gas cooler, dedusting by a low-temperature bag filter, and waste gas decomposition in a catalytic reactor.

The melting furnace is a high-temperature, gasifying/melting shaft furnace with a supply of oxygen provided by pressure swing adsorption (PSA). In this type of furnace, waste is thermally decomposed and gasified by the heat exchange between the waste and the high-temperature gas arising from the combustion of coke and ash and from waste that is melted at high temperatures in the lower part of the furnace. The combustible portions of the waste are gasified in the single-shaft furnace through four zones: (1) the drying zone (the uppermost zone), (2) the thermal decomposition/gasification zone, (3) the solution reaction zone, and (4) the high-temperature combustion and melting zone (the lowermost zone). The high gas

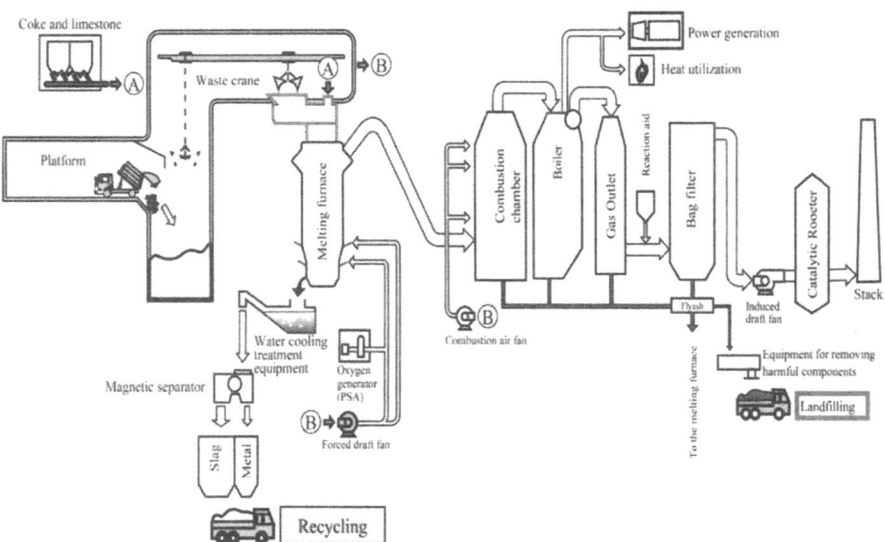

Fig. 5.48. Basic flow of direct coke-bed melting system

temperatures in the thermal decomposition/gasification zone the solution reaction zone are features of the direct melting furnace, and the gasification efficiency is high compared to thermal decomposition by partial combustion of waste alone.

The gas generated by thermal decomposition in the melting furnace is discharged through the drying zone, usually at temperatures reduced to between 400 and 500 °C as a result of removing the moisture from the waste. From the heat balance, the temperature of the gas that enters the drying zone is estimated to be ≥ 1000 °C. Therefore, the waste that enters the thermal decomposition/gasification zone from the drying zone is thermally decomposed and gasified by high-temperature gases of ≥ 1000 °C. The solution reaction converts the carbon in the waste into CO gas by the contact reaction with the high-temperature CO_2 gas that rises from the high-temperature combustion zone in the lower part of the furnace. Although the gas temperature in the high-temperature combustion and melting zone (the lowermost zone) becomes ≥ 1800 °C, as the oxygen supplied by the PSA oxygen generator is consumed by combustion, the gas temperature falls to about 1000 °C due to the endothermic reaction in the solution reaction zone. The solution reaction is another feature of the high-temperature gasification of the shaft furnace that contributes to the increased gasification of the waste.

In the lower part of the melting furnace, waste and coke are combusted at an oxygen ratio of 0.36, generating 33.6% of the whole quantity of heat as high-temperature gas (≥ 1800 °C) necessary for stable melting (Fig. 5.49).

The sensible heat of the high-temperature gas is recovered as the latent heat of gases by the solution reaction, and the gas temperature is lowered to about 1000 °C. In the drying zone, a further 10% of the sensible heat becomes latent heat of water evaporation, and the gas is discharged at an appropriate temperature

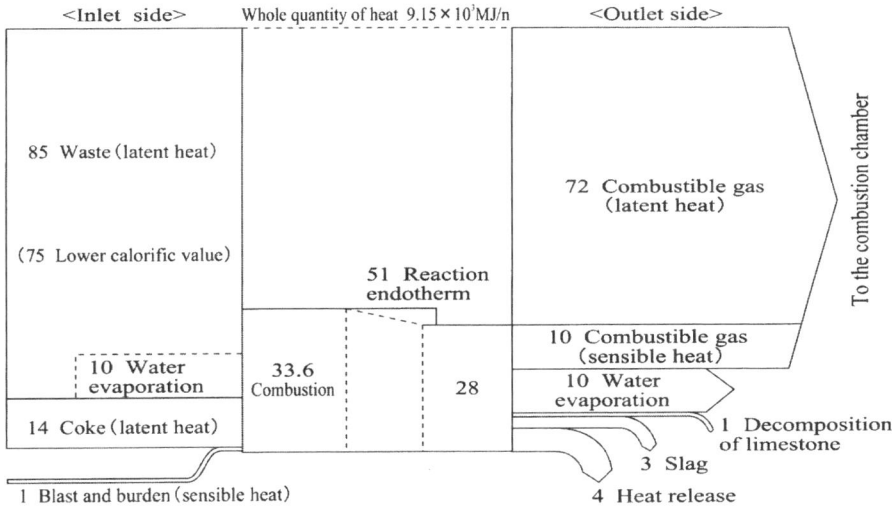

Fig. 5.49. Heat balance of direct coke-bed melting system

of about 400 °C. The solution reaction in the melting furnace is favorable for both melting and gasification because it converts the sensible heat of the high-temperature gas necessary for stable melting into latent heat that can be utilized in the downstream combustion chamber. In terms of the total quantity of heat in the heating furnace, 72% of the heat quantity is converted into the latent heat of combustible gases by consuming 28% of the whole quantity of heat as sensible heat. In this system, the gasification of combustibles and complete melting of incombustibles are achieved in the melting furnace; therefore, further melting is unnecessary and controlled combustion of the generated gas is possible in the downstream combustion chamber.

The combustible gases (CO, H_2, CH_4) generated in the direct melting furnace are injected into the combustion chamber through a main burner to produce a diffused and mixed combustion with the primary and secondary combustion air. Because only gases are combusted, this system offers excellent control of combustion, adequately adapting to variations in waste quality and achieving very stable combustion. The gas temperature at the outlet of the combustion chamber is 900 to 950 °C and the CO concentration is stable, with an average of 10 ppm. The dioxin concentration was sufficiently lower than 0.1 ng-TEQ/Nm^3 at the outlet of the combustion chamber, and the same at the inlet of the stack.

Plasma Melting [215]

Plasma melting is a process of melting incineration residues by way of electricity. Figure 5.50 shows the process flow and material balance when incineration fly ash from MSW incinerated in a fluidized bed system is melted by a plasma melting system.

The results are derived from Clean Center Mima, which was completed in 1997. This fluidized bed system can incinerate 5.5 tons of MSW per hour with a calorific value of 1810 kcal/kg. The 360 kg/h of incineration fly ash that is generated from the fluidized bed incineration is melted by plasma. Thus, 288 kg/h of melted slag is obtained. In plasma melting, arc energy produced by a voltage difference between a main electrode and a bottom electrode transforms a nitrogen jet at the edge of the main electrode into plasma. The heat of the plasma then melts the incineration fly ash, which overflows as molten slag. The maximum output of the plasma torch is 600 kW, and consumptions reach 870 kWh / t-fly ash (equivalent to about 50 kWh / t-waste). PCDDs/DFs included in the waste gas from the melting process amount to 0.1 ng-TEQ/Nm^3, and hydrogen chlorides are 500 ppm. Both are treated as waste gas and are sent back to the waste gas management system in the incineration system.

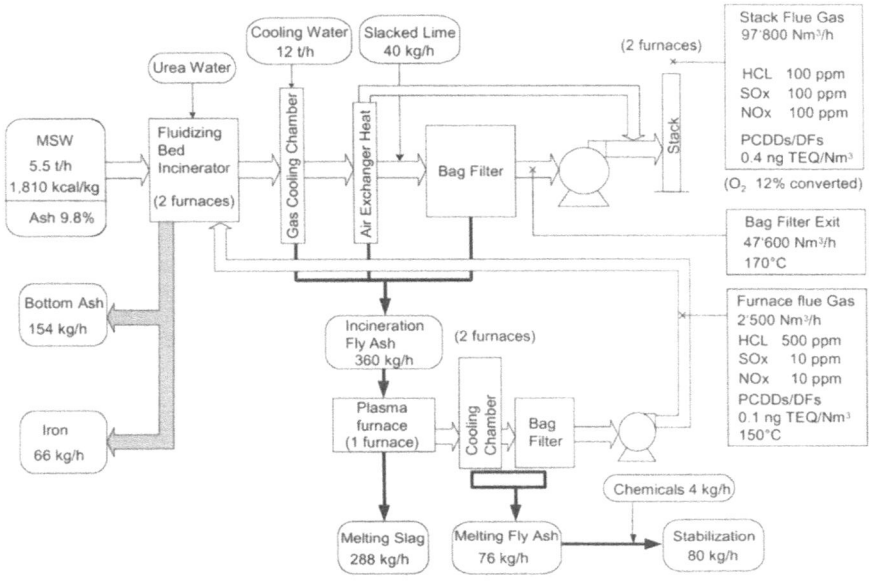

Fig. 5.50. Basic flow of plasma melting system

Rotary Drum Gasification/Melting [241]

This system uses low-temperature pyrolysis of MSW with a high-temperature combustion of the solid and gaseous products of the pyrolysis process in an ash-melting furnace (Fig. 5.51).

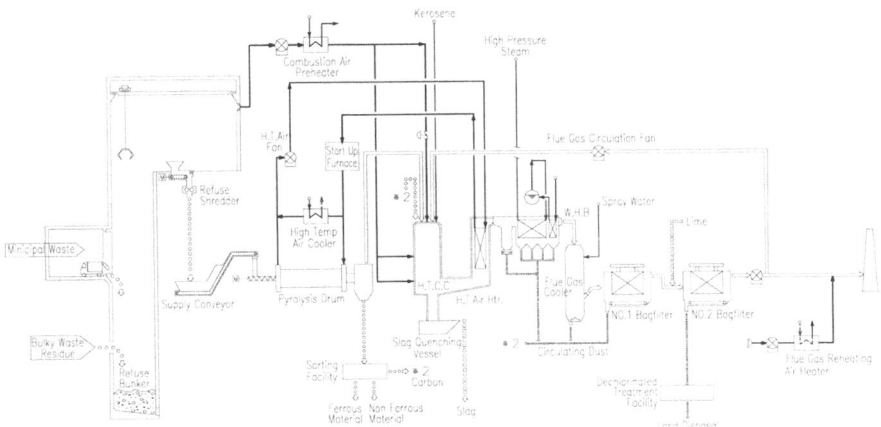

Fig. 5.51. Process flow diagram of rotary drum gasification/melting

Yame West is the first commercial pyrolysis, gasification, and ash melting processing plant for MSW in Japan. The Yame West Clean Center is designed to handle 220 t/d of MSW and 50 t/d of bulky waste. After recovery of recyclable materials from the bulky waste, both the MSW and the bulky waste residue are delivered to a refuse bunker. A crane feeds the waste into a biaxial shearing/crushing unit, which reduces the waste to a top size of around 200 mm. The crushed waste is then fed, via a waste conveyor, to the inlet hopper of a screw feeder into a rotating pyrolysis drum.

The crushed waste undergoes low-temperature drying and pyrolysis in the rotary drum. The MSW is heated indirectly by hot air, which passes through a number of heat transfer tubes arranged along the length of the drum. The pyrolysis drum is effectively sealed against air input, with the result that the metals, both ferrous and nonferrous, are unfused and nonoxidized, and can be recovered in a state suitable for material recycling. The residence time of the waste in the pyrolysis process is around 1 h. The system is relatively insensitive to variations in the quality of the feed waste, allowing stable operation of the downstream equipment, principally the high-temperature combustor, the heat recovery boiler, and the flue gas clean-up equipment. The products of the pyrolysis process are

- pyrolysis gas, which is carried forward directly to the high-temperature combustor;
- solid residues, which comprise char, inert solids, and metals.

The solid residues from the pyrolysis drum are separated from the pyrolysis gas, and passed on to the solids-handling facility. The hot solids are cooled and then sorted through a series of screens. The ferrous and nonferrous metals are recovered for recycling in a clean, unfused, nonoxidized form, which finds ready markets. In contrast, the aluminum recovered from conventional MSW incinerators tends to be fused and contaminated with dirt particles, and the recovered ferrous metal is oxidized. After metals recovery, the solid residues, which contain some combustibles, char, and inert material, are crushed to a top size of 1 mm, and conveyed pneumatically to the high-temperature combustor. The heat balance of the process is presented in Table 5.12.

These data illustrate the stable operation of the process and the ability to melt all of the ash without the need for supplementary fuel. In this case, the net calorific value of the MSW was 7.12 MJ/kg (as received, 1700 kcal/kg). Operational data from the Yame West Clean Center has indicated that operation of the process without supplementary fuel can be achieved with MSW that has a net calorific value as low as 6.28 MJ/kg (as received, 1500 kcal/kg).

The process material balances for the Yame West Clean Center are presented in Table 5.13.

Table 5.12. Energy balance for a rotary drum gasification/melting plant

Heat Input			Heat Output		
Item	MJ/h	%	Item	MJ/h	%
Waste input	31'475.4	100.00	Flue gas losses	3'098.9	9.85
Boiler feedwater	5'160.1	16.39	Circulating flue gas	902.5	2.87
Circulating flue gas	662.6	2.11	Fly ash	62.4	0.20
Circulating ash	16.7	0.05	Recovered steam	27'826.0	88.40
Cooling water	597.8	1.90	Continuous blowdown	78.7	0.25
Combustion air	45.2	0.14	Slag	359.6	1.14
Slag tap heating burner	272.1	0.86	Ferrous and nonferrous materials	1.3	0.00
			Cooling water	1'095.9	3.48
			Heat losses	2'862.8	9.09
			Hot air losses	1'941.9	6.17
Total	38'229.9	121.45	Total	38'229.9	121.45

Table 5.13. Material balance of a rotary drum gasification/melting plant (December 21st, 1999 to July 31st, 2000)

	Quantities [t]	Percentage of MSW throughput [%]	Mass ratio	Volume ratio	Apparent specific gravity [t/m^3]
MSW	20'477	-	-	-	0.25
Acidic gas clean-up residues	728	3.56	1/28	1/101	0.9
Slag	1'568	7.66	-	-	1.2
Ferrous metal	80.0	0.39	-	-	2.5
Non ferrous metal	31.7	0.15	-	-	0.6

All of the recovered metals and the slag are sold as recycled materials. Only the acidic gas-clean-up residues have to be disposed of in landfills, and this represents a landfill volume of only about 1% of that of the original MSW. This compares very favorably with the performance of conventional MSW incineration processes, where the landfill volume requirement for disposal of ash residue is around 20% of that of the volume of received MSW.

At the Yame West Clean Center, all of the recovered metals are of high quality and are sold. Commercial arrangements have been made with Nihon Hodo Co. Ltd. for the sale of all of the slag generated from the plant for use as aggregate for road building.

5.3.3 Persistent Chemicals in High-Temperature Melting Processes

Contents of Hazardous Substances in Incineration Residues

I have summarized the content of hazardous substances contained in the fly ash from MSW incineration, the incineration bottom ash, the melted slag, and the fly ash generated during melting (hereinafter referred to as "melting fly ash") [189]. We studied the heavy metal Pb, contained in high concentrations in incineration residues, and the persistent toxic substances, PCDDs/DFs. Figure 5.52 shows the distribution of Pb content (mg/kg-dry) in fly ash, bottom ash, melted slag, and melting fly ash; Table 5.14 lists the statistical values (number of samples, maximum value, minimum value, average value, median value, and mode value).

Because Pb is evaporated once during the high-temperature waste incineration and again during the residue melting process, Pb is highly concentrated in the fly ash and melting fly ash, while its concentration is smaller in the slag and bottom ash. On the basis of an examination of the Pb content in the soil of a Japanese city, a Pb content of above 600 mg/kg is regarded as contaminated. When this is compared with the median values for slag and bottom ash - 92 mg/kg and 555 mg/kg, respectively - the greater portion of the slag samples and about half of the bottom ash samples show levels below this guideline value.

Table 5.14. Some statistical values of Pb and PCDDs/DFs in incineration residues

Component	Statistic value	Fly ash	Bottom ash	Slag	Melting fly ash
Pb	N.	137	39	33	40
[mg/kg]	AVE.	2146	886	145	28'351
	MAX.	8'290	3'700	1'240	100'000
	MIN.	91.9	85.0	0.005	2'200
	MED.	1'650	555	92	19'000
	MOD.	1'290	1'000	100	19'000
PCDDs/DFs	N.	38	39	18	21
[ng-TEQ/g]	AVE.	6.54	0.20	0.001	0.33
	MAX.	100	1.50	0.010	3.50
	MIN.	0.069	0.00070	10^{-5}	0.0041
	MED.	2.0	0.032	10^{-5}	0.016
	MOD.	1.7	0.016	10^{-5}	0.25

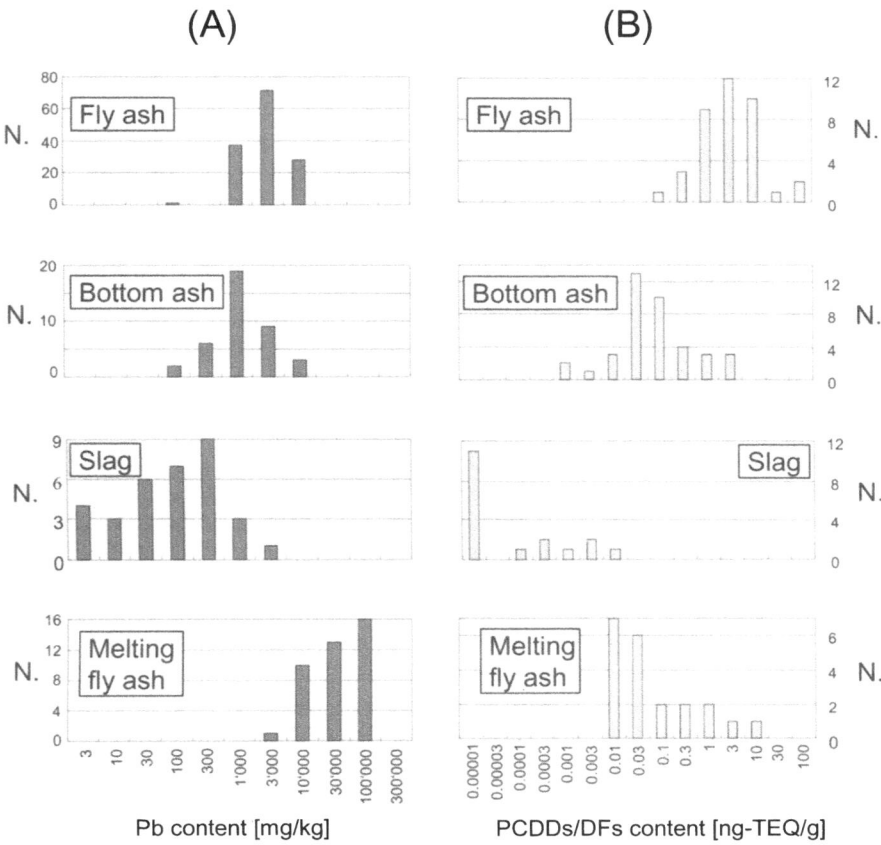

Fig. 5.52 (A): Distribution of Pb content in incineration residues. **(B)**: Distribution of PCDD/DF content in incineration residues (a value of 0.03 means that $0.03 < N. \le 0.1$)

Figure 5.52B shows the distribution of PCDD/DF contents (ng-TEQ/g-dry) and Table 5.14 lists the statistical values. Fly ash contains the highest concentrations of PCDDs/DFs. Incineration bottom ash and melting fly ash contain about the same level. Slag has the lowest concentrations. The average values for slag and bottom ash are 0.001 ng-TEQ/g and 0.20 ng-TEQ/g, respectively. The median values are <1E-5 ng-TEQ/g and 0.032 ng-TEQ/g, respectively. Slag exhibits an overwhelmingly low value. The maximum allowable concentration of PCDDs/DFs in the soil before some ameliorative measures need to be taken is 1 ng-TEQ/g in Japan. Most of the fly ash samples show a level above this concentration. On the other hand, the greater portion of bottom ash and melting fly ash samples show a value below 1 ng-TEQ/g. The melted slag provides a concentration at the background level of soil. Fly ash exhibits a high level of heavy metal content as well as PCDD/DF content. This has, again, demonstrated the need for the proper management of fly ash.

Leaching of Hazardous Substances from Incineration Residues

Figure 5.53 shows the concentrations of Pb - represented as a function of pH [190] - in the leachate from various batch-type leaching tests of slag, fly ash, combined ash, and bottom ash.

Here, JLT46 denotes the Japanese leaching test method for soil where the test is conducted by filtering with a membrane filter with a nominal pore size of 0.45 μm after shaking the test sample for 6 h. The test uses samples with a particle size of ≤ 2 mm, distilled water, and L/S 10. JLT13 is the same as JLT46 except that the sample particle size is ≤ 5 mm and a glass fiber filter (1-μm nominal pore size) is used. The pH-dependent test is a test where HNO_3 or NaOH is used to control each different final pH. Solutions of HNO_3 and solutions saturated with carbon dioxide gas are used as solvents for the Ini-pH4 method and CO_2 method, respectively, and the samples are shaken for 6 h after the initial pH is adjusted to 4. The Open and pH4-stat tests are leaching tests where the sample is stirred for 6 h. In the Open method the leachate is open to the air.

In the pH4-stat test method, a pH controller and nitric acid are used to ensure that the pH of the leachate is maintained at 4. Independently of the type of leaching protocol, both, slag and incineration residues show a pH-dependent leaching behavior characteristic of Pb as an amphoteric metal: the leaching concentration of Pb is high in the low-pH and high-pH areas. This shows that the pH value is one

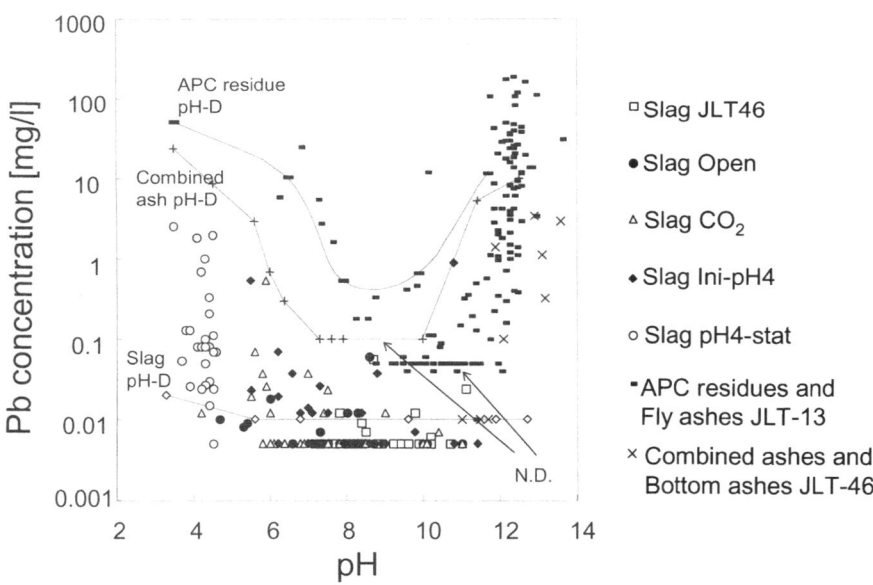

Fig. 5.53. Leaching behavior of Pb from residues in various batch leaching tests

of the most important factors in the leaching of Pb. Therefore, for adequate residue management, it is important to keep track of such leaching behavior [68, 244].

Compared to the incineration residue, a much smaller concentration of Pb leaches from the slag at any pH. With respect to the leachate from slag, the pH is slightly lower in the Ini-pH4 method and CO_2 method than in that of the JLT46 test using distilled water. The Pb leaching concentration tends to be slightly higher. Nevertheless, the leaching concentrations are extremely low. Furthermore, in cases where the pH of the slag is acidic and in cases where the pH of the bottom ash is slightly alkaline, the leaching concentrations of Pb are almost the same. This shows that heavy metal leaching depends on pH. Moreover, it is also important to keep track of the pH level that the residue assumes in the environment according to the actual environmental conditions and properties of the residues such as acid neutralization capacity. This is an important factor in heavy metal leaching. It is reported that, when the carbonate content of clay is high, the pH buffering capacity is high and the Pb adsorption capacity is also high.[259]

It is also reported that neutral leachates obtained at disposal sites are due to the reduction in the pH values of the leachates themselves, rather than due to a reduction in the pH of the residues.[155] High buffering capacity can be expected from aged bottom ash and other residues where carbonate contents are high. Regarding the leaching of heavy metals, it is also important to study the transition of pH values in the environment from the viewpoint of this acid neutralization capacity.

Figure 5.54 shows the concentration distribution of PCDDs/DFs leaching from fly ash and fly ash treated by cement and heavy metal stabilizer.

It contains leachate data based on the JLT13 test (L/S 10 using distilled water). Comparing the mode value of the PCDD/DF content of fly ash (1.7 ng-TEQ/g) and the mode value of PCDDs/DFs leached from the fly ash (0.11 pg-TEQ/L)

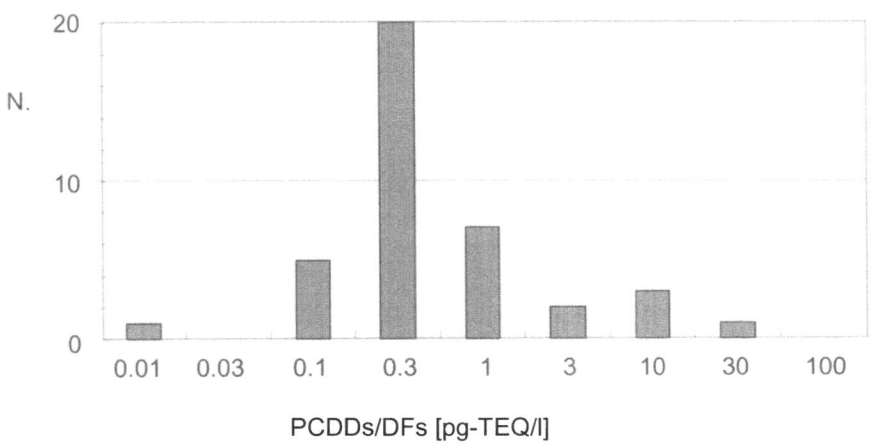

Fig. 5.54. Distribution of PCDD/DF concentrations leaching from fly ash

shows that a very small amount—10^{-5}% to 10^{-4}%—of PCDDs/DFs is leached from fly ash.

PCDDs/DFs leaching from fly ash was studied by Carsch et al.[50]. In an 8-day batch-leaching test using 100 g of fly ash per 1 L of distilled water, 20 ng/L of HpCDD and 40 ng/L of OCDD were detected. In a 2-week leaching test using 1 kg of fly ash per 10 L of distilled water, 20 ng/L of HpCDD and 30 ng/L of OCDD were detected. On the other hand, Berbenni et al. consider the leachate of fly ash from the ESP to be less than 20 ng/L in terms of the concentration of each homologue [39]. The concentration range of PCDDs/DFs leaching from fly ash detected in the present study was 0.03 to 30 pg-TEQ/L, which corresponds to about 0.003 to 3 ng/L in terms of total concentration.

Although Berbenni et al. did not detect such concentrations, the leaching concentrations detected in the present study are quite feasible from the viewpoint of solubility—the solubilities of dioxins obtained by Friesen et al. are 4.4 ng/L for 1,2,3,4,7,8-HxCDD, 2.4 ng/L for 1,2,3,4,6,7,8-HpCDD, and 0.4 ng/L for OCDD [82] Nevertheless, the leaching rate of PCDDs/DFs from fly ash is low. This is because of the strong capacity for adsorption into ash particles. Also, the median value of the concentrations of PCDDs/DFs leaching from fly ash is 0.22 pg-TEQ/L. When compared to the Japanese environmental standard for water quality of 1.0 pg-TEQ/L, 80% or more of the samples meet this standard. Only one sample exceeded the proposed standard value of 10 pg-TEQ/L for waste water. Although oil, solvents, surfactants, and humic substances are all thought to facilitate the leaching of PCDDs/DFs, [187, 206] a survey of these factors and the leaching behavior from bottom ash and melting fly ash containing little carbon content is not sufficient. It is important to also examine ash types and other factors that accelerate leaching, and to investigate the leaching test methods, including the filter pore size to determine the impact of fine particulates. It is also essential to evaluate the levels of trace quantities of PCDDs/DFs, taking into consideration biological concentrations in living systems taken up from the aquatic environment.

Substance Flow Accounting in Melting Residues

For the purposes of substance flow accounting in residues from MSW incineration, input is the amount of hazardous substances contained in the waste that is put into the incineration facility, while output is the amount of hazardous substances contained mainly in the emission gas, bottom ash, fly ash, and waste water. When a melting facility is also installed, output is the emission gas, melting fly ash, slag, and waste water. Table 5.15 shows the flow of Pb and PCDDs/DFs in the residue of waste incineration and ash melting when 1 t of MSW is incinerated. Here, the contents of Pb and PCDDs/DFs in the various residues are based on the average values in Table 5.14 from the viewpoint of total amounts of each that are released. In the case of waste incineration, a total of 197 g-Pb/t-waste is released from bottom ash and fly ash. Even if this is melted, a total of 133 g-Pb/t-waste is released from the slag and melting fly ash.

Here, the total amount of Pb released from 1 t of MSW is such that the total amount of Pb released from bottom ash and fly ash added together should be almost the same as that released from the slag obtained by melting them and the melting fly ash added together. This is because there is not much flow of Pb to emission gas and the metal. If they are different, this is because part of the bottom ash may contain some fly ash. Incineration fly ash contains 65 g of Pb for 1 t of waste. When the residues are melted, about twice the weight of Pb, 112 g, is concentrated in the melting fly ash; on the other hand, the amount of Pb present in the slag is much decreased.

In the case of PCDDs/DFs, 1 t of incinerated waste produces a total of 226 µg-TEQ from the bottom ash and fly ash. By contrast, when these residues are melted, a total of 1.48 µg-TEQ/t-waste is produced from the slag and melting fly ash; in other words, dioxin release is reduced by 99% or even more. The volume of emission gas generated during waste incineration is assumed to be 5000 Nm^3/t-waste. Thus, the concentration of PCDDs/DFs in the emission gas is 0.5 µg-TEQ/t-waste if the emission standard of 0.1 ng-TEQ/Nm^3 is assumed. It can again be confirmed that the residues account for a high percentage of the total PCDDs/DFs released by incineration. Therefore, it is evident that treatment by melting is a very effective method for the achievement of a target value of 5 µg-TEQ/t-waste for total PCDDs/DFs released during MSW management.

Table 5.15. Substance flow of Pb and PCDDs/DFs in the residues

Residue	Residue released for one ton of waste [kg]	Pb content (average) [mg/kg]	PCDDs/DFs content (average) [ng-TEQ/g]
Bottom ash	150	886	0.20
Fly ash	30	2'150	6.54
Melting fly ash	4	28'000	0.33
Slag	160	145	0.001

Residue	Pb released for one ton of waste [g]	PCDDs/DFs released for one ton of waste [µg-TEQ]
Bottom ash	132.9	30
Fly ash	64.5	196.2
Melting fly ash	112	1.32
Slag	23.2	0.16

Utilization of Melted Slag

The main use of recycled melted slag is likely to be as aggregate for road pavement. Road building requires large quantities of materials of a variety of qualities; this allows us to utilize wastes according to specific purposes. There is approximately 1.2 million km of roads in Japan, accounting for 3% of the gross area of Japan, which is almost equal to the area of land used for housing [106]. The pavement depth ranges from a few centimeters to about 1 m; the total volume is nearly 5 billion m^3. Road surfaces paved with asphalt are designed to last 5 years, roads paved with asphalt have a design life of 10 years, and roads paved with concrete have a design life of 20 years; the average road life is approximately 10 years.

Although natural materials, such as crushed rock and sea sand, have been used for pavement materials so far, supplies of these natural resources can easily become exhausted and we are urged to control their use. Therefore, melted slag has gradually been promoted for use as an aggregate for road building. In addition to the economic profit of slag, this type of use would also help to avoid costs to dispose incineration residues at the landfill sites. Furthermore, we should not forget the significance of slag as an alternative to natural resources, which was impossible to measure in terms of economic value.

For melted slag to be used as a pavement aggregate, it must first meet the following required specifications: (1) particle size range is within the given specification, (2) abrasion loss by Los Angels abrasion testing machine is lower than 50%, (3) plasticity index is lower than 4 or 6, and (4) the modified California Bearing Ratio (CBR) is higher than 80%. These indexes are prescribed in tests for civil engineering materials.

The characteristics of melted slag can be controlled by its method of manufacture, particularly the speed of cooling from the high-temperature melting conditions. In the plasma melting plant mentioned above, the melted slag is cooled by laying it on a mold conveyer [215]. The cooling speed of the molten slag can be changed by varying the operating speed of the conveyer and by supplying water to the mold. Table 5.16 lists the physical characteristics of slag when the slag-cooling speed is changed.

In general, the faster the cooling speed, the smaller the diameter of slag particles and the smaller the modified CBR values; this means higher strength values for materials for road construction.

Furthermore, on the premise of utilizing the slag, fewer hazardous substances, such as heavy metals, leach out from the slag to the aquatic environment. The results of leaching tests of plasma-melted slag indicate that concentrations of cadmium, lead, arsenic, and selenium are each less than 0.001 mg/L; lower than the 0.01 mg/L that is the standard for recycling in Japan. Leaching of hexavalent chromium is under 0.005 mg/L (standard: 0.05 mg/L) and mercury is below 0.0005 mg/L (standard: 0.0005 m/L).

Table 5.16. Changes in physical characteristics by slag cooling speed

	Water-crushed slag	Air-cooled slag	Air-cooled slag	Specification of sub base material
Slag cooling speed	Rapid	Medium	Slow	
Mean particle diameter [mm]	10	16	20	
Specific gravity of surface dry	2.803	2.808	2.810	
Modified CBR [%]	24.5	29.7	38.4	Over 20
Absorption ratio [%]	0.463	0.590	0.555	
Abrasion loss by Loss Angels machine [%]	19	21.7	19.9	Under 50
Soundness [%]	0.5	0.8	1.0	Under 20

5.4 Characteristics, Behavior and Durability of High Temperature Materials

Didier Perret, Kaarina Schenk, Marc Chardonnens, Peter Stille, and Urs Mäder

Foreword

For the purposes of long-term environmental protection and sustainable development, the Swiss authorities have decreed an obligation to incinerate the entirety of non-recyclable municipal solid wastes (MSW), effective as of January 2000 [3]. Today, the conditions to achieve this requirement, *i.e.* the total incineration capacity of the country, are technically fulfilled for the next 25-30 years via a network of 28 incineration plants, which can absorb *ca.* 3×10^6 t MSW/yr. The mass of MSW collected and incinerated every year throughout the country leads to the production of secondary residues from incineration, namely bottom ash (BA; *ca.* 750×10^3 t/yr), fly ash (FA; *ca.* 46.5×10^3 t/yr) and filter cake (FC; *ca.* 11.625×10^3 t/yr). Because of their high content in toxic metals and their high leachability, these secondary residues cannot be recycled for civil engineering applications, and must therefore be deposited in landfills as stabilized or inert material, depending on the properties of the residues [4, 24, 25].

It is therefore pertinent to address the potentialities of the advanced thermal treatment technologies which are said to produce inert secondary residues, hereafter referred to as high temperature materials, which might be reused as a secondary raw material resource. This has been the aim of an in-depth study [167] initiated by the Swiss Agency for Environment, Forests and Landscape (SAEFL/BUWAL), which is in charge of the technical, economical and social aspects of waste policy for the country. This section surveys the main scientific issues resulting from the determination of the characteristics (physical, chemical and microscopic features) and behavior (leachability during corrosion) of 23 secondary materials produced by various high temperature treatment processes. On the basis of these issues, but without consideration of the energetic compatibility and social sustainability of such processes, possible answers are provided for a better future in waste management.

5.4.1 Frame of the Study

The Swiss Technical Ordinance on Waste [3] requires a revision which is informed by consideration of long-term sustainability and environmental impacts of wastes and secondary residues. Within this frame, the SAEFL has addressed basic questions on the long-term durability of the materials originating from alternative high temperature treatment technologies (hereafter referred to as HT materials and HT processes). These questions are summarized as follows:

- Taken as a generic class of residues, do the intrinsic characteristics and long-term behavior of the so-called HT materials exhibit substantial improvements over conventional BA, FA and FC residues?
- If so, is it possible to distinguish between the different types of HP processes and their input material in order to design general rules for process optimization toward HT materials in compliance with the concept of sustainability?
- Which guidelines can be derived from the results to revise the TOW in the direction of a more sustainable management of waste in Switzerland?

HT Materials and Standards

To answer these questions, a series of 23 by-products originating from 16 various HT treatment processes developed by 10 different companies in Switzerland, Germany, France and Italy, were extensively studied. HT materials were produced under realistic conditions (*e.g.* from wastes representative of average MSW) and were sampled so as to avoid interpretation biases.

The classification of HT samples is based on the input material from which they were produced; this makes no assumption of the various technologies used to produce HT materials. Two main families of processes, schematically depicted in Figure 5.55, can be distinguished on the basis of their dominant input material:

- In-line processes: These include technologies which operate at higher than normal temperature during incineration.

- Post-processes: These include technologies which treat the residues obtained by conventional thermal treatment of MSW. Two additional sub-categories are given, depending on the dominant type of residues:
  - Post-processes for BA: treat mostly the BA fraction of residues.
  - Post-processes for FA: treat mostly the FA fraction of residues.

This study does not present the exhaustive physico-chemical status of HT treatment processes of wastes and their residues, but, to the best of our knowledge, it is the first of its kind to embrace such a broad range of technologies and input materials (MSW, BA, FA, or mixtures thereof with additives).

For comparative and normalization purposes, the study was also performed on 3 HT standards, SON68, R2bis and R3. SON68 is the non-radioactive surrogate of the nuclear high-level waste (HLW) glass R7T7 designed in France to inertize fission products of nuclear power plants. This borosilicate glass is one of the most studied HLW glasses [18, 52, 77, 162] and is thus an ideal analytical standard and control for our bulk chemical analyses and corrosion tests. R2bis and R3 are HT materials produced in France by the HT treatment of FA under different conditions [57, 58].

HT samples and standards were identified by physical, chemical and microscopic analyses (static picture), their behavior was assessed under corrosive conditions (dynamic behavior), and their long-term durability was estimated with a thermodynamic model (thermodynamic stability).

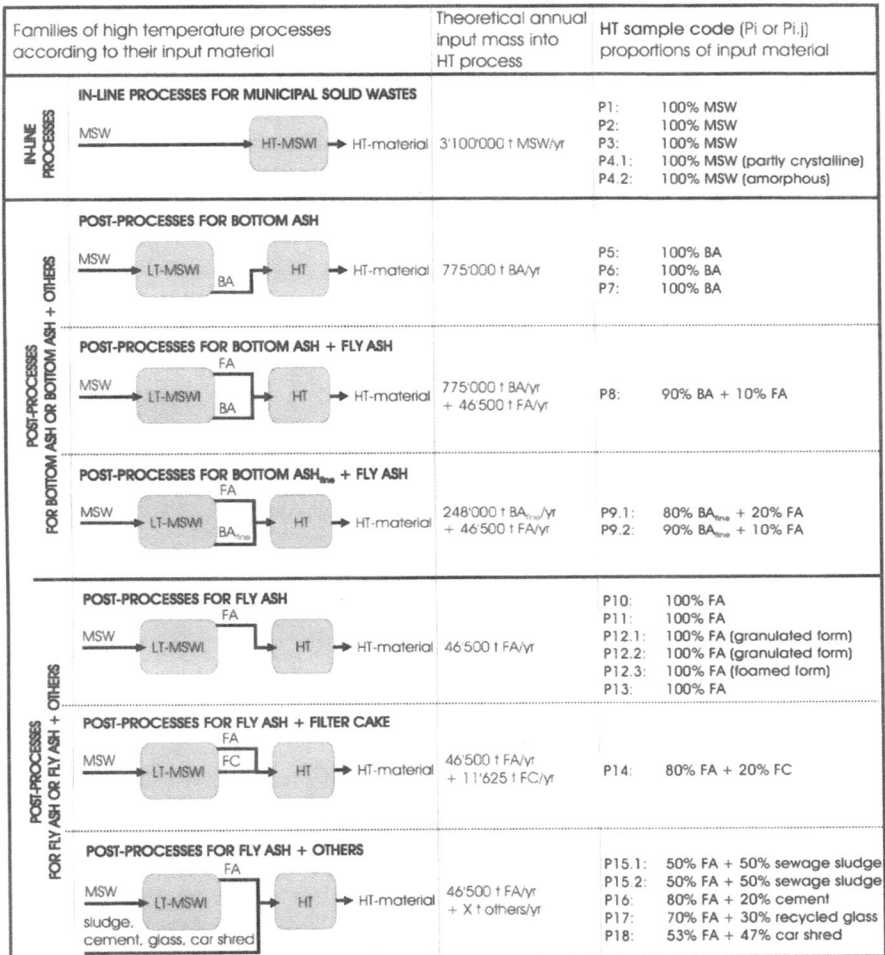

Families of high temperature processes according to their input material	Theoretical annual input mass into HT process	HT sample code (Pi or Pi.j) proportions of input material
IN-LINE PROCESSES FOR MUNICIPAL SOLID WASTES	3'100'000 t MSW/yr	P1: 100% MSW P2: 100% MSW P3: 100% MSW P4.1: 100% MSW (partly crystalline) P4.2: 100% MSW (amorphous)
POST-PROCESSES FOR BOTTOM ASH	775'000 t BA/yr	P5: 100% BA P6: 100% BA P7: 100% BA
POST-PROCESSES FOR BOTTOM ASH + FLY ASH	775'000 t BA/yr + 46'500 t FA/yr	P8: 90% BA + 10% FA
POST-PROCESSES FOR BOTTOM ASH_fine + FLY ASH	248'000 t BA_fine/yr + 46'500 t FA/yr	P9.1: 80% BA_fine + 20% FA P9.2: 90% BA_fine + 10% FA
POST-PROCESSES FOR FLY ASH	46'500 t FA/yr	P10: 100% FA P11: 100% FA P12.1: 100% FA (granulated form) P12.2: 100% FA (granulated form) P12.3: 100% FA (foamed form) P13: 100% FA
POST-PROCESSES FOR FLY ASH + FILTER CAKE	46'500 t FA/yr + 11'625 t FC/yr	P14: 80% FA + 20% FC
POST-PROCESSES FOR FLY ASH + OTHERS	46'500 t FA/yr + X t others/yr	P15.1: 50% FA + 50% sewage sludge P15.2: 50% FA + 50% sewage sludge P16: 80% FA + 20% cement P17: 70% FA + 30% recycled glass P18: 53% FA + 47% car shred

Fig. 5.55. Classification of HT samples on the basis of the input material from which they were produced. In-line processes directly transform MSW into HT material. Post-processes transform residues of MSW incineration (BA, FA, FC, or mixtures of them with possible additives). The codes Pi or Pi.j given to the HT samples denote (i) the process type for a given input material, and when applicable (j) the variations in the operating conditions of this process (*e.g.* different treatment temperatures, or quenching rates, or proportions of input materials). The theoretical annual input mass into HT process is the theoretical mass of input material that should be handled by the given process to treat the total amount of MSW or residues, on the basis of the Swiss waste picture in 2000.

5.4.2 Characteristics of HT Materials: The Static Picture

With respect to the status of knowledge acquired over decades on the characteristics and stability of nuclear high level waste glasses [33, 71, 234, 235, 247], many physical (morphology, crystallinity), microscopic (surface and inner structure) and chemical features (proportions in major and trace elements) of an HT material may play an important role in its long-term behavior. These features are presented and discussed in the next sections.

Morphology

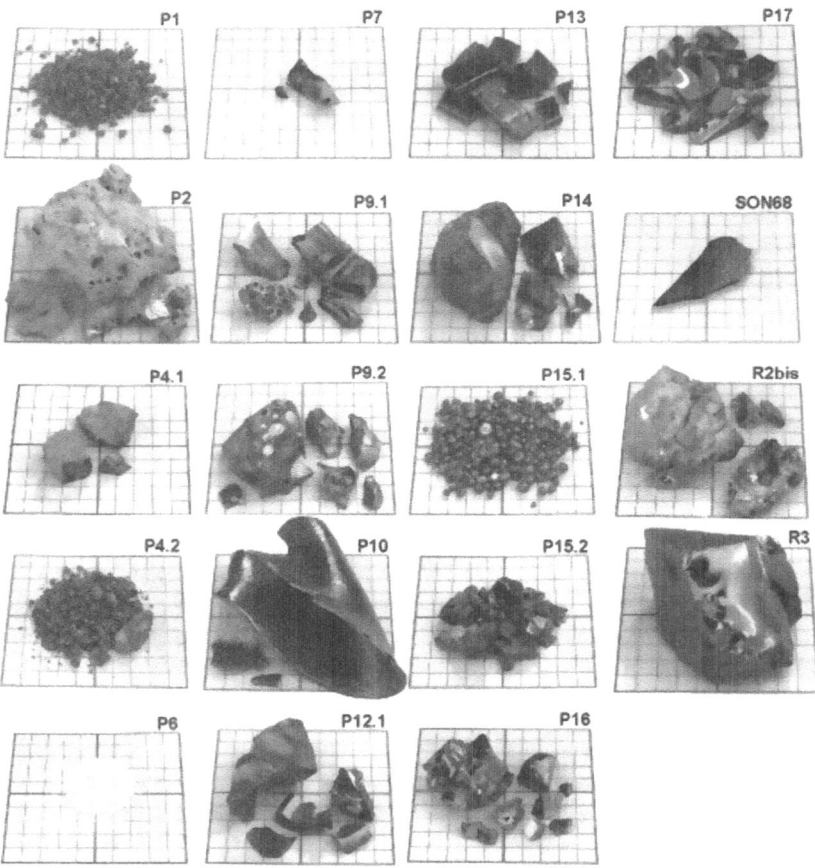

Fig. 5.56. Examples of the various morphologies of the studied HT materials in their original state. The samples are on a 10 cm² scaling grid with 1 cm² subdivisions. Note the visual dissimilarities between samples prepared from the same HT processes but under slightly different conditions (*e.g.* P4.1 and P4.2; P9.1 and P9.2; P15.1 and P15.2)

Figure 5.56 shows that HT materials exhibit a broad range of morphologies. They range from millimeter-sized grains or granules to thin or thick plates and regularly shaped beads, and to medium or large irregular blocks; their visual appearance extends from homogeneous (glassy aspect) to highly heterogeneous (non glassy or vitrocrystalline). The final morphology of HT materials is affected by the quenching techniques which are used in the different HT treatment processes, but it is not influenced by the input material (MSW, BA, FA, FC, additives). Nevertheless, it is expected that the larger the size of the HT materials the more stable they should be when subjected to corrosion.

Crystallinity

HT materials have been produced at high temperatures (> 1000 °C, except for P6 and P11, obtained at 900 °C) favoring the build up of a silicate melt which may partly crystallize upon slow quenching, or contain relicts of mineral phases inherited from the input material. The degree of crystallinity of the HT material or the proportion of crystalline inclusions in the silicate matrix is a key parameter for classification purposes; it can be determined by semi-quantitative X-ray diffraction.

HT materials containing less than 2 % crystalline components are considered to be vitreous (*i.e.* they have a glassy matrix); HT materials are considered to be vitrocrystalline when they contain negligible amounts of crystalline components, and as crystalline when their crystalline components become significant. Although difficult to quantify unambiguously, HT samples never exhibited purely crystalline characteristics; thus, HT materials belong exclusively to the families of vitreous and vitrocrystalline materials (vitrocrystalline samples are assigned an asterisk, *e.g.* P2*, to distinguish them from vitreous samples). Figure 5.57 shows typical diffraction patterns of HT samples. It can be seen that vitrocrystalline samples still exhibit a background pattern indicative of silicate glass.

Among the phases identified in vitrocrystalline HT samples, quartz is ubiquitous; albeit ($NaAlSi_3O_8$) and melilite (gehlenite) are also frequently identified, but to a lesser extent. Other exotic phases are present from time to time: diopside ($CaMg(SiO_3)_2$), pyroxenes, forsterite (Mg_2SiO_4), portlandite, Fe-Mg-spinels, Fe-Ti-oxides.

It is difficult to draw unambiguous conclusions on the presence of crystalline components in HT materials; crystalline inclusions may be depleted or enriched in heavy metals, and their individual leachability will be either beneficial or detrimental to the overall stability of the HT material [17, 205, 226]. Nevertheless, the silicates and metal oxides identified in HT materials tend to be more resistant to corrosion than silicate glass is. In addition, the metal-embedding characteristics of our HT materials are expected to be stronger than the conventional residues of incineration (BA, FA, FC).

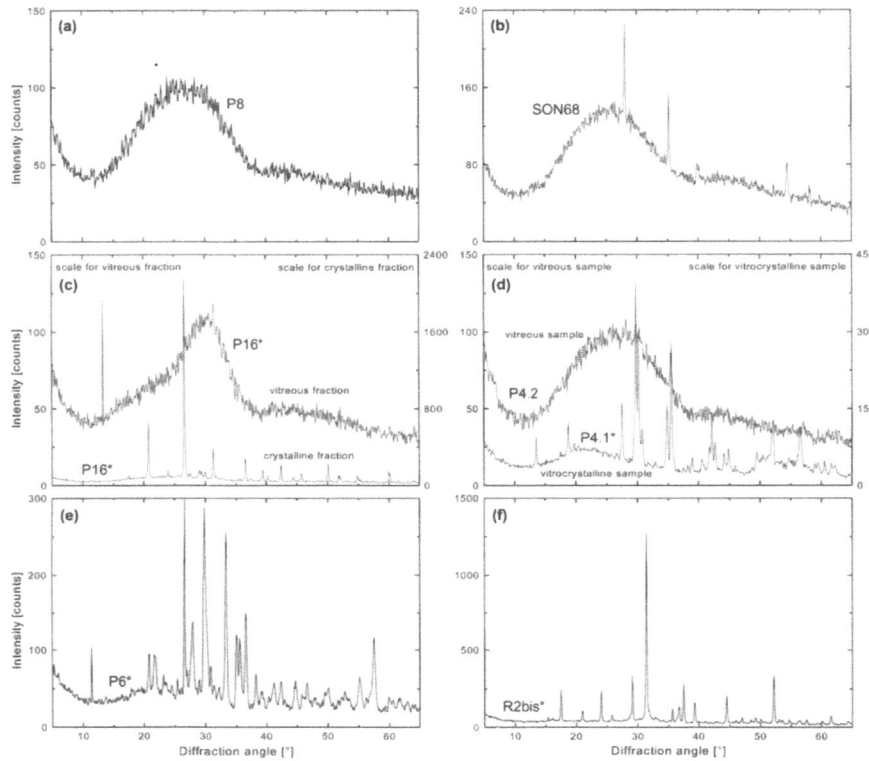

Fig. 5.57. Typical X-ray diffractograms of 100-125 µm ground HT materials, showing either purely vitreous characters (a, b), contrasted vitreous/vitrocrystalline characters (c, d), or purely vitrocrystalline characters (e, f). In Figure 5.57c, the sample P16* (partly vitreous, partly vitrocrystalline) shows large scale heterogeneities. In Figure 5.57d, samples P4.1* (vitrocrystalline) and P4.2 (vitreous) originate from the same HT process, but were obtained from variations in the quenching step

Surface and Inner Structure

The specific surface area S_{spec} of HT materials is a key parameter that directly influences their chemical reactivity at the solid-solution interface, and, in turn, their long-term durability. S_{spec} is partly controlled by the proportion of crystalline inclusions in the glassy matrix of HT materials and partly by the HT process itself, in particular the quenching step.

S_{spec} was determined for 100-125 µm ground HT materials. Taking the standard HLW glass SON68 as the reference for durability with respect to S_{spec} (S_{spec}(SON68) = 383 cm^2/g ; by comparison, S_{spec} = 230 cm^2/g for 100 µm glass spheres) one can discriminate three families of HT materials: Samples with S_{spec}

similar to SON68 (*i.e.* $S_{spec} < 600$ cm^2/g ; S_{spec}:S_{spec}(SON68) < 1.5), samples with S_{spec} larger than SON68 (*i.e.* $S_{spec} = 600$-1000 cm^2/g), and samples with S_{spec} much larger than SON68 (*i.e.* $S_{spec} > 1000$ cm^2/g). With the exception of P1 ($S_{spec} = 607$ cm^2/g), vitreous HT materials exhibit a low specific surface area, close to that of SON68. Alternatively, HT materials P12.1*, P16* and R2bis* also display a low S_{spec} though they are vitrocrystalline. Vitrocrystalline HT materials P2*, P6* and P11* are characterized by very high values of S_{spec} (3600, 8200, respectively 7000 cm^2/g); they are expected, *a priori*, to exhibit a higher reactivity during the initial stages of corrosion. Nevertheless, these values of S_{spec} reflect ground samples, not the initial status of HT materials.

The surface structure and bulk inclusions of 100-125 µm ground HT materials were characterized through scanning electron microscopy. A close examination of micrographs (Fig. 5.58) indicates that there is fairly good agreement between S_{spec} and the apparent roughness of HT materials, regarding their extent in surface and bulk mineral inclusions (an indication of the inhomogeneity of the samples).

With respect to surface roughness, HT materials were classified as smoother than, similar to, or rougher than the standard SON68. Surprisingly, HT samples exhibit fewer mineral inclusions than the standard SON68, or at most the same proportions of inclusions, whatever their vitreous or vitrocrystalline state. Nevertheless, the combination of the roughness and inclusion features confirms that vitrocrystalline HT materials are less favorable than SON68, with the exception of sample P16* (roughness similar to SON68 and fewer inclusions than SON68). As previously observed, there is no relationship between the surface and inner structure of HT materials (S_{spec}, roughness, inclusions) and the family of processes by which they were produced.

Semi-quantitative and detailed petrographic analyses of surface and bulk mineral inclusions indicate the presence of a very large palette of phases, the most abundant ones being spinels, alloys, plagioclase, quartz, gehlenite, melilites, pyroxene, silicates and metallic inclusions. Many of them exhibit metal enrichment (in particular Cr, Cu, and Zn, which are the dominant trace elements in most HT samples).

Major Elements

The composition of HT materials (Fig. 5.59) plays a dominant role in their durabiliy [93, 205]. Although subtle changes in the proportion of certain elements may affect their characteristics, it has been shown that silicon, the dominant network-forming element, is the most critical parameter for the durability of glasses. Aluminum may also influence durability, acting either as a network-forming element in tetrahedral sites or as a network-modifying element in octahedral sites. By contrast, calcium (and to a lesser extent magnesium) is a network-modifying element which detrimentally affects glass durability.

surface
features

bulk
features

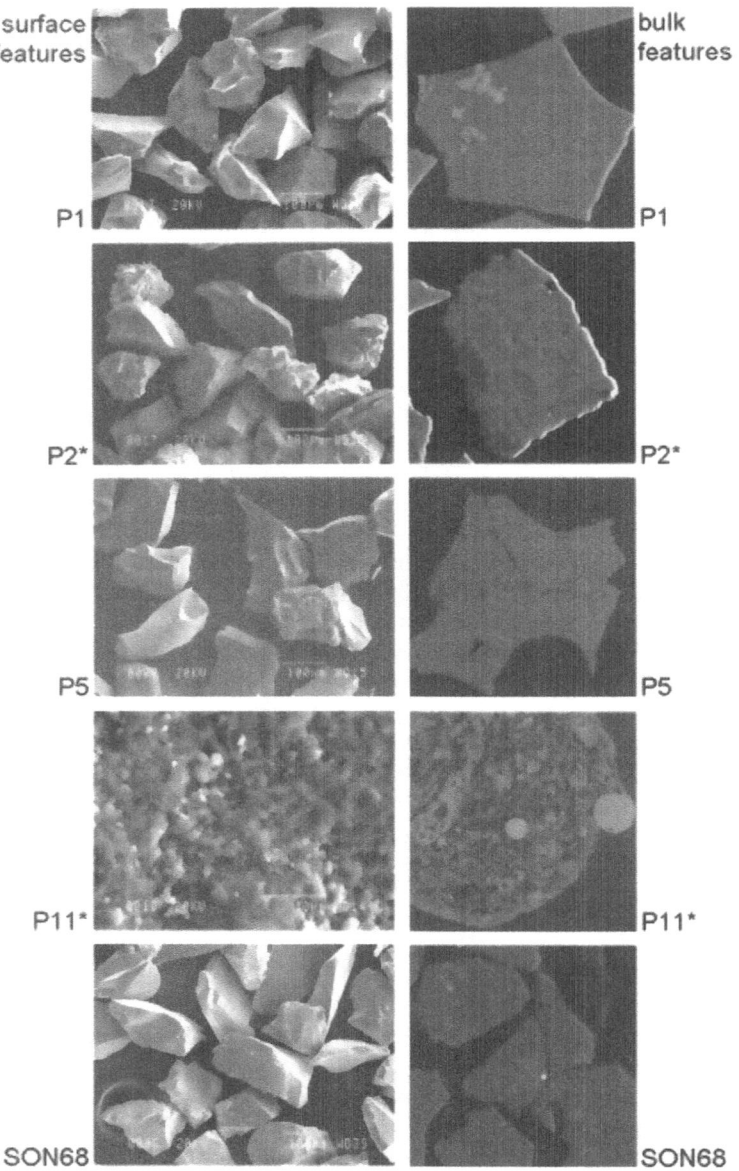

P1

P1

P2*

P2*

P5

P5

P11*

P11*

SON68

SON68

Fig. 5.58. Scanning electron micrographs of 100-125 µm ground HT materials. Left: Surface features (apparent roughness; SEM in secondary electrons mode). Right: Bulk features (density of mineral inclusions; SEM in backscattered electrons mode on resin-embedded and polished samples)

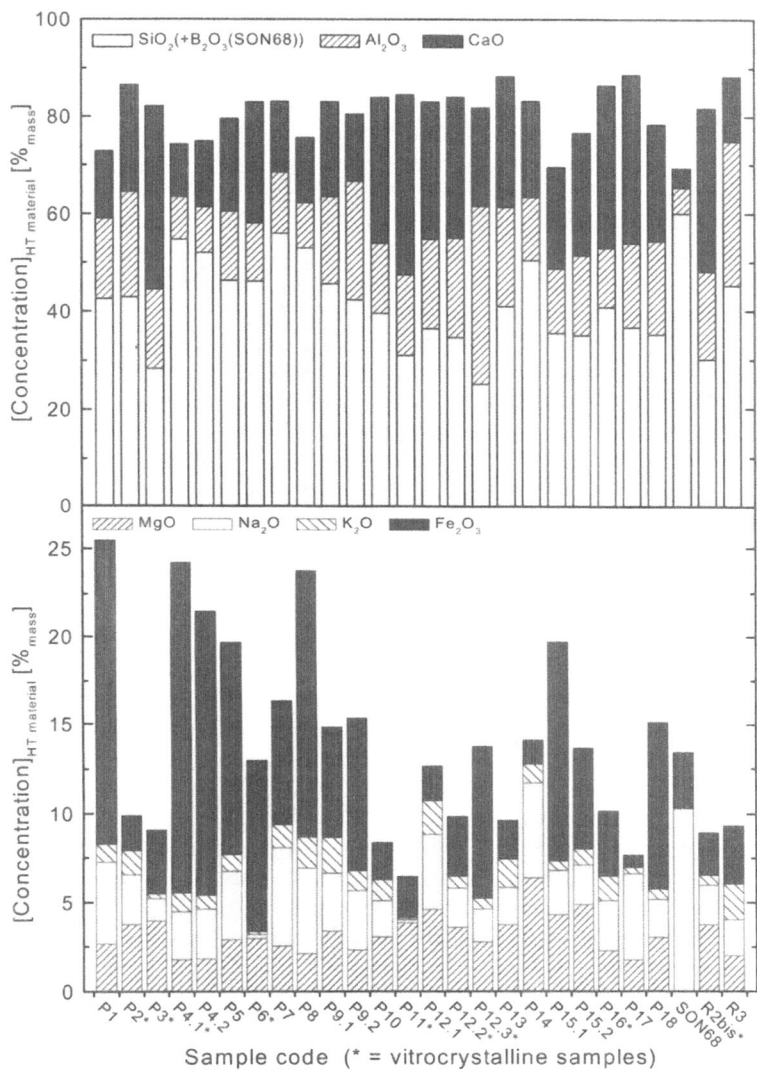

Fig. 5.59. Proportions of the major constituents in HT materials. Top: SiO$_2$ (network-forming), Al$_2$O$_3$ (network-forming or modifying) and CaO (network-modifying); for the HT standard SON68 (borosilicate glass), SiO$_2$ and B$_2$O$_3$ (added during vitrification for better control on the plasticity of the melt) have been summed up. Bottom: MgO, Na$_2$O, K$_2$O and Fe$_2$O$_3$

Our HT materials have physical and microscopic characteristics similar to glasses, and we expect them to exhibit contrasting features connected with their Si, Al and Ca content; in particular, SiO_2-rich HT materials should have a longer lifetime when subjected to corrosion.

For all samples and standards, $[SiO_2 + Al_2O_3] = 45\text{-}75$ %; this is an indication that HT materials should exhibit a high durability, at least the vitreous ones. There is no relationship between proportions of major constituents and crystallinity, though CaO and SiO_2 are inversely related. This is true in particular for HT materials produced by post-processes for fly ash, where the ratio $[SiO_2]:[CaO]$ tends to be much lower (1.4 for P10-P18) than for in-line processes (2.94 for P1-P4.2) or post-processes for bottom ash (2.92 for P5-P9.2).

MgO, Na_2O, K_2O and Fe_2O_3 exhibit large variations from one material to another: their sums range between 6 % and 25 %, with no influence from process family or crystallinity. It is known that alkali-poor glasses are more durable than alkali-rich ones, because the latter build up phase-separated glasses with reduced resistance to corrosion [205, 224]; however, these constituents should not drastically influence the durability of HT materials, as the sum of their concentrations never exceeds 7 %, except for the standard HT material SON68 ($[Na_2O + K_2O] = 10$ %).

Trace Elements

The concentrations of heavy metals in HT materials with regard to environmental impacts and possible recycling are of utmost importance. Of course, a high amount of metal in an HT material does not necessarily translate into a large release of the metal under corroding constraints.

Overall, the concentrations of metals in a HT material are a good indication of the efficiency of the HT process in solubilising them and embedding them in the matrix of the material during melting. In addition, $[\text{metal}]_{HT\ material}$ is a useful comparative parameter with respect to the existing trigger values of the TOW, even if these apply exclusively to residues of conventional incineration (BA, FA, FC).

Figure 5.60 shows the concentrations of nine environmentally relevant metals in HT samples and standards. There is indeed no correlation between the families of HT processes and the metal contents $[\text{metal}]_{HT\ material}$, although several HT processes for fly ash accumulate larger amounts of the most volatile elements Cd, Zn, Pb and Sb.

For comparative purposes, Table 5.17 highlights the limited number of HT materials that would be admitted for disposal in landfills for inert materials on the exclusive basis of the Swiss legislation. Most HT materials contain between 1 and 5 metals that exceed the TOW limits. Indeed, it would be misleading to assess the critical sustainability of HT materials on the exclusive basis of their metal content without determining their release rate during corrosion.

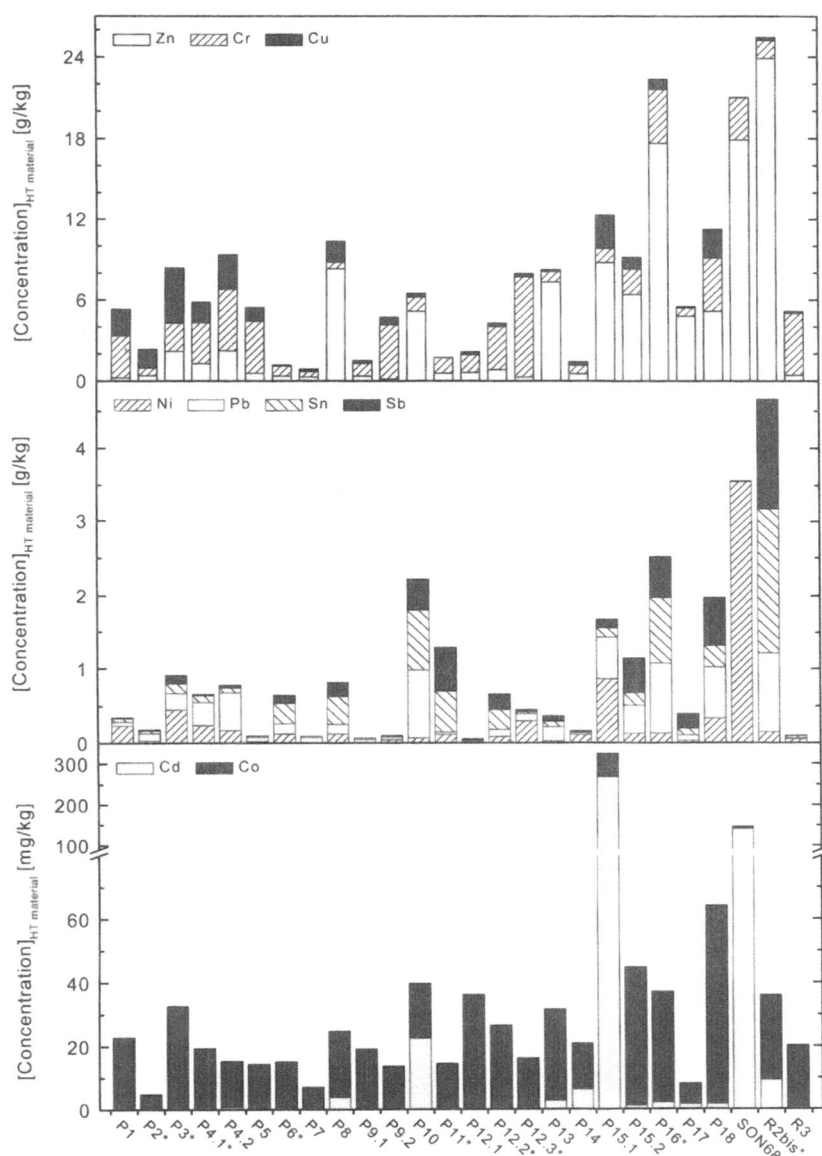

Fig. 5.60. Concentrations of nine environmentally relevant metals in HT materials

Table 5.17. Matrix of admissibility for disposal of HT materials in landfills for inert materials, according to the maximum allowed concentrations of heavy metals (TOW trigger value ; in parentheses). For HT materials that would not be admitted, the numbers indicated are the ratio [Element]$_{HT\ material}$:[Element]$_{TOW\ limit}$. Cr, Sn, Sb, Co, which are not listed in the TOW, are not mentioned here

HT material	Admissible in landfills for inert material	Elements above the maximum concentration allowed by TOW				
		Zn (1 g/kg)	Cu (0.5 g/kg)	Ni (0.5 g/kg)	Pb (0.5 g/kg)	Cd (0.01 g/kg)
P1	no		4 ×			
P2*	no		2.8 ×			
P3*	no	2.2 ×	8.2 ×			
P4.1*	no	1.3 ×	3 ×			
P4.2	no	2.2 ×	5.1 ×		1.02 ×	
P5	no		2.1 ×			
P6*	yes					
P7	yes					
P8	no	8.3 ×	3.1 ×			
P9.1	yes					
P9.2	no		1.1 ×			
P10	no	5.2 ×			1.8 ×	2.3 ×
P11*	yes					
P12.1	yes					
P12.2*	yes					
P12.3*	yes					
P13	no	7.4 ×				
P14	yes					
P15.1	no	8.8 ×	5 ×	1.7 ×	1.2 ×	26.7 ×
P15.2	no	6.4 ×	1.8 ×			
P16*	no	17.6 ×	1.5 ×		1.9 ×	
P17	no	4.8 ×				
P18	no	5.2 ×	4.3 ×		1.4 ×	
SON68	no	17.9 ×		7.1 ×		14.1 ×
R2bis*	no		23.9 ×		2.1 ×	
R3	yes					

The Global Static Picture of HT Materials

Taken as a whole, HT materials do not show obvious trends with respect to their physical, microscopic or chemical characteristics. It is thus difficult to draw general conclusions pertaining to the quality of these materials, because of the wide diversity of their parent processes (in-line or post-processes) and input materials. Nevertheless, the following differentiations can be summarized:

- HT materials are either homogeneous and vitreous, or heterogeneous and vitrocrystalline. The former should be less reactive under corroding conditions, while the mineral phases of the latter could account for contrasting reactivities. Purely crystalline materials were not identified.
- HT materials either exhibit a smooth surface with a small specific surface area and few mineral inclusions, or a rough surface with a large S_{spec} and a high density of inclusions. With some exceptions, these characteristics fit well together with the other physical characteristics (*i.e.* a vitreous material is homogeneous, with a small S_{spec}, a smooth surface and only few mineral inclusions, if any).
- With respect to the TOW limiting values for admissibility in landfills for inert materials, HT products either contain acceptably low proportions of toxic metals or too high proportions of these elements. It is not possible to differentiate between HT materials on the basis of their major constituents, in particular SiO_2, CaO and Al_2O_3, which account for *ca.* 70-90 % of the total mass of samples, as there is no reference material that could be used as a boundary for classification purposes. However, HT materials with a high ratio $([SiO_2] + [Al_2O_3]):[CaO]$ should theoretically exhibit a moderate reactivity.

Although a clear relationship between all physical, microscopic and chemical characteristics is difficult to establish, a combination of the measured parameters into a single descriptive figure would help to find plausible explanations for the expected behavior and durability of HT materials under corroding conditions. This is the purpose of the ternary diagram depicted in Figure 5.61.

The diagram combines all parameters discussed previously, *i.e.* vitreous or vitrocrystalline state, low or high specific surface area, surface smoothness or roughness, presence of few or many mineral inclusions, and a content in toxic metals below or above the TOW limiting values. Several minerals and rocks or anthropogenic materials are also indicated for comparison. Remarkably, HT materials belong to the domain of compositions that are able to form a silicate structure (area above the larnite - gehlenite - mullite boundary).

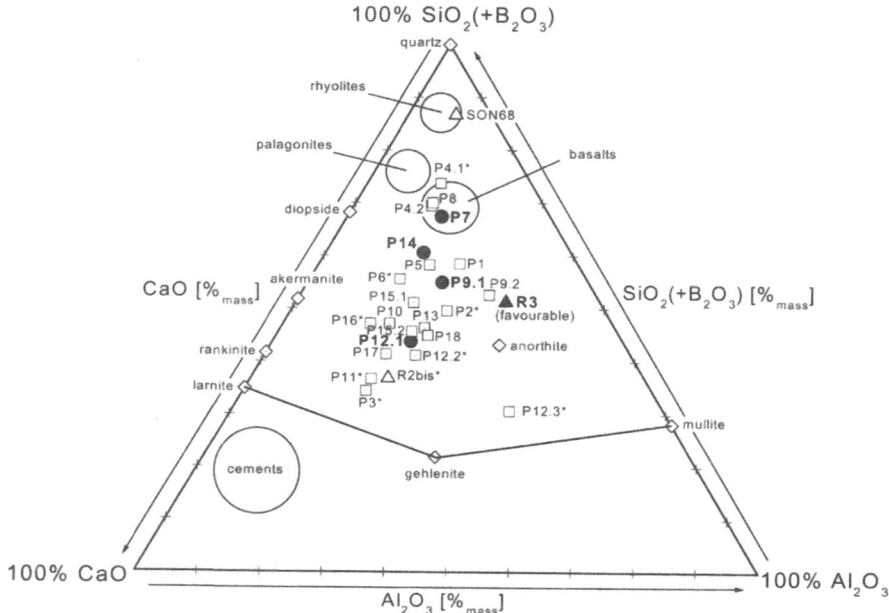

Fig. 5.61. Ternary SiO_2 - Al_2O_3 - CaO diagram of HT materials (solid circles and dotted squares), HT standards (solid and open triangles), rocks (rhyolites, palagonites, basalts ; areas), minerals (dotted diamonds) and cements (area). The line joining the minerals larnite, gehlenite and mullite is the boundary between silicate networks and crystalline structures (ortho-silicate). For HT materials, it is considered that $[SiO_2]$ + $[Al_2O_3]$ + $[CaO]$ (+ $[B_2O_3]$ for SON68) = 100 %. HT materials exhibiting an exclusively favorable static picture are depicted with a solid circle ; HT materials with at least one less favorable characteristic are depicted with a dotted square

A detailed analysis of data indicates that vitrocrystalline HT materials are distributed across the lower part of the diagram (with the exception of P4.1*). This is also roughly the case for HT materials which either present a high S_{spec}, a rough surface, or many mineral inclusions. On the other hand, HT materials containing toxic metals above the TOW limits are spread throughout the diagram, without relationship to their physical or microscopic characteristics.

In the ternary diagram of Figure 5.61, HT materials have globally favorable characteristics (dotted circles) when all their physical, microscopic and chemical characteristics are propitious to a lower reactivity and a higher durability (*i.e.* vitreous <u>and</u> low S_{spec} <u>and</u> smooth <u>and</u> only few inclusions <u>and</u> no toxic metal above the TOW limits). On the other hand, and under worst-case conditions, HT materials with at least one of their characteristics classified as detrimental are considered less favorable.

Only a limited number of HT materials (P7, P9.1, P12.1, P14 and the HT standard R3) exhibit consistently favorable static properties. It must however be kept in mind that this restrictive classification is not exact, as it is based on qualitative operator-dependent parameters (smoothness, density of inclusions), quantitative parameters (crystallinity, S_{spec}), and TOW trigger values which were not originally designed for HT materials.

In conclusion, HT materials cannot be clearly distinguished according to their origin. The large variability in input materials, thermal processes and quenching modes makes it difficult to extract obvious parameters controlling the final static picture of these products. The static picture of HT materials is therefore not sufficient to establish pertinent guidelines for their recycling or disposal into landfills. Nevertheless, the determination of the physical, microscopic and chemical characteristics of HT materials forms a sound basis for comparison and interpretation of their behavior under conditions of corrosion, as discussed in the next section.

5.4.3 Behavior of HT Materials: The Dynamic Picture

As already stated, there is no *a priori* relationship between the toxic metal content of HT materials and their long-term behavior, and it is thus not possible to forecast the evolution of a given type of HT material disposed of in a landfill or reused in civil engineering applications on the sole basis of its concentrations of metals.

For instance, a CaO-rich HT material will develop high pH values during corrosion, which may accelerate dissolution of the glassy matrix on a short-term basis; under these conditions, however, the apparent release of metals may diminish as a consequence of metal hydroxide precipitation. In addition, a high pH may favor the long-term formation of secondary zeolites, acting as a physico-chemical barrier against further corrosion. Likewise, an HT material with a high density of mineral inclusions shall have a higher reactivity under corroding constraints, but some of these mineral phases may be efficient metal scavengers. Consequently, one relies on the direct determination or estimation of the long-term behavior of HT materials under typical situational conditions to assess their compliance with respect to environmental impacts.

With nuclear HLW glasses, several corrosion experiments have been performed over a span of decades under real conditions of leaching (e.g. [33, 56, 71]). For a given HLW glass formulation, these long-term experiments highlight the consequences of specific constraints on the release of radionuclides and other species in the environment. However, for obvious practical reasons, one cannot consider testing a given HT material over a period of decades prior to making a decision about its final destination. As a consequence, one relies on the estimation of the long-term behavior of HT materials by means of accelerated corrosion tests, as is also the case for nuclear HLW glasses [18, 27, 33, 162].

Accelerated corrosion tests usually offer accurate insight into the mechanisms of matrix corrosion and kinetics of species release. However, these tests can only give rough estimates of the long-term behavior of the tested materials, because they are always performed under conditions far from realistic corrosion (*e.g.* high temperature, high concentrations of leachants, high reactive surfaces, *etc.*). One must thus keep in mind that accelerated corrosion tests, whatever their set-up, offer responses for situations of worst-case conditions.

Conditions of Accelerated Corrosion

In order to obtain the most accurate picture of the behavior of HT materials, the experimental set-up of the accelerated corrosion was performed under carefully controlled and reproducible conditions. For experimental reasons and because of the diversity of the HT materials (small granules, fragments, large blocks) a specific corrosion test, hereafter referred to as the Strasbourg test, was developed for this study; its main features are:

- Type of test: Static (no flow-rate, no stirring, closed vessel);
- Leachant: Ultrapure water (pH evolves freely during corrosion);
- Sample preparation: 100-125 μm ground HT materials;
- Ratio sample:leachant: *ca.* 50 mg:100 mL;
- Temperature: 90 °C;
- Duration: 1 day, 3 days, 10 days;
- Post-corrosion measurements: Analysis of the leachates for their pH and content in major, minor and trace elements; determination of the microscopic surface features of the corroded samples.

The chemical analyses of leachates are expressed either in absolute concentration $[C_i]_{leachate}$ (*i.e.* concentration of element i in leachate, in [g/L]), or in apparent normalized loss NL_i (NL_i [g/m^2] = {$[C_i]_{leachate} \times V_{leachant}$} / {$[C_i]_{HT\ material} \times S_{HT\ material}$}, with $V_{leachant}$ = volume of leachant [L], $[C_i]_{HT\ material}$ = concentration of element i in the HT material [g element/g sample], and $S_{HT\ material}$ = surface of HT material exposed to leachant [m^2]).

Absolute concentrations are useful in the comparison of the release of one given element from different HT materials, while apparent normalized losses are helpful in the comparison of the releases of different elements from one given HT material; in this context, NL_i values provide access to the mechanisms of corrosion (*e.g.* differentiation between congruent and selective releases from one HT material). To a certain extent, apparent normalized losses also help in comparing the mechanistic behaviors of different samples for one given element.

Surface Features after Corrosion

The micrographs in Figure 5.62 are a pertinent starting point in the discussion of the physico-chemical consequences of the corrosion of HT materials. These micrographs exhibit the surface structure of HT materials prior to corrosion and after 10 days corrosion. A qualitative estimate of the modifications undergone by samples during corrosion was also performed by a comparison of the apparent roughness, the density of mineral inclusions and the analysis of the mineral phases present at their surface, prior to and after corrosion.

Corroded HT materials exhibit almost systematically an amorphous gel layer and crystalline secondary phases. In some instances, pits and holes with preferential corrosion pathways are visible. Although experiments were performed without stirring, several samples show traces of spallation of the gel layer, suggesting that corrosion may be a discontinuous process. It is expected that the growing gel layer acts as a diffusion barrier and reduces further hydrolysis of the silicate network. Additionally, the gel layer is supposed to be chemically more durable than the silicate network, thus reducing the overall rate of corrosion. However, exfoliation of the gel layer leaves a fresh and reactive surface, which may momentarily accelerate corrosion, at least locally; this phenomenon suggests that mechanical constraints could partly govern the premature disaggregation of HT materials.

The petrographic analysis of secondary mineral phases identified at the surface of corroded HT materials yields a large, yet not exhaustive, variety of species, and a high sample-to-sample variability, suggesting that the surface reactivity of HT materials is influenced by the composition of their silicate network. The most abundant minerals identified are aluminosilicates, calcium phosphates, iron-rich phases, minerals enriched with magnesium, and even zeolites. The distinction between primary and secondary minerals is indeed difficult for HT materials showing a high heterogeneity (*e.g.* P2*, P6*, P11*). Nevertheless, no assumption can be made as to the role of these mineral phases, which may either act as scavengers of trace metals or enhance their release in solution. It is, however, expected that secondary mineral phases increase the surface heterogeneity of HT materials, and hence their specific surface area.

Release of Major Elements

The apparent matrix dissolution of HT materials, expressed as the absolute concentrations of the major elements Si (network-forming), Al (network-forming or modifying) and Ca (network-modifying) released in the leachates after 1 day, 3 days and 10 days of corrosion, is shown in Figure 5.63. Information on the true release of elements, *i.e.* their extraction from the silicate network, is not readily accessible because significant proportions of many elements released are usually immobilized into secondary minerals at the surface of corroded HT materials, or precipitated at the surface or in solution during corrosion, and also because most glasses have been observed to develop a gel layer which partly scavenges the extracted elements [61, 62, 89, 148, 172].

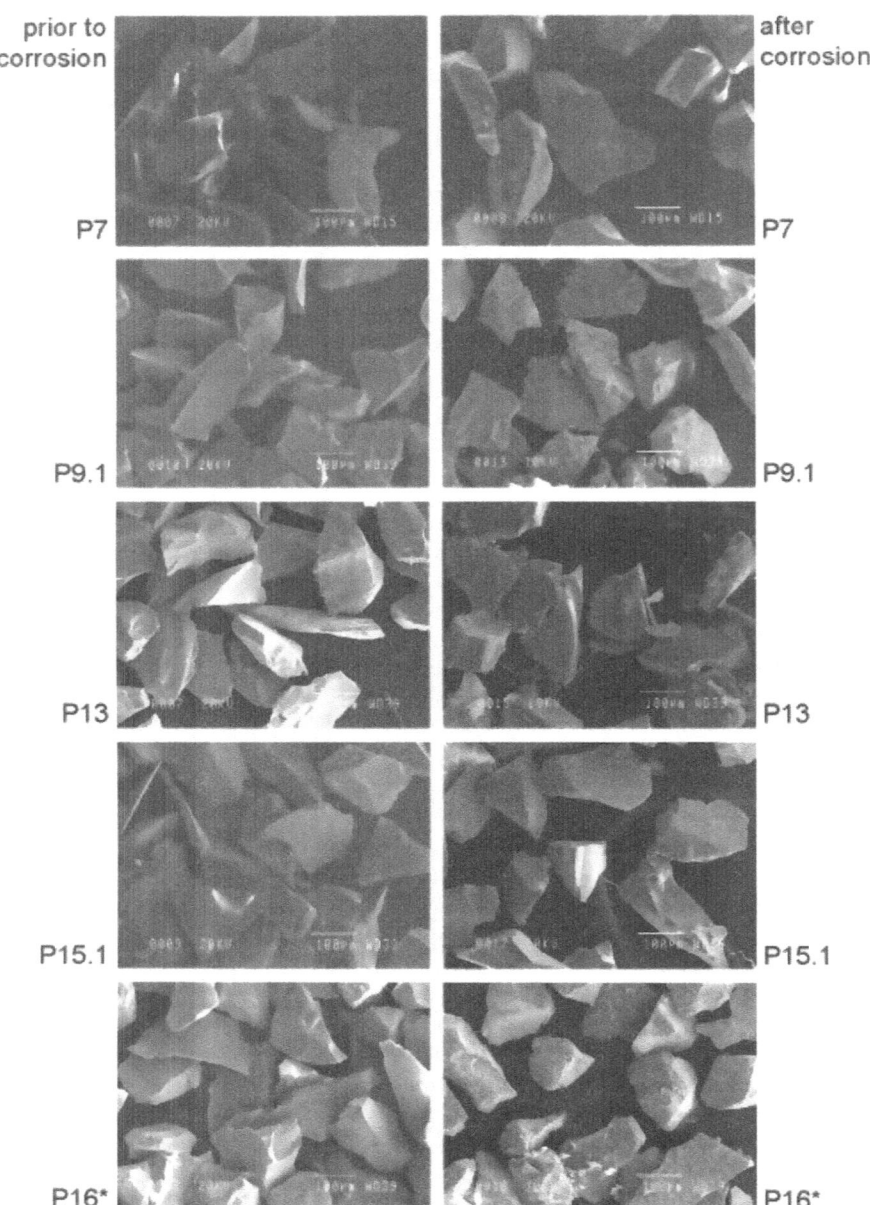

Fig. 5.62. Scanning electron micrographs (SEM in secondary electrons mode) of the surface features of 100-125 μm ground HT materials. Left: Prior to corrosion. Right : After 10 days corrosion in H_2O at 90 °C

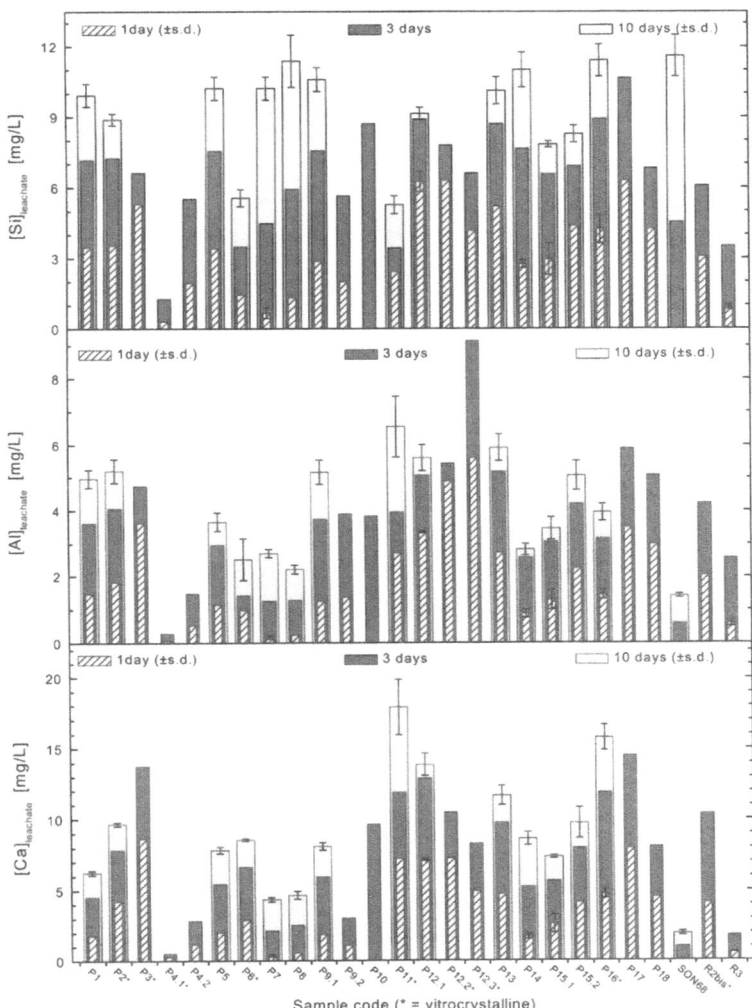

Fig. 5.63. Concentrations of Si, Al and Ca measured in the 0.2 μm filtered leachates of the Strasbourg test after 1 day, 3 days and 10 days corrosion of *ca.* 50 mg ground HT materials in 100 mL H$_2$O at 90 °C. For duplicate analyses, the minimum and the maximum concentrations are also displayed as error bars

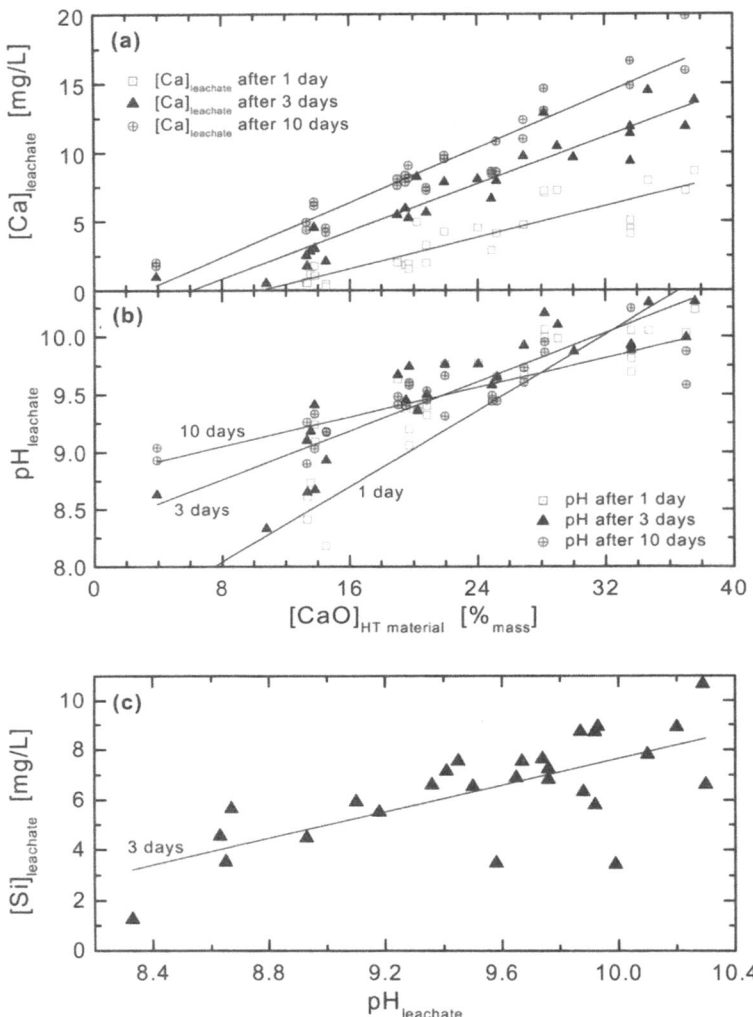

Fig. 5.64. Relationships between strongly correlated parameters measured in the leachates. (a) $[Ca]_{leachate}$ *vs.* $[Ca]_{HT\ material}$ shows the direct influence of the amount of calcium in the matrix of HT materials on the release of this element after 1 day, 3 days and 10 days corrosion. (b) $pH_{leachate}$ *vs.* $[Ca]_{HT\ material}$ shows that the pH of the leachate is mostly, but not exclusively, controlled by the the amount of Ca in the matrix of HT materials. (c) $[Si]_{leachate}$ *vs.* $pH_{leachate}$ (after 3 days corrosion) shows that the dissolution of the matrix is pH-dependent

The concentrations of major elements released in solution increase with corrosion time. The trend is, however, different from sample to sample, and from element to element. After 1 day of corrosion, the release of silicon ranges from 0.4 mg/L (P7) to 6.7 mg/L (P12.1), but the difference is less pronounced after 3 days (1.3 mg/L (P4.1*) to 10.7 mg/L (P17)), and even less after 10 days (4.9 mg/L (P11*) to 12.5 mg/L (P8, SON68)); this indicates that the release of Si stabilizes over time. Similar observations can be made for the release of aluminum and calcium, but the global sample-to-sample fingerprints of Si, Al and Ca do not superimpose; this suggests that matrix dissolution follows specific pathways (*e.g.* different solution chemistries and/or secondary phases formed). These differences cannot be explained in terms of process families (in-line processes, post-processes) or crystallinity of the HT materials.

Figure 5.64 shows the strong relationships which exist between $[Ca]_{HT\ material}$, $[Ca]_{leachate}$, $pH_{leachate}$ and $[Si]_{leachate}$. In fact, calcium, a network-modifying element, is easily extracted from the silicate network of HT materials (Figure 5.64a), and the released amounts of Ca are directly related to the proportion of Ca in the HT materials. This alkaline-earth is present in fairly large amounts in samples (11 % (P4.1*) to 38 % (P3*), except for SON68 (4 %)), and its release primarily governs the pH of the leachates (Figure 5.64b) as given by the following reactions:

$$\{CaO\}_{HT\ material} + 2H_2O \rightarrow Ca^{2+} + 2OH^-$$
$$\{CaSiO_3\}_{HT\ material} + 2H_2O \rightarrow Ca^{2+} + H_4SiO_4 + 2OH^-$$

$pH_{leachate}$ increases with time, but the effect is more pronounced between 1 and 3 days than between 3 and 10 days, indicating that pH stabilizes rapidly during matrix dissolution, as is observed for Ca, Al and Si.

Silicon, in turn, dissolves under the influence of alkaline conditions (Figure 5.64c), as expected from the mechanisms of glass corrosion described in [56, 205], which shows that the higher the pH, the less stable the silicate network :

$$\{\equiv Si\text{-}O\text{-}Si(OH)_3\}_{HT\ material} + OH^- \rightarrow \{\equiv Si\text{-}O^-\}_{HT\ material} + H_4SiO_4\ \text{(hydrolysis of}$$
the external silicate network)

$$\{\equiv Si\text{-}O\text{-}Si\equiv\}_{HT\ material} + OH^- \rightarrow \{\equiv Si\text{-}O^-\}_{HT\ material} + \{HO\text{-}Si\equiv\}_{HT\ material}$$
(hydrolysis of the internal network)

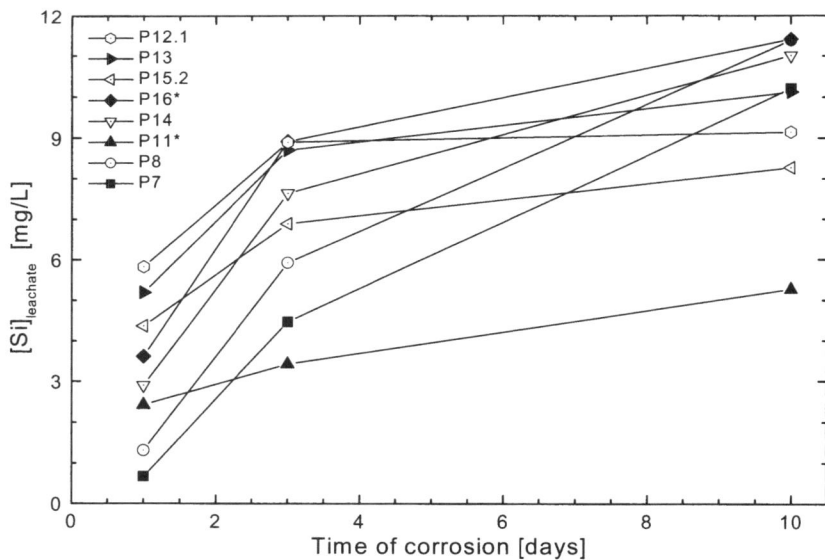

Fig. 5.65. Evolution of the concentration of Si in the leachates of several HT materials corroded during 1 day, 3 days, and 10 days. The temporal fingerprint is different from sample to sample

Mechanistically, corrosion of HT materials by pure water leads to the release of Ca, which in turn increases the pH of the leachate, with a concomitant enhancement of matrix dissolution and Si release; the apparent rate of these mechanisms decreases with time but is sample-dependent, as exemplified in Figure 5.65. These observations suggest that Ca-poor HT materials (*i.e.* Si-rich HT materials) are less prone to corrosion, and thus more durable, than Ca-rich HT materials, without a strong influence of other physico-chemical parameters. However, the variable short-term behavior of HT materials, at least with respect to the dominant component SiO_2, makes it difficult to accurately predict their long-term stability with regard to corrosion, or even to make a classification of HT materials on the basis of their dynamic picture.

Although the evolution of the major elements Al, Ca, Na and K in the leachates does not superimpose onto the evolution of Si, their trend is globally similar to that of Figure 5.65: Concentrations in the leachates stabilize over time. On the other hand, Mg exhibits a different behavior: After an increase between 1 and 3 days of corrosion, $[Mg]_{leachate}$ drastically decreases, indicating the formation of Mg-rich secondary phases, either in solution or at the surface of the corroded HT materials. This is in agreement with the identification of abundant Mg-rich secondary mineral phases at the surface of samples after corrosion.

Release of Trace Elements

The time profiles of the major elements indicate that the strong element-to-element and sample-to-sample differences follow no specific trend with respect to the physical, microscopic and chemical characteristics previously discussed. This is also clearly observed for the trace elements Cr, Co, Ni, Cu, Zn, Cd, Sn, Sb and Pb, which show highly contrasting fingerprints in leachates, both in terms of element-to-element and sample-to-sample dissimilarities. Table 5.18 summarizes the average and maximum concentrations of these trace elements measured in the leachates, along with the maximum amount of concentrations permitted according to the TOW leaching test used for the disposal of wastes and residues of incineration in landfills for inert materials.

Indeed, the Strasbourg test developed for this study and the TOW leaching test cannot be directly compared: The former is performed in ultra-pure water at 90 °C during a period of 10 days, with a ratio mass:volume = 50 mg:100 mL, $i.e.$ under more aggressive conditions than the latter, which is performed in CO_2-saturated water at room temperature [2]. Nevertheless, the measured concentrations of trace elements released during corrosion are remarkably low (Co, Cd, Sn, Pb : below 1-5 µg/L ; Cr, Ni, Cu, Zn, Sb : below 10-100 µg/L) and systematically beneath the maximum concentrations allowed by Swiss regulations. By comparison, untreated bottom ash or fly ash usually leads to much higher metal concentrations in leachates when corroded under the conditions of the TOW leaching test.

Table 5.18. Average and maximum concentrations of metals measured in the leachates after 1 day, 3 days, 10 days corrosion of HT materials under the conditions of the Strasbourg test (50 mg ground HT material in 100 mL H_2O at 90°C, without stirring). HT standards (SON68, R2bis, R3) are omitted. For comparison, the maximum concentrations of metals allowed by the TOW test for disposal of residues in landfills for inert materials are given

Eleme nts	1 day corrosion		3 days corrosion		10 days corrosion		TOW limit
	$[M^{n+}]_{mean}$ [µg/L]	$[M^{n+}]_{max}$ [µg/L]	$[M^{n+}]_{mean}$ [µg/L]	$[M^{n+}]_{max}$ [µg/L]	$[M^{n+}]_{mean}$ [µg/L]	$[M^{n+}]_{max}$ [µg/L]	$[M^{n+}]_{max}$ [µg/L]
Cr	b.d.l.	b.d.l.	b.d.l.	b.d.l.	30	40	50
Co	0.18	0.40	0.05	0.14	0.04	0.09	50
Ni	b.d.l.	b.d.l.	b.d.l.	b.d.l.	14	22	200
Cu	b.d.l.	b.d.l.	b.d.l.	b.d.l.	14	19	200
Zn	19	25	37	113	27	47	1000
Cd	0.06	0.10	0.04	0.10	1.1	2.7	10
Sn	0.20	0.64	0.20	0.65	0.13	0.69	200
Sb	2.9	10	6	26	7	38	no limit
Pb	0.27	0.74	0.37	2.2	0.29	2.5	100

b.d.l. below detection limit

Although a detailed mechanistic interpretation of the results is not possible, trace metals extracted from the silicate network can be scavenged by the amorphous gel layer or by secondary minerals which form at the surface of the HT materials, thus explaining the very low concentrations in the leachates. This is of course a favorable behavior of HT materials with regard to the initiation of toxic metals into vitreous and vitrocrystalline silicate networks.

Without making any distinction between elements, processes, or duration of the leaching experiments, the release of major, minor and trace elements in leachates is not congruent: Apparent normalized release rates r_{norm} ($r(i)_{norm} = dNL_i/dt$) range between $ca.$ 10^{-4} $g/m^2 \cdot d$ and $ca.$ 30 $g/m^2 \cdot d$. Alkali elements (Na and K, but not Rb) are released at higher rates ($r(i)_{norm}$ around 0.5-1.8 $g/m^2 \cdot d$) than network-forming and network-modifying elements (Si, Al, Ca, Mg, Sr, Ba ; $r(i)_{norm}$ around 0.2-0.6 $g/m^2 \cdot d$); indeed, the former are mostly released in solution without side reactions, while the latter are partly re-incorporated into secondary minerals, thus highlighting the complex solution and secondary phase chemistries involved and the non-congruency of the mechanisms of release [18, 88]. Fe and Mn ($r(i)_{norm}$ around $2 \cdot 10^{-2}$ $g/m^2 \cdot d$) are almost exclusively precipitated into secondary minerals during matrix dissolution, leaving no Fe or Mn in the leachates. Trace elements exhibit highly contrasting behaviors, and are either retained in crystalline phases (*e.g.* Cr-, Ni- and Co-spinels, Zn-melilite) or dissolved in the leachate. Apparently, process families influence the normalized release rates of major elements (in-line processes > post-processes for FA > post-processes for BA).

What is probably the most important feature observed during corrosion is the possibility to discriminate HT materials with respect to the initial stages of network hydrolysis: As shown in Figure 5.66, short-term release rates are highly differentiated from one sample to another, but they rapidly decrease to small, similar values ($r(i)_{norm}$ around 0.2 $g/m^2 \cdot d$). This suggests that the physico-chemical characteristics of HT materials influence their initial reactivity, but that the dissolution rates also converge to low limiting values, whatever the nature of the HT materials.

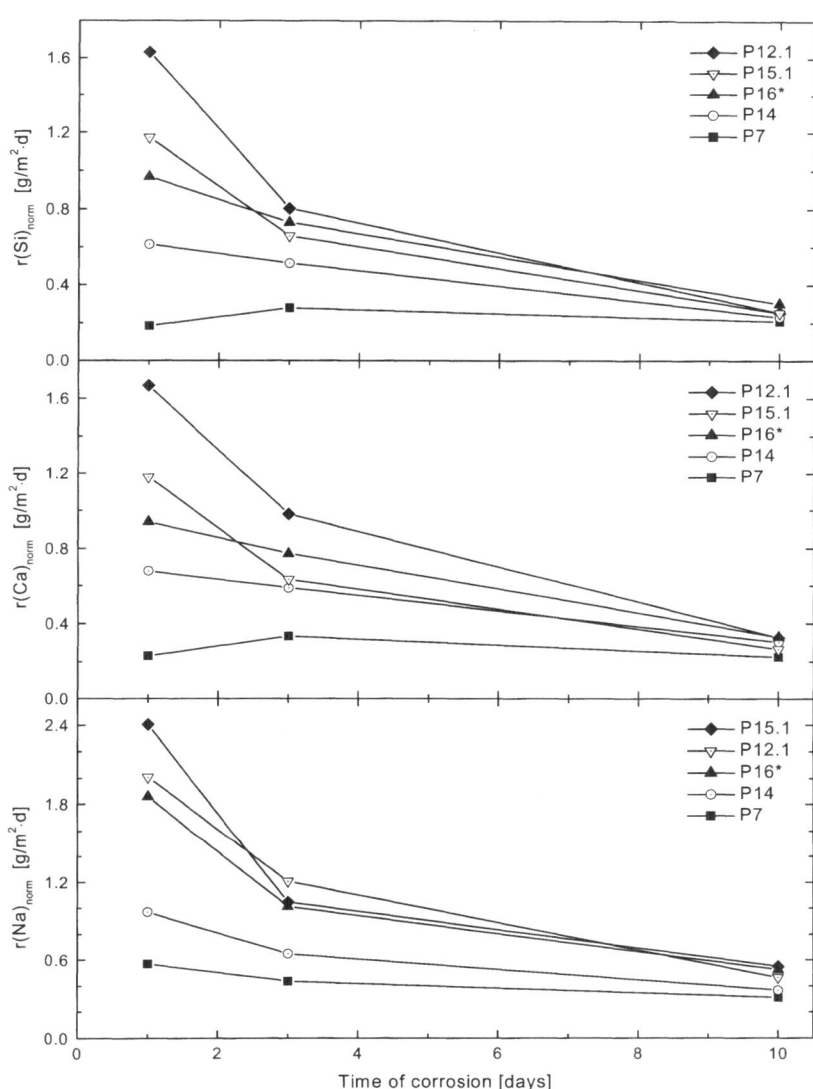

Fig. 5.66. Temporal evolution of the apparent normalized release rates of some major elements in HT materials. Although the initial rates (*i.e.* after 1 day) are different from sample to sample, the rates measured after 10 days tend to a limiting value, whatever the HT material

The Global Dynamic Picture of HT Materials

Figure 5.67 compares HT materials with respect to their corrosion behavior. Materials with a normalized release rate $r(Si)_{norm} \leq SON68$, such as P1, P2*, P4.1*, P6*, P7, P11* and R3, are considered to have a consistently favorable dynamic picture.

Although the classification is based on relative terms (comparison of $r(Si)_{norm}$), the following considerations can be taken into account for a clearer characterization of HT materials with respect to their behavior during corrosion: Microscopically, several materials develop a gel layer and secondary mineral phases, but their effect is not known in detail. Short-term corrosion (*i.e.* initial release rate) shows large differences from sample to sample and from element to element; this allows for differentiation between fast-reacting and slow-reacting HT materials during the first steps of matrix hydrolysis. Overall, release rates rapidly decrease towards low values (r_{norm} around 0.2 $g/m^2 \cdot d$ for Si, Al, Ca after 10 days of corrosion). Finally, the release rates of trace elements are highly scattered, but their concentrations in leachates are systematically and remarkably very low; even under the drastic conditions of the Strasbourg test, $[metal]_{leachate}$ are below the TOW limits for admissibility into landfills for inert materials.

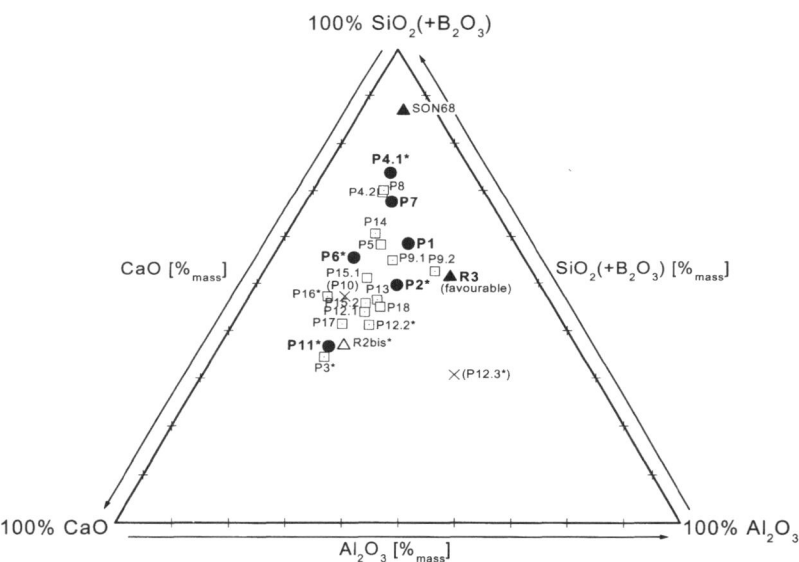

Fig. 5.67. Ternary SiO_2 - Al_2O_3 - CaO diagram of HT materials (solid circles, dotted squares and crosses) and HT standards (solid and open triangles). HT materials exhibiting a favorable dynamic picture (*i.e.* with $r(Si)_{norm} \leq SON68$) are depicted with a solid circle. HT materials P10 and P12.3* are depicted with crosses, because their normalized release rate could not be calculated (S_{spec} missing). The classification is based on the 3 days corrosion experiment, which was performed on the whole set of samples

The results of the corrosion tests show that the behavior of HT materials during corrosion is in agreement with observations on nuclear glasses and analog materials. Under the conditions of the Strasbourg test, the concentrations of major, minor and trace elements measured in solution are very low (< 20 mg/L for major elements and < 100 µg/L for trace metals after 10 days corrosion at 90 °C). HT materials comply with the conditions of the TOW leaching test for admissibility into landfills for inert materials, even when their content in toxic metals is above the TOW limits (see 5.4.2).

The chemistry of the leachates at the beginning of corrosion is governed by the composition of HT materials, but it evolves rapidly towards similar patterns of behavior. Corrosion is controlled by the pH of the leachate, which is in turn in direct correlation to the amount of CaO present in HT materials. It follows that SiO_2-rich/CaO-poor HT materials are less prone to corrosion, even during the initial stages of alteration.

As already stated, the legitimacy of the actual TOW regulation for the assessment of wastes and residues (*i.e.* the TOW limits on the content of toxic metals and their release during leaching tests) may be addressed. For instance, HT materials like P1, P2* and P4.1* release only minute amounts of toxic metals in the leachates of the Strasbourg test, but could not be admitted into a landfill for inert materials because they contain excessively high proportions of several metals.

5.4.4 Durability of HT Materials: The Thermodynamic Picture

Undoubtedly, the assessment of the long-term stability of HT materials requires realistic conditions of corrosion (*e.g.* burial, landfill disposal, reuse in road construction or other civil engineering applications) during an appreciable amount of time (months to years) in order to allow for kinetically slow mechanisms to take place. This approach is, however, technically difficult to perform, and the alternative of accelerated corrosion experiments is a convenient way to overcome the limitations of real-time alteration experiments. Nevertheless, accelerated corrosion, with its cohort of operational parameters (high temperature, high reactive surface, *etc.*) forcing the solid-solution far from equilibrium, may either overestimate or underestimate in an uncontrolled manner the dissolution rates when extrapolated to realistic conditions.

Another way to estimate the relative durability of HT materials is to make use of thermodynamic models, amongst which the strongly documented thermodynamic assessment of glass hydration is certainly the most appropriate and convenient. The approach requires only a limited number of parameters (composition of HT material; thermodynamic hydration constants of the constituents) for the calculation of the relative stability of an HT material, and its application to a large palette of glasses and analog materials makes it practical for comparative purposes.

In this context, the seminal work of Paul and Newton [159, 160, 165, 166] on the thermodynamics of hydration of glasses and vitreous materials opens strong opportunities for the estimation of the durability of HT materials on the basis of their composition. According to the concept of Paul and Newton, a glass is a homogeneous assembly of silicate polyhedra of the different constituents present. When subjected to water corrosion, the glass hydrates in proportion to its overall thermodynamic characteristics. The approach is based on simplifying assumptions, *i.e.* the release of species from the glass matrix is a congruent process and no secondary products form at the surface of the glass. The overall free energy of hydration ΔG_{hydr} represents the thermodynamic ability of the glass to hydrate spontaneously in water, and it can be estimated as the molar-weighted sum of the free energies of hydration $\Delta G_{hydr}(i)$ of the individual constituents of the glass.

This concept was later refined and successfully used to predict the long-term stability of nuclear HLW glasses that must be stored in deep geological repositories [33, 71, 111-113, 174, 175, 221]. In addition, it has been observed empirically that there is a fairly good relationship between calculated ΔG_{hydr} of several glass types (HLW, natural, ancient and commercial glasses) and the apparent normalized release rate of major elements during accelerated corrosion experiments. However, the approach had never been used before to predict the relative stability of materials originating from the high temperature incineration of municipal solid wastes or their residues [167, 168].

Application of the Concept of Glass Hydration to HT Materials

A correct definition of glass or vitreous material is required to apply the concept of glass hydration to HT materials. During the high temperature incineration of municipal solid wastes or their residues, the elements present in the melt combine in a liquid phase. Provided that the quenching (cooling down of the melt) is rapid, the liquid phase transforms into a more or less homogeneous solid. This solid is characterized by a 3-dimensional network of tetrahedrons MO_4^{4-} (M = Si, Al, B ; network-forming elements), with network-modifying elements (Ca, Mg, Na, K, Fe) and trace elements being dissolved or intercalated into the matrix. This network is analogous to the homogeneous assembly of silicates in a glass, rather than simply the sum of its oxides.

Therefore, provided that the raw composition of a vitreous HT material is known (and expressed as the oxides of the constituents; *e.g.* SiO_2, Al_2O_3, CaO, Na_2O, K_2O, Fe_2O_3, *etc.*), an assembly of silicates and residual oxides can be built (*e.g.* $CaSiO_3$, $MgSiO_3$, Na_2SiO_3, K_2SiO_3, *etc.*), each having its own free energy of hydration $\Delta G_{hydr}(i)$. In the frame of the present study, the concept of the free energy of hydration of glasses was applied to the vitreous HT materials and extended to the vitrocrystalline HT materials, considering that the structural inhomogeneities (mineral inclusions) of the latter induce only insignificant distortions of the assembly of silicates.

Provided that the hydration of the individual silicates and residual oxides governs the overall dissolution of the HT material, one expresses the free energy of hydration as $\Delta G_{hydr} = \Sigma(X_i \times \Delta G_{hydr}(i))$, where X_i is the molar fraction of the constituent (i) in the HT material, and the individual values of $\Delta G_{hydr}(i)$ being available in the literature for most silicates and oxides, or calculated as $\Delta G_{hydr}(i) = -RT \times \ln(K_{hydr}(i))$. The more negative values of ΔG_{hydr} correspond to materials that hydrate more spontaneously, *i.e.* that dissolve easily in water. The thermodynamic approach thus consists of estimating the durability of HT materials on the basis of their thermodynamic propensity to corrode, and to relate this durability to the release rate determined experimentally.

Thermodynamic Stability of HT Materials

After conversion of HT materials (expressed as oxides) into the silicates $CaSiO_3$, $MgSiO_3$, Na_2SiO_3, K_2SiO_3, $BaSiO_3$, $ZnSiO_3$, $ZrSiO_4$ and residual oxides SiO_2, Al_2O_3, CaO, Fe_2O_3, MnO_2, TiO_2 (+ B_2O_3, Li_2SiO_3, Cs_2O_3, Nd_2O_3, $SrSiO_3$, NiO, MoO_3, La_2O_3, Ce_2O_3 for the HT standard material SON68), their $K_{hydr}(i)$ are extrapolated for $T = 90\ °C$ [18], and their free energy of hydration is calculated as $\Delta G_{hydr}(i) = -RT \times \ln(K_{hydr}(i))$ (for details on the approach, see Perret *et al.*, 2002). The overall free energy of hydration is then calculated as explained above. $K_{hydr}(i)$ values for trace elements are either unknown or subject to large variabilities. These trace elements were therefore not included in the calculation of ΔG_{hydr}, except for Ba, Zn, Zr (known $K_{hydr}(i)$; non-negligible concentrations in several samples). It is estimated that the error in ΔG_{hydr} caused by the missing contribution of trace elements is negligible, except for Cr (up to 7500 mg/kg) and Cu (up to 2500 mg/kg) in some samples.

Figure 5.68 shows the dependency of the calculated ΔG_{hydr} of HT materials on the pH of the leachate, over a much wider range of pH values (1 to12) than the ones measured during the 1-10 days corrosion experiments. The results obtained for vitrocrystalline HT materials are shown for comparative purposes, assuming that their crystallinity is not large enough to impair the overall free energy of hydration. Taken collectively, the thermodynamic stability reaches a maximum at around pH = 5-7.5. As expected from theoretical considerations and experimental observations [166], the hydration becomes more spontaneous under alkaline conditions than under acidic conditions.

Although the observed sample-to-sample variability in ΔG_{hydr} is large ($\Delta G_{hydr,max} = +0.1$ kcal/mol for P4.1*, the most stable sample; $\Delta G_{hydr,min} = -12.6$ kcal/mol for P10, the least stable sample), the average stability of HT materials is comparable to that of the durable HLW glass SON68; this is a positive indication of the long-term stability of HT materials. Overall, HT materials produced by post-processes for FA are less stable ($\Delta G_{hydr} = -9.6$ kcal/mol $\pm$ 2.0) than the ones produced by in-line processes (-4.7 kcal/mol $\pm$ 3.1) or post-processes for BA (-5.0 kcal/mol $\pm$ 1.5). This can be accounted for by the higher amount of CaO in HT materials produced from FA.

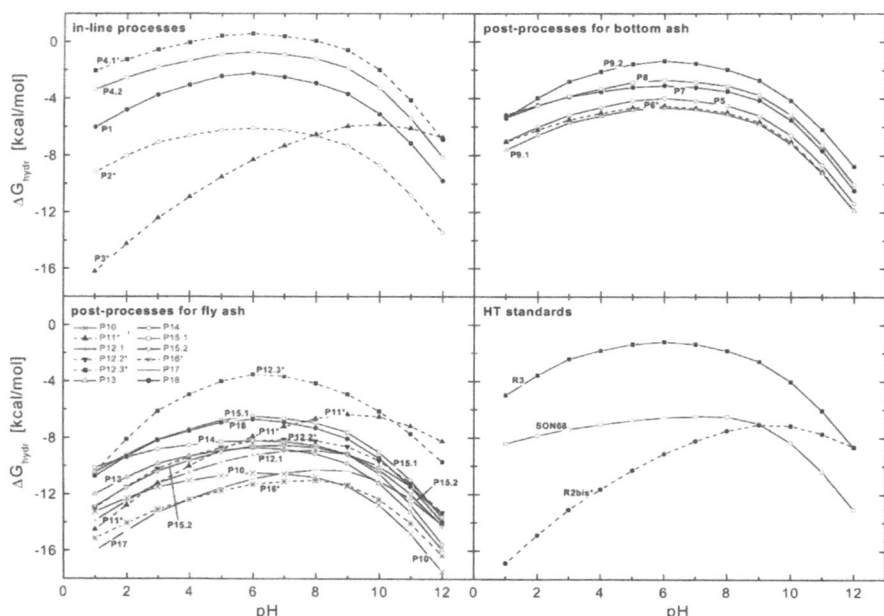

Fig. 5.68. Calculated free energy of hydration ΔG_{hydr} of the HT materials in function of the pH. Solid curves apply to purely vitreous HT materials, for which the model of Paul and Newton was originally designed. Dotted curves apply to vitrocrystalline HT materials. The pH values of the leachates after 1 day, 3 days and 10 days range between 6.7 and 10.3

The overall free energy of hydration of a given HT material is primarily governed by the constituents present in abundant proportions and exhibiting large $\Delta G_{hydr}(i)$ values. $CaSiO_3$ (17-63 %mol ; $\Delta G_{hydr}(i) = -16$ kcal/mol), and to a lesser extent Na_2SiO_3 and $MgSiO_3$, are the species which contribute mostly to the reduction in thermodynamic stability of HT materials. By comparison, the stabilizing contribution of the oxides SiO_2, Al_2O_3, CaO, Fe_2O_3 and MnO_2 is weaker than the destabilizing contribution of most silicate polyhedra. Nevertheless, calculations clearly show that HT materials containing the largest possible proportions of network-forming elements and the lowest possible proportions of network-modifying elements exhibit the highest thermodynamic stabilities.

Combining Rates of Corrosion and Thermodynamic Stabilities

As previously stated, there exists a clear relationship between the overall free energy of hydration of glasses from different origins and their release rate during corrosion experiments. This empirical relationship was obtained by means of an accelerated corrosion test, the MCC-1 test [33, 66]. This test was designed to assess the leachability of nuclear HLW glasses; it is a static test performed on glass cubes in water, at 90 °C over a period of 28 days.

Due to the intrinsic nature of several of the HT materials, it was not possible to prepare polished monoliths for all samples, and thus the MCC-1 test could not be used. Nevertheless, the MCC-1 and Strasbourg tests can be compared in terms of leaching efficiency ($S_{HT\ material}/V_{leachant}$)×t ($S_{HT\ material}$ = surface exposed to leachant; $V_{leachant}$ = volume of leachant; t = duration of the corrosion experiment). For the MCC-1 test, (S/V)×t = 2.8 d/cm. For the Strasbourg test, (S/V)×t = 0.2-2 d/cm (1-10 days corrosion, S_{spec} = 400 cm^2/g). Thus, both tests can be compared, as their leaching efficiency is in the same order of magnitude.

The validation of the ΔG_{hydr} approach for HT materials is given in Figure 5.69. Apparent normalized release rates of Si, measured after 1, 3, and 10 days of corrosion, are plotted in function of the calculated values of ΔG_{hydr}. The results of the MCC-1 test obtained on a series of 115 glasses of different origins [174] are superimposed for comparison.

Whatever the duration of the Strasbourg test, Figure 5.69 exhibits a strong relationship between r(Si)$_{norm}$ and ΔG_{hydr}. The HT standard SON68 (Strasbourg test) yields results very close to the ones obtained for nuclear HLW glasses (MCC-1 test), confirming that both tests are comparable for similar samples and short-term corrosion. A rough difference of 10 kcal/mole between two HT materials translates into an approximate one order of magnitude difference between their dissolution rates. This demonstrates that the thermodynamic approach of the free energy of hydration, though limited by simplifying assumptions, can be applied for the discrimination of HT materials during the initial stage of matrix hydrolysis. After 3 days and 10 days corrosion, the influence of the differences in ΔG_{hydr} on the release rates vanishes. This suggests that the behavior of HT materials depends on the conditions of corrosion, and evolves toward limited and uniform matrix dissolution (ca. 0.25 g/m^2·d, calculated from the release of Si after 3 and 10 days).

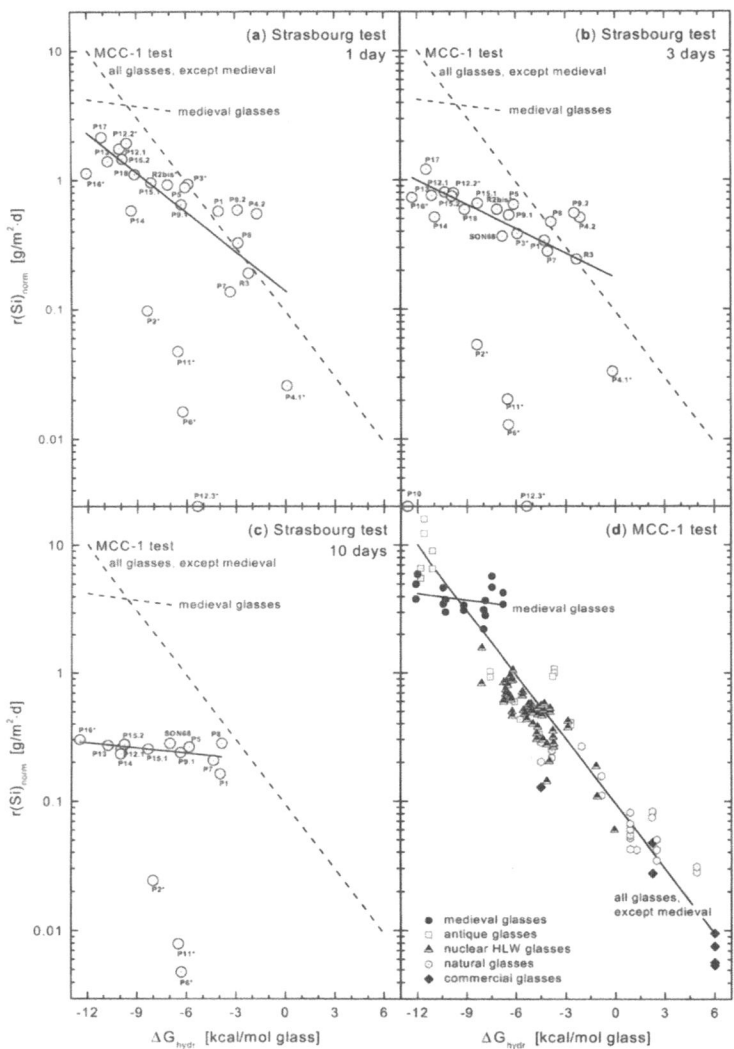

Fig. 5.69. Relationship between the normalized apparent release rate of Si in HT materials, $r(Si)_{norm}$, measured after (a) 1 day, (b) 3 days, respectively (c) 10 days of corrosion under the conditions of the Strasbourg test, and the free energy of hydration of HT materials, ΔG_{hydr}, calculated for the pH values measured after 1, 3, 10 days corrosion. The linear fits exclude P2*, P6* and P11* (S_{spec} too large), respectively P10 and P12.3 (S_{spec} missing). (d) Literature-extracted results [174] of $r(Si)_{norm}$ *vs.* ΔG_{hydr} obtained for 115 glasses of different origins corroded under the conditions of the MCC-1 test (reproduced as dashed lines in plots (a-c))

Figure 5.69 also indicates that the technical discrepancy between the MCC-1 test (blocks) and the Strasbourg test (powder) complicates a comparative interpretation. Blocks corroded under the conditions of the MCC-1 test react under a regime of initial corrosion, while the Strasbourg test facilitates the formation of Si-rich secondary phases, leading to a regime of quasi-equilibrium. This explains the differences in slopes between both tests. However, HT materials (Strasbourg test ; 10 days corrosion) and medieval glasses (MCC-1 test) exhibit similar flat slopes, probably because of the high proportions of Mg in both types of materials; indeed, Mg favors the formation of protective secondary phases on the surface of corroded glasses [61]. Overall, ΔG_{hydr} of HT materials lies within the range of the thermodynamic stabilities of nuclear HLW glasses.

The Global Thermodynamic Picture of HT Materials

It is difficult to extract simple information from the results of the thermodynamic approach, mostly due to the wide diversity of HT materials. Nevertheless, HT materials can be categorized with respect to their free energy of hydration, in comparison to the one of the durable HT standard SON68. In this respect, Figure 5.70 highlights the most and least durable HT materials; it must however be remembered that the results for vitrocrystalline samples are semi-quantitative.

The long-term durability of HT materials is estimated to range between 10^3 years and 10^4 years. This estimation is based on the observed durability of medieval and antique glasses ($\geq 10^3$ years; [226, 261-263], and on the extrapolated durability of nuclear HLW glasses ($\geq 10^4$ years; [263]). Figure 5.69 indicates that medieval, antique and nuclear HLW glasses can be considered (thermodynamically) pertinent analogs of HT materials. The results also show that there is no thermodynamic discrimination of HT materials in relation to their parent process.

Although Si and Ca govern the thermodynamic picture of HT materials, the calculation of ΔG_{hydr} involves the contribution of many other silicates or oxides with different hydration behaviors. The thermodynamic approach provides valuable information for HT process developers on one hand, and for waste management and environmental policy on the other: Subtle changes to the final composition of waste materials (*e.g.* more Si, Al, and in particular Mn; much less Ca) drastically increase their overall thermodynamic stability without impairing their ability to trap toxic metals. For example, 1 %$_{mol}$ MnO_2 in the HT materials would contribute to +1.3 kcal/mol of their overall free energy of hydration, counteracting the destabilizing influence of, *e.g.*, 8 %$_{mol}$ $MgSiO_3$.

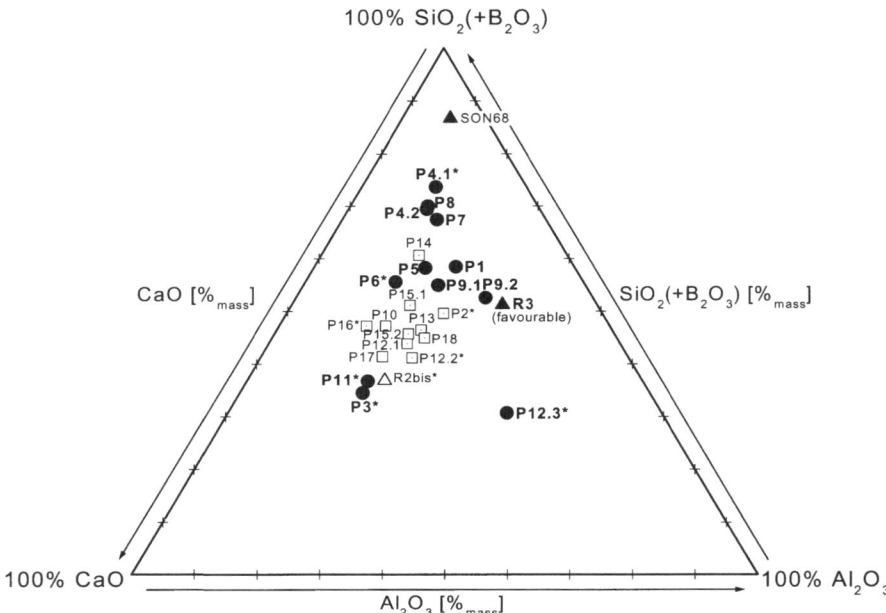

Fig. 5.70. Ternary SiO$_2$ - Al$_2$O$_3$ - CaO diagram of HT materials (solid circles and dotted squares) and HT standards (solid and open triangles). HT materials exhibiting a favorable thermodynamic picture (*i.e.* with $\Delta G_{hydr} \geq$ SON68) are depicted with a solid circle. For each HT material, ΔG_{hydr} was calculated for the pH value measured at the end of the 3 days corrosion experiment

5.4.5 From Facts to Policy

The global survey of the physico-chemical characteristics (static picture), the behavior under aggressive conditions of accelerated corrosion (dynamic picture), and the estimated long-term durability (thermodynamic picture) of HT materials produced from different sources (MSW, BA, FA, FC, or combinations of them) offers the following results and information.

Static Picture

- HT materials cannot be distinguished on the basis of their parent process or on the basis of the input material (MSW, BA, FA or combinations of them). Globally, the process types (in-line or post-processes) do not noticeably influence the final physico-chemical characteristics of the products, provided that they are operated at higher temperatures than conventional MSW incinerators.

- The size and aspect of HT material is mostly dictated by the quenching technique.
- HT materials are either vitreous or vitrocrystalline; none of them are crystalline. This characteristic is governed by the quenching technique.
- Most HT materials exhibit a favorably low specific surface area (< 600 cm^2/g for the 100-125 μm ground fraction) and are expected to have a poor reactivity towards leachants. Specific surface area and vitreous/vitrocrystalline state are not directly linked.
- With some exceptions, HT materials are microscopically smooth; mineral inclusions, either inherited from the starting wastes and residues, or produced during quenching, are relatively rare.
- Network-forming (Si), intermediate (Al, Fe) and network-modifying (Ca, Mg, Na, K) elements govern the status of HT materials. The sum of SiO$_2$ (25-56 %), Al$_2$O$_3$ (9-37 %) and Fe$_2$O$_3$ (1-19 %) ranges between 48 % and 82 %, while network-modifying elements are present in moderate amounts (CaO = 11-38 %; MgO = 2-6.5 %; Na$_2$O = 0.2-5.5 %; K$_2$O = 0.1-2 %; sum = 16-43 %).
- HT materials contain significant amounts of toxic metals, which is an indication of the favorable ability of HT processes to embed these metals in the vitreous or vitrocrystalline matrix. However, the high levels of metals in several HT materials would make them inappropriate for disposal into landfills for inert materials with respect to actual Swiss regulations.

Dynamic Picture

- During corrosion of HT materials in water (Strasbourg accelerated test), the pH of leachates increases but stabilizes rapidly. pH is mostly governed by the release of Ca in solution, which in turn directly correlates to the amount of Ca in HT materials.
- High pH values (pH > 9.5) enhance the hydrolysis of the silicate network. SiO$_2$-rich/CaO-poor ([SiO$_2$]:[CaO] > 2) samples develop lower pH during corrosion and exhibit a higher stability than CaO-rich samples.
- The leaching of major, minor and trace elements is not congruent, and the behavior of HT materials is highly differentiated during the initial stage of corrosion under the conditions of the Strasbourg test (1 day), but element releases systematically decrease toward small, undifferentiated values.
- Alkali (Na, K) are released at higher rates (r(i)$_{norm}$ = 0.5-1.75 g/m^2·d) than alkaline earths (Ca, Mg, Sr, Ba), network-forming (Si) and intermediate (Al) elements (r(i)$_{norm}$ = 0.2-0.6 g/m^2·d). This reflects the differentiated incorporation of these elements into secondary minerals formed during corrosion.
- The maximum releases of toxic trace metals during the batch corrosion experiments are systematically below the maximum limits allowed by Swiss regulations. In most instances, [metal]$_{leachate}$ is < 100 μg/L (Cr, Ni, Cu, Zn, Sb),

and even below 5 µg/L (Co, Cd, Sn, Pb); on this basis, all HT materials would be admissible into landfills for inert materials.

- Microscopically, the majority of HT materials develop a protective amorphous gel layer during corrosion, sometimes partly exfoliated, with visible pits and holes; the formation of secondary mineral phases completes the apparent surface modifications caused by corrosion.

Thermodynamic Picture

- The theoretically maximum thermodynamic stability of HT materials is calculated to be around pH = 5-7 (*i.e.* HT materials would have the highest stability if $pH_{leachate}$ could be confined around 5-7); experimentally, most HT materials are slightly less stable ($pH_{leachate}$ = 6.7-10.3 during the Strasbourg test).

- Taken collectively, the thermodynamic stability of HT materials compares with the one of the standard nuclear HLW glass SON68 (ΔG_{hydr} = -0.16 kcal/mol for the most stable, -12.6 kcal/mol for the least stable, -6.8 kcal/mol for SON68), which is a favorable indication of their long-term durability.

- SiO_2-Al_2O_3-rich and CaO-MgO-poor HT materials exhibit the highest thermodynamic stabilities. In this respect, post-processes operating fly ash as their main input material produce on average HT materials with lower stabilities than post-processes for bottom ash and in-line processes for municipal solid wastes.

- For medium-term corrosion (Strasbourg test, 10 days corrosion) the combined dynamic and thermodynamic behavior of HT materials resembles that of medieval glasses. On shorter terms (1 day corrosion), HT materials can be compared to nuclear HLW glasses. From these results, the estimated durability of HT materials spans between 10^3 years and 10^4 years.

Assessment of HT Materials: A New Paradigm

As can easily be concluded from the information herein, HT materials, despite differences in their static, dynamic and thermodynamic pictures, behave collectively favorably in comparison to bottom ashes or fly ashes produced by conventional low temperature MSW incinerators. Whatever their status and reactivity, HT materials release very low amounts of toxic metals, at least in batch mode under the aggressive conditions of the Strasbourg test. Thus, the information acquired from the characterization of our HT materials allows one to derive general guidelines for the quality of a globally favorable HT product originating from the high-temperature treatment of municipal solid wastes or their residues:

- No process family yields high-temperature materials with a consistently superior quality;
- A tendency towards superior properties is expected for HT materials with a high macroscopic homogeneity, a low specific surface area, and a high Si:Ca ratio or a high Si content;

- The production of HT materials should be performed at a high temperature (T > 900 °C) to guarantee a vitreous or vitrocrystalline state;
- HT processes should have an energy consumption efficient enough to meet the requirements of sustainable development; according to the principles of risk prevention, HT processes should lead to HT materials from which metals have been extracted.

With regard to the establishment of environmental guidelines for the possible disposal or reuse of HT materials, we are faced with the following dichotomy: On the one hand, HT materials have a high propensity to embed toxic metals in their matrix, and most of them would thus not fulfill the requirements of the existing TOW with respect to their metal content. On the other hand, HT materials are very stable during corrosion, and their low metal releases, which are much lower than the TOW limits, allow them to comply with the requirements of the existing TOW with respect to their leachability in toxic metals.

Hence, the existing Swiss regulation on waste management, originally designed for bottom and fly ashes and their impact on the environment, is no longer appropriate for the proper redirection of HT materials. Provided that high temperature incineration technologies would be gradually introduced in Switzerland, the assessment of the ultimate fate of their products should thus be based on updated regulations, taking into account the high efficiency of HT materials in inertizing toxic metals over long-term periods.

Our results show that these materials have favorable static, dynamic and thermodynamic characteristics; these results should be considered as a sound environmentally oriented database to be used within the framework of the revision of the Swiss regulations on the sustainable management of wastes and their residues.

Undoubtedly, other considerations should also be taken into account in order to assess the complete picture of such materials. For example, energy fluxes required to operate HT processes and the economical and environmental impacts of elevated energy consumption must be estimated. The possibility of optimizing HT processes toward the fine tuning of the final characteristics of HT materials (*e.g.* use of stabilizing additives) must be explored. Redirection of HT materials to civil engineering purposes (*e.g.* foundation layers for road construction; additives for the production of cement or concrete, see section 3.5) must also be assessed in terms of ecological, technical and economical feasibility (*e.g.* liberation of metals under various conditions of use, mechanical stability, cost-effective use of surrogate materials). Finally, the environmental definition of reusable *vs.* non-reusable HT materials needs to be translated on a local and regional scale and for short- to long-term periods of time, in terms of predicted fluxes of toxic substances released to the ecosystems by these materials, in comparison to other natural and anthropogenic fluxes.

5.5 Separate Treatment of Hazardous MSW Components

Charles Keller

Household waste contains waste fractions, which are highly enriched with heavy metals, such as batteries and electronic scrap. The flood of electronic products is still increasing. Although prepaid taxes encourage the consumer to bring these goods back to the stores after use for separate collection, a large portion is still disposed of with mixed waste. It is assumed, therefore, that heavy metal concentrations will also increase in MSW in the future. Additionally, other toxic wastes, such as Automotive Shredder Residues (ASR) from shredding cars, are mixed with MSW for treatment in the incinerators.

Automotive Shredder Residues (ASR) from shredding cars are rich in organics which are partly toxic. Additionally, ASR contain metals, which are about one order of magnitude more concentrated than in average Swiss MSW. In Switzerland ASR is currently mixed with the fractions of MSW which are sent to incineration. Via this method, the heavy metals are transferred to the MSWI ashes. Considering various aspects of sustainable ASR processing, ASR should no longer be co-incinerated. In Switzerland, a dedicated plant will be built to process the total annual accumulation of ASR, about 60'000 tons/year. An interesting side effect of this new technology is that it can process all filter ashes produced by the Swiss MSW incinerators (Switzerland does incinerate 100% of MSW).

The following chapter discusses alternatives for the treatment of wastes such as ASR (section 5.5.1), batteries (section 5.5.2) and electronic wastes (section 5.5.3). Separate collection and treatment is essential for the successful recovery of the heavy metals. A combination of mechanical dismantling and thermal treatment processes allows for efficient recovery.

5.5.1 Optimized Disposal of Automotive Shredder Residues (ASR)

Introduction

Automotive shredder residues (ASR), also known as fluff, SLF (shredder light fraction), or RESH (residues from shredding), are the lighter materials separated from scrap with the airflow during shredding of carcasses (wrecks, depolluted and dismantled size-diminished end-of-life-vehicles (ELV)). The heavy fraction, mainly composed of iron (scrap) can be recycled in the iron and steel industry as secondary raw material. Depending on the content of non-ferrous metals and plastics, this fraction is usually separated further manually, with sieve drums, with magnetic separators and with float-and-sink plants.

ASR is still mostly dumped in landfills in Europe today, despite its unsuitable composition for this. Due to the high content of organic materials, chemical and biological degradation reactions are triggered then. This results in the evolution of gas with offensive smells and high concentrations of pollutants in the effluent leachate of a landfill [30]. Consequently, it is necessary to destroy the organic pollutants contained therein (unless sustainable alternatives are found) via thermal treatment and to immobilize the metals contained in the inorganic material [120].

As shown in Table 5.19, the accumulated quantity of ASR with complete incineration of municipal solid waste (MSW) in a region corresponds to approximately the accumulated quantity of (dry) electrostatic filter ash (EFA) from municipal solid waste incinerators (MSWI). EFA was previously disposed of mostly alone or together with slag (bottom ash) from MSWI directly or used in road construction. In Switzerland, in 2000 around 4.6% of EFA were disposed of together with slag from MSWI, 56% were disposed of as untreated EFA in underground depositories and the rest was mixed with hydraulic binder (mainly cement) in order to immobilize the heavy metals in landfills [49].

Table 5.19. Amounts of ASR in relation to municipal solid wastes and their residues in Switzerland (year 2000; data of ASR from [9], others from [49])

Waste category	Quantity [t/y]	Quantity [kg/cap./y]	Quantity [%] relative to MSW total (= 100)
Total ASR	61'200	9	2.4
Share incinerated	60'800	8	2.4
MSW total	2'540'000	353	100
Share incinerated	2'260'600	314	89
Total waste incinerated	2'800'000	389	110
Share of slag (bottom ash)	640'000	89	25
Share of electrostatic filter ash (EFA)	62'000	9	2.4
Share of residues from wet scrubbers	9'200	1	0.4

Legal Issues

The disposal of ASR is regulated mostly in addition to the general laws on waste disposal and disposal of ELV.

- In the European Union (EU), the ELV Directive 2000/53/EC [5] obligates in its articles 5, 6 and 7 the member states to ensure that operators of ASR disposal set up a collection network for ELV.
- The delivery of a car to an authorized treatment facility is free for the last owner (consumer). The car manufacturers are financially responsible for these take-back-systems.
- All authorized facilities comply with current environmental regulations (Waste Directive) and treat the cars according to the ELV Directive's Appendix 1.
- Recycling businesses achieve 85% of recovery as of 1 January 2006 (including max. 5% energy recovery) and 95% of recovery (including max. 10% energy recovery) as of 1 January 2015.

The implementation of this ELV Directive into national law of the EU member states is based on two different strategies [5, 264]. The Netherlands, Sweden, Denmark and Finland are willing to organize a nationwide service of ELV recovery based on the current operators of ASR disposal. The environmental standard of businesses, the will of all participants in administration and industry to do their best and available information of existing disposal structure is quite uniform and advanced. The other countries intend to reinforce competition between those involved. There are great differences on these aspects between countries such as Germany with advanced legislation and other ones in this group. The discrepancies in fact will even grow if EU is extended to Eastern Europe during the next years. This will also have increasing economically undesired side-effects (ecological dumping) on competitors from different EU member states with their own environmental standards in fact.

The EU ELV Directive also prohibits and restricts the use of lead, cadmium, mercury and hexavalent chromium in new cars. In the long-term, this ban helps to improve the efficiency of ELV and ASR disposal as well as to reduce the impact on occupational health and environment. But in the short- and medium-term, there is little influence on the disposal situation, because the mean lifetime of an automobile is approximately 8 to 15 years [229].

Several regulations of the EU ELV Directive such as the mentioned reuse and recycling quota are disputed. For example, a life cycle analysis of a car shows that the disposal phase of ELV is overall not decisive [229], especially with respect to its energy balance, but instead the operation phase of the car, i.e. the fuel consumption per 100 km (mainly determined by design and engineering) and the driving behavior (length of routes driven, way of driving). The automobile industry fears that recycling goals have been set so high that currently they can only be achieved with steel car bodies, which will make the production of cars with lightweight construction (e.g. aluminum instead of iron) and of biomass-based raw materials more difficult and hence have a negative ecological effect [195]. In any case, new types of design and engineering of cars will greatly change the existing disposal structures [76].

In Japan, the very limited landfilling capacity (approximately two to three years of annually deposited wastes) and increasing landfilling costs led to an advanced proposed ELV legislation [22]. It obligates automobile manufacturers and importers to take back and dispose air bags and chlorofluorocarbons from dismantlers as well as to take back and recycle ASR from shredders. They can do so directly or by commissioning such services to specialized facilities. The new automobile holder and/or owner has to pay for these services to the fund managing public corporation. After implementation of this system, ASR recyclers registered by the government can receive subsidies for their environmentally compatible recycling work.

Generally, there is much more landfill space in the USA than in Europe and Japan. Correspondingly, there is less governmental control and intervention concerning ELV and ASR disposal, although more and more governmental regulations have to be fulfilled by recyclers [225]. The American Automotive Recyclers Association (ARA) has defined a standard procedure It includes the draining of all fluids of ELV, removing hazardous materials for reuse or disposal for material such as batteries, dismantling reusable parts and recovery of the bulk consisting of steel and aluminum for the refining industry.

Quantities and Composition

The composition of ASR is very complex (Fig. 5.71, Tables 5.20 and 5.21) and varies substantially. On one hand, the used sorting and crushing procedures of ELV as well as their composition are important. For example, a car contains approx. 66% iron/steel, 6% light metals (especially aluminum), 2% copper, 1% lead, 1% zinc, 9% plastics, 5% rubber, 4% glass and 6% other materials [54, 96, 171]. On the other hand, the ASR composition is determined to a large extent by other waste mixed in via the shredder such as larger household appliances (e.g. cook stoves) and technical appliances from industry (e.g. equipment housing made of steel). These goods also processed during shredding of carcasses constitute up to 50% of the quantity in Germany [229] and 30 to 40% in Switzerland [49].

The content of ash-forming matter in ASR is approximately 50%. Due to the high contents of plastics, the calorific value is very high at an average of 13 MJ/kg (range from 7 to 26 MJ/kg, MSW: around 12 MJ/kg) [219]. For certain fractions like ASR heavy fraction coming from a float-and-sink plant with a high rubber share, the calorific value is considerably higher at approx. 20 MJ/kg.

The average heavy metal content in ASR is substantially higher than in MSW (Table 5.21). This applies especially to copper and nickel with contents of more than ten times that in MSW. Also substantially higher contents of iron, copper and nickel occur in ASR than in EFA. But only the copper content in ASR is similar to the corresponding content in economically explorable ores.

The relatively high chlorine content (mainly from PVC) in ASR is significant for its further disposal. Without corresponding precautionary measures during the ASR incineration and the cooling of the flue gas, the high chlorine contents in ASR can result in increased formation of toxic polychlorinated dibenzo-dioxins and furans (PCDD/Fs). Similar effects occur by bromine in ASR, present as flame retardant. High contents of alkali metals, chlorine/bromine and sulphur in ASR also will give corrosion problems in thermal ASR treatment plants.

In addition, ASR contains an average of approx. 9.7 mg/kg polychlorinated biphenyls (PCB) (range of 5-14 mg/kg) and 16.3 mg/kg polycyclic aromatic hydrocarbons (PAH) (range of 10-28 mg/kg) [219] as relevant organic pollutants (similar values of Swiss ASR: 9.3 mg PCB per kg, 35.7 mg PAH per kg and 0.014 μg PCDD/Fs toxicity equivalents per kg [119]).

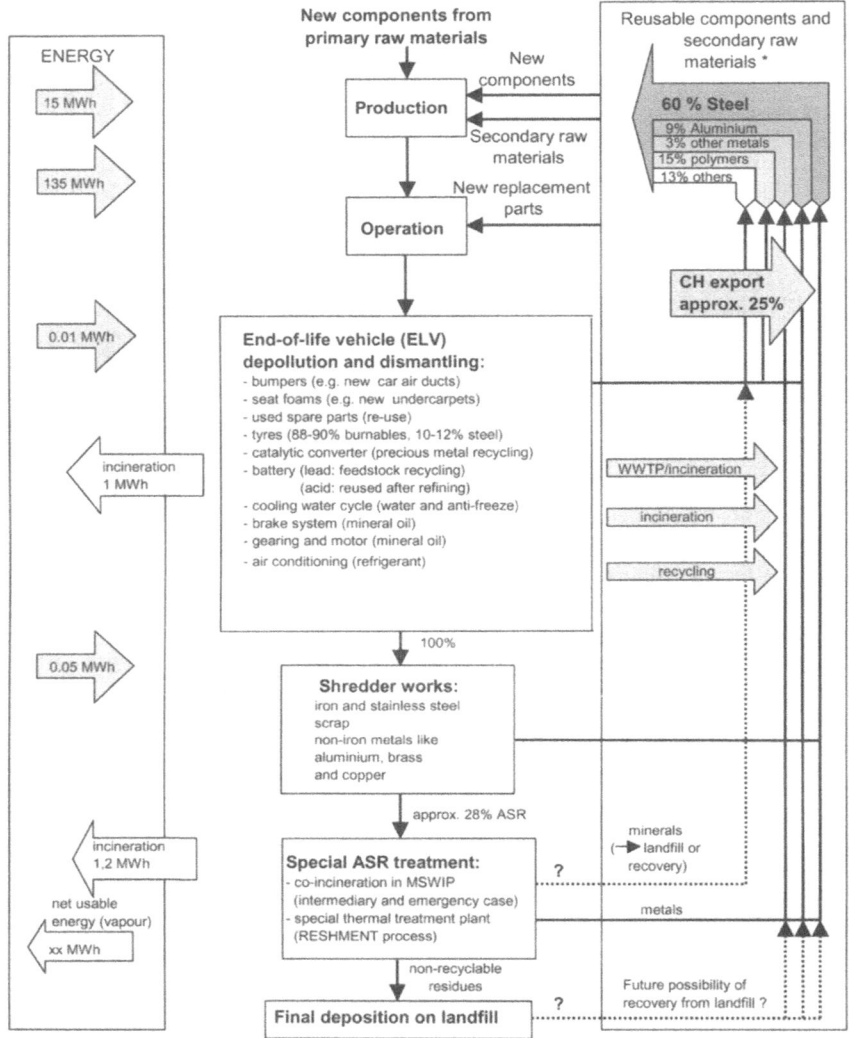

Fig. 5.71. Disposal path of ELV (based on [229], modified and expanded by the author) *WWTP* = Waste Water Treatment Plant; * Percentages before implementation of the special ASR treatment with Reshment process. The net usable energy of the Reshment process depends on the co-treated wastes and their amounts.

Table 5.20. Materials in ASR [199]

Parameter	Contents in ASR
Inorganic solids	
- dust, soil and rust	10 - 20%
- inert material (glass and ceramics)	3 - 16%
- metals	ca. 20%
Organic solids	
- plastics (thermoplasts, polyurethane, PVC, duroplasts)	30 - 48%
- elastomers (incl. tires)	10 - 32%
- fibers (textiles, wood, paper, cardboard)	4 - 26%
- lacquer	3 - 10%
Liquids	
- oils and water	15 - 17%

Table 5.21. Contents on some chemical elements in ASR in relation to corresponding values of mixed municipal solid waste (MSW) and electrostatic filter ash (EFA) and explorable ores (EO)

Chemical element	Absolute contents in ASR mean (range) [219][a]		Absolute contents in mixed MSW [38]	Absolute contents in EFA mean [29]	Absolute contents in EO mean [29]
Unity	[mg/kg]		[mg/kg]	[mg/kg]	[mg/kg]
Iron	141'000	(50'000-240'000)	27'000	34'000	500'000
Aluminum	20'000	(10'000-25'000)	11'000	90'000	300'000
Copper	11'400	(3'300-30'000)	780	3'000	12'500
Zinc	9'000	(2'000-13'000)	1'600	26'000	45'000
Lead	5'100	(300-14'000)	700	9'000	35'000
Chromium	1'200	(340-1'300)	360	1'100	300'000
Nickel	1'200	(400-2'800)	110	250	15'000
Manganese	1'000	(400-1'400)	220	840	200'000
Tin	66.7	(25-90)	71	660	5'100
Cadmium	61.2	(40-80)	11	400	17'500
Mercury	2.1	(1-3)	2.9	8	862'000
Silicium	77'000	(45'000-110'000)	37'000	128'000	no data
Calcium	40'000	(30'000-60'000)	24'000	60'000	no data
Magnesium	8'700	(8'000-10'000)	3'200	34'000	20'000
Sodium	7'100	(1'000-12'000)	5'100	50'000	400'000
Potassium	2'700	(1'500-3'000)	2'300	43'000	no data
Chlorine	18'000	(5'000-30'000)	6'900	59'000[b]	600'000
Barium	5'780	[119]	430	no data	no data
Nitrogen	9'000	(2'000-18'000)	no data	no data	no data
Sulfur	6'000	(1'000-14'000)	1'300	30'000	no data

[a] ASR with co-treated other materials at shredder works; German data in accordance with Swiss data from [121] and [21], [b] strong variation of concentrations for different processes

Objectives of Processing ASR

The objectives for sustainable ASR processing should be formulated based on its complex composition, its relatively small energy content in comparison to the energy consumption in the operating phase of a car (Fig. 5.71) and considering valid legal framework conditions such as the reuse and recycling quota of EU legislation on ELV.

It is primarily a question of removing organic pollutants and of immobilizing the inorganic contaminants, if they cannot be recycled. The excess energy released during thermal treatment of ASR should be used efficiently in the form of electricity and heat for external use or by direct heat transfer to vitrification or separation steps in the process.

Most promising for the disposal of are combinations of mechanical and thermal process steps [118]. The respectively suitable treatment method also depends on the disposal structures already existing in a region.

The concept for disposing the complete ASR in Switzerland from 1991 [120] is given as an example:

- As a short-term objective, the ASR is to be mineralized in existing incineration plants instead of dumping it on landfills.
- As a medium-term objective, specialized thermal treatment plants with sufficient capacity to treat all ASR from the considered region have to be constructed and commissioned.
- The long-term objective is to construct new cars which are easily dismantlable to a great extent, whereby all dismantled pieces can be reused, recycled or finally deposited on a landfill as inert materials without need for special control and possible harm to environment.

Finally the Swiss ELV disposal concept of 1991 set up an organizational and financial framework. It enabled to evaluate more advanced treatment methods of ASR, to finance the short-term solution of ASR disposal (the co-incineration of ASR in MSWI instead of its deposition on landfills) and to invest into the future Swiss special ASR treatment plant (see case study below). The financing system is based on a recycling fee charged on every imported car by Swiss automobile importers and administered by a foundation (Fig.5.72). The dismantler is mainly financed by the last owner of a car and the revenues from wrecks. The additional costs of a shredder plant operator at delivering ASR to a MSWI instead to landfill are roughly paid by the mentioned foundation. Other important but not sufficient revenues of a shredder plant operator come from the sale of recovered scrap.

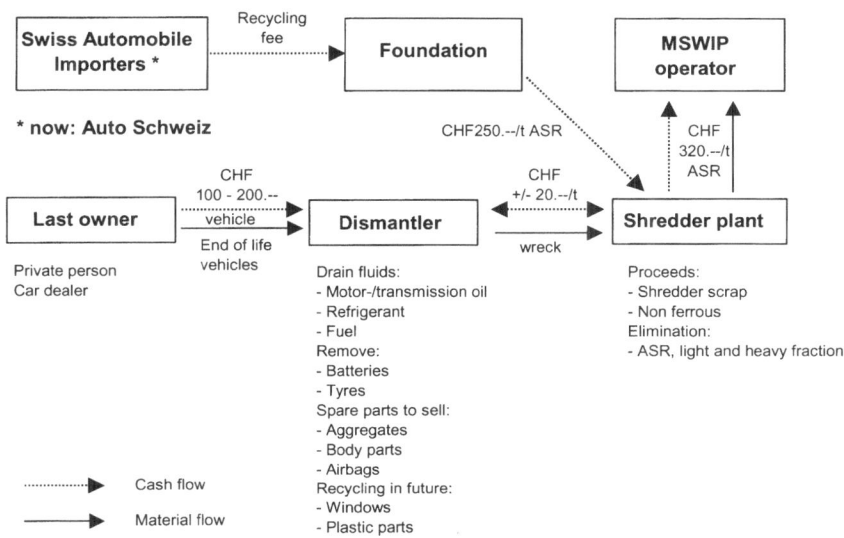

Fig. 5.72. Cash and material flows at disposal of Swiss ELV [55]

Co-Incineration of ASR in MSWI as an Intermediary Short-Term Measure for ASR Processing

Comprehensive experiments were conducted on a technical scale for co-incineration of ASR and MSW in various Swiss MSWI in 1993. The share of ASR in MSW was increased incrementally to approx. 10% in a Swiss MSWI during a week [122]. Based on the positive test results, extended experiments were conducted with the co-incineration of a total of approx. 100 t ASR (5% in MSW) [123]. In addition to the material flows of heavy metals, also the flows of some important organic pollutants during co-incineration of ASR such as PCB, PCDD/Fs and PAH were investigated in flue gas and solid residues.

The results show that at least short-term ASR co-incineration does not need – besides sufficient homogenization of ASR and MSW – significant changes of the MSWI operation. There is no relevant change of emissions of organic pollutants either. The effect of an increased heavy metal quantity due to the incinerated ASR is acceptable. But the following preliminary factors still remain in force [64]:

- The MSWI must have incineration with a good burning out of flue gas and incineration residues, i.e. low total organic carbon contents, modern flue gas cleaning and an efficient energy recovery system.
- The ASR part of the effectively incinerated waste has to be limited to an average of less than 10% and a highly homogeneous mixture has to be guaranteed.

- The composition of the products from co-incineration (such as slag and electrostatic filter ash) must be periodically controlled prior to their further treatment and disposal.
- Slag from co-incineration of ASR with MSW may not be recycled, but instead it must disposed of on a controlled landfill.

Until 2002, a percentage of up to 6% shredder residue in the total waste amount led to no severe problems at co-incineration in most of the involved Swiss and German MSWI plants. Additional tests at the German MSWI of Würzburg with increasing amounts of ASR up to 31% showed that German legal requirements concerning air pollution control (17. BlmSchV) and leaching behavior of slag (German leaching procedure DEV S4) are fulfilled [127]. No significant increase of heavy metals and PCDD/Fs concentrations in the cleaned flue gas and the leachate from solids occurred. The heavy metal concentrations of the solid residues increased in accordance with the previous observations made in Swiss MSWI to the extent depending on the already existing heavy metal content in the co-incinerated MSW. However, in fact the percentage of ASR in the Swiss and German MSWI fed with Swiss ASR does not exceed significantly 5 to 6% of the totally incinerated waste capacity. Due to the described composition of ASR (especially its high contents of alkali metals and chlorine) caking and elevated corrosion rates cannot be excluded in the furnace, boiler and flue gas system of a MSWI if ASR is co-incinerated with MSW during a long time.

Generally, co-incineration of ASR in MSWI is ecologically superior to the previous ecologically insufficient and currently unacceptable deposition on landfills. But it must clearly be identified as only an intermediate, short-term solution. Additionally, co-incineration of ASR will be an alternative disposal path to prevent the need of landfilling ASR when the future specially designed thermal treatment plants cannot be operated at their full capacity in emergency cases.

Technologies for Special Treatment of ASR (with and without other Similar Wastes)

A long-term solution for disposing of ASR of a whole country has not been implemented anywhere yet. The construction of such an innovative treatment plant for ASR would entail substantial but manageable technical and economic risks. In addition, long, drawn-out approval procedures are usually connected with such plants. For example, a plant must be integrated into an existing transport and disposal structure, and the population in the surrounding area must be convinced of the environmental compatibility of such a hazardous waste treatment plant [125].

A great number of different ASR treatment processes have been developed and offered for sale on the market over the past 10 years. Many of the technical processes offered failed due to economic underperformance.

In Germany, there was a widespread evaluation of mechanical treatment processes for ASR in 1999 [248]. They produce fractions, which can be recycled, dumped in landfills without problems or – as a fraction with high calorific value but low pollutant content – utilized for pyrolysis, as a reducing agent in blast furnace and/or for energy recovery. The top rating was awarded to the WESA-LSF process [193] equipped with rotary shears for preliminary shearing and a single-shaft rotor as second shearing level. Construction and operation of a pilot plant for a capacity of 4 tons per hour of ASR was subsidized by the German ARGE-Altauto. Since that time, another interesting largely mechanical process has been introduced by the GS Gesellschaft für Umwelt- und Energie-Serviceleistungen mbH, Singen (Germany), with two rotor impact mills [200].

Among the thermal processing procedures offered in Europe, the processes of the following companies should be listed primarily. All these processes have been used for thermal treatment of more than 100 t ASR on a technical scale:

• Alstom/Ebara (fluidized bed gasification and ash melting process, called Twin Rec) [22],
• Babcock Borsig Power CT Environment (smelting cyclone technology called Reshment process)[196],
• Citron SA (Oxyreducer process in the plant in Le Havre (France))[45] and
• Thermoselect (high temperature waste treatment with ash smelting, production of synthesis gas and its shock cooling)[223].

In Japan, several ASR treatment plants are already operating, including a plant with an annual capacity of approx. 100'000 t/a in Aomori built by Ebara (Table 5.22). A considerable feature of this plant is its optional co-treatment of ASR with other wastes, e.g. sewage sludge. In addition to minimizing residues from the process for landfill, energy, ferrous metals and a glass granulate (fulfilling the German leachate limits of DEV S4) are recovered. The required recycling and recovery quota of EU legislation for 2015 are met.

Table 5.22. Experiences of thermal ASR treatment in Japan [22]

Technology	Capacity [t/y]	Startup	Supplier	Operator / Shredder
Pyrolysis, gasification and gas cracker	20'000	10/2001	Toshiba	Yamanaka
Pyrolysis drum and ash melting system	30'000	10/1998	Takuma	Kanemura
Drum type gasification and copper smelting resource	53'000	04/2002	-	Dowa mining
Copper smelting resource	78'000; capacity increase planned	11/2000	-	Onahama smelting and refining
Twin Rec fluidized bed gasification and ash-melting system	100'000	02/2000	Ebara	SEINAN

Please refer to the cited technical publication for further details.

In the following case study, the Reshment process is presented in more detail. This process has been selected in 2001 as the most promising technology for a dedicated plant co-processing ASR and EFA in Switzerland.

Case Study: The Reshment Process

The Reshment process is seen as the medium-term technology for ASR processing in Switzerland and as the best solution to the processing of electrostatic filter ash from Swiss MSWI. The process (block diagram in Figure 5.73 and material flux analysis in Figure 5.74) is based on a thermal treatment of ASR under reducing conditions at high temperatures to free volatile heavy metals such as zinc, lead and cadmium. The process conditions are set such that immediate smelting of the inorganic mineral constituents occur. Energy contained in the ASR can be used at high temperatures with direct heat transfer to co-processed material [196].

A smelting cyclone is at the core of the process. Such smelting cyclones have been tested and used in various smelting operations of the mining industry. Because of the low residence time of the materials to be treated at extremely high turbulence in the cyclone, the material must have a fairly small particle size (< 5 mm).

Therefore a mechanical size reduction of the ASR is necessary before feeding to the cyclone. This step with grinding and magnetic separation also allows preliminary removal of bulky material, which consists to a high degree of economically interesting metals (like iron, copper and aluminum) for recovery from the process, subsequent refining in the metal industry and use as a secondary raw material. Besides the mechanical protection of the thermal equipment and the cost-saving use of energy by prohibition of melting otherwise easier separable metals, this preliminary metal recovery step also diminishes the corresponding leachable concentrations in the molten product coming from the cyclone during deposition on landfill.

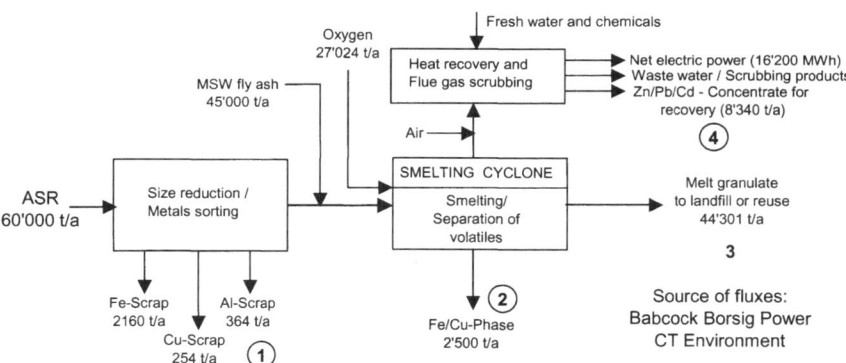

Fig. 5.73. Simplified block diagram of the Reshment process by Babcock Borsig Power CT Environment [197]

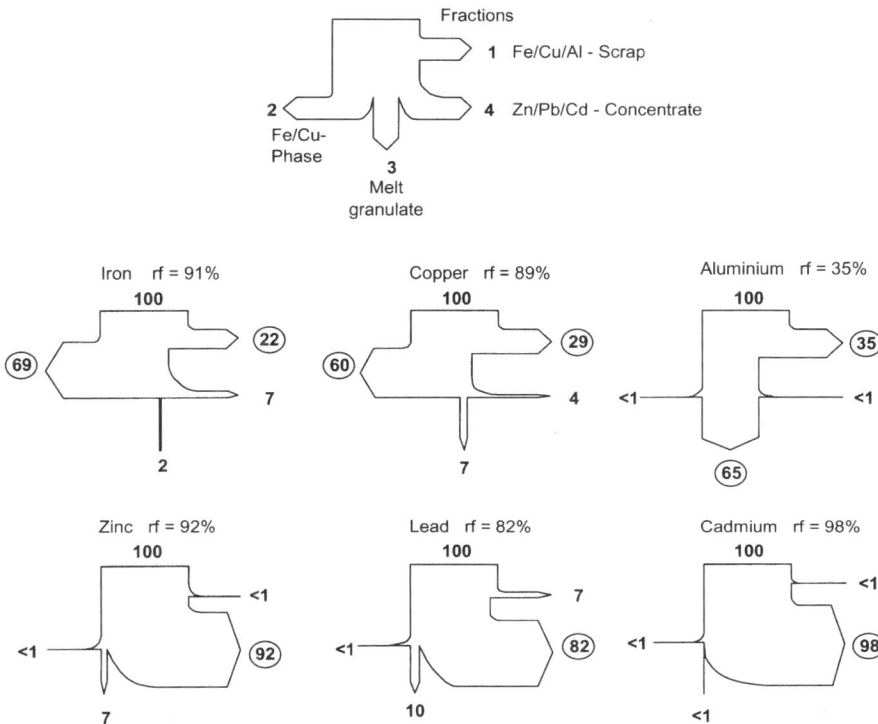

Definition of recovery factors rf (%): sum of transfer coefficients of the fractions for feedstock recycling

Fig. 5.74. Material flux analysis of the Reshment process based on data from [197]

The remaining mixture after the mechanical pre-treatment consists mainly of fine powder containing organic materials, metals and metal oxides as well as glass and ceramics. It is processed in the smelting cyclone by the addition of oxygen at approximately 2000 °C. The organic substances (also contaminants such as PCB, PCDD/Fs and PAH) are completely dissociated there. Oxidized iron (rust) in the feed is not reduced but remains in the molten slag, improving the leaching behavior of the final product during deposition on landfills [260].

The remaining fractions of reduced iron and copper are melted as well and separated from the mineral material by gravity separation. Other non-volatile heavy metals such as chromium also are dissolved therein, consequently diminishing the corresponding contents in the co-produced melt granulate. It shows very low concentrations of the legally limited contaminants in the leachate and fulfills the Swiss laws concerning leachate concentrations for waste deposition on a landfill for inert material.

Additionally, the process meets the advanced targets of the EU ELV Directive concerning reuse and recycling for the year 2015 in combination with the Swiss standard procedure of ELV de-pollution and shredding (mostly with steel recovery).

The start of construction of the Swiss Reshment plant (with about 105'000 t annual capacity) is planned for 2003.

An important advantage of this process is its combination of tried and tested technologies. Therefore there is a minimum of uncertainties associated with this new technical process. The determining factor in Switzerland for the choice of the Reshment process was its high flexibility for co-treatment of other problematical wastes. This is true especially for wastes with low calorific value but high contents of different heavy metals and high disposal costs.

Such wastes are, for example, EFA, galvanic sludge and some other problematic powdery materials. Calculations made by the supplier [197] as well as the data of waste amounts and compositions in Tables 5.19 and 5.21 indicate that not only the whole amount of ASR in Switzerland but also most of EFA from MSWI can be treated in the planned ASR thermal treatment plant. Hereby it is assumed that the ratio of ASR and EFA in plant design can reach up to 1:1. After operation start-up of the new thermal ASR treatment plant, ASR and most of EFA will be disposed of completely in Switzerland. At least underground depositories abroad and immobilization of heavy metals with cement prior to landfilling are no longer necessary for EFA disposal.

Conclusions: Are all Objectives of ASR Disposal Met After Implementation of the Reshment Process?

Environmentally compatible disposal of ASR is feasible today using combinations of mechanical and thermal processes. Even the ambitious re-use and recycling quota of EU regulations per 2015 can be met by various ASR treatment processes.

Thanks to suitable utilization of disposal plants for ASR, other wastes such as EFA can be co-treated in an environmentally compatible way. Their present treatment costs can be significantly reduced. But contrary to the start-up of separation of recyclable components during dismantling and of scrap during shredding, additional costs are produced during future thermal treatment of ASR. These result in negative overall returns on ELV disposal. Consequently, this future environmentally more advanced ELV disposal system than today requires sufficient governmental and/or private industry regulation concerning the organization and financing of the disposal.

With the operation start-up of the planned Swiss ASR processing plant, an important milestone is being set towards sustainable handling (incl. disposal) of cars and ASR as well as of the co-treated wastes. As an alternative disposal path in emergency situations (if there are breakdowns at the planned special ASR treatment plant) the co-incineration of ASR in MSWI retains its importance [124].

In spite of this, a lot remains to be done, bearing in mind that the creation of an ASR processing plant is only a medium-term objective in the Swiss concept of ELV disposal from 1991 [120]. However, material recycling beyond the metal reclamation from ELV and ASR, especially the recycling of plastics, only makes sense if there is a market for the recycling products, additional costs are acceptable and the recycling proves to be more favorable ecologically than production using virgin raw materials. Correspondingly, the manufacture of new cars with substantially less fuel consumption has priority over recycling of used plastic products for savings in primary energy carriers such as crude oil, gas and coal.

The principal long-term goal of an environmentally friendly car operation and disposal system remains a further reduction of the environmental impact especially via sustainable transport of persons and goods. Insofar as this is a question of technical measures (e.g. light weight construction of cars), further progress is certainly possible in the near future [169]. It will be more difficult to change the attitudes of individual users of cars, which tend to offset the successful implementation of technical developments towards lower emissions.

5.5.2 Disposal and Thermal Treatment of Spent Batteries

General Aspects

Batteries can be divided into primary batteries (which are used once only) and rechargeable secondary batteries (accumulators). Spent batteries, which are sometimes very small and easily overlooked, represent only a small proportion of overall domestic waste. The annual amount of batteries disposed of in Switzerland is estimated to be about 5,000 t, compared to a total domestic waste amount of 4.3 Mil. t/a (1997)[231], i.e. only about 0.1%. Similar figures also apply to other industrialized countries. Batteries nevertheless represent a significant source of heavy metals found in domestic waste and therefore need special attention.

Measurements taken by the waste disposal authority of the Swiss city of Zurich [135] found that about 10% of the zinc, 85% of the cadmium and 67% of the nickel in the waste intended for incineration stem from batteries, despite the fact that a separate collection of batteries, with a recycling quota of about 50%, is a currently functional system. Similar figures would apply to mercury if its content in most batteries had not been gradually reduced recently to practically zero [80] and a functioning return system for the remaining mercury batteries had not been installed [115]. These figures are even more alarming considering that certain heavy metals (such as cadmium) are inclined to concentrate in some incineration and flue gas cleaning residues [38, chapter 3].

The first initiatives to specifically regulate battery disposal (Directive 91/157: Scope, member state programs, implementation record; Directive 93/86/EEC: Marking requirements) were undertaken by the EU in 1991 and 1993. Directive 98/101/EEC stipulated the elimination of remaining mercury content in the most commonly used carbon-zinc batteries by 1 January 2000. A proactive industry response by the European Portable Battery Association consists of a two-stage plan to have a collection and recycling scheme ready for all batteries by 2004. It includes the following:

- collection of all types of batteries;

- collection targets of 75% for all consumer batteries;
- decisions about the collection and treatment system to be implemented in every member state;
- taxes for collection and treatment of spent batteries should not be regulated by EU, but by the member states [31].

Nickel-cadmium accumulators contribute significantly to the Cd loads in domestic waste. They have not been phased out despite the fact that it is usually technically possible to replace them, for example, by nickel-metal hydride batteries.

In Japan, a new battery recycling law was enacted for the collection and recycling of small-size secondary batteries (i.e. portable rechargeable batteries) in June of 2000. It also imposed obligations on manufacturers of batteries and equipment for the collection and recycling of spent batteries. As in Europe, the involved industry is collectively organizing a collection and disposal system that should combine ecological and economical targets [84]. In the USA many retailers also participate in the education and separate collection programs of the non-profit organisation Rechargeable Battery Recycling Corporation (RBRC)[70]. Japan and the USA have technically advanced battery treatment plants like those in Europe.

Consumption, Composition and Corresponding Environmental Impact

Until now, metal-hydride and lithium tonnage has been relatively small (Table 5.23) in comparison to other types of batteries, but it will increase significantly during the coming years and will demand additional specific disposal capacities that do not yet exist [99].

Table 5.23. Annual consumption of batteries in Germany 1994 [99]

Battery use	Battery type / system	Million pieces	Metric tons	Percentage [%]	Tendency in consumption
primary	acid Zn/MnO_2	335	13'000	50.8	slightly decreasing
primary	alkaline Zn/MnO_2	322	9'323	36.4	slightly increasing
primary	Zn/HgO	6	18	0.1	strongly decreasing
primary	Zn/O_2	15	21	0.1	increasing
primary	Zn/AgO_2	15	27	0.1	decreasing
primary	Li-systems	30	60	0.2	increasing
secondary	NiCd (steel)	81	3'095	12.1	invariable
secondary	NiMeH	n.a.[1]	n.a.[1]	-	increasing
secondary	small Pb/PbO_2	0.1	50	0.2	Insignificantly variable
	SUM	804.1	25'594	100	

n.a. not available
[1] significant amounts since 1995

With regard to primary batteries, the extensive removal of mercury from the carbon-zinc batteries in previous years leaves only the mercury button cells with particular ecological problems caused by mercury. Lithium containing batteries demand specific safety measures for their disposal because of their high reactivity.

Secondary batteries have the ecological advantage of multiple use (rechargeability), which may in some cases offset the impact caused by the toxic metals (Pb, Cd) they contain. Nickel-metal hydride and lithium batteries in particular have relatively few ecologically problematic constituents (Table 5.24). However, mainly because of their high reactivity, these batteries must also be collected separately from other waste and disposed of in special plants. The nickel in these batteries could be recovered, which would significantly decrease the nickel contents in domestic waste, as the above mentioned study [135] indicates.

Table 5.24. Main components of commonly used batteries [99] and especially for metal hydride batteries according to [183]

Battery type/sytem		Zn/MnO_2 -acid	Zn/MnO_2 -alkaline	$Zn/$ HgO	$Zn/$ O_2	$Zn/$ AgO_2	$Li/$ MnO_2	NiCd	NiMeH (*)	$Pb/$ PbO_2
	Zn	20	20	10	30	10	-	-	-	-
	Mn	25	30	-	-	-	30	-	-	-
	Hg	-	-	30	1	1	-	-	-	-
	Ag	-	-	-	-	30	-	-	-	-
	Ni	-	-	-	-	-	-	20	39	-
Metals	Cd	-	-	-	-	-	-	15	-	-
[%]	Li	-	-	-	-	-	2	-	-	-
	Fe	20	20	40	45	40	50	45	8	-
	Pb	-	-	-	-	-	-	-	-	65
	others	-	-	-	-	-	-	-	15	-
	SUM	65	70	80	76	81	82	80	62	65
	KOH	-	5	3	4	3	-	5	-	-
	NH_4Cl $ZnCl_2$	5	-	-	-	-	-	-	-	-
Electrolytes [%]	organic	-	-	-	-	-	9	-	-	-
	H_2SO_4	-	-	-	-	-	-	-	-	8
	H_2/O_2	-	-	-	-	-	-	-	18	-
	H_2O	10	10	6	8	6	-	10	-	17
Plastics, paper, cardboard, coal [%]		20	15	11	12	10	9	5	20	10

* prismatic cells

The analysis of the constituents of the most important types of batteries in Table 5.24 shows that most of the economically recyclable constituents are in metallic form in the cathode and anode. The substances in the electrolytes are, on the other hand, mainly substances which cannot be recycled (with exception, e.g. sulphuric acid in lead-acid batteries). The organic constituents, particularly the casing (approximately 5 to 20% plastic, paper, cardboard, coal) can at least be used for energy recovery.

Generally, in addition to an expected increase in the use of batteries with less ecologically problematic substances in future, the separate collection of batteries and their disposal in suitable specific treatment processes will continue to be necessary.

Survey of Used Technologies

Practically all the technical treatment methods for used batteries (Table 5.25) consist of the following main processing steps:
- Separate collection of batteries, where possible, at the source of waste generation (for reduction of further separation and treatment costs during the conditioning of the plant feed);
- Mechanical detection and removal of unsuitable battery types from feed and its homogenisation (possibly with the addition of other waste with well defined composition, and crushing for achieving relatively constant and efficient operation conditions);
- Thermal treatment in plants specifically designed for batteries or suitable for co-treatment with other similar wastes (destruction of organic substances and eventual energy recovery);
- Separation of the main metallic components to the greatest possible extent (for ecological and economical reasons), with perhaps some cleaning steps and closing the loop by subsequent treatment along with other materials in the metal refining industry;
- Treatment of the produced flue gas, waste water and solid residue in accordance with the ecological demands by legislation.

The types of treatment available vary greatly and depend mainly on the type of battery. Few of the thermal processes used can be applied to the treatment of a wider range of battery compositions. Even for those, pre-treatment is necessary to obtain homogenous process conditions over a long time and for stable operation conditions. Also, only in this way is it possible to achieve a high yield and purity for the recovered substances of interest. Particularly strict requirements for the composition due to environmental regulations apply if batteries form only a small proportion of the feed but have a significantly higher content of problematic constituents than the other input (ores) (example: strictly limited cadmium content of batteries in waelz process).

Table 5.25. The relevant battery treatment processes already existing on an industrial scale (co-treatment of other wastes not excluded) [69] and for lead acid batteries [161]

Suited battery fraction	Process type	Operator	Plant location	Recycling capacity
Roughly sorted battery mixtures	Oxyreducer (IN-METCO) process	Citron SA	Rogerville (near to Le Havre) (F)	23'000 t/a [1]
Zinc containing batteries (Hg free)	Imperial smelting process	MIM Hüttenwerke Duisburg GmbH	Duisburg (D)	>3'500 t/a (10 t/d)
	Waeltz process	B.U.S. AG	Freiberg (D) and other EU locations	Several thousands t/a each
	DK (blast furnace process) [2]	DK Recycling and Roheisen GmbH	Duisburg (D)	several thousands t/a
	Electric arc furnace (steel making)	Several in USA, NL, UK, and Spain	e.g. Alblasserdam (NL), Sheerness(UK), Alabama (USA)	1-3% of input (several thousands t/a)
Zinc containing batteries (with and without Hg)	Batrec (Sumitomo) process	Batrec Industrie AG	Wimmis (CH)	3'500–5'000 t/a
	Electric arc furnace[3]	VALDI	Feurs, Le Palais (F)	1.5 t/h
	Revatech/Erachem process	Revatech S.A., Erachem Europe (B)	Liège (B) Tertre (B)	3'000–4'000 t/a
	DMA	Chemtec (A)	Simmering (A)	3'000 t/a
Button cells, mercury batteries	Vacuum thermal Recycling	GMR (D)	Leipzig (D)	750–1'000 t/a
		NQR (G)	Lübeck (D)	n.a.
	MRT process	Trienekens AG (D)	Grevenbroich (D)	n.a.
Nickel containing batteries (1) NiCd (2) NiCd, ...NiMH (3) NiMH	(1) Distillation, pyrolysis	SAFT AB (S)	Oskarshamn (S)	1'500 t/a
	(2) Distillation, pyrolysis	S.N.A.M. (F)	St. Quentin-Fallavier, Viviez (F)	4'000–8'000 t/a
	(2) Vacuum destillation	ACCUREC (D)	Mülheim a/d Ruhr (D)	2'500 t/a
	(2) Rotary hearth furnace process	INMETCO (USA)	Ellwood City (USA)	3'000 t/a
	(3) NIREC recycling process	NIREC Recycling GmbH (D)	Dietzenbach (D)	>1'000 t/a
Lithium batt.	ToxCo process	ToxCo (Ca)	Trail (Ca)	3'500 t/a
Lead batteries	Crushing & drainage of whole spent batteries with the following process: - shaft furnace	some examples: VARTA Recycling GmbH	Krautscheid (D)	n.a.
	- short rotary furnace	Metallhütten-Gesell-schaft Schumacher	Rommerskirchen (D)	n.a.
	- short rotary furnace with battery preparation before refining the lead	BSB Recycling GmbH	Braubach (D)	n.a.
	Bath smelting processes before refining the lead: - QSL process	some examples: Berzelius GmbH	Stolberg (D)	n.a.
	- Sirosmelt process	Metaleurop Weser Blei GmbH	Nordenham (D)	ca. 90'000 t/a

[1] Permitted, otherwise much higher; [2] Foundry pig iron production; [3] Ferro alloy production

There are also alternative hydro-metallurgical processes, such as Metek [182]. These processes have as yet not been used in technical plants, but have been implemented in pilot plants with a throughput of up to about 1 t/h. Ecologically and economically most interesting are the expected significant reduction of energy consumption and the expected efficient separation of specific metals from metallic solutions. Whether these expectations can realistically be fulfilled has still to be proven by extended tests in technical plants.

With few exceptions, the battery recycling capacity achieved in disposal plants in practice thus far is relatively small and lead to high capital cost. Most plants are operated in developed countries with strict environmental and safety regulations requiring substantial investment into environmental protection equipment. Process-integrated environmental protection measures instead of expensive end-of-pipe solutions are therefore very much in demand. However, these are usually only to a limited extent technically possible because the more specific the central thermal treatment process is in practice, the narrower the spectrum of treatable wastes will be (see examples of operating plants given below).

Ecological requirements of battery processing consist of high recycling quota and compliance with regulations regarding waste gases, waste water and landfill residues as well as high yield and high purity of recyclable products. Economic requirements consist of financially acceptable disposal costs which are still competitive compared with alternative methods. To meet all requirements within reason, the following strategies (sometimes combined) are used in practice (as an illustration, compare the following process example from [208] and [129] and see processes in Table 5.25):

- The application, variable depending on the operating conditions, of specific flexible thermal processes which are at the same time suitable for a wide spectrum of wastes;
- Treatment temperatures which are as low as possible (smelting processes are particularly expensive and technically demanding) and the avoidance, as far as possible, of the oxidation of metals present in elementary form;
- High throughput of batteries for minimization of the process costs per ton of waste;
- Concentration on batteries mainly as feed in possible combination with the co-disposal of other metal-rich wastes with high disposal costs, or alternatively the co-disposal of batteries as valuable special waste in suitable, very large-capacity ore or secondary raw material processing plants (e.g. waelz process);
- The optimal incorporation of the battery disposal plant, both geographically and logistically, in an existing waste disposal infrastructure with assured supply channels at cost-efficient prices for battery feed and assured sales channels of the secondary raw materials obtained [81].

Case Studies

Industrial process example 1:	Batrec (adapted Sumitomo) process [129, 250] (Fig. 5.75)
Plant location:	Wimmis (Switzerland)
Capacity:	3'000-4'000 t/a
Cost:	CHF 4'000 to 5'000 per ton of treated household batteries
Start of operation:	1994
Treated batteries:	Zinc carbon batteries (also mercury-containing).
Proven capability of treatment for other wastes (examples):	Mercury-containing mineral covering, metal hydroxide sludges, zinc containing dusts, broken fluorescent tubes, activated carbon filters, dental wastes.
Principle of treatment:	Pyrolysis of the organic part of the batteries in a shaft furnace, followed by reduction of the metallic parts in a smelting furnace, whereby metals are either molten (Fe, Mn), or evaporated and recovered in a splash condenser (Zn).
Process conditions:	400 to 750°C (shaft furnace) and 1500°C (smelting furnace).
Recovered products:	Zinc, ferromanganese, mercury.
Other products:	Slag from smelting furnace (sludge from waste water treatment will be reprocessed in the shaft furnace), cleaned flue gas and waste water.

Industrial process example 2:	Citron (adapted INMETCO) process [208](see Fig. 5.76)
Plant location:	Rogersville (near Le Havre, France)
Capacity:	130'000 t/a
Cost:	CHF 1'200 to 2'200 per ton of treated household batteries
Start of operation:	1998
Treated batteries:	Roughly sorted batteries (e.g. nickel metal hydride, alkaline, silver oxide, sodium sulphur, zinc carbon, lithium). Button cells are separately treated in a distillation unit.
Proven possibility of treatment for other wastes (examples):	Automotive shredder residues, soil from contaminated industrial sites and different kinds of sludges and dusts containing even high amounts of heavy metals
Principle of treatment:	Rotary hearth furnace process: the organic components are pyrolyzed and the resulting carbon monoxide is oxidized to carbon dioxide in the gaseous phase inside of the furnace. The reducing atmosphere within the waste layer reduces the metal oxides and hydroxides. The metals with relatively low evaporation points (Zn, Pb, Cd, Hg) are vaporized and the gaseous phase of the furnace re-oxidized (except Hg) into fine solid particles before they are recuperated in a gravity chamber. The metallic mercury is washed out in a flue gas treatment unit with quencher, electrostatic precipitator and scrubber.
Process conditions:	Temperatures up to 1'350°C (oxy-reducer unit).
Recovered products:	Zinc oxide, steel scrap, manganese concentrate, mercury, other metals in feed elementary or as corresponding oxides depending on specific feed and process conditions.
Other products:	Sludge to suitable landfills, cleaned air and waste water.

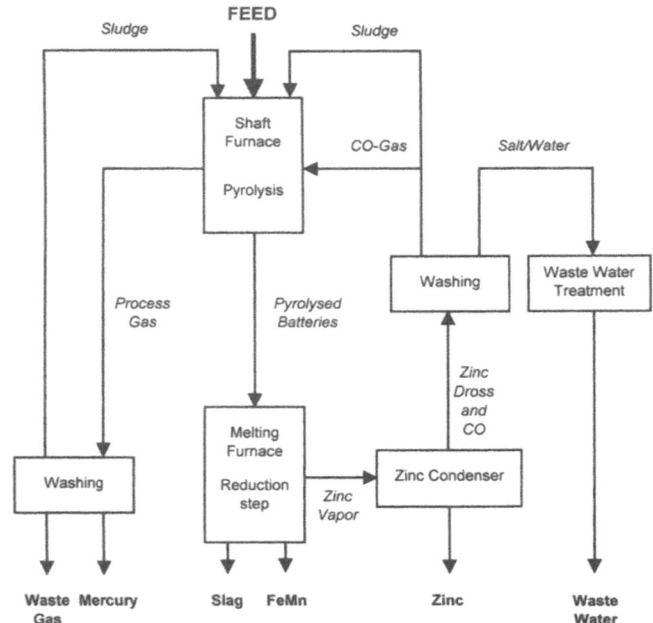

Fig. 5.75. Flowsheet of the BATREC process [47]

The Oxyreducer™ Technology

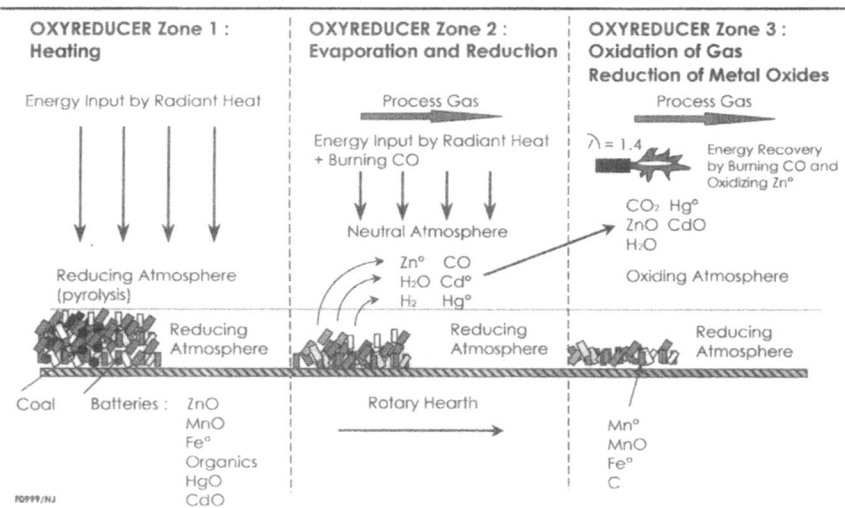

Fig. 5.76. Scheme of the Oxyreducer© technology [10]

Conclusions and Outlook

Thermal processes in practical operation are currently available for the environmentally compatible disposal of all types of batteries. The valuable metallic materials contained in the batteries are, for the largest part, recovered at a purity required by the processing industry. The electrolytes and casings cannot be completely recycled. One of the few exceptions is sulphuric acid in lead-acid batteries.

An even greater challenge for new battery recycling plants than the technical problems is the cost of recycling. Due to their smaller size, specific battery processing plants are more costly than disposal plants for domestic waste. Necessary strategies that ensure success in this case, such as assured waste deliveries, sophisticated logistic systems and flexible highly specific disposal systems with large throughput have been noted. In addition, because of the high investment costs, the existing or planned battery processing plants must be established and used in reality if they are funded (e.g. by pre-paid disposal fees on consumer prices, suitable information campaigns).

Future technological developments in electrical energy storage should consider recycling aspects at the beginning of product development. Care must be taken to ensure that new overall advantages (e.g. improved service life and energy-saving technology) are not outweighed by the disadvantages in the use of more toxic and chemically reactive substances. Therefore the impact of new battery technologies should be ecologically assessed as early as possible, before the production and distribution of batteries as consumer products.

5.5.3 Disposal of Waste Electrical and Electronic Equipment (WEEE)

General Overview

In the European Union (EU), electrical and electronic scrap is called "Waste Electrical and Electronic Equipment" (WEEE). It comprises a wide variety of waste materials (Table 5.26 and 5.28) and corresponding chemical constituents. It ranges from small everyday household objects, such as mobile telephones, very often discarded together with normal household waste, to computer main frames and entire telephone switchboards in offices. A characteristic feature of all these pieces of equipment is that they contain electrical and electronic components such as cables, circuit boards and cathode ray tube (CRT) screens. In many cases the components containing harmful substances have a relatively short life span owing to rapid technological progress (see Ch. 8). Electronic components are also integrated in increasing amounts in conventional machinery such as washing machines, cars and telecommunication equipment.

The plants existing today for the disposal of this waste are large-scale shredders and incineration plants for combustible waste. Increasingly complete electrical machines, such as washing machines, end-of-life vehicles (ELV) or copying machines should be pre-treated so that only the harmless but bulky remains, such as casings made of metal, plastics, ceramics, glass, paper and cardboard are sent to shredders and incinerator plants. This means separating electrical and electronic components and other potentially harmful materials. The design and further development of plants with sufficient capacity for disassembly and treatment of WEEE which are also able to produce waste fractions suited for the intended special recycling and disposal plants, therefore become an important issue.

Legal aspects

The currently planned European Union regulations for disposal of WEEE [12] are closely connected to the reduction of hazardous substances [6] in future market products. In 2001 the European Parliament undertook the first review of a directive regarding WEEE and reduction of hazardous substances. The directive is expected to finally be adopted in 2002. The following are the most important regulations contained in this directive and indicate the future direction (based on commission texts) of the WEEE legislation in Europe [75, 184]:

- Returning of WEEE must be free of charge for consumers;
- Member states ensure adequate number of collection facilities;
- Collection target of 4 kg per person per year;
- Producers are obligated to set up systems to treat WEEE at end-of-life;
- Mandatory selective treatment for certain substances and components;
- Ambitious recovery and recycling targets (Table 5.26) 5 years after the directive has come into effect (CE=consumer electronics, CRT=cathode ray tube, IT=industrial technology)

Table 5.26. Recovery and recycling targets of EU regulation concerning WEEE

Type of WEEE	Rate of recovery	Rate of reuse and recycling
Large household appliances	80 %	75 %
CE products with no CRT	60 %	50 %
IT Products with no CRT	75 %	65 %
Gas discharge lamps	-	80 %
Products with a CRT	75 %	70 % (recycling)

As a preventive measure, these WEEE regulations are amended by stipulations regarding permissible materials (qualifying date: January 1, 2008) – with the exception that their use may be expressly allowed for lack of reasonable alternatives in clearly defined cases. This applies, for instance, to lead, mercury, cadmium, hexa-valent chromium, PBB (polybrominated biphenyls) and PBDE (polybrominated diphenylethers).

The effective implementation of this regulation will hopefully lead to a rapid alignment of the standards and actual state of the WEEE disposal in the member states of the European Union. By now, some of the states have implemented appropriate systems for covering disposal costs as well as extensive and well functioning collection organisations and disposal infrastructure. States with already mandated free of charge return systems are, Belgium, the Netherlands, Norway, Sweden and, as a non-EU member state, Switzerland. Equally important are suitable financial incentives for being environmentally conscious as well as a corresponding knowledge and awareness on the part of the waste producers (section 7.3.). Because of the increasing global trade in goods, an internationally coordinated modus operandi is indispensable, at least among the industrial states, particularly in the area of reduction of hazardous substances.

The necessity of increased international collaboration in the reduction of hazardous substances and WEEE disposal is shown in Table 5.27.

Table 5.27. Comparison of the WEEE disposal in the EU, USA and Japan [43] (for current life-cycle design practices of multinational companies see [212])

Issue	Japan	Western Europe (e.g. Germany, the Netherlands and Sweden) (cf. also[65])	USA (cf. also [138])
Collection	Each municipality has its own policy. Some let people drop off used appliances on specific dates and locations free-of-charge. Others have them picked up at end users' homes on request with a charge varying from $2 to $15.	A collection infrastructure is present in most communities. Collection services are usually free of charge, and paid for by local taxes.	Most cities have pick-up services for large appliances. Usually, no additional fees have to be paid for this service; it is covered by local taxes.
Disassembly	Little or no disassembly is done. However, some companies have their own pilot plants.	Disassembly as a step prior to mechanical processing is done (on a commercial basis) at a number of recycling facilities.	Little or no disassembly is done. Some disassembly takes place to recover reusable components from IT equipment.
Processing	Appliances are shredded only for retrieving iron. Cities have different policies for the treatment of fractions. Plastics are usually landfilled. Small appliances are incinerated.	Many different product categories are mechanically processed. Remaining plastics fractions are usually incinerated for energy recovery. Precious metals, copper and aluminium are often recovered. Landfill is increasingly restricted.	Some shredding operations exist for appliances with raw material value. Still, the majority of small appliances is landfilled. 75% of large appliances are recycled.

With the planned WEEE directive, the European Union and its member states, as well as other European non-EU member states with similar legislation, will have highly ecologically advanced regulation with regard to WEEE disposal.

Case Study 1: SWICO: The Swiss WEEE Collection, Organisation and Financing System

Of practical interest is the situation in Switzerland where the recycling rate of WEEE today amounts to 65 %. An additional 25 % of the appliances are exported as reusable, second-hand products.

One of the reasons for this situation is most likely that in addition to a well developed disposal infrastructure (cf. case study 2 of this chapter) and a legally regulated WEEE disposal that is already in effect (according to which the consumer is generally obliged to return end-of-life electrical and electronic appliances to the retailer, who is in turn obliged to take it back), there exists a credible, economical collection system organized by SWICO (Swiss business association of information, communication and organisation technologies). The system was developed in close cooperation between the government and the most important trade and industrial companies involved [see e.g. 227]. Therefore, a corresponding simple but effective financing model was designed to minimize the total costs.

The financing of the WEEE disposal is based on pre-paid disposal charges on new electronic products sold over the counter (e.g.: TV set: CHF 15.-, stereo system: CHF 10.-, walkman: CHF 2.-). These appliances can be returned to the retailer or to central collection sites without additional charge. By payment of the pre-paid disposal charge at the time of the purchase, the consumer is guaranteed by SWICO (under governmental supervision) that the purchased equipment is disposed of by the licensed recycling operator at its end-of-life in an ecologically 'clean' way according to environmental regulations in effect. The adherence to the regulations is periodically checked and certified by technical specialists of EMPA, the Swiss governmental Material Testing and Research Institute.

Every six months the sellers of new electronic equipment hand over the proceeds from the disposal charges to the operating organisation, SWICO. The latter reimburses its licensed recycling firms for each processed electronic set. This method also makes it possible to keep an adequate statistical record of the flow of materials and thereby to direct material flows optimally.

This disposal system, which up to now has only been in effect for office and data processing equipment such as personal computers, provides many advantages for the waste disposers, the equipment manufacturers, the recycling operators and the state. This system was therefore assigned to apply as of January 1, 2002 to consumer electronics, which are provided mostly by foreign manufacturers and trading firms. Furthermore, as of January 1, 2003, an analogue system of organisation, financing and control of the disposal of electrical household equipment such as vacuum cleaners, mixers, dishwashers and cookers will be introduced by the applicable trade association. Previously, a fee had to be paid upon the return of used household appliances for disposal; this will now be done away with.

Quantities and Composition of WEEE

The quantities of WEEE per inhabitant according to Table 5.28 are relatively small in comparison to the quantity of municipal solid wastes. Because of the higher content of harmful substances and potentially recyclable valuable materials in WEEE, this waste is nevertheless of great importance in an ecological materials economy. According to German statistics, about 60 % of WEEE comes from consumer goods and the rest from offices and the industry.

The comparison of three studies [94, 104, 251] is summarised in Table 5.29, showing the appliances and equipment divided into the various material groups after the disassembly of the sets.

Iron/steel forms the largest class of materials with around 40% of the WEEE (mainly contained in casings), followed by plastics (ca. 20%) and the remaining materials groups, each with about 10%. The latter groups contain technically and economically easily recoverable materials such as non-ferrous metals, in particular copper, and glass (especially in CRT).

CRT currently make up about 56% of the weight of TV sets. For computer screens the corresponding figure is about 39%. Other materials of importance in these appliances, as far as weight is concerned, are (using TV sets as an example, in descending order) wood, iron, plastics, circuit boards, transformers and capacitors. The CRTs themselves in turn are made up of a large variety of materials.

Table 5.28. WEEE quantities (whole appliances with casing and auxiliaries, inclusive of direct factory deliveries to waste disposal operators) per inhabitant in Germany, data BVSE [1], also released by German Umweltbundesamt

Equipment category (EDP = Electronic data processing, PC = personal computer)	Annual amount of WEEE [tons/year]	Annual amount of WEEE [kg/inhabitant]	Composition of WEEE [%]
Large household appliances	570'000	6.9	31
Small household appliances	130'000	1.6	7
TV-sets (without monitors)	320'000	3.9	17
Small consumer electronics units	80'000	1.0	4
Sub-total of consumer goods	1'100'000	13.4	59
EDP/PC private	30'000	0.4	1
EDP/PC industrial	80'000	1.0	4
Communication technology	140'000	1.7	8
Office machines	110'000	1.3	6
Sub-total of information, office and communication technology	360'000	4.4	19
Industrial electronics	360'000	4.4	19
Medicinal and laboratorial technical equipment	50'000	0.6	3
Sub-total of industrial goods	410'000	5.0	22
Total	1'870'000	22.8	100

Table 5.29. Composition of WEEE according to disassemblable materials groups

Category	Total [%] of WEEE category	Thereof					
		Iron [%]	Non-iron metals [%]	Plastics [%]	Circuit boards 1 [%]	Others 1 [%]	CRT2 [%]
Large household appliances ("white goods")	29-30	63	6	14	1	16	-
Small household appliances	7-8.3	24	16	52	2	6	-
TV/Monitor	13-20	8	5	10	9	12	56
Commercial appliances	4.4-8	25	15	37	17	6	-
Computer	3-7.8	30	20	30	15	5	-
Consumer electronics	4-6.3	30	13	32	9	16	-
Rest (large industrial & office sets)	28-31	50	15	18	15	2	-
Grand total (range of values)	100	40.3 -42.2	10.4 -11.4	20.0 -21.2	8.6 -8.7	9.1 -9.5	7.4- 11.2

[1] Glass, wood
[2] Mainly glass and metals

Composition of Circuit Boards

The major part by weight of populated circuit board waste is made up of semiconductor components (33%), followed by capacitors (24%), unpopulated circuit boards (23%), resistances (12%), switches and other materials (8%) [23].

The circuit boards are composed of a wide variety of materials (Table 5.30). The metal content, 28% (copper: 10-20%, lead: 1-5%, nickel: 1-3%), is considerably higher than that in CRT. The content of the remaining most important materials is: plastics 19%, bromine (especially as flame-retardant) 4%, glass and ceramics 49% [63]. In addition, there are also considerable quantities of toxicologically relevant substances, such as the flame-retarding "synergist" antimony trioxide, which is suspected to be carcinogenic, as well as precious metals (in decreasing order: silver, platinum metals like palladium, gold, to a total of approx. 0.3-0.4 %) [63].

Table 5.30. Elementary composition of circuit board waste and comparison with corresponding contents in ores (in order to assess the suitability thereof for use as raw materials) and in the earth's crust (in order to assess the quality of inert materials, i.e. possibility of being able to deposit them in landfills in an environmentally compatible way)

Chemical element	Content in circuit boards of PC [23] [mg/kg]	Content in earth's crust [151] [mg/kg]	Concentration factor[1] [59] [-]	Content in explorable ores */**/*** [mg/kg]	Relative content in circuit boards to earth's crust [-]	Relative content in circuit boards to explorable ores [-]
Fe	108'000	50'000	6	300'000	2	0.36
Al	48'000	81'300	4	325'200	0.5	0.15
Cu	37'000	70	160	11'200	529	3.3
Sn	31'350	40	5'000	200'000	784	0.16
Pb	30'000	16	3'000	48'000	1'875	0.63
Br	27'000	1.6	-	-	16'875	-
Mn	21'500	1'000	350	350'000	22	0.06
Zn	14'500	132	600	79'200	110	0.18
Sb	4'500	1	-	-	4'500	-
Ni	3'160	80	125	10'000	40	0.32
Cl	1'860	314	-	* 600'000	6	0.003
Na	1'840	28'300	-	* 400'000	0.07	0.005
Cr	1'610	200	3'000	* 600'000	8	0.005
Cd	395	0.15	-	* 4'400	2'632	0.09
Co	83	23	-	-	4	-
Hg	9	0.5	-	* 2'000	18	0.005
Ag	ca. 1'000	0.1	-	** 500	10'000	2.0
Pt	ca. 200	0.01	-	*** 10	20'000	20
Au	ca. 5	0.005	-	** 5	1'000	1.0

Without notation: calculated by multiplication of concentration in earth's crust [151] with corresponding concentration factor [59], with notation: * contents from [28], ** contents from[13], *** contents from [185]
[1] Concentration factor for economically explorable ores relative to earth's crust

Aside from the special case of precious metals, copper in circuit boards is economically the most interesting metal for recovery and use as a secondary raw material because of its relatively high content compared to the corresponding content in explorable ores and its economic value. Copper recovery is also of ecological importance because of the relatively high copper content in comparison to earth's crust composition and the corresponding low acceptable concentration limits, e.g. in Swiss regulations concerning leachate from landfills because of its high toxicological impact especially on ruminants as well as some water organisms [154]. No other metals in circuit boards occur in higher contents without further refining processes than in corresponding ores. Nevertheless the contents of some chemical elements are much higher than in earth's crust, especially of some harmful chem.-ical elements like bromine, antimony, cadmium and lead.

In addition to these inorganic elements, the following important organic compounds are found in circuit boards: isocyanates and phosgene from polyurethanes, acrylic and phenolic resins, epoxides and phenols, such as chip glues [23]. Therefore the circuit boards cannot be considered inert materials and are not suited for direct deposition in landfills.

Composition of Cathode Ray Tubes (CRT)

One characteristic of CRT is the high proportion of glass, amounting to around 90%. Around 60.7 % is screen (panel) glass and 29.3%, cone (funnel) glass together with neck and solder glass [201]. Further characteristics are the almost complete absence of organic constituents and, unlike circuit boards and other electronic components, the relatively small amount of metals (in total 10%).

Screen glass contains about 6% strontium oxide and 10% barium oxide [201]. For cone glass the corresponding figures are about 0.3% strontium oxide, 1.3% barium oxide and 22.7% lead oxide. In glass these toxic chemical elements are contained in immobilised form and therefore are of little concern. But in leachable form they are toxic. When the waste glass is used as a secondary raw material, it is therefore important that they remain in immobilised form. This means that the distribution during the lifetime of the materials is controlled and the ecologically correct disposal thereof is ensured.

The phosphorescents only amount to 0.02-0.04% by weight [201] but contain materials which are partly problematic from an ecological point of view or valuable as secondary raw materials, such as zinc sulphide, cadmium, yttrium, gold, silver and rare earth metals such as europium.

Measures Required in Order to Achieve Ecologically Sound WEEE Disposal

A comprehensive procedure with respect to all areas of waste management is required to improve the current unsatisfactory waste disposal situation in many locations regarding electrical and electronic appliances. The following waste management priorities (ecologically, from highest to lowest level) generally apply ([128] with some specific illustration by the author):

1) reuse product as a whole (e.g. second-hand computer),

2) reuse of (even functional) subassemblies and components (e.g. disks and microchips),

3) material recycling (e.g. metals, glass and perhaps some pure plastics fractions),

4) energy recovery from organic materials (plastics, cardboards and papers)

5) adequate treatment of residual fractions for final deposition as an inert material in a controlled landfill.

To fulfil these objectives and keep the mentioned WEEE composition in consideration, the following three principal processing steps must occur:

- disassembly;

- fractionation of the disassembled pieces in various crushing, sorting and classification operations;
- refining with recovery of reusable and recyclable materials, as well as environmentally compatible disposal of the other WEEE fractions produced.

The first two processing steps of WEEE disposal are of decisive importance for a high recovery of valuable components and secondary raw materials. These processing steps must be geared to the already existing regional recycling and disposal facilities for the materials fractions produced as well as to the general economic situation.

Disassembly of WEEE

In the disassembly stage the materials delivered to the plant are recorded (weight, type of appliance, origin, estimated quantity of reusable and hazardous materials), their suitability for reuse assessed (e.g. necessary removal of radioactive substances, batteries, mercury-containing materials, etc.), disassembled as far as required and the individual components forwarded to the respective internal and external special processing plants. This operation is mostly automated to only a low degree and therefore cost-intensive. Efforts are being made to increase the degree of automation. However, if at all technically feasible, this process would entail considerable investments and a guaranteed supply of materials to be processed.

The disassembly can also be economically successfully carried out in de-centralised areas with a high rate of waste produced, i.e. not only at the location of the processing plant, but e.g. at manual separation stations. An important factor in the disassembly operation is the training of personnel in order to avoid high costs further down-stream caused by operators failing to identify foreign, harmful, haz-ardous and reusable materials. The major fractions obtained from the disassembly operation, arranged according to the further processing steps thereof, are:

- reusable components and component groups such as certain electronic components;
- larger quantities of mainly unmixed materials which can therefore be forwarded to refining (e.g. metals) or to recycling (e.g. combustible materials such as wood);
- parts containing hazardous materials (e.g. batteries, selenium-containing copier drums, cable scraps) inclusive of foreign materials brought in from outside, which all require special treatment (reuse, disposal);
- composite materials to be processed separately such as monitors and circuit boards.

The degree to which the disassembly and associated production of fractions (degree of disassembly) should be effected must be decided by the composition of the original materials, the available internal and external processing plants, as well as by economical considerations (e.g. if differentiation between circuit boards having a low and a high content of precious metals is feasible).

Processing of Disassembled Circuit Boards

Basically, one can distinguish between the following currently rival processing methods which can also be combined with each other:

- dry or wet mechanical processes (only with physical separation of materials);
- chemical processes with conversion of materials (among others hydrometallurgically, electrochemically);
- thermal processes (with addition or removal of heat, with and without conversion of materials) (e.g. pyrolysis and smelting metallurgy).

Circuit Boards. Mechanical Processing Methods (Including Electromagnetic Separation Techniques)

The mechanical processing of circuit board scraps consists of various crushing and separation stages, one after the other. In a first stage, a pre-crushing operation is carried out, mostly in shredder or cutting plants, whereby larger pieces of iron are separated out by a magnetic separator. This is followed by a post-crushing operation to < 1 mm size, whereby composite materials can be further divided up into fractions of different composition using suitable, highly specific separation methods.

The crushing methods generally used are shredders, hammer mills and granulators [23, 201]. Typical dry separation and classification methods (without dust separation and exhaust air cleaning) are screens (oscillating and drum screens), wind and air-suspension sifters, magnetic separators, magnetic drum screens and cyclonic separators. Commonly used wet separation and classificatory methods are flotation plants and hydro-cyclones.

The attached case study 2 of the Swiss company Immark Ltd. shows that merely by the skilful introduction of such mechanical-electromagnetic separation methods, a high degree of purity of unmixed materials and a high proportion of reusable materials can be achieved, even with large variations in the incoming materials. The personnel must be highly observant when receiving materials, as even small, erroneously directed quantities of hazardous materials can endanger the reusability of the produced fractions. A high degree of know-how is also required when selecting the correct equipment and the processing conditions to suit the feed characteristics. Under these conditions the mechanical-magnetic WEEE processing method alone is very efficient and is also successfully applied at many locations in Central Europe [78].

Circuit Boards. Chemical Processes

Mainly hydrometallurgical and electrochemical methods are in use in the chemical processing of circuit boards as described below.

a) **Hydrometallurgical (wet chemical) processes.** The processing of circuit boards is normally carried out in three stages [201]:

- Pre-treatment consisting of:
  Mechanical crushing of the circuit boards aimed at exposing as large a surface as possible for the subsequent etching process;
  Separation of solder and components fixed to the circuit boards by heating the boards
- Dissolving the remaining metals by etching and screening off the non-soluble remainder consisting of plastics, ceramics and glass
- Selective separation of the individual metals in the solution obtained, e.g. by precipitation, extraction, filtration, ion exchange and electrolysis.

As regards its complexity the hydrometallurgical treatment method for circuit boards can be compared with mechanical treatment methods, but requires additional chemicals and waste water treatment. Moreover, it produces sludge which is difficult to dispose of.

Examples: The MR Recycling, Neustadt (Wied), process

b) Electrochemical processes. Electrolysis is particularly suited to the treatment of preliminary processed circuit boards with a high content of precious metals and to selective separation of interesting metals. The separation of precious metals from circuit boards is increasingly uneconomical because of the continuing decreasing content in the WEEE and without the possibility of raising the concentration of precious metals-containing components in the disassembly stage. The electrochemical processes are therefore increasingly only of economic interest for use in special cases with fractions having a high content of precious metals and in separation plants which also process other materials with a high content of precious metals.

Examples: The Degussa AG and W.C. Heräus GmbH processes

Circuit Boards. Thermal Processes

a) General aspects. The earlier processing method of combustion of circuit boards in the open air, which was occasionally practiced in Europe, is now forbidden, with good reason, because of the release of organic hazardous substances, e.g. strongly carcinogenic PCDD/Fs and PBDD/Fs (polychlorinated and polybrominated dibenzo dioxins and furans) from chlorine-containing plastics, PCB (polycyclic polychlorinated biphenyls), PAH (polycyclic aromatic hydrocarbons) and bromine. Large quantities of metals with a relatively low boiling point, such as mercury, cadmium, zinc and lead are also released into the environment in this case. When chlorine and bromine are present these metals become even more volatile [246].

The municipal solid waste incineration plant (MSWI) is not suited either to the disposal of whole circuit boards as the heavy metal contents in the circuit board scrap is much higher than in domestic waste. Additionally, ecologically relevant recovery of metals from the incineration residues (except iron of inferior quality from slag or bottom ash) has not been practised in the past until now. Furthermore, it has not been proven on a technical scale that in carrying out a more extensive processing of the slag, e.g. for separation of copper, the same purity can be achieved from circuit boards as is the case, for example, of using the described mechanical fractionation in the following case study of this chapter.

Furthermore, the use of WEEE fractions as a substitute for coal, for instance in blast furnaces and in cement plants, is a realistic proposition only in the case of plastics fractions with high heating value [137]. In cement works, care must be taken that the heavy metal content in the cement does not rise above acceptable limits and the heavy metals are not released in an uncontrolled manner into the environment. Additionally, sometimes the content of chlorine and bromine in wastes is a technical limit for the substitution of coal as a primary energy source.

Thermal classification processes used for various plastics materials based on their melting point, and for materials containing plastics-metals composites based on cryogenic methods are not discussed here. These are special processes which require a preliminary fractionation.

Only pyrolysis and smelting metallurgy will be discussed in more detail here, both being thermal processes generally suited to the treatment of WEEE

b) Pyrolysis. The pyrolysis of circuit boards consists of heating under oxygen exclusion at 500 to 700°C. Under these conditions the plastics materials are thermally decomposed without oxidising the metals, which are mainly present in elementary form. During the process, gas and oil are formed which can be used for heat generation and as raw materials in the chemical industry. Furthermore a metal-enriched solid residue is obtained. Using further mechanical and magnetic separation methods it is at least possible to recover the metals present in larger quantities, such as iron, copper, manganese, etc. A glass-like slag containing the remaining metals is formed as a by-product of this process.

Pyrolysis is a technically proven and fully developed alternative to mechanical treatment processes. It is however a considerably more expensive process, except in the case of large plants with very high throughput and, in addition to circuit boards, which also handle other metal-containing waste. The existing pyrolysis plants are furthermore found in locations more centralised than the mechanical processing plants, which are usually of smaller capacity and the installation of which is economically and logistically more justified. The transport costs for pyrolysis plants are therefore generally higher, except for those located close to the source of the waste.

In cases with similarly extensive fractionation, pyrolysis and mechanical treatment processes are comparable as far as the ecological aspects go, i.e. a decision regarding the selection of the disposal process to be used in a concrete case must be based on a detailed investigation of the local and regional environmental implications and established on the basis of an ecological balance.

Examples: The Pyrocom plant, Bernau (D), the pyrolysis of Ruhrkohle Oel und Gas GmbH, Bottrop (D), and the atypical Inmetco based Oxyreducer process of Citron SA, Le Havre (F) as a process at more elevated temperatures like gasification and with both oxidation and reduction steps [46].

c) Smelting metallurgy. A realistic alternative to the pyrolysis described above is the co-processing of circuit boards in plants where favourable local conditions exist (e.g. treatment plants for copper scrap and copper ore). The circuit boards are first disintegrated in shredder or cutting plants. Subsequently they can be treated in the same way as the other input into the reactor and converter of the copper works. Organic hazardous materials are fully destroyed at high temperature (1200°C) and residence time in the converter, and energy is recovered with high efficiency. The metals contained in the scrap and ore, consisting almost exclusively of non-precious metals such as copper, are oxidised at the prevailing process conditions and are removed from the smelt. They can then be separated with various selective mechanical and chemical methods.

This is an entirely feasible and ecologically compatible technological solution in modern copper works with associated gas cleaning systems as well as process plants for slag (iron separation) and dust filter (in which enrichment of more volatile metals, such as lead, tin and zinc, takes place). If copper alone, i.e. about 10% of the circuit board, is recovered, this process is in fact superior to disposal on landfills but clearly should be considered ecologically inferior to the dry-mechanical process described with its relatively high percentage of recyclable fractions.

Examples: The processes developed by the Norddeutsche Affinerie (D), the Hüttenwerke Kayser AG (Lünen, D), the Montanwerke Brixlegg (A) [116], the Metallgiesserei Velmede GmbH, Bestwig-Velmede (D), and the Boliden Mineral AB (Sweden) [134].

Processing of Cathode Ray Tubes (CRT)

The processing of CRTs is, in the first place, aimed at the recycling of glass. The recycling of metals, provided this is technically and economically feasible, is of secondary importance.

There are a number of processing methods which differ widely from one another with respect to the proportion of glass that can be recycled (Table 5.31). The situation regarding utilisation of toxic materials (strontium, barium, lead) will in longer terms be mitigated, as more and more the CRTs are being replaced by liquid crystal displays (LCD), containing a lower level of harmful substances (and which are energetically more efficient).

The processing of the glass to secondary raw materials must in any case be done by dry or wet mechanical methods [176], and the phosphorescents must be separated out. Although in the past in Europe these materials have been disposed of in underground depositories and hazardous waste incineration plants, these materials are potential secondary raw materials which can be recovered and reused. Technical methods for this purpose exist (e.g. the Norwegian HAS System from HAS Consult A/S [201].

Table 5.31. Review of fundamental methods for CRT processing [201]

Type of process	Process	Characteristics	Assessment of possibilities of reuse
Mechanical	Shredders or breakers	Simple, no thorough separation of the types of glass	Problematic because of Phosphorescents, at the most down-cycling (lead recovery) of the glass
	Knocking off the cone glass	Costly, low level of clean separation of the types of glass	Limited reuse, only down-cycling
	Cutting	Costly, difficult to achieve thorough separation of the types of glass	At least limited reuse (screen glass or cone glass certainly possible)
Chemical	Wet chemical separation	High chemical consumption	High-quality reuse after uccessful separation of the different types of glass
Thermal	Thermal breaking-off of the cone half	Complete separation of the components difficult to achieve	At least pure fractionation of the screen or cone glass, down-cycling with lead recovery possible
	Thermal separation in the furnace	Complete separation of the components possible	High-quality recycling products

Processing of Other Fractions from the Disassembly and Processing of WEEE

Special processing methods exist today for most of the materials contained in WEEE, such as batteries (section 5.5.1), fluorescent tube lamps [179], selenium drums in copiers (see case study of the company Immark in this chapter) and cables. They result in an extensive recovery of reusable materials and immobilisation of harmful materials (inorganic compounds) and/or destruction of harmful materials (organic compounds). Metallic and certain mineral constituents in the WEEE can be refined at reasonable cost after the disassembly and possible fractionation in a WEEE processing plant located in metal and glass works and reused as secondary raw materials.

However, reuse and recycling of fractions of plastics materials is more problematic and comparable to conditions when treating ELV (end-of-life vehicles) and ASR (automotive shredder residues). Firstly, these fractions are mostly mixed and also not easily identifiable in the disassembly process. Secondly, in particular after the WEEE processing, they are additionally contaminated with foreign materials such as metals and minerals. It must therefore be decided on the basis of the feasibility of identifying, disassembling and separating the materials into reusable plastics fractions, which of the following recovery cycles, listed in descending order with respect to their reusable value, should be chosen:

- Reuse as components in new products;
- Recycling of the materials for direct use as raw materials (e.g. after re-granulation or melting of thermoplastics);
- Chemical (feedstock) recycling for use as raw materials (e.g. after cracking);
- Thermal recycling (energy recovery) in suitable thermal treatment plants.

Typical thermal processing methods for utilisation of the plastics as materials are pyrolysis (typically at 700-800°C and at oxygen exclusion, cf. [130]), gasification (production of synthesis gas containing carbon monoxide and hydrogen at elevated temperature and pressure with addition of water and oxygen, cf. [153]), hydrogenation (temperature about 500°C at elevated pressure and in a hydrogen atmosphere, cf. [177]) and solvolysis (hydrolysis [addition of water], alcoholysis [addition of alcohols] with methanolysis and glycolysis as well as aminolysis [addition of amines], cf. [34]).

In no case should the plastics be deposited in larger quantities on landfills, as some of the organic decomposition products and inorganic additions (e.g. cadmium as softener) greatly endanger the quality of the groundwater. Additionally, the heating value of the plastics is not taken advantage of, which results in unnecessary consumption of non-renewable energy (gas, oil, coal). Depending on the local legislation and the market situation regarding the energy sources otherwise used, the potential savings in production costs in thermal processing plants, such as blast furnaces and cement factories, are not utilised. Extensive German and Swiss experience shows that a large proportion of the coal otherwise required can be saved [137], making use of existing technology, as for instance in cement plants. But a sufficiently low content of foreign materials in the plastics must be ensured. This solution functions to the benefit of the population and the environment and – as in a win-win situation – also to the economic advantage of the plant operator.

Case Study 2: WEEE Processing of the Swiss Company Immark

Based on information from the Swiss company Immark Ltd (Table 5.32), this case study shows that a very high recycling rate can be achieved with state-of-the-art technology in private industry, i.e. without governmental subsidies, but only on the basis of a guaranteed disposal price per ton of WEEE, taking into account the unavoidable transport and processing costs. Based on the total recovered WEEE in Switzerland according to the first case study of this chapter, the company IMMARK shows a market share of about 50%.

Table 5.32. WEEE Processing by IMMARK Ltd

Brief description of the company (Fig. 5.77)	
WEEE processed:	Computers (PC and main frame), TV/computer screens, mobile telephones, circuit boards.
Beginning of operation:	1986
Processing capacity:	Kaltenbach: about 8'000 t/a, Regensdorf: about 20'000 t/a.
Area of activity:	Throughout Switzerland.
Recycling partner for large companies:	Compaq, IBM, Philips, Siemens-Nixdorf etc.
Services:	- Collection of WEEE from large-scale collecting areas. - Disassembly of WEEE at customers' locations.
Locations:	4 regional disassembly locations in Switzerland
Disassembly and processing steps:	1. Disassembly and delivery to a special processing plant: Inspection, sorting and disassembly of the appliances delivered; among components separated are the following: capacitors (possibly containing PCB); all battery types; mercury-, cadmium- and selenium-containing components; LCD-displays (containing azo dyes); beryllium containing special telecommunication devices; other disassembleable components containing foreign, harmful, process-disturbing as well as directly reusable (e.g. some electrical components), refinable (e.g. metals) and disposable bulk materials (i.e. packaging materials and casings of combustible materials). 2. Mechanical-electromagnetic separation of disassembled WEEE and composite materials (Figs. 5.77 and 5.78).
Percentage of recycled materials:	66%, consisting of - Pre-crushing and primary separation - Result$^©$-Plant (further crushing to finer materials and separation into recyclable fractions and such which can be disposed of at existing incineration plants).

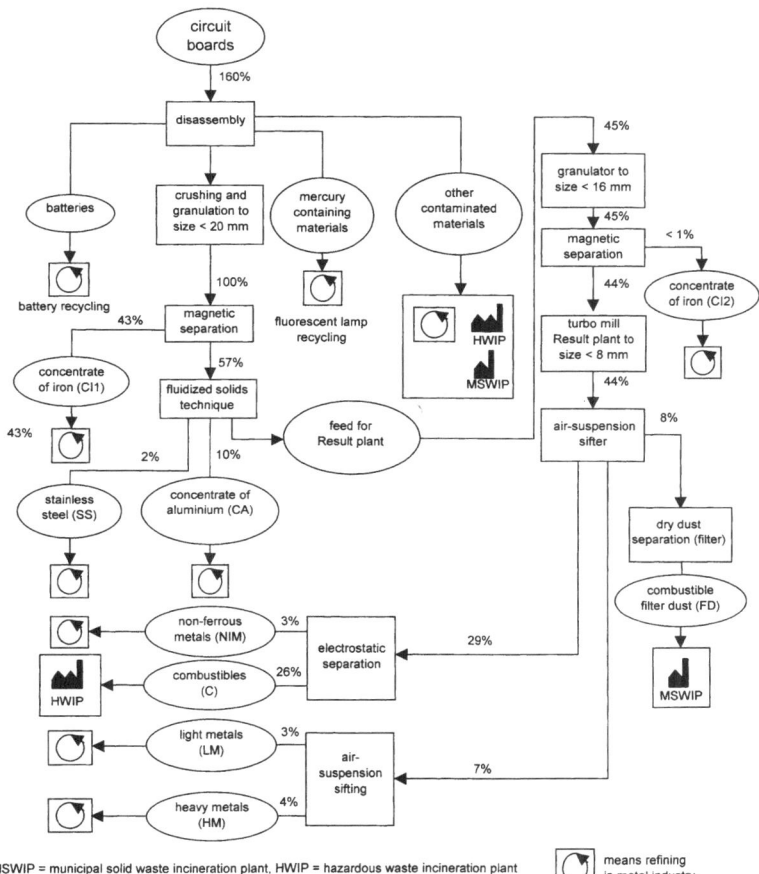

Fig. 5.77. Flow sheet showing processing of WEEE and circuit boards at the IMMARK Company in Kaltenbach (Switzerland) (Feed: 70% electronics from office, information and communication technology and 30% consumer electronics [78])

In addition to the high proportion of reusable materials, the following features are worth mentioning:

- The fully closed material cycle (no production of non-disposable fractions or fractions intended for dumping on landfills);
- The autonomous economy (complete coverage of the investment and operating costs by the fixed disposal price, established by the above described Swiss collecting organisation SWICO of the first case study in this chapter);
- The full compliance with all governmental safety and environmental regulations.

Table 5.33. Approximate composition (%) of the fractions from WEEE processing in the IMMARK plant at Kaltenbach (same feed as in Figure 5.77)

Fraction	Total percentage of feed	Fe content	Heavy metal content without Fe and Cu	Cu content	Light metal content (like Al and Mg)	Residual fraction content[1]
Concentrate of iron	43	91.8	5	2	0.2	1
Stainless steel	2	44.8	15	10	0.2	30
Concentrate of aluminium	10	0.2	10	10	69.8	10
Concentrate of iron	1	96.6	0.2	1	0.2	2
Combustible filter dust	8	0.2	0.5	0.5	0.5	98.3
Non-ferrous metals	3	0.2	39.9	39.9	10	10
Combustibles	26	0.2	0.2	0.5	0.5	98.6
Light metals	3	0.5	0.5	0.5	98.2	0.3
Heavy metals	4	0.5	7	91.3	0.2	1

[1] Mainly combustibles and glass

Refrigerators were until recently disassembled in an environmentally correct way i.e. with extensive removal of the harmful CFC gas (ChloroFluoroCarbons, which when released cause a breakdown of the ozone layer) from the appliances before shredding. This part of the work is now carried out by a different company. Large electrical appliances, such as cookers, are also disassembled, and hazardous components removed and processed in the company's own processing plant. The bulky metal casings are sent to large-scale shredders and to wet mechanical separation plants for extensive recovery of metals.

The results of material flux analysis in Figure 5.78 are base on the flow sheet in Figure 5.77 and the data of fractions composition in Table 5.33 delivered by IMMARK Ltd. and adapted by the author.

The results from the material flux analysis illustrate the following:

- the overall recovery rate is around 66%;
- generally, the recovery rate of the different investigated metal categories is very high (>98% for iron, 98% for non-ferrous and non-cupreous heavy metals, 97% for copper, >98% for light metals) if all the metals delivered to refining industry are recycled;
- most of the recycled metals are concentrated in only two fractions from WEEE processing;
- beside the already mentioned high overall recovery rates, the noted metals in the fractions are only concentrated to the extent demanded by the refining industry and to the extent that it is economical for further processing of fractions;
- most of the residual materials are concentrated in only two fractions which are suitable as a secondary fuel in waste incineration plants with heat recovery;

nevertheless the organic components are also useful in the fractions delivering to the metal refining industry because there thermal processes occur with direct heat transfer from thermal destruction of organics to the thermal metal refining.

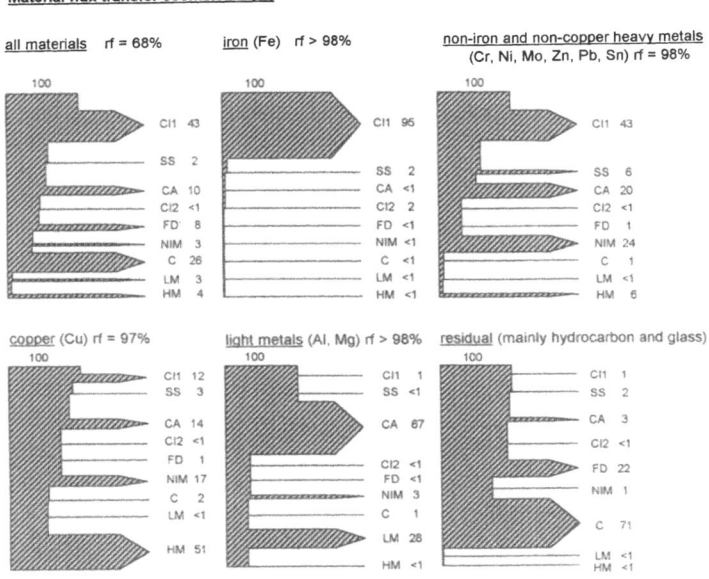

Fig. 5.78. Material flux analysis (distribution of the several fractions from WEEE processing) and recovery rates of the most important WEEE components. Definition of recovery factor rf (%): Sum of transfer coefficients of the fractions for feedstock recycling, namely, Cl1, SS, CA, Cl2, NIM, LM, and HM

Conclusions

Various highly selective processes already currently exist for WEEE disposal in an environmentally compatible manner with extensive reuse of components, recycling of materials (especially metals and glass) and energy recovery from combustible fractions (energy recovery). These processes are also economical, provided that suitable collection, logistics and financial systems exist as described. Additionally it is essential that an appropriate legislation is in effect, that the awareness of the waste producers is increased, that the personnel handling the processing of the waste are adequately trained and finally that a sufficiently extensive infrastructure (MSWI, landfills) to deal with waste disposal exists. Provided that these conditions are met, and if this system is professionally carried out, there will be no harmful emissions for man and environment.

Further improvements to the above WEEE processes are possible. A particularly interesting possibility in connection to this is co-processing with other heavy metal-containing and organic-rich materials of complex composition, such as automotive shredder residues (ASR).

From an ecological and economical point of view, it is essential that the large differences in the infrastructure related to the disposal of WEEE which currently exist between various countries be eliminated as soon as possible. Their disposal practices should be brought into line with the standards described in this chapter, not only within the European Union but also world-wide.

On the basis of the results achieved from eco-balances and considerations of economic costs made up to now and the resulting ecological gains, it is evident that there is still a great need for further action in research, development, legislation and implementation (life-cycle design [212]) of measures to be taken for the reduction or total ban [181] of hazardous substances in market products currently used. For some harmful materials, there are already – or will likely be in the next few years - fully adequate alternatives (e.g. lead-free solder [158], halogen-free circuit boards [40]). In recent years there have been considerable ecological improvements in this respect (e.g. lower content of harmful substances and energy consumption in LCD units).

References

1. Bundesverband für Sekundärrohstoffe und Entsorgung e.V., Information Paper: Electronics Scrap Recycling - Facts, Data and Process, bvse-recycconsult, Bonn, 1998.
2. SAEFL / BUWAL (1988). Entwurf für eine Richtlinie zur Durchführung des Eluat-Tests für Inertstoffe und endlagerfähige Reststoffe. Berne, Switzerland.
3. SAEFL / BUWAL (1990). Technical Ordinance on Waste December 10, 1990 (TOW; SR 814.600); Technische Verordnung über Abfälle vom 10. Dezember 1990 (TVA; SR 814.600). Berne Switzerland.
4. BUWAL (1998). Die Rückstände der Verbrennung; Flugasche und Filterkuchen. Umwelt-Materialien #100. BUWAL, Bern, Switzerland.
5. EU Directive 2000/53/EC of the European Parliament and the Council of 18 September 2000 on End-of Life Vehicles. European Union
6. EU Directive on the Restriction of the Use of Certain Hazardous Substances in Electrical and Electronic Equipment (2000/159 COD). European Union
7. EU Directive on Waste Incineration 2000/76/EC. European Union
8. Finnish Standard for Recover Fuels, 5875/2000. Finland
9. IGEA-Stiftung, RESH-Entsorgung 2000, 5.2.2001.
10. M.U.T.Umweltpreis, Wettbewerbsprojekt CITRON SA, Kostengünstige und umweltfreundliche Entsorgung schadstoffreicher Abfälle mit der OXYREDUCER™-Technologie, Basel, 2000.
11. Osada M, Murahashi K, Shibaike H (2000) Characteristics of Gasifying Combustion in MSW Direct Melting System, Proc.of ICIPEC, June 8-10 Seoul Korea, 101-107
12. Proposal of the EU Commission COM (2000) 347 for a Directive on Waste Electrical and Electronic Equipment (2000/158 COD). European Union
13. Uhren-Bijouterie Weissbach, Laupen, Switzerland (2002) personal communication

14. Anonymous (2001) The Management of Residues from Thermal Processes. In: Patel N, Cordon G, Howlett L (eds) IEA Bioenergy task 23 Energy from Thermal Conversion of MSW and RDF

15. Anonymous (2001) The Status of Waste Incineration in Europe. Waste management world November-December 2001:p. 63

16. TVA (1993) Technische Verordnung über Abfälle, SR814.600, Berne, Switzerland. (see also [3])

17. Adams PB (1992) Predicting Corrosion. In: Clark DE, Zoitos BK (eds) Corrosion of Glass, Ceramics and Ceramic Superconductors; Principles, Testing, Characterization and Applications, Noyes Park Ridge U.S.A.

18. Advocat T (1991) Les mécanismes de corrosion en phase aqueuse du verre nu-cléaire R7T7; approche expérimentale, essai de modélisation thermodynamique et cinétique. PhD thesis, University Louis Pasteur Strasbourg

19. Ahrens-Botzong, Redmann (1993) Das Schwel-Brenn Verfahren, Brennstoff Wärme Kraft

20. Alakangas E (2000) Recent Developments in Bioenergy Sector in Europe. EU AFB Altener report. EU

21. Amt für Umweltschutz und Energie des Kantons Basel-Landschaft (1995) Bericht über die Untersuchung der stofflichen Zusammensetzung und das Auswaschverhalten von Rückständen aus einer Shredderanlage (RESH) im Hinblick auf deren Entsorgung in Reaktordeponien. Umweltschutzlabor, Liestal

22. Ando G, Steiner C, Selinger A, Shin K (2002) Automobile Shredder Residue Treatment in Japan - Experience of 95'000 t ASR Recycling and Recovery Available for Europe with TwinRec. International Automobile Recycling Congress, Geneva Switzerland

23. Angerer G, Bätcher K, Bass P (1993) Verwertung von Elektronikschrott - Stand der Technik, Forschungs- und Entwicklungsbedarf, Abfallwirtschaft in Forschung und Praxis, Band 59. Erich Schmidt Verlag, Berlin

24. Anonymous (1990) Durchführung von Pilotversuchen zur Aufbereitung und Entsorgung von Filterstaub aus Kehrichtverbrennungsanlagen. Schlussbericht. ABB-EAWAG-EMPA-KEZO, Baden Dübendorf Switzerland

25. Anonymous (1992) Emissionsabschätzung für Kehrichtschlacke. Schlussbericht. Auftraggebergemeinschaft für das Projekt EKESA, Switzerland

26. Antal MJ, Allen SG, Schulman D, Xu X (2000) Biomass Gasification in Supercritical Water. Ind. Eng. Chem. Res. 39:4040-4053

27. Atassi H (1989) Evaluation de la résistance à la corrosion en solution aqueuse de quelques verres silicatés. PhD thesis, University Louis Pasteur Strasbourg

28. Baccini P, Brunner PH (1985) Behandlung und Endlagerung von Reststoffen aus Kehrichtverbrennungsanlagen. Gas-Wasser-Abwasser 65:403

29. Baccini P, Gamper B (1993) Deponierung fester Rückstände aus der Abfallwirtschaft - Endlagerqualität am Beispiel Müllschlacke. ETH-EAWAG-vdf Hochschulverlag, Zurich

30. Baccini P, Zimmerli R, Krähenbühl M (1985) Bestimmung und Beurteilung der physikalisch-chemischen Eigenschaften von Abfällen der Autoverwertung Ostschweiz AG. Nr. 30-323. EAWAG-Projekt, Dübendorf

31. Barlow R (2001) Industry Response to European Battery Legislation. Proceedings. International Congress for Battery Recycling, Montreux

32. Barner HE, Huang CY, Johnson T, Jacobs G, Martch MA, Killilea WR (1992) Supercritical Water Oxidation: An Emerging Technology. J. Hazard. Mater. 31

33. Bates JK, Bradley CR, Buck EC, Cunnane JC, Ebert WL, Feng X, Mazer JJ, Wronkiewicz DJ, Sproull J, Bourcier WL, McGrail BP, Altenhofen MK (1994) High-Level Waste Borosilicate Glass: A Compendium of Corrosion Characteristics. Technical report. US-DOE, Washington USA

34. Bauer G (1991) Alkoholyse - Chemisches Recycling für PUR und gemischte Kunststoffabfälle. Kunststoffe 90 81:301

35. Beckmann M, Scholz R (1999) Energetische Bewertung der Substitution von Brennstoffen durch Ersatzbrennstoffe aus Abfällen bei Hochtemperaturprozessen zur Stoffbehandlung. ZKG International 52:287-303; 411-419

36. Beckmann M, Horeni M, Keldenich K (2002) Herstellung und Einsatz von Ersatzbrennstoffen-Möglichkeiten zur Prozessoptimierung der gesamten Verfahrenskette. VDI-GET Tagung Ersatzbrennstoffe in der Energietechnik. VDI-Verlag GmbH Düsseldorf, Dortmund

37. Belevi H (1993) Was können Stoffflussanalysen bei der Bewertung der thermischen Behandlung leisten? Bericht 1033. VDI-Verlag GmbH, Düsseldorf, pp 261-276

38. Belevi H (1998) Environmental Engineering of Municipal Solid Waste Incineration. Habilitation. Verlag der Fachvereine, Dübendorf

39. Berbenni P, Nobili F, Basetti A, Farneti A, Barbaresi U (1989) Detoxification of Dusts Derived from Urban Solid Wastes Incineration Plants. Recycling International:1550-1560

40. Bergendahl CG (2000) Electronics goes Halogen-Free: International Driving Forces and the Availability and Potential of Halogen-Free Alternatives. Proceedings. IEEE International Symposium on Electronics and the Environment, San Francisco USA, p 54

41. Biollaz S, Grotefeld V, Künstler H (1999) Separating Heavy Metals by the VS-Process for Municipal Solid Waste Incinerators. In: Barrage A, Edelmann X (eds) Recovery, Recycling, Reintegration R'99 Conference. EMPA, Geneva

42. Boie W (1957) Vom Brennstoff zum Rauchgas. In: Feuerungstechnisches Rechnen mit Brennstoffkenngrößen und seine Vereinfachung mit Mitteln der Statistik. B.G.Teubner Verlagsgesellschaft,, Leipzig

43. Boks C, Nilsson J, Masui K, Suzuki K, Rose C, Lee BH (1998) An International Comparison of Product End-of-Life Scenarios and Legislation for Consumer Electronics. Proceedings. IEEE International Symposium on Electronics and the Environment, Oak Brook Illinois USA, p 19

44. Brandl, Hellweg, Schmidt, Stucki, Wochele (1999) Forschung für eine nachhaltige Abfallwirtschaft. Paul Scherrer Institut, Villigen

45. Brüggler M (2002) ASR Recycling in Europe. International Automobile Recycling Congress, Geneva Switzerland

46. Brüggler M (2002) Citron's Oxyreducer Process - A Solution to Pyrolyse WEEE. Proceedings. International Electronics Recycling Congress, Davos Switzerland

47. Burri R (1995) The Wimmis Project. Proceedings. International Battery Recycling Congress, Lucerne

48. SAEFL / BUWAL (1998) Die saubere Kehrichtverbrennung: Mythos oder Realität? Schriftenreihe Umwelt

49. SAEFL / BUWAL (2000) Schweizerische Abfallstatistik.. http:// www.umwelt-schweiz.ch

50. Carsch S, Thoma H, Hutzinger O (1989) Leaching of Polychlorinated Dibenzo-p-Dioxins and Polychlorinated Dibenzofurans from Municipal Waste Incinerator Fly Ash by Water and Organic Solvents. Chemosphere 15:1927-1930

51. Chandler AJ, Eighmy TT, Hartlen J, Hjelmar O, van der Sloot HA, Sawell SE, Vehlow J, Kosson DS (1995) Municipal Solid Waste Incinerator Residues. Studies in Environmental Science. Elsevier, Amsterdam

52. Cheron P, Chevalier P, Do Quang R, Tanguy G, Sourrouille M, Woigner S, Senoo M, Banka T, Kuramoto K, Yamaguchi T, Shimizu K, Fillet C, Jacquet-Francillon N, Godard J, Dussossoy J-L, Pacaud F, Charbonnel J-G (1995) Examination and Testing of an Active Glass Sample Produced by COGEMA. Mat. Res. Soc. Symp. Proc., Scientific Basis for Nuclear Waste Management XVIII, pp 353, 55-

53. Chien Y-C, Wang HP, Lin K-S, Yang YW (2000) Oxidation of Printed Circuit Board Wastes in Supercritical Water. Wat. Res 34:4279-4283

54. Christen D (1997) Zusammensetzung von Altfahrzeugen nach Stoffgruppen und ihre Entwicklung. IGEA, Bern

55. Christen D (2002) Elimination of Shredder Residue: The Swiss Approach. International Automobile Recycling Congress, Geneva Switzerland

56. Clark DE, Zoitos BK (1992) Corrosion of Glass, Ceramics and Ceramic Super-conductors; Principles, Testing, Characterization and Applications. Noyes Publications, Park Ridge

57. Colombel P (1996) Etude du comportement à long terme de vitrifiats de REFIOM. PhD thesis, University of Poitiers

58. Colombel P, Godon N, Vernaz E, Thomassin J-H (1997) Mécanismes d'altération des vitrifiats de REFIOM et analogie avec d'autres verres industriels ou naturels. In: Cases J-M, Thomas F (eds) Procédés de Solidification et de Stabilisation des Déchets, Nancy France

59. Cox PA (1995) The Elements on Earth - Inorganic Chemistry in the Environment. Oxford University Press, Oxford

60. Crooker PJ, Ahluwalia KS, Fan Z, Prince J (2000) Operating Results from Supercritical Water Oxidation Plants. Ind. Eng. Chem. Res. 39:4865-4870

61. Crovisier J-L, Honnorez J, Eberhart J-P (1987) Dissolution of Basaltic Glass in Seawater: Mechanism and Rate. Geochim. Cosmochim. Acta 51:2977-

62. Crovisier J-L, Vernaz E, Dussossoy J-L, Caurel J (1992) Early Phyllosilicates Formed by Alteration of R7T7 Glass in Water at 250°C. Appl. Clay Sci. 7:47-

63. Demmelhuber S (1994) Elektroschrott und Recycling: Völlig falsche Entsorgung. Funkschau 14

64. Disler W, Keller C (1997) Co-Incineration of Non-Metallic Automobile Shredder Waste (RESH) in Solid Waste Incineration Plants. p. V.69 proceedings. R'97 recovery, recycling, re-integration, congress, Geneva Switzerland

65. Dowdell DC, Adda S, R. N, Laurent D, Glazebrook B, Kirkpatrick N, Richardson L, Doyle E, Smith D, Thurley A (2000) An Integral Life Cycle Assessment and Cost Analysis of the Implications of Implementing the Proposed Waste from Electrical and Electronic Equipment (WEEE) Directive. Proceedings. IEEE International Symposium on Electronics and the Environment, San Francisco (USA), p 1

66. Ebert WL, Mazer JJ (1994) Laboratory Testing of Waste Glass Aqueous Corrosion: Effects of Experimental Parameters. Mat. Res. Soc. Symp. 333. Scientific Basis for Nuclear Waste Management XVII, pp 27

67. Eckert CA, Busch D, Brown JS, Liotta CL (2000) Tuning Solvents for Sustainable Technology. Ind. Eng. Chem. Res. 39:4615-4621
68. Eighmy TT, Crannell BS, Butler LG, Cartledge FK, Emery EF, Oblas D, Krazanowski JE, Eusden (Jr) JD, Shaw EL, Francis CA (1997) Heavy Metal Stabilization in Municipal Solid Waste Combustion Dry Scrubber Residue Using Soluble Phosphate. Environmental Science and Technology 31:3330-3338
69. Engels B (2001) Collection & Recycling of Spent Batteries, Thermal Treatment of Spent Batteries - Results of a Research Project on Behalf of the German Federal Environmental Agency. Proceedings. International Congress for Battery Recycling, Montreux
70. England N (2001) Activities of the Rechargeable Battery Recycling Corporation (RBRC) - Country report USA. Proceedings. International Congress for Battery Recycling, Montreux
71. Ewing RC (1996) Glass as a Waste Form and Vitrification Technology: Summary of an international workshop. National Research Council, National Academy Press, Washington
72. Fang Z, Kozinski JA (2002) A Study of Rubber Liquefaction in Supercritical Water using DAC-Stereomicroscopy and FT-IR Spectrometry. Fuel 81:935-945
73. Fang Z, Xu S, Kozinski JA (2000) Behavior of Metals during Combustion of Industrial Organic Wastes in Supercritical Water. Ind. Eng. Chem. Res. 39:4536-4542
74. Faulstich M, Freudenberg A, Köcher P, Kley G (1992) RedMelt-Verfahren zur Wertstoffgewinnung aus Rückständen der Abfallverbrennung. In: Faulstich M (ed) Rückstände aus der Müllverbrennung. EF-Verlag, Berlin, pp 703-726
75. Feehan T (2002) Proposed Directives on WEEE and Reduction of Hazardous Substances. Proceedings. International Electronics Recycling Congress, Davos Switzerland
76. Feillard P (2002) ELV Directive Implementation in Europe. Proceedings. International Automobile Recycling Congress, Geneva Switzerland
77. Fillet S (1987) Mécanismes de corrosion et comportement des actinides dans le verre nucléaire R7T7. PhD thesis, University Montpellier II
78. Bundesministerium für Land- und Forstwirtschaft, Umwelt und Wasserwirtschaft der Republik Oesterreich (2000) Schriftenreihe des BMLFUW Band 7/2000
79. Frankenhaeuser M (2002) European Standardization of Solid Recovered Fuels In: Power Production from Waste and Biomass IV. Proceedings. VTT symposium series, Espoo Finland, p in press
80. Fricke J (2001) Collection and recycling of portable batteries in Germany - How to achieve high quantities. Proceedings. International Congress for Battery Recycling, Montreux
81. Fricke JL, Knudson N (2001) Entsorgung von Gerätebatterien. In: Kiehne H-A (ed) Gerätebatterien, Kontakt & Studium. expert verlag, Renningen-Malmsheim
82. Friesen KJ, Sarna LP, Webster GRB (1985) Aqueous Solubility of Polychlorinated Dibenzo-p-Dioxins Determined by High Pressure Liquid Chromatography. Chemosphere 14:1267-1274
83. Frisch MA (1997) Supercritical Water Oxidation. In: Freeman HM (ed) Standard Handbook of Hazardous Waste Treatment and Disposal, 2nd (edn). McGraw-Hill, New York, p 8.176
84. Fujimoto K (2001) Status of New Battery Legislation & Recycling Activities in Japan. Proceedings. International Congress for Battery Recycling, Montreux

85. Göttlicher R, Anton P (1990) Reststoffe der Müllverbrennung. AbfallwirtschaftsJournal 2 Nr.6
86. Goudriaan F, Peferoen DGR (1990) Liquid Fuels from Biomass via a Hydrothermal Process. Chemical Engineering Science 45:2729-2734
87. Goudriaan F, van de Beld B, Boerefijn FR, Bos GM, Naber JE, van der Wal S, Zeevalkink JA (2001) Thermal Efficiency of the HTU© Process for Biomass Liquefaction. In: Bridgwater AV (ed) Progress in Thermochemical Biomass Conversion. Blackwell Science Oxford
88. Grambow B (1985) A General Rate Equation for Nuclear Waste Glass Corrosion. Mat. Res. Soc. Symp. Proc. 44. Scientific Basis for Nuclear Waste Management VIII (C.M. Jantzen, J.A. Stone and R.C. Ewing, Eds), pp 16-
89. Grambow B (1994) Remaining Uncertainties in Predicting Long-Term Performance of Nuclear Waste Glass from Experiments. In: Barkatt A, Van Konynenbourg RA (eds) Mat. Res. Soc. Symp. Scientific Basis for Nuclear Waste Management XVII
90. Granatstein DL (2001) Technoeconomic Assessment of Fluidized Bed Combustors as Municipal Solid Waste Incinerators: A Summary of Six Case Studies. In: Patel N, Cordon G, Howlett L (eds) IEA Bioenergy task 23 Energy from Thermal Conversion of MSW and RDF
91. Granatstein DL, Hesseling WFM (2001) Robins Resource Recovery Facility, Robins, Illinois, USA. In: Patel N, Cordon G, Howlett L (eds) IEA Bioenergy task 23 Energy from Thermal Conversion of MSW and RDF
92. Griffith JW, Wofford III WT, Griffith JR (1999) Apparatus for Oxidizing Undigested Wastewater Sludges. US Patent 5 888 389, USA
93. Gutmann R (1996) Thermal Technologies to convert Solid Waste Residuals into Technical Glass Products. Glastech. Ber. 69:285-
94. Hanke M, Ihrig C, Ihrig DF (2001) Stoffbelastung beim Elektronikschrott-Recycling, Schriftenreihe der Bundesanstalt für Arbeitsschutz und Arbeitsmedizin, Gefährliche Arbeitsstoffe GA 56, Schriftenreihe. Wirtschaftsverlag NW, Verlag für neue Wissenschaft GmbH, Bremerhaven
95. Hannes JP, Wachenhausen M (1999) Mitverbrennung von "Refuse Derived Fuel" in kohlegefeuerten Kraftwerkskesseln. Bericht 1492. VDI Verlag GmbH, Düsseldorf, pp 381-386
96. Härdtle G, Markek K, Bilitewski B, Gorr C (1994) Altautoverwertung. Grundlagen - Technik - Wirtschaftlichkeit - Entwicklungen. Müll und Abfall 32
97. Hazlebeck DA, Spritzer MH, Downey KW (1995) Supercritical Water Oxidation of Chemical Agents, Solid Propellants and Other DOD Hazardous Wastes. In: Holm FW (ed) Proceedings Workshop on Advances in Alternative Demilitarization Technologies. Science Applications International Corporation, McLean Virginia, USA, Hyatt Regency Reston Virginia
98. Heck L, Schweppe R (2001) Flame Retardant Degradation by SCWO, In: Proceedings of the Exploratory Workshop on Supercritical Fluids as Active Media: Fundamentals and Applications. European Science Foundation, Valladolid Spain
99. Hiller F, Giercke R, Kiehne H-A (1998) Entsorgung von Gerätebatterien, Primärbatterien und Kleinakkumulatoren, Kontakt & Studium. expert verlag, Renningen-Malmsheim
100. Hiraoka M, Sakai S (1994) The Properties of Fly Ash from Municipal Waste Incineration and its Future Treatment Technologies. Journal of the Japan Society of Waste Management Experts 5:3-17

101. Hirth M, Wieckert CH, Jochum J, Jodeit H (1989) A Thermal Process for the Detoxification of Filter Ash from Waste Incinerators. Recycling International 2:1561-1566

102. Hirth T, Schweppe R, Jähnke S, Bunte G, Eisenreich N, Krause H (1996) Degradation Processes in Sub- and Supercritical Water. In: Rudolf von Rohr P, Trepp C (eds) 3rd International Symposium on High Pressure Chemical Engineering. Elsevier Amsterdam, Zurich Switzerland

103. Hoenig V (1998) Emissionen bei Einsatz von Sekundärbrennstoffen in Drehofenanlagen der Zementindustrie. Wissenschaftliche Konferenz "Einsatz von industriel-len und kommunalen Abfällen im Zementherstellungsprozess", Opole-Jarnoltòwek, pp 45-54

104. Holst H (1995) Aufkommen und Zusammensetzung von Elektro- und Elektronikschrott aus Hausmüll und hausmüllähnlichen Gewerbeabfällen. Landesumweltamt Nordrhein-Westfalen, Germany, Verwertung von Elektro- und Elektronikgeräten:

105. Huckauf H (1988) Stand und Möglichkeiten der rationellen Energieanwendung beim Zementklinkerbrand. ZKG INTERNATIONAL 41:153-157

106. Inoue T (2001) The Actual Conditions of Using Melting Slag for the Aggregate for Pavement. Urban Cleaning 54:346-352

107. Jakob A (1997) Complete Heavy Metal Removal from Fly Ash by Heat Treatment: Control Mechanisms and Parameters, Recovery, Recycling. Re-integration World Congress, Geneva

108. Jakob A, Mörgeli R (1999) Detoxification of Municipal Solid Waste Incinerator Fly Ash, the CT-FLURAPURTM Process. In: Gaballah, Hager, Solozabal (eds) REWAS'99, San Sebastian, Spain

109. Jakob A, Mörgeli R (1999) Removal of Heavy Metals from the Municipal Solid Waste Incinerator Fly Ash: The CT-Fluapur Process. In: Barrage A, Edelmann X (eds) Recovery, Recycling, Reintegration R'99 Conference. EMPA, Geneva

110. Jakob A, Stucki S, Struis R (1996) Complete Heavy Metal Removal from Fly Ash by Heat Treatment. Environ. Sci. Technol. 30:3275-3283

111. Jantzen CM (1984) Effects of Eh (oxidation potential) on Borosilicate Waste Glass Durability. Adv. Ceram. 8:385-

112. Jantzen CM (1988) Prediction of Glass Durability as a Function of Glass Composition and Test Conditions: Thermodynamics and Kinetics. Proc. 1st Int. Conf. Advances Fusion Glass 24:1-

113. Jantzen CM, Plodinec MJ (1984) Thermodynamic Model of Natural, Medieval and Nuclear Waste Glass Durability. J. Non-Cryst. Solids 67:207-

114. Johanson NW, Spritzer MH, Hong GT, Rickman WS (2001) Supercritical Water Partial Oxidation In: Proceedings of the 2001 DOE Hydrogen Program Review. NREL/CP-570-30535

115. Jordi H (1995) A Financing System for Battery Recycling in Switzerland. Proceedings. International Battery Recycling Congress, Lucerne

116. Kaltenböck FJ, Sauer E (1985) Electronics Scrap Processing at Brixlegg Copper Works. Metall 39 11. Vereinigte Metallwerke Ranshofen-Berndorf AG, p 1047

117. Kehl P, Scharf K-F, Scur P, Wirthwein R (1998) Die Betriebsergebnisse aus den ersten 30 Monaten mit der neuen Ofenlinie 5 im Zementwerk Rüdersdorf. ZKG INTERNATIONAL 51, pp 410-426

118. Keller C (1999) Nachhaltige RESH-Entsorgung durch kombinierte mechanische und thermische Behandlung zusammen mit den Betreibern der Shredderwerke und Kehrichtverbrennungsanlagen. In: Forschung für eine nachhaltige Abfallwirtschaft - Ergebnisse des Integrierten Projekts Abfall im Schwerpunktprogramm Umwelt des

schweizerischen Nationalfonds 1996-1999. Paul Scherrer Institut, Villigen Schweiz, p 59

119. Keller C (1999) Thermal Treatment of Automotive Shredder Residues (ASR) under Reducing Conditions - Experiences from Investigations on Technical and Laboratory Scale. Conference proceedings p. II.187. Re'99, recovery, recycling, re-integration, Geneva Switzerland

120. Keller C, Elektrowatt Ingenieurunternehmung AG (1991) Ueberprüfung und Ergänzung des Konzeptes zur Entsorgung nichtmetallischer Shredderabfälle. Hauptbericht. Vereinigung Schweizerischer Automobil-Importeure (VSAI), Bern

121. Keller C, Elektrowatt Ingenieurunternehmung AG (1993) Bestimmung der aktuellen Zusammensetzung von RESH - Untersuchungsergebnisse über die Glührückstands- sowie Gesamtchlor- und Schwermetallgehalte bei verschiedenen Shredderanlagen. Stiftung für umweltgerechte Entsorgung von Motorfahrzeugen, Bern

122. Keller C, Elektrowatt Ingenieurunternehmung AG (1994) Stofffluss-Voruntersuchung der Verbrennung von nichtmetallischen Shredderabfällen (RESH) zusammen mit Kehricht in der KVA Horgen. Stiftung für umweltgerechte Entsorgung von Motorfahrzeugen, Bern

123. Keller C, Elektrowatt Ingenieurunternehmung AG (1995) Mitverbrennung nichtmetallischer Shredderabfälle von Altautos (RESH) in Kerichtverbrennungsanlagen - Zusammenfassung der Ergebnisse anlässlich der Stofffluss-Hauptuntersuchungen in der Kehrichtverbrennungsanlage Bazenheid vom 11. bis 30. September 1995. Stiftung für umweltgerechte Entsorgung von Motorfahrzeugen, Bern

124. Keller C, Schebdat K (2000) Entsorgung: Shredder-Leichtfraktion von Altautos - Mechanische und thermische Behandlung. Umweltfocus:26

125. Keller C, Schebdat K (2000) Sustainable Disposal of Automotive Shredder Residues - Disposal of Light-Density Automotive Shredder Residues (ASR) by Means of a Staged, Purposeful Approach. UTA Technology and Environment Int Edition 9:31

126. Kinni J (2002) RDF co-combustion in fluid bed boilers. In Power production from waste and biomass IV. Proceedings, VTT syposium series, Espoo Finland, in press

127. Kleppmann F (2001) Energetic Recovery of Automobile Shredder Residual Waste (RESH) in the Refuse-Fired Cogeneration Plant (MHKW) Würzburg. Proceedings. International Automobile Recycling Congress, Geneva Switzerland

128. Knoth R, Hoffmann M, Kopacek B, Kopacek P (2001) A Logistic Concept to improve the Reusability of Electric and Electronic Equipment. Conference Record. IEEE International Symposium on Electronics & the Environment, Denver USA, p 115

129. Krebs A (2001) The Future Way for Battery Recycling - BATREC recycling system. Proceedings. International Congress for Battery Recycling, Montreux

130. Kühl TPR (1996) Charakterisierung und Enthalogenierung von kondensierten Produkten aus der Pyrolyse von Elektronikschrott und flammfesten Kunststoffen. PhD thesis, Universität-Gesamthochschule

131. Künstler H, Grotefeld V (1996) Untersuchung zum VS-Verfahren im Dauerbetrieb. Abschlussbericht. Küpat AG, Zurich

132. Künstler H, Klukowski C, Grotefeld V (1994) Der VS-Kombi-Reaktor der Firma KÜPAT AG. In: Reimann DO (ed) Beiheft zu Müll und Abfall, Berlin

133. Kurkela E, Nieminen M (2002) Gasification of Waste Derived Fuels. In: Power Production from Waste and Biomass IV. Proceedings. VTT Symposium series, Espoo Finland, in press

134. Lehner T (2002) Recycling of Electronic Scrap at the Rönnskar Smelter. Proceedings. International Electronics Recycling Congress, Davos Switzerland

135. Lemann M (1995) Heavy Metals in MSWI-Residues. Proceedings. International Battery Recycling Congress, Lucerne

136. Lin S-Y, Suzuki Y, Hatano H, Harada M (2002) Developing an Innovative Method, HyPr-RING, to produce Hydrogen from Hydrocarbons. Energy Conversion and Management 43:1283-1290

137. Locher D, Baier H, Gardeik HO (1999) Utilization of Secondary Materials in Cement plants - Wishful Thinking and Reality. R'99 recovery, recycling, re-integration. EMPA, Geneva

138. Lopez JD (2002) Electronics Asset Management Initiatives by the US Government. Proceedings. International Electronics Recycling Congress, Davos Switzerland

139. Ludwig C, Schuler AJ (1998) 19838383. Patent application, Germany

140. Ludwig C, Wochele J, Stucki S (2000) Recycling Zinc from Municipal Solid Waste, Re-cycling, Re-covery, Re-integration. R'2000, Toronto

141. Ludwig C, Schuler AJ, Wochele J, Stucki S (1999) Measuring the Evaporation Kinetics of Heavy Metals: A New Method, Recovery, Recycling, Re-integration. Report. R'99 4th World Congress, February 2-5, II, pp 205-210

142. Ludwig C, Schuler AJ, Wochele J, Stucki S (2000) Measuring Heavy Metals by Quantitative Thermal Vaporization. Journal of Water Science and Technology 42:7-8:209-216

143. Ludwig C, Lutz H, Wochele J, Stucki S (2001) Studying the Evaporation Behavior of Heavy Metals by Thermo-Desorption Spectrometry. Fresenius' J. Anal. Chem. 371:8:1057-1062

144. Lutz H (2002) Detoxification of Filter Ashes from Waste Incinerators. Ph.D. thesis, ETH Zurich

145. Lutz H, Ludwig C (2002) The Potency of Additives for Thermal Treatment of Waste Incinerator Filter Ash. Conference Proceeding. Recovery, Recycling, Re-integration World Congress, Geneva Feb. 12-15

146. Lutz H, Ludwig C, Struis R (2001) On-Line Monitoring of Heavy Metal Evaporation Using Inductively Coupled Plasma Optical Emission Spectrometry. ICP Information Newsletter 27:3:167-172

147. Mäkinen T, Sipilä K, Hietanen L, Heikkonen V (2000) Waste Magement Strategies in Helsinki area. In: Helsink metropolitan council (ed) Reports from Helsink metropolitan council YTV-C

148. Malow G (1982) The Mechanisms for Hydrothermal Leaching of Nuclear Waste Glasses: Properties and Evaluation of Surface Layers. In: Lutze W (ed) Scientific Basis for Nuclear Waste Management V

149. Maniatis K (2002) Overview of Waste to Energy Aspects and RES In: Power Production from Waste and Biomass IV. Proceedings. VTT symposium series, Espoo Finland, in press

150. Mark FE, Rodriguez J (1999) Energy Recovery of Greenhouse PE Film: Co-Combustion in a Coal Fired Power Plant. Sonderdruck. APME, Brussels

151. Mason B (1952) Principles of Geochemistry. John Wiley & Sons, New York

152. Mauch W, Reitemann T (1993) Kumulierter Energieaufwand verschiedener Verfahren zur Restmüllbehandlung. In: Fricke K, Thomè-Kozmiensky K-J, Neumüller (eds) Integrierte Abfallwirtschaft im ländlichen Raum. EF-Verlag für Energie- und Umwelttechnik GmbH, Berlin, pp 225-247

153. Menges G, Lackner V (1992) Gaserzeugung aus Kunststoffabfällen. In: Menges G, Michaeli, W, Bittner, M. (ed) Recycling von Kunststoffabfällen. Carl Hanser Verlag, München

154. Merian EH (1984) Metalle in der Umwelt - Verteilung, Analytik und biologische Relevanz. Verlag Chemie, Weinheim

155. Mizutani S, Sakai S, Takatsuki H (1999) Effects of CO_2 in the Air on Leaching Behaviour of Heavy Metals from Alkaline Residues. Proceedings. Waste Stabilization and Environment 99, Lyon Villeurbanne, pp 125-127

156. Modell M (1989) Supercritical Water Oxidation. In: Freeman HM (ed) Standard Handbook of Hazardous Waste Treatment and Disposal. Mc Graw-Hill, New York, p 8.153

157. Modell M, Mayr ST, Kemna A (1995) Supercritical Water Oxidation of Aqueous Wastes. Preprint of paper IWC-95-50. 56th Annual International Water Conference, Pittsburgh PA USA

158. Murphy CF, Pitts GE (2001) Survey of Alternatives to Tin-Lead Solder and Brominated Flame Retardants. Conference Record. IEEE International Symposium on Electronics & the Environment, Denver USA, p 309

159. Newton RG (1985) The Durability of Glass. Glass Technol. 26:21-

160. Newton RG, Paul A (1980) A New Approach to Predicting the Durability of Glasses from their Chemical Compositions. Glass Technol. 21:307-

161. Niklas H (1990) Die Rückgewinnung von Wertstoffen aus Altakkumulatoren (Recycling). In: Hiller F, Hartinger L, Kiehne H-A, Niklas H, Schile R, Steil H-U (eds) Die Batterie und die Umwelt, Kontakt & Studium. expert verlag, Renningen-Malmsheim

162. Nogues J-L (1984) Les mécanismes de corrosion des verres de confinement des produits de fission. PhD thesis, University Montpellier II

163. Palonen J, Nieminen JY, Berg E (1998) Thermie demonstrates Biomass CFB Gasifier at Lahti. Modern Power Systems February 1998

164. Park Y, Reaves JT, Curtis CW, Roberts CB (1999) Conversion of Tire Waste Using Subcritical and Supercritical Water Oxidation. J. Elastomers and Plastics 31:162-179

165. Paul A (1977) Chemical Durability of Glasses. J. Mater. Sci. 12:2246-

166. Paul A (1982) Chemistry of Glasses. Chapman and Hall

167. Perret D, Stille P, Shields G, Crovisier J-L, Mäder U (2000) Long-Term Stability of HT Materials. Report 4. SAEFL, Switzerland

168. Perret D, Crovisier J-L, Stille P, Shields G, Mäder U, Advocat T, Schenk K, Chardonnens M (2002) Thermodynamic Stability of Waste Glasses compared to Leaching Behaviour. Appl. Geochem.:

169. Peters T (2001) Design for Recycling as Integrated Engineering Tool. Proceedings. International Automobile Recycling Congress, Geneva Switzerland

170. Peters WA, Griffith P, Harris JG, Herzog HJ, Howard JB, Latanision RM, Smith KA, Tester JW (1994) Supercritical Water Oxidation for Wastes Cleanup: Enabling Research for Practical Application. Proceedings Paper S-6. The First International Conference on Solvothermal Reactions, Takamatsu Japan, p 1

171. Petersen M, Kleba I (2000) PURer Leichtbau - Leichtbaukonzepte mit Polyurethan im Automobilbau. Kunststoffe 90 3:136

172. Petit J-C, Della Mea G, Dran J-C, Magonthier M-C, Mando PA, Paccagnella A (1990) Hydrated Layer Formation During Dissolution of Complex Silicate Glasses and Minerals. Geochim. Cosmochim. Acta 54:1941

173. Pilz S (1999) Modeling, Design and Scale-Up of an SCWO Application Treating Solid Residues of Electronic Scrap Using a Tubular Type Reactor - Fluid Mechanics, Kinetics, Process Envelope. In: Dahmen N, Dinjus E (eds) VDI-GVC High Pressure Chemical Engineering Int. Meeting. Forschungszentrum Karlsruhe GmbH, Karlsruhe Germany

174. Plodinec MJ, Wicks GG (1994) Application of hydration thermodynamics to in-situ test results. In: Barkatt A, Van Konynenbourg RA (eds) Scientific Basis for Nuclear Waste Management XVII, pp 145-

175. Plodinec MJI, 755-. (1984) Stability of Radioactive Waste Glasses assessed from Hydration Thermodynamics. Scientific Basis for Nuclear Waste Management. Mat. Res. Soc. Symp. Proc. 26, pp 755-

176. Quade J-U (2002) To get New Picture Tubes from Old Ones. Proceedings. International Electronics Recycling Congress,, Davos Switzerland

177. Rauser G (1992) Verfahren zur hydrierenden Verflüssigung von Kunststoffabfällen. In: Menges G, Michaeli W, Bittner M (eds) Recycling von Kunststoffabfällen. Carl Hanser Verlag, München

178. Reimann DO (1994) Beiheft. Müll und Abfall 31

179. Reimer B (2002) Lamp Recycling - State of the Art and Trends. Proceedings. International Electronics Recycling Congress, Davos Switzerland

180. Ristola P (2002) New Approach to Recycling and Waste-to-Energy in Paper Production, Urban Mill, In: Power Production from Waste and Biomass IV. Proceedings. VTT symposium series, Espoo Finland, in press

181. Rollet P (2002) Present Trends on Dangerous Substances Ban in Electronics. Proceedings. International Electronics Recycling Congress, Davos Switzerland

182. Rosenberg A (2001) Battery Recycling at METEK Metal Technology in Israel. Proceedings. International Congress for Battery Recycling, Montreux

183. Ruetschi P, Meli F, Desilvestro J (1995) Nickel Metal Hydride Batteries - The Clean Batteries of the Future? Proceedings. International Battery Recycling Congress, Lucerne

184. Rummler T (2002) Future Requirements for Electronics Recycling in Germany. Proceedings. International Electronics Recycling Congress, Davos Switzerland

185. Saager R (1984) Metallische Rohstoffe von Antimon bis Zirkonium. Bank Vontobel, Zurich

186. Sakai S, Tejima H, Kimura T (1996) Cycle Technologies and Strategies on MSW Incineration Residue. Air & Waste Management Association VIP-53:737-749

187. Sakai S, Urano S, Takatsuki H (1998) Leaching Behavior of Persistent Organic Pollutants (POPs) in Shredder Residues. Chemosphere 37:2047-2054

188. Sakai S, Hiraoka M, Takeda N, Tsunemi T (1990) Sewage Sludge Melting Process: Preliminary System Design and Full-Scale Plant Study. Water Science and Technology 22:392-338

189. Sakai S, Mizutani S, Uchida T, Yoshida T (1997) Substance Flow Analysis of Persistent Toxic Substances in the Recycling Process of Municipal Solid Waste Incineration Residues. In: Waste Materials in Construction Putting Theory into Practice. Elsevier, Amsterdam

190. Sakai S, Mizutani S, Uchida T, Yoshida T, Sato T (1999) Leaching Behaviors of Melting Slag of Municipal Solid Waste and it's Secondary Products. Proceedings. Waste Stabilization and Environment 99, Lyon Villeurbanne, pp 489-494

191. Sako T, Sugeta T, Otake K, Sato M, Tsugumi M, Hiaki T, Hongo M (1997) Decomposition of Dioxins in Fly Ash with Supercritical Water Oxidation. J. Chem. Eng. Japan 30:744-747

192. Sander HJ (2000) Mitverbrennung von Ersatzbrennstoffen aus aufbereiteten Siedlungsabfällen. Bericht 430403. VDI-Bildungswerk, Düsseldorf

193. Sattler H-P (2001) Processing Shredder Residues with WESA-SLF, the Final Step for End-of-Life Vehicles. Proceedings. International Automobile Recycling Congress, Geneva Switzerland

194. Savage PE, Gopalan S, Mizan TI, Martino CJ, Brock EE (1995) Reactions at Super-critical Conditions: Applications and Fundamentals. AIChE J. 41:1723-1778

195. Schäpper S (2001) Side-Effects of the Recycling Quotas of the EU End-of-Life-Vehicle Directive on Lieghtweight Construction Concepts and on the Use of Biomass Based Materials in Car Production. Proceedings. International Automobile Recycling Congress, Geneva Switzerland

196. Schaub M (2002) The Reshment Process - Evolution in ASR Processing. International Automobile Recycling Congress, Geneva, Switzerland

197. Schaub M (2002) Rohstoffliches Recycling am Beispiel der RESH-Verwertung mit dem Reshment -Prozess. ISWA-Frühjahrstagung - Stoffliches Recycling in der Schweiz - sind die Grenzen erreicht? EMPA, Dübendorf Switzerland

198. Schaub MRaR (2001) The Reshment Process: Maximized Reuse. Report from ASR Processing. 1st Int. Automobil Recycling Congress, March 5-7, Geneva

199. Schenk (1998) Altautomobilrecycling - Technisch-ökonomische Zusammenhänge und wirtschaftspolitische Implikationen. Gabler Verlag Deutscher Universitätsverlag, Wiesbaden

200. Schicht E, Müller G, Schons G (2000) Size Reduction and Disintegration of Recycling Material in BHS Rotor Impact Mills. Aufbereitungstechnik 41:93

201. Schlögl M (1995) Recycling von Elektro- und Elektronikschrott. 1st edn., Vogel, Würzburg

202. Schmieder H, Dahmen N, Schön J, Wiegand G (1996) Industrial and Environmental Applications of Supercritical Fluids. In: van Eldik R, Hubbard CD (eds) Chemistry under Extreme or Non-Classical Conditions. Wiley and Spektrum, New York

203. Schmieder H, Abeln J, Boukis N, Dinjus E, Kruse A, Kluth M, Petrich G, Sadri E, Schacht M (2000) Hydrothermal Gasification of Biomass and Organic Wastes. J. Super-crit. Fluids 17:145-153

204. Scholz R, Beckmann M, Schulenburg F (2001) Abfallbehandlung in thermischen Verfahren - Verbrennung, Vergasung, Pyrolyse, Verfahrens- und Anlagenkonzepte. Teubner, BG, Stuttgart, Leipzig, Wiesbaden

205. Scholze H (1991) Glass, Nature, Structure, and Properties. Springer-Verlag, New York

206. Schramn KW, Wu WZ, Henkelmann B, Merk M, Xu Y, Zhang YY, Kettrup A (1995) Influence of Linear Alkylbenzene Sulfonate (LAS) as Organic Cosolvent on Leaching Behavior of PCDDs/DFs from Fly Ash and Soil. Organohalogen Compounds 24:513-516

207. Schulenburg F (2000) Energetische Bewertung thermischer Abfallbehandlungs-anlagen unter Berücksichtigung verschiedener Prozessführungen. PhD thesis 1.Auflage, Technische Universität

208. Schutz B (2001) Economical Aspects of Battery Recycling. Proceedings. International Congress for Battery Recycling, Montreux

209. Sealock LJ, D. C. Elliott, E. G. Baker, R. S. Butner (1993) Chemical Processing in High-Pressure Aqueous Environments. 1. Historical Perspective and Continuing Developments. Ind. Eng. Chem. Res. 32:1535-1541

210. Selinger A (2002) TwinRec - Energy and Material Recovery from Car Shredder Residues. Conference Proceeding. R'02, Recovery, Recycling, Re-integration World Congress, Feb. 12-15, Geneva

211. Serikawa RM, T. Usui, T. Nishimura, H. Sato, S. Hamada, H. Sekino (2002) Hydrothermal Flames in Supercritical Water Oxidation: Investigation in a Pilot Scale Continuous Reactor. Fuel 81:1147-1159

212. Shapiro KG, White AL (1999) Life-Cycle Design Practices at Three Multinational Companies. Proceedings. IEEE International Symposium on Electronics and the Environment, Danvers Massachusetts USA, pp 116-121

213. Shaw RW, Cullinane MJ (1998) Destruction of Military Toxic Materials. In: Meyers RA (ed) Encyclopedia of Environmental Analysis and Remediation. John Wiley & Sons, New York, pp 2821-2836

214. Shaw RW, T. R. Brill, A. A. Clifford, C. A. Eckert, E. U. Franck (1991) Supercritical Water: A Medium for Chemistry. Chem. Eng. News 69

215. Shimizu Y, Nakazawa S, Nikaido H (1999) The Operating Situations of MSW Floating Bed Incineration + Plasma Melting System of Fly Ash. Proceedings. 20th National Urban Cleaning, pp 245-247

216. Smith A, Brown K, Ogilvie S, Rushton K, Btaes J (2001) Waste Management Options and Climate Change. Final report. European Commission, GD Environment,ED21158R4.1

217. Smith KA, Griffith P, Harris JG, Herzog HJ, Howard JB, Latanision RM, Peters WA, Tester JW (1995) Supercritical Water Oxidation: Principles and Prospects. Proceedings. 56th Annual International Water Conference, Pittsburgh PA

218. Smith RLJ, Fang Z, Inomata H, Arai K (2000) Phase Behavior and Reaction of Nylon 6/6 in Water at High Temperatures and Pressures. J. Appl. Polym. Sci. 76:1062-1073

219. Spanke V (1998) Aufbereitung der Shredderleichtfraktion aus der Altautoverwertung. Entsorgungspraxis 5:26

220. Specht E, Jeschar R (1990) Beurteilung von Industrieöfen bei Wärmerückgewinnung. In: Energieverfahrenstechnik If (ed) Seminar zu Methoden der Energieeinsparung bei Industrieöfen. Technische Universität Clausthal

221. Sproull JF, Marra SL, Jantzen CM (1994) High-Level Radioactive Waste Glass Production and Product Description. In: Barkatt A, Van Konynenbourg RA (eds) Scientific Basis for Nuclear Waste Management XVII, pp 15-

222. Stahlberg R (1993) Thermoselect - Energie- und Rohstoffgewinnung durch thermische-chemische Stoffwandlung im geschlossenen System. VDI-Bericht 1033

223. Stahlberg R, Kaiser W, Nyhuis G, Mattson N, Drost U (2002) The Thermoselect High Temperature Recycling Technology for Automotive Shredder Residue - Results and Perspectives. International Automobile Recycling Congress, Geneva Switzerland

224. Stanworth JE (1950) Physical Properties of Glass. Clarendon Press, Oxford England

225. Steinkuller W (2001) ARA (Automotive Recyclers Association) Activities / Country report USA. Proceedings. International Automobile Recycling Congress, Geneva Switzerland

226. Sterpenich J (1998) Altération des vitraux médiévaux, contribution à l'étude du comportement à long terme des verres de confinement. Ph.D. thesis, Université Henri Poincaré

227. Strässle R (2001) Vorgezogene Gebühr - gratis entsorgen. Umwelt Focus 6:36
228. Struis R, Ludwig C, Lutz H, Scheidegger A (2002) Heavy Metal Recovery from Fly Ash: A Zn K-edge EXAFS study. to be published in Environ. Sci. Technol.
229. Stücheli A (2000) Technologien und Wirtschaftlichkeit von Recycling und Entsorgung von Altautos. Bericht. Zürcher Hochschule, Winterthur
230. Swallow KC, Killilea WR (1992) Comment on "Phenol Oxidation in Supercritical Water: Formation of Dibenzofuran, Dibenzo-p-dioxin, and Related Compounds". Environ. Sci. Technol. 26:1849-1850
231. Swiss Federal Statistical Office SAftE (1997) The Environment in Switzerland 1997 - Facts, Figures, Perspectives. EDMZ, Bern
232. Takeda N, Hiraoka M, Sakai S, Kitai K, Tsunemi T (1989) Sewage Sludge Melting Process by Coke-Bed Furnace: System Development. Water Science and Technology 21:925-935
233. Tester JW, Holgate HR, Armellini FJ, Webley PA, Killilea WR, Hong GT, Barner HE (1993) Supercritical Water Oxidation Technology - Process Development and Fundamental Research. In: Tedder DW, Pohland FG (eds) Emerging Technologies in Hazardous Waste Management III, ACS Symposium Series 518. American Chemical Society Washington DC, Atlanta GA USA
234. Thomassin J-H (1995) Apport des analogues naturels, industriels ou archéologi-ques à la connaissance du comportement à long terme des vitrifiats de déchets toxiques. Rapport final. University of Poitiers, France
235. Thomassin J-H (1996) Rapport CRP OESIP 96-1. Rapport CRP OESIP 96-1. University of Poitiers, Poitiers France
236. Thomé-Kozmiensky KJ (1994) Deglor-Elektroschmelzverfahren ABB Umwelttechnik. In: Thermische Abfallbehandlung. EF-Verlag, Berlin, pp 626-627
237. Thornton TD, LaDue III DE, Savage PE (1991) Phenol Oxidation in Supercritical Water: Formation of Dibenzofuran, Dibenzo-p-dioxin, and Related Compounds. Environ. Sci. Technol. 25:1507-1510
238. Traber D (2000) Petrology, Geochemistry and Leaching Behaviour of Glassy Residues of Municipal Solid Waste Incineration and their Use as Secondary Raw Material. PhD Thesis, Bern University
239. Traber D, Mäder UK, Eggenberger U (2002) Petrology and Geo-Chemistry of a Municipal Solid Waste Incinerator Residue Treated at High Temperature. Schweiz.Mineral.Petrogr. Mitt. 82:1-14
240. Traber D, Mäder U, Eggenberger U, Simon F-G, Wieckert C (1999) Phase Chemistry Study of Products from the Vitrification Processes AshArc and Deglor. Glastech. Ber. Glass Sci. Tech. 72(3):91-98
241. Tsukada T (2001) Operation Report of Pyrolysis Gasification and Melting System for Municipal Solid Waste - Performance of Mitsui Recycling 21 System.
242. Tuhkanen S, Pipatti R, Sipilä K, Mäkinen T (2001) The Effect of New Solid Waste Treatment Systems on Greenhouse Gas Emissions. In: Greenhouse Gas Control Technologies. In: Williams D, Durie J, McMullan RAP, Paulson CAJ, Smith AY (eds) Proceedings of the Fifth International Conference on Greenhouse Gas Control Technologies (GHGT-5). CSIRO Publishing, Collingwood
243. Umweltbundesamt (1998) Möglichkeiten der Kombination von mechanisch-biologischer und thermischer Behandlung von Restabfällen. Förderkennzeichen: 1471114. Im Auftrag des Bundesministeriums für Bildung und Forschung BMBF

244. van der Sloot HA, Heasman L, Quevauviller P (1997) Harmonization of Leaching/Extraction Tests. Elsevier, Amsterdam, p 281
245. Vehlow J (2000) Low Cost, Low Pollution - Current Objectives in Waste Combustion. International Conference on Combustion, Incineration/Pyrolysis and Emission Control (ICIPEC), Korea Science and Technology Center, Seoul
246. Vehlow J, Bergfeldt B, Hunsinger H, Jay K, Mark FE, Tange L, Drohmann D, Fisch H (2002) Recycling of Bromine from Plastics Containing Brominated Flame Retardants in State-of-the-Art Combustion Facilities. Technical paper. APME, EBFRIP, Forschungszentrum, Karlsruhe
247. Vernaz E, Dussosoy J-L (1992) Current State Knowledge of Nuclear Waste Glass Corrosion Mechanisms: The Case of R7T7 Glass. Appl. Geochem.:13-
248. VKE/APME (1998) Bewertung von Aufbereitungsverfahren für die Verwertung der Shredderleichtfraktion Phase II. technisch/wirtschaftliche Bewertung ausgewählter Verfahren. Fichtner, Stuttgart
249. Vogel F, Hildebrand F (2002) Catalytic Hydrothermal Gasification of Woody Biomass at High Feed Concentrations. paper 123. 4th International Symposium on High Pressure Process Technology and Chemical Engineering, EFCE, Venice Italy
250. Weber A, Burri R (1996) Flexibility of the Batrec process. Proceedings. International Battery recycling Congress, Cannes
251. Wissing FJ (1995) Lösungskonzepte der deutschen Elektroindustrie für die Verwertung und Entsorgung elektrotechnischer und elektronischer Geräte. In: Schimmelpfeng L, Huber, R. (ed) Elektrik-, Elektronikschrott, Datenträgerentsorgung - Möglichkeiten und Grenzen der Elektronikschrott-Verordnung. Springer, Berlin
252. Wochele J, Stucki S (1994) Fate of Heavy Metals in Refuse Incineration. PSI Annual Report An.V
253. Wochele J, Stucki S (1998) Labor-Simulierung des Festbettes von Müllverbrennungsanlagen aufgrund der Ähnlichkeitsgesetze. VDI-Bericht 1390. VDI-Tagung Braunschweig, pp 287-295
254. Wochele J, Stucki S (1999) Similarity Laws for the Tubular Furnace as a Model of a Fixed-Bed Waste Incinerator. In: Chem. Eng. Technol. 22. WILEY-VCH Verlag GmbH, Weinheim
255. Wochele J, Ludwig C, Stucki S (1999) Model Experiments on Heavy Metal Evaporation in MSW Incineration Plants. Conference Proceeding. R'99, Recovery, Recycling, Re-integration World Congress, Geneva Feb.99
256. Xu X, Antal Jr MJ (1998) Gasification of Sewage Sludge and Other Biomass for Hydrogen Production in Supercritical Water. Env. Prog. 17:215-220
257. Yamada K, Akai Y, Matsubayashi Y, Yamaguchi Y (1998) Organic Solid Waste Treatment with High-Temperature, High-Pressure Water. Presented paper. RECOD 98: 5th International Conference on Recycling, Conditioning and Disposal, Nice France
258. Yamaguchi H, Shibuya E, Kanamaru Y, Uyama K, Nishioka M, Yamasaki N (1996) Hydrothermal Decomposition of PCDDs/PCDFs in MSWI Fly Ash. Chemosphere 32:203-208
259. Yong RN, Warkentin BP, Phadungchewit Y, Galves R (1990) Buffer Capacity and Lead Retention in Some Clay Materials. Water, Air, and Soil Pollution 53:53-67
260. Zeltner C (1998) Petrologische Evaluation der thermischen Behandlung von Siedlungsabfällen über Schmelzprozesse. PhD thesis, ETH Zurich

261. Gillies KJS, Cox GA (1988) Decay of medieval stained glass at York, Canterbury and Carlisle; II. Relationship between the composition of the glass, its durability and the weathering products. Glasstech. Ber. 61:101

262. Macquet C, Thomassin JH (1992) Archaeological glasses as modelling of the behaviour of buried nuclear waste glasses. Appl. Clay Sci. 7:17

263. Jollivet P, Nicolas M, Vernaz E (1998) Estimating the alteration kinetics of the French vitrified high-level waste package in a geologic repository. Nucl. Tech. 123:67

264. Schäpper S (2002) A vehicle producer's remarks regarding national implementations of EU ELV Directive. International Automobile Recycling Congress, Geneva Switzerland

265. BCR (1984) Commission of the European Communities, The Certification of the Contents (mass fraction) of Cd, Co, Fe, Hg, Ni, Pb, Zn, Sb, Se, Tl an Cr in a Sample of a City Waste Incinerator Ash, Community Bureau of Reference, BCR No. 176, Report EUR 9664 EN, pp. 62-63, 1984

6 Ecology: Which Technologies Perform Best?

Stefanie Hellweg, Gabor Doka, Goeran Finnveden, and Konrad Hungerbühler

What is more important, reducing long-term toxic emissions to the environment or optimizing the energy efficiency of waste treatment plants? Are emissions of nitrogen oxide more harmful than non-methane volatile organic compounds (NMVOC)? How can financial resources be invested most efficiently to reduce environmental impact? These and a long list of similar questions must be addressed when making decisions related to waste management; for example, when choosing between different disposal technologies. This book proposes and depicts many waste treatment processes and strategies that attempt to optimize the environmental and/or economical performance of waste treatment. However, since optimizing one feature of a technology is usually done at the cost of another (e.g., preventing toxic emissions may increase the energy demand), it is usually not evident at first sight which technology is most appropriate in a given situation. While the previous chapters focused on the technical description of technologies, the following chapters deal with their assessment and, thereby, address the questions posed at the beginning of this paragraph.

Environmental assessment can be performed using various tools (section 6.1). One of these tools is Life-Cycle Assessment (LCA). LCA analyzes all interactions with the environment during the complete life cycle of a product or service and quantifies the resulting potential impact on the environment (section 6.2). LCA is useful in waste management because it allows for the analysis of (waste product) systems such as package systems for a certain good, waste treatment or recycling processes, or complete waste management strategies. It can be used to identify improvement potentials from an environmental perspective or to respond to people's concerns (chapter 7).

The results of a case study comparing various treatment technologies show that long-term leaching of heavy metals from landfills is a major problem of end-of-pipe technologies such as sanitary landfill, mechanical-biological treatment, and incineration (section 6.3). However, since these emissions can occur in the far future, even many thousand years from now, whether or not these emissions are perceived as a serious problem depends entirely on the time frame considered. This temporal dimension raises the question of whether future environmental impacts should be weighted equally to current impacts or whether they should be discounted (section 6.4). While, in fact, many people attach less weight to damages in the distant future, we bear responsibility for future generations according to the goals of sustainability and ethical considerations. Therefore, a discount rate close or equal to zero should be applied if ethical standards are to be met.

In many cases, economic and environmental costs are inversely correlated. For instance, sanitary landfill is a relatively cheap waste treatment option (disregarding potential costs of aftercare), but the environmental impact potential is high (Section 6.3). In order to improve the environmental performance, a number of alternative treatment options are available (chapters 3 to 5). The choice of alternative technologies is usually influenced by ecological, economic, and social aspects. Very often, only a limited amount of financial resources is invested in technologies which have the goal of achieving the highest environmental benefit possible. One way to identify the most efficient investment option is to divide the net environmental benefit (e.g. measured with the help of LCA) by the net present value of the investment. This indicator, the ecological-economic efficiency, is applied to four end-of-pipe technologies in section 6.3.2.

6.1 Assessment Tools for Waste Treatment Systems

Waste treatment has long been recognized as a key environmental issue [2, 13, 25, 97]. Given this environmental relevance, there is an increasing demand to analyse and compare the environmental performance of products, technologies or waste policies. Such work can be done with environmental assessment tools. There are various tools that can be used with respect to waste treatment and management [101]. Table 6.1 gives an overview of some of these tools and their main fields of application. These tools are complementarily used for different purposes. For instance, the risk of accidents of a new incineration technology is best assessed in a risk assessment study (not with a substance/material flow analysis or a life-cycle assessment); while the question as to where a certain substance (e.g., Cd) is pro-

Table 6.1. A selection of environmental assessment tools [8] and their fields of application

Assessment tool	Purpose	Application (example)
Environmental Audit	Evaluating the current ecological performance and setting targets for the future environmental performance of a company.	Auditing a recycling plant (situation report and targets).
Environmental Impact Assessment (EIA)	Analysis of the potential environmental impacts of a new plant or activity related to its location (prognosis).	Choosing a new landfill site and estimating its impact on the local environment.
Life-Cycle Assessment (LCA)	Comparison of products/ activities with the same function, identification of weaknesses within the life cycle of a product or process; identification of improvement potentials, ecolabelling.	Comparison of different package systems. Identification of key pollutant processes when treating 1 kg of average MSW. Ecolabelling of packages.
Risk Assessment	Coupling quantitatively the impacts of an event or chain of events with their probability of occurrence.	Assessment of possible failures and their probability of MSW incinerators; risks of subsurface deposits.
Substance / Material Flow Analysis	Identification of problem flows and causes of environmental problems in a region; monitoring; incitement for substance regulations; evaluation of abatement measures.	Analysis of Cd flows and storage within a region to find the most promising reduction potentials.

duced, used, stored, and disposed of can be answered with a material flow analysis.

In this chapter, we will analyze and compare various waste treatment options. We have chosen the tool Life-Cycle Assessment (LCA) for this purpose because it can be applied for the assessment of many emissions and the use of resources during the complete life cycle ('cradle-to-grave').

6.2 An Introduction to Life-Cycle Assessment

Life-Cycle Assessment (LCA) is used to study the environmental aspects and potential impacts throughout a product's life from raw material acquisition through production, to use and disposal (i.e. from cradle to grave) [3]. The term 'product' refers to material products as well as to service functions, for example treating a certain amount of solid waste. LCA serves to compare different products or processes, to identify environmental key issues, and to optimize processes.

The LCA methodology developed significantly during the 1990s. There is an international standard for LCA, the ISO norms [3], which provides a framework, though only limited guidance concerning methodology. There are also several national guidelines for LCA, for example in Denmark [44] and the Netherlands [41]. Essentially, this methodology can be used when LCA is applied to waste management, although various aspects of the methodology may come more acutely into focus [22, 31]. LCA consists of four phases (Fig. 6.1) [3]:

1. In the *Goal and Scope Definition*, the purpose of the study and the boundary conditions are discussed.
2. The *Life-Cycle Inventory* phase (LCI) comprises the gathering of emission and resource use data.
3. In the *Life-Cycle Impact Assessment* phase (LCIA), potential environmental impacts of these emissions and of resource consumption are quantified.
4. In the *Interpretation* phase, conclusions are drawn.

The four phases of LCA are iterative steps, e.g., the goal and scope definition might be revised in the process of data collection (LCI) or impact assessment (LCIA).

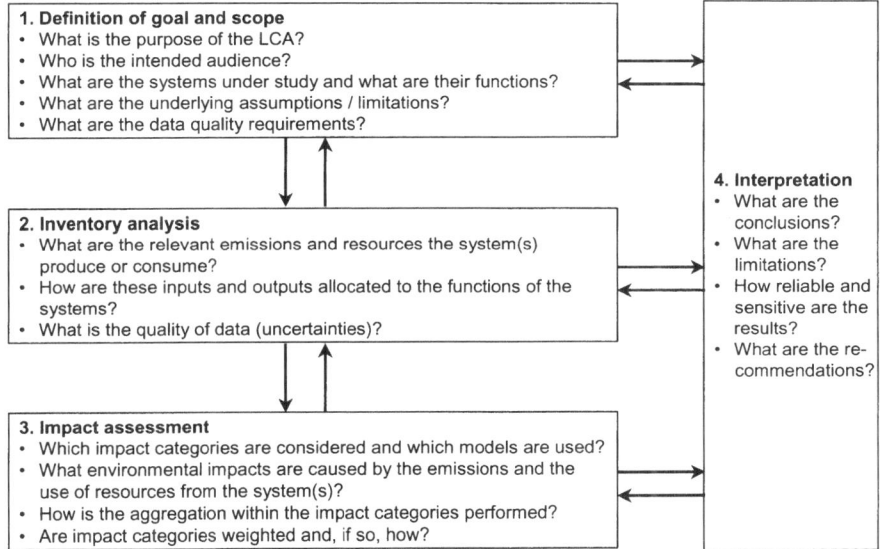

Fig. 6.1. LCA framework [3] and the questions answered in each phase. The arrows in both directions between the different phases indicate that LCA is an iterative technique.

6.2.1 Definition of the Goal and Scope

In this phase of LCA, the purpose and the boundary conditions of the study are defined. Some of the items to be described are the function(s) of the system to be studied, system boundaries, allocation procedures, underlying assumptions, limitations, and data quality requirements (Fig. 6.1) [3]. Of these issues, the definition of the functional unit is of crucial importance because it provides the reference to which all inputs and outputs are normalized. The functional unit captures the function(s) of the system according to the goal of the study. In Table 6.2 we present several examples for adequate functional units in different problem situations.

Table 6.2. Definition of the functional unit – examples.

Example LCA case	Functional unit
Comparison of waste treatment technologies	Treatment of a certain amount of waste with a defined composition (e.g., 1 kg of mixed MSW)
Comparison of beverage container systems	1 piece of container carrying 1 liter of beverage (not 1 kg of package, since, e.g., a glass bottle is heavier than a PET bottle)
Waste transport by train or truck	Transport of 1 ton of waste over 1 km

In theory, all processes from 'cradle to grave' should be considered. This means that all inputs should be flows that are drawn from the environment and that outputs should be flows that are released to the environment without further human transformation [31]. However, those system elements that are identical in all systems can be disregarded. For this reason, an LCA comparing waste treatment technologies does not need to consider the generation of waste, but can start right away with its treatment (Table 6.2).

The appropriate system boundaries may depend on the goal of the study (e.g., [22]). The goals of an LCA can have several dimensions. A first fundamental dimension is concerned with whether the study is change-oriented (prospective) or descriptive (retrospective) [6, 36, 90, 96]. If the study is change-oriented, it analyses the consequences of a choice; ideally the data used should reflect the actual changes taking place, and may depend on the scale of the change and the time over which it occurs. With regard to time, a distinction can be made between a very short time frame (less than a year), short (years), long (decades) or very long (centuries). Studies that are not change-oriented might be/are called environmental reports. In such studies the appropriate data should reflect what actually happened in the system in the past.

6.2.2 Life-Cycle Inventory Analysis (LCI)

The inventory analysis involves data collection and calculation procedures to quantify relevant inputs (= resources) and outputs (= emissions) [3]. This phase is started by setting up flowcharts of all systems under study. Since all processes 'from cradle to grave' shall be considered in LCA, these flowcharts are usually very complex and the gathering of data can therefore be hard work. Potential data sources are primary data (measurements at one or several plants), literature, or expert judgment. Moreover, there are public databases that offer averaged inventory data for a large number of processes (e.g., [37, 79]). After data collection, all data needs to be normalized to the functional unit.

If the system produces more than one output product or service, the functional unit needs to be harmonized for all systems under study. Obviously, it would be unfair to compare an alternative X (which produces the service A) directly to an alternative Y (which produces the service A, B, and C) and conclude that X is less of a burden than Y, not heeding the fact that X generates less services than Y. Figure 6.2 presents an example in which grate incineration (MSWI) is compared to the PECK technology (Section 5.2.4). The principal function in both cases is the treatment of waste. Both systems additionally produce energy and heat. However, in contrast to the grate incinerator, the PECK technology recovers Cu, Zn, Pb, and a mineral product which can be used as secondary materials. In order to make the two systems comparable, the MSWI system needs to be enlarged by adding the environmental impacts of alternative processes that produce these materials (e.g., an industrial plant producing Cu from ore). Such complementary processes are called *reference or complementary systems* and the whole process of harmonizing the functional unit *system enlargement* in LCA terminology [3]. An important prerequisite for system enlargement is that the co-product of the system must be of equal quality to the primary product produced by the reference system.

When modeling waste incineration processes, an important question to address is how the emissions and resources are assigned to the different ingredients or elements contained in the waste input. The ISO norms [3] propose, in this case, that the total environmental burden be split up according to physical-chemical causalities. For instance, the emissions of Cd from a waste incinerator are a function of the Cd contained in the waste input [3]. Such a dependency exists concerning many other elements than Cd. This relationship between waste input and emission output permits an individual assessment of different types of waste, i.e., it is possible to differentiate between the incineration of 1 kg of paper and 1 kg of metal. Unfortunately, it is not sufficient only to know that all substances entering the system with the waste input also leave it; one needs to know whether they do so as emissions to air, water, or soil. Since emissions to different environmental compartments have different impacts, this is a relevant problem and requires thorough modeling in the inventory phase.

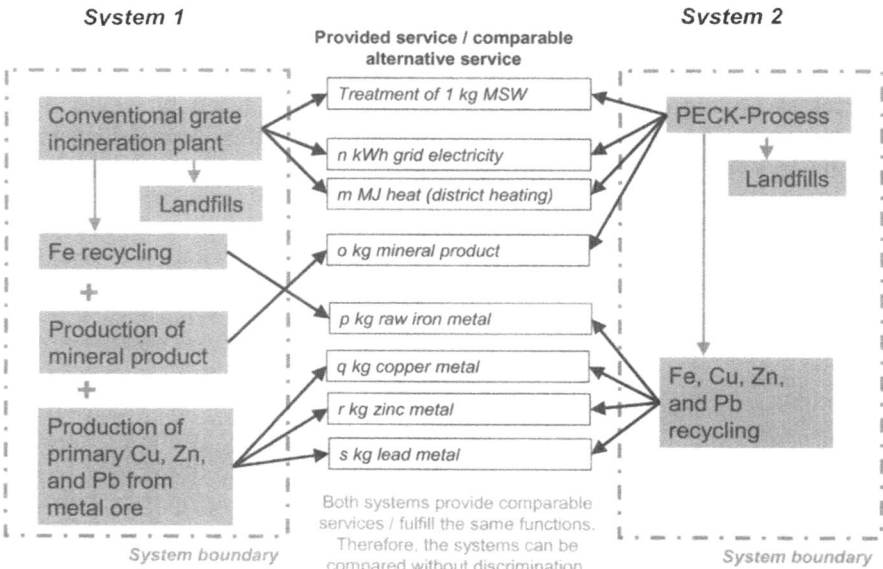

Fig. 6.2. Services produced by the grate and the PECK technologies (it is assumed here that the amounts of energy produced are identical). The functional unit is the superset of functions of all systems. Therefore, the system 'grate incineration' needs to be enlarged to be comparable to the system PECK. A '+' indicates system enlargement.

When the data for all processes are available and referenced to the functional unit, the emissions produced and the resources consumed can be aggregated. For instance, the Cd emissions to air from the incineration process can be added to the Cd emissions of the remaining processes of the system (e.g., transport, production of ancillary products). The final result is a table of all resources consumed and all emissions produced by the complete system (Fig. 6.3, left).

6.2.3 Life-Cycle Impact Assessment (LCIA)

The Life-Cycle Impact Assessment (LCIA) aims at understanding and evaluating the magnitude and significance of the potential environmental impacts of a product system. It is divided into several elements. In the ISO norms [3], some of these elements are described as mandatory and some as optional.

The first mandatory element is a selection of impact categories, indicators for these categories, and models to quantify the contributions of different resource inputs and emissions to the impact categories. Table 6.3 presents some examples for impact categories considered in different LCIA methods. In practice, however, a shorter list of impacts is normally considered in current LCAs [32]. The second mandatory element (classification) is the assignment of the inventory data to the

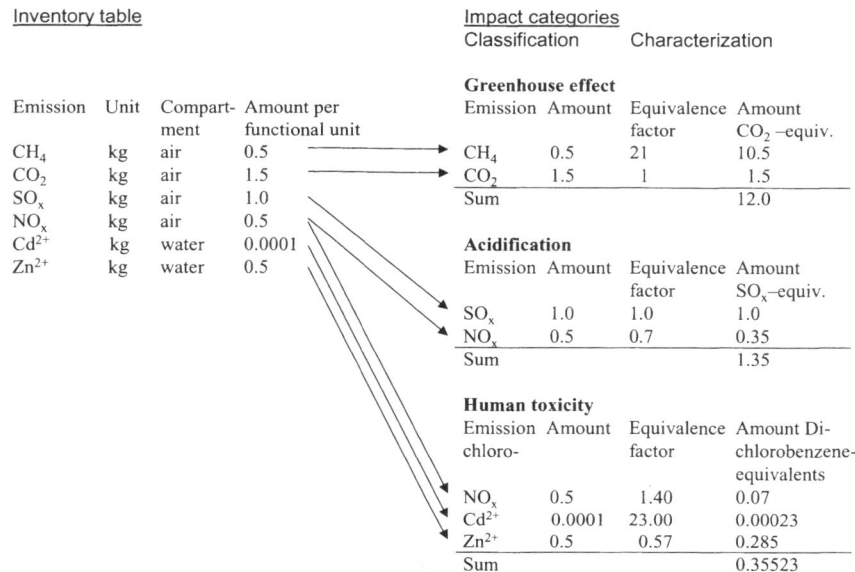

Fig. 6.3. Classification and characterization of inventory data (simplified example with fictitious numbers).

impact categories. The third mandatory element (characterization) is the quantification of the contributions of the inventory data to the chosen impacts. For instance, all emissions contributing to global warming are expressed in CO_2-equivalents. Fig 3 shows an example for the classification and characterization of inventory data.

There are also several optional elements that can be included in the LCIA depending on the goal and scope of the study [3]. Normalization relates the magnitude of the impacts in the different categories to reference values; an example of a reference value is the total contribution to an impact category by a nation. Grouping includes sorting and possibly ranking the indicators. Weighting aims at converting and possibly aggregating indicator results across impact categories resulting in a single result, sometimes with a monetary measure. The final element is a data quality analysis.

In the last 20 years several LCIA methods have been developed. These methods differ by their definition of impact categories, consideration of environmental compartments, number of emissions and resources considered, level of aggregation, and their weighting approach. Table 6.3 shows a selection of commonly used methods.

Table 6.3. A selection of commonly used LCIA methods.

LCIA method	Impact categories considered	Compartments considered	Weighting approach
CML [21]	Abiotic depletion, land use, global warming, ozone layer depletion, human toxicity, freshwater aquatic and marine ecotoxicity, freshwater sedimental and marine sedimental ecotoxicity, terrestrial ecotoxicity, photochemical oxidation, acidification, eutrophication, radiation, odour.	Air, fresh and marine water, agricultural and industrial soil, resources.	No weighting between impact categories
Eco-indicator 99 [40]	Damages to resources: extraction of minerals, extraction of fossil fuels; the measuring unit is MJ surplus energy needed to extract resources in the future. Damages to human health: Carcinogenic effects, respiratory effects, climate change, ionizing radiation, ozone layer depletion; the measuring unit is disability adjusted life years (DALY). Damages to ecosystem quality: ecotoxicity, acidification and eutrophication, land occupation, land conversion; the measuring unit is potentially disappeared fraction of species) (PDF)*m^2*y.	Air, water, soil, resources.	Damage oriented approach. The weighting between the three different damage categories human health, ecosystem health, and resource consumption is based on panel studies. Different archetypes for the diversity of people's preferences are suggested [52] (see text)
EDIP [44]	Global warming, stratospheric ozone depletion, photochemical ozone formation, acidification, nutrient enrichment, ecotoxicity, human toxicity, resource consumption, further impact categories for working environment.	Air, fresh water, and soil; resources.	The weighting factors are based on Danish political reduction targets for emissions and the supply horizon for resources.
EPS [87]	Human health (life expectancy, morbidity, nuisance), ecosystem production capacity (crop, wood, fish/meat, base cation capacity of soil, water), abiotic stock resources (e.g., oil, coal, metal ores, gravel), bio-diversity.	Air, fresh water, and soil; noise; resources, land use.	Monetarization of all impacts using market prices where available; otherwise the Willingness to Pay approach was used.
Method of Ecological Scarcity [15]	Global warming, ozone depletion, primary energy consumption; all other emissions are not characterized prior to weighting.	Air, water soil.	Weighting based on political reduction targets (international and Swiss).

CML'01 is a partially aggregating method, while all the other methods mentioned in Table 6.3 are fully aggregating. Different techniques are used for weighting. For instance, the Eco-indicator 99 method uses different cultural archetypes, the 'Hierarchist', the 'Egalitarian' and the 'Individualist', that feature different biases and beliefs concerning value judgments, moral, and ethical questions of individuals and societies [52]. While any real-life individual is usually a mixture of these archetypes, they bring structure and consistency into the difficult topic of subjective value judgment. Important distinguishing features are, e.g., that the Individualist has a risk-seeking and optimistic personality: He/she has a short-term perception of problems and belittles damages to future generations. Only damages that are actually observable are of concern to the Individualist. On the other hand the Egalitarian has a risk-avoiding personality: He/she has a long-term perception of problems and judges damages to future generations at least as important as damages to the present generation. Apart from observable damages the Egalitarian also heeds suspected or uncertain environmental damages. The third archetype, the Hierarchist, takes an intermediary position between the Individualist and the Egalitarian: The Hierarchist believes in the systems established by society and tries to moderate or consolidate extreme positions. Damages to the future and present generations are equally important to the Hierarchist. As a firm believer in the hierarchic, scientific system, the Hierarchist heeds damages that are proven and scientifically established, but discounts damages of an uncertain or merely suspected nature. Goedkoop et al. [40] recommend using the Hierarchist perspective as default and the Individualist and Egalitarian perspective in sensitivity analysis.

It is advisable to check the characteristics of the LCIA method before use. For instance, the former Eco-indicator 95 method [39] did not consider emissions to soil. Therefore, the application of this method to certain problems, such as agricultural production, would not make sense. It is recommendable to apply different adequate methods to the same problem to reinforce the results or reveal uncertainties. International activities have begun to review LCIA methods with the aim of suggesting the best available practice in the area [92, 93].

6.2.4 Interpretation

Life-cycle interpretation is a phase where the findings from the inventory analysis and the impact assessment are combined in order to reach conclusions and recommendations consistent with the goal and scope of the study [3]. This phase is comprised of three parts:

1. Identification of significant issues. Such 'hot spots' might be relevant process steps of the system as well as impact categories or emissions/resources that proved to be particularly relevant in the LCI or LCIA phase.

2. Evaluation considering completeness of data, sensitivity, and consistency checks.

3. Conclusions and recommendations.

The interpretation is therefore the phase in which the results of the LCA are summarized and evaluated.

6.3 Case Study: An LCA of Waste Treatment Processes

In the following we will use the LCA methodology to identify and compare environmental key issues of some of the technologies described in the previous chapters. We will not discuss waste prevention or reduction schemes in this chapter (as proposed in Section 3 1). It is clear that mere prevention of waste is ecologically beneficial, because all environmental burdens from the production, use, and disposal phase are avoided at the same time. Waste reduction by substitution of goods or substances or material recycling, on the contrary, does not necessarily lead to a reduction in burden, even if the waste volume or mass is decreased. For instance, the new product might cause other or more toxic effects than the old product. However, because scenarios for substitution are manifold they would require comprehensive analysis beyond the scope of this book. In the present study, we therefore restrict the analysis to the comparison of waste treatment processes.

6.3.1 System Description (Goal and Scope and Inventory Analysis)

Definition of the Goal and Scope

The goal of this LCA is to identify environmental "hot spots" and compare some of the waste treatment options described in the previous chapters using environmental criteria. The analysis is performed with respect to mixed waste of an average composition. The results can be used to aid in the decision of which waste treatment is most suitable for the treatment of MSW in a given region. The environmental data is coupled to economic costs to facilitate the decision-making.

The time frame is indefinite in order to consider all environmental impacts. However, the results are also shown for a shorter time period of approximately 100 years (*surveyable time period* [33]). Since the technologies can be set up anywhere in the world, the analysis is, in principle, not bound to a certain country. However, the waste composition refers to typical values for Switzerland. The data in [105] (transfer coefficients) provide the possibility of adjusting these input parameters to the specific situation of other countries. Transport distances and data about the gathering of waste were also averaged data from Switzerland [104]. However, these figures were assumed roughly equal with respect to all technologies (slight differences are due to the transport of solid output products to processing plants), so that the final results of the technology comparison are not sensitive

Table 6.4. Waste treatment technologies studied in the LCA and data sources.

Waste treatment technology	Technology description	Main data sources[a]
MSW sanitary landfill	Section 2.2	[11, 33]
Mechanical-Biological (Pre-)treatment (MBP) prior to landfill	Section 4.1	[95], assumptions
Thermal treatment (grate incineration and PECK technology)	Sections 3.3, 5.2.1 and 5.2.4	[27, 49, 104]

[a] Data for background processes such as production of ancillary products, recycling of valuables, transport, and infrastructure were taken from [23, 24, 37, 38, 57, 58, 70, 104].

to transport. The waste treatment processes considered and the data sources used are shown in Table 6.4.

The data quality varies widely from one technology to the other. Sanitary landfills and regular grate-incineration plants are in operation all over the world and numerous measurements are available for short-term time horizons. However, the long-term behavior beyond 100 years of any kind of landfill is still a matter of speculation. Mechanical-biological treatment has been implemented at full scale for only several years and the availability of measurements is limited to few studies. The uncertainties are largest for the PECK technology because only some of its components have been tested separately (Fig. 6.5). The complete assembly with feedback loops has not been implemented in pilot plants so far and measurements were limited to the composition of some solid products from trial runs of some of the components (e.g., slag and the analysis of metal particles).

Technology Description and Functional Unit

The technologies considered in this study are only briefly presented in flowcharts (Fig. 6.4) since they have already been described in earlier chapters of this book (Table 6.4). For the sake of visibility, Figure 6.4 only indicates the main processes while many of the background processes are not shown, although they were also considered. The small empty boxes in Figure 6.4 indicate that such processes have been taken into account by using data from a public database [37].

The principal function of all technologies is the treatment of waste. Therefore, the functional unit includes the treatment of 1 kg of waste material. The average composition of MSW to be disposed of is shown in Table 6.5.

To make a fair comparison of the environmental burdens of the different technological options, the services delivered by the assessed systems must be equal (system enlargement, Section 6.2.2). Since several co-products, such as electricity, heat, and metals, are produced in different quantities by the considered technologies, the systems need to be enlarged. However, system enlargement is only applied if the quality of the secondary product is equal to the primary product and

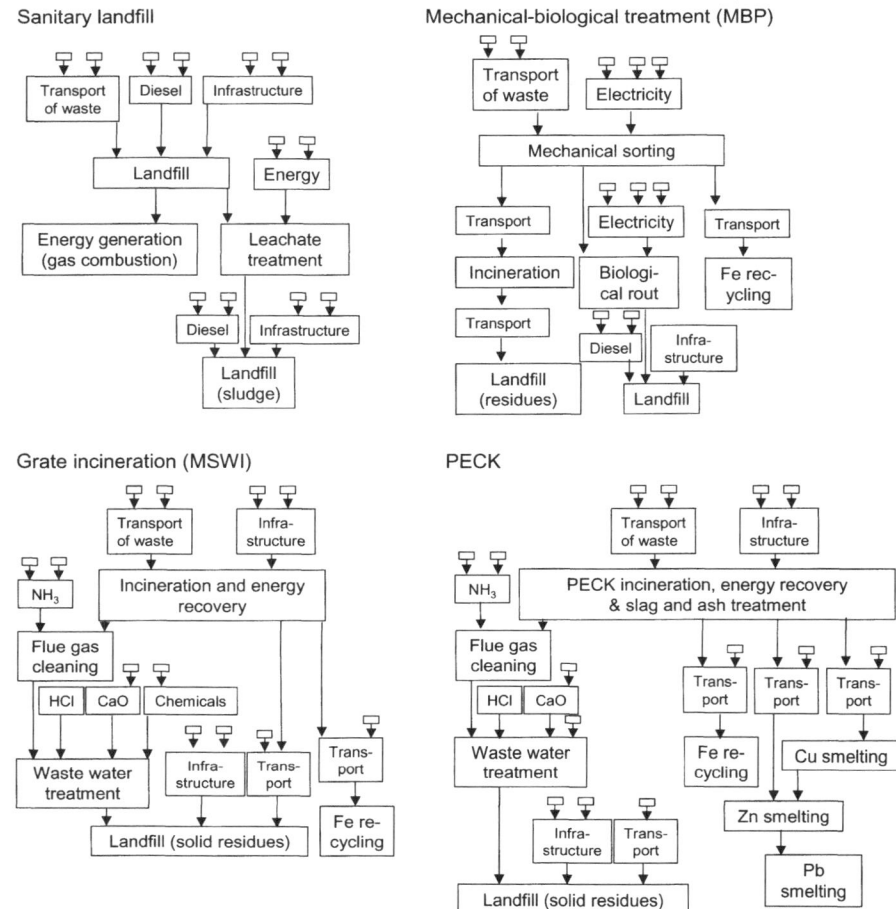

Fig. 6.4. Flowcharts of the processes considered. The empty boxes indicate prior background processes that have been considered (use of data from a public database [37]).

if the secondary product is marketable. For instance, gypsum produced by the flue gas cleaning of incineration plants was not considered a valuable co-product because the gypsum market is already glutted in many countries, especially in Europe. The functional unit and the reference systems used for system enlargement are shown in Table 6.6.

The functional unit is a superset of all services provided by the four treatment technologies considered (Table 6.6). All technical options examined in this study must fulfill all services of the functional unit. Table 6.6 shows that all systems need to be complemented by regular industrial processes producing primary products (system enlargement). The choice of the reference system for system enlargement can have a decisive influence on the results [28]. Especially concerning the generation of heat and electricity, this choice can be important since the environmental impacts differ between different energy sources. As suggested in Section 6.2.1, the choice of the reference system can depend on the goal of the study. Here an average mix for electricity and natural gas for heat was chosen to show results referring to the general situation in Europe. One should bear in mind that these reference systems might be inadequate for some countries, which constitutes a source of uncertainty if the data is not adapted to the site-specific conditions.

Table 6.5. Composition of average waste and sewage sludge used in this study (Swiss conditions [9, 14, 104] and Table 2.4, Section 2.1). The data is presented with three digits; however, the level of significance is lower. The lower heating value of mixed waste was assumed to be 11.7 MJ/kg and of sewage sludge 10.9 MJ/kg.

Element	Mixed waste [%]	Sludge [%]	Element	Mixed waste [%]	Sludge [%]
H_2O	22.9	5.00	Cu	0.121	0.0418
O	25.7	32.9	Hg	0.000144	0.00741
H	4.83	6.18	Mn	0.0259	0.00
C	33.4[a]	21.7	Mo	0.000196	0.00
S	0.112	0.924	Ni	0.0107	0.00622
N	0.312	2.57	Pb	0.0502	0.0332
P	0.0894	1.71	Sb	0.00226	0.00
B	0.000719	0.00	Sn	0.00734	0.00
Cl	0.687	0.380	V	0.000922	0.00
Br	0.00136	0.00	Zn	0.131	0.149
F	0.00563	0.000529	Si	4.85	7.26
Ag	0.0000714	0.00	Fe	3.00	8.63
As	0.0000625	0.00	Ca	1.41	5.99
Ba	0.0149	0.00	Al	1.24	2.41
Cd	0.00117	0.00950	K	0.206	0.209
Co	0.000135	0.00119	Mg	0.338	0.380
Cr	0.0315	0.0133	Na	0.514	3.53

[a] Consisting of approximately 15,0 % fossil and 18.4 % biotic carbon.

Table 6.6. Functional unit for the technology comparison concerning average MSW (service cluster), services provided by the treatment technologies, and reference systems used for system enlargement. The term 'SE' indicates that the system needs to be enlarged because the quantity of service provided by the respective technology is not sufficient to fulfill the functional unit.

Services/ functions	Functional unit	Amount provided by sanitary landfill	Amount provided by MBP	Amount provided by grate incineration	Amount provided by PECK-Process	Reference system for system enlargement
Disposal of waste (average composition)	1 kg	1 kg	1 kg	1 kg	1 kg	-
Co-disposal of sewage sludge[a]	13 g	13 g (SE)	13 g (SE)	13 g (SE)	13 g	Co-disposal in the respective treatment plant
Generation of electricity	1.8 MJ	0.6 MJ (SE)	0.8 MJ (SE)	1.8 MJ	1.7 MJ (SE)	European grid electricity (UCPTE) [b]
Generation of heat	4.2 MJ	1.2 MJ (SE)	1.8 MJ (SE)	4.2 MJ	3.8 MJ (SE)	Industrial natural gas furnace[b]
Production of pig Fe	26 g	- (SE)	26 g	20 g (SE)	19 g (SE)	Production of primary pig Fe[b]
Production of Cu	0.46 g	- (SE)	- (SE)	- (SE)	0.46 g	Production of Cu[b]
Production of Zn	0.91 g	- (SE)	- (SE)	- (SE)	0.91 g	Production of Zn[b]
Production of Pb.	0.44 g	- (SE)	- (SE)	- (SE)	0.44 g	Production of Pb[b]
Production of mineral material	161 g	- (SE)	- (SE)	- (SE)	161 g	Construction sand from natural sources[b]

[a] The PECK technology needs sewage sludge for the treatment of filter ash. Therefore, all other systems are required to treat the same amount of sewage sludge as well.
[b] Production from average industrial sources; [37].

Modeling the Technologies

The technologies considered exist in different technological versions and not all necessary data was available. Therefore, approximations and assumptions had to be made.

It was assumed that the landfills for MSW, the output of MBP, and incineration residues are not covered by an impermeable layer. The reason is that such coverings, if they are applied at all, only have a life expectancy of a couple of decades, and it is highly improbable that they would be continuously renewed in the future. The infiltration rate was therefore assumed quite high, 400 mm/y, which is reasonable at humid sites. The thickness of all landfills was assumed to be 20 meters.

The model for *sanitary landfills* is described in Finnveden et al. [33] and based on Björklund [11]. Emissions from landfills are separated into emissions to water and to gas, and also in emissions occurring during the surveyable time period (corresponding to approximately one century) and those that occur during the remaining time period. Leachate purification and landfill fire data from Fliedner [35] have been added.

Landfills were assumed to have a collection system for landfill gas operating during the surveyable time period, with an efficiency of 50%. The gas collected was assumed to be used for electricity and heat production. Gas that is not collected passes through the soil where 15% of the methane was assumed to be oxidized to CO_2. The leachate from the landfill was assumed to be collected and treated before being released to recipients. It was assumed that 80% of the leachate could be collected during the surveyable time period and that the remaining 20% is directly emitted. The collected leachate was assumed to be transported to a municipal wastewater treatment plant. Sludge from the leachate treatment was assumed to be landfilled and emissions from the sludge landfill were considered in the model. Landfill fires can produce significant amounts of hazardous substances. In the model, emissions of chlorobenzene, chlorinated dioxins, PAH, PCB and Hg from landfill fires during the surveyable time period were included. It was assumed that 25% of the produced pollutants are emitted from the landfill.

Concerning the *mechanical-biological treatment (MBP)*, sound data was only available for the short-term emissions of one German plant [95]. We extrapolated from this data that approximately 2.7% of the total wet waste mass (Swiss composition) is recovered as iron scrap with a Fe-content of 96% (corresponding to 90% of the total Fe in the waste), 25.5% are separated as a high calorific light fraction for energy recovery (containing approximately 60% plastics), while the rest (71.8%) is biologically treated. The composition of Fe scrap was taken from literature [51] (the Cu content was updated according to [103]) and is shown in [105]. Because the content of oxides is rather low, we assumed that the scrap could be recycled without pretreatment in an electric arc furnace. We assumed that the high-calorific light fraction is burned in a modern grate incinerator. The biological digestion lasts 16 weeks in the plant considered [95]. In order to construct transfer coefficients (quantifying the relation between in- and outputs, [105]), we estimated the composition of the input waste of the considered plant from literature [5, 95]. The MBP product is deposited in landfills. The organic short-term emissions were taken from Wallmann [95]. As recommended in this study [95], we assumed that heavy metal emissions in the leachate during the first 100 years are identical to those of slag landfills [104], while other emissions to water such as sulfates, ammonia, nitrates, phosphor, fluorides, and chlorides were set equal to those of MSW sanitary landfills [11, 104]. In analogy to the MSW landfills, we assumed that all heavy metals are released in the long run. The TOC transfer coefficient to the water was set equal to that of slag landfills [95, 104] (small differences in the inventory data are due to the differences in density of slag and MBP output, which alters the ratio between amount of leachate and landfill material). Since oxygen enters the landfill, we assumed that no CH_4 emissions are formed and released after the first 100 years [95].

The *grate technology* is the most commonly used waste incineration technology in the world. Therefore, abundant data was available for the modeling (e.g., [9, 49, 67, 104]). The variations in emissions from one incineration plant to another are large. We assumed that the incinerator is of a modern standard and that it is equipped with wet flue gas cleaning, because it would be unfair to compare average or obsolete plants to new technologies, such as PECK. For the Fe recovered from slag, a composition according to [51, 91, 103] was assumed. Because of the high oxide content of the scrap (up to 20% [51]), we assumed that this Fe scrap needs to be treated in a blast furnace. Further aspects of the modeling can be consulted elsewhere [49, 104].

The three main components of *PECK* (staged incineration, Fluapur ash treatment and mechanical slag treatment, Fig. 6.5 and Section 5.2.4) were modeled based on information on solid outputs from experimental studies and trial runs. The emissions to air, water, and expenditures from the flue gas treatment were adapted from the grate incinerator above, but augmented by a mercury trap. Wastewater treatment sludge from the gas cleaning was assumed to be landfilled. Mercury is deposited in an underground deposit site and no further emissions are assumed. The composition of the Cu and Fe products were extrapolated from the analysis of metal particles in the raw Küpat slag [91]. The composition of the zinc hydroxide product from the filter ash treatment was based on measurements from laboratory studies. The metal products from PECK are expected to be recyclable in secondary Cu and Zn smelters and Fe blast furnaces (Fig 1 in [105]). PECK concentrates Pb in the zinc product. Lead is isolated as a side product in the zinc smelting. To assess the specific emissions released during the recycling process of PECK output metals, transfer coefficient models of the smelters and furnaces were developed from LCI data of modern plants [19, 23, 24, 38, 57, 58, 70]. The amount of Pb entering the lead smelter was estimated from the Pb content in the Zn product and the model of the Zn smelter. However, there was no information available on the content of the trace metals. As an approximation we assumed that the lead input is equal to the average input of Pb smelters. Further burdens from the manufacture and use phase of recycled metals were not considered. More details on the modeling are contained in Doka [27].

All treatment technologies were modeled based on physical-chemical causalities between waste input and output to air, water, and solid output material/residue (Section 6.2.2). This relation between input waste and output was represented by transfer coefficients. The sum of transfer coefficients for each given substance equals 100%. Table 6.7 displays some of these transfer coefficients for illustration. Two out of 33 elements considered are shown as representatives for typical emissions to air (S) and to water by landfill leaching (Cu). The complete list of transfer coefficients can be deduced from [27, 33, 49] and is given in [105].

In Table 6.7 the transfer coefficients are only shown for one process step. The other steps of the system, for instance, the landfills for the MBP product, are modeled in the same way using transfer coefficients [105] and taking the composition of the solid outputs (slag, filter ash, MBP output) from Table 6.7 as input. This procedure is relatively simple for the conventional treatment methods (sanitary landfill, MBP, grate incineration), because the process steps sequentially follow each other. On the contrary, the PECK process involves two feedback loops that need to be modeled (Fig. 6.5 and Section 5.2.4): First, metal-depleted mineral remains from the filter ash treatment are fed back into the Küpat incinerator. Second, a copper-rich transient product from the slag treatment is fed into the filter ash treatment. Based on the measurements of trial and experimental runs of the Küpat, Fluapur and Eberhard parts of PECK, a complete PECK model was synthesised by heeding mass balances and the feedback between Fluapur and Küpat. The PECK model was recursively calculated using average waste input compositions until all transfer coefficients approached a constant final value. Table 6.7 shows these final transfer coefficients for S and Cu.

In some cases, a clear causality between the waste input and the output to the environment cannot be observed: for instance, when emissions are formed due to process conditions such as combustion temperature. Concerning the incineration process, such emissions are, for example, CO, CH_4, NMVOC, NH_3, benzene, toluene, HCB, PCB, benzapyrene, dioxins, and partially NO_x. In the current work, these process-related emissions were allocated as a total to 1 kg of input waste, irrespective of the waste composition.

The amounts of ancillary products needed were calculated according to the target substances in the waste [104]. For example, the consumption of lime, $Ca(OH)_2$, was calculated as a function of acid equivalents neutralised in the wet scrubber. The inventory data for the production of ancillary products was taken from Frischknecht et al. [37]. All data concerning infrastructure were taken from Zimmermann et al. [104].

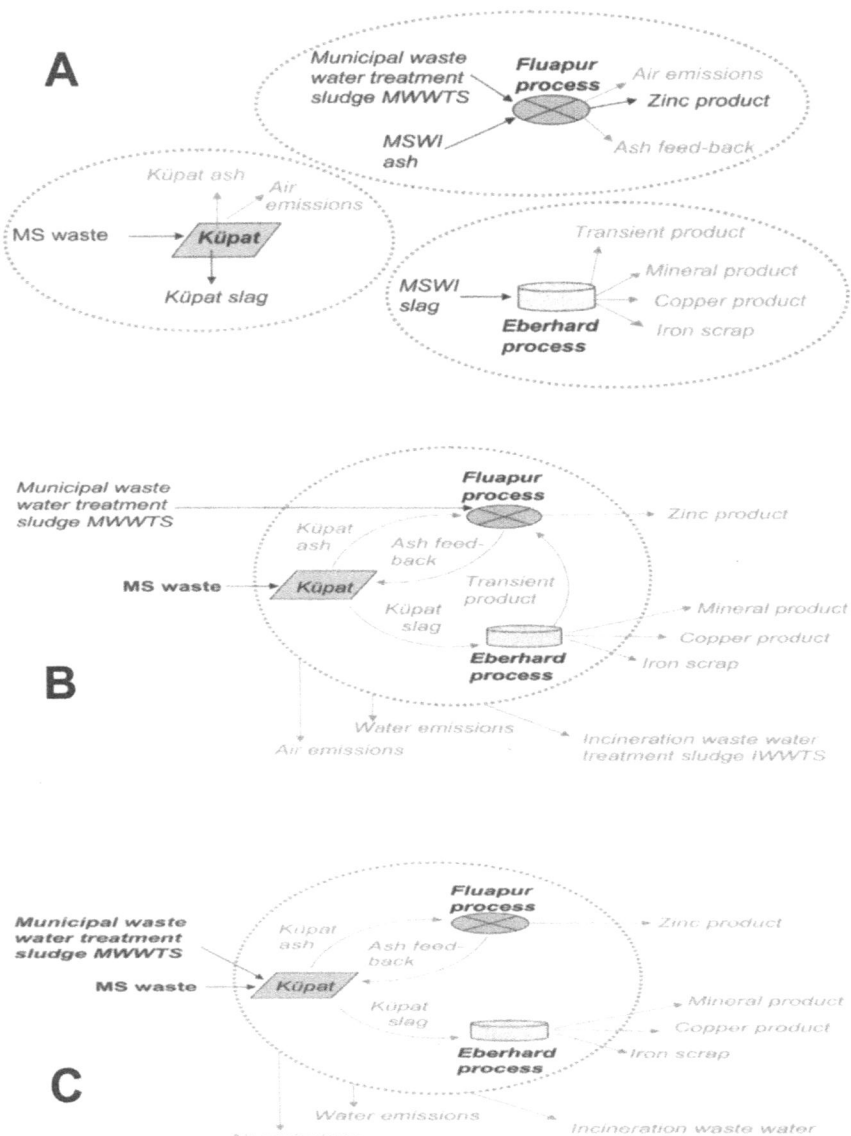

Fig. 6.5. The three components of the PECK process – Küpat, Fluapur and Eberhard processes – as they have been tested in pilot plants and laboratory tests (A), the intended assembly with feedback loops (B) and the process as modeled in this LCA (C). All measured material compositions are printed in black.

Table 6.7. Transfer coefficients are used to calculate the system outputs as a function of inputs. Two out of 33 elements considered are shown as representatives for typical emissions to air (S) and to water by landfill leaching (Cu) (example figures). The product of waste composition and transfer coefficients provides the composition of the output. The numbers quantifying the input and output composition are rounded to two significant digits.

MSW sanitary landfill

Technology	Element	Amount in waste	To air, <100 y	To water, <100 y	To air, >100 y	To water, >100	To air, <100 y	To water, <100 y	To air, >100	To water, >100
			Transfer coefficients TC [%]				Output O = C * TC [g/kg]			
MSW sanitary landfill	S	1.1	5.1	1.0	-	93.9	0.056	0.011	-	1.0
	Cu	1.2	-	0.0042	-	99.996	-	0.0050	-	1.2

Mechanical-biological treatment (only rout)

Element	Amount in waste	To air[a]	To water[b]	Rout output	To air[a]	To water[b]	Rout output
S	1.1	0.08	-	99.92	$8.8 \cdot 10^{-4}$	0.0	1.1
Cu	1.2	0.0006	-	99.9994	$7.2 \cdot 10^{-6}$	0.0	1.2

Modern grate incinerator with wet flue gas cleaning[c]

	Element	Amount in waste	To air[a]	Slag	Other residues	Waste water	To air[a]	Slag	Other residues	Waste water
Inert fraction C_{inert}		Transfer coefficients for inert materials TC_{inert}					Total output: O=C_{inert}*TC_{inert}+C_{burn}*TC_{burn}			
	S	0.07	0.00	100.00	0.00	0.00				
	Cu	0.55	0.00	100.00	0.00	0.00				
Burnable fraction C_{burn}		Transfer coefficients for burnable materials TC_{burn}								
	S	1.04	0.213	55.374	36.24	7.136	0.0022	0.65	0.39	0.074
	Cu	0.65	0.01	80.069	19.27	0.00	$6.5 \cdot 10^{-5}$	1.1	0.13	0.0

PECK technology[c,d]

Element	Amount in waste	To air[a]	Cu scrap	Waste water sludge	Mineral product	To air[a]	Cu scrap	Zn product	Mineral product
S	1.1	0.20	0.29	65.66	33.56	0.0022	0.0032	0.72	0.37
Cu	1.2	0.0004	37.45	0.81	16.08	$4.8 \cdot 10^{-6}$	0.45	0.0097	0.19

[a] Purified gas.

[b] It was assumed that the process water is circulated (wetting of rout input) [95].

[c] Concerning grate incineration, the waste composition vector was split: one for inert and another for burnable materials. These two compositions are multiplied by transfer coefficients for inert and burnable waste, respectively. The sum of the inert and burnable output is the resulting output. The PECK technology works at higher temperatures and homogenizes the waste to a larger extent. Therefore, the differentiation between inert and burnable input materials is not necessary.

[d] Not all output products are shown concerning PECK. Therefore, transfer coefficients do not add up to 100%.

Results of the Inventory Analysis

Table 6.8. Selected results for the landfill, rout, incineration and recycling/production process steps of metals (infinite time horizon; the complete list of emissions and resources is contained in [105]). Only direct exchanges with the biosphere are shown; the indirect emissions and resource use from the production of ancillary products, transport, energy, and infrastructure are not considered in the table.

Compartment	Resource, emission, process	Unit	Landfill	MBP	MSWI	PECK
Resource	Cu	kg	2.74E-04	2.74E-04	2.74E-04	0.00E+00
Air	CH_4 Methane	kg	4.05E-02	1.86E-04	1.45E-05	1.45E-05
Air	CO_2 [a]		1.09E+00	1.22E+00	1.22E+00	1.24E+00
Air	Cu	kg	1.15E-07	9.85E-08	5.36E-06	6.35E-06
Air	Hg	kg	7.61E-09	1.05E-08	5.84E-09	1.46E-09
Air	NMVOC	kg	4.19E-02	1.29E-04	3.69E-05	3.64E-05
Air	NO_x as NO_2	kg	4.47E-04	7.65E-05	2.65E-04	2.64E-04
Air	PAH	kg	4.01E-08	9.92E-08	2.95E-08	2.95E-08
Air	R11 CFC	kg		1.32E-06		
Air	R12 CFC	kg		3.42E-07		
Air	SO_x as SO_2	kg	1.92E-04	2.99E-05	7.43E-05	4.43E-05
Air	Dioxins/furans [b]	ng	4.52E-01	7.73E-01	3.10E+00	3.11E+00
Water	Ammonia as N	kg	1.28E-03	9.72E-05	3.44E-05	7.68E-07
Water	Fluorides	kg	5.02E-06	5.66E-05	6.14E-05	9.38E-06
Water	Ion Sb	kg	2.26E-05	2.26E-05	1.75E-05	2.15E-06
Water	Ion Cd	kg	1.19E-05	1.18E-05	1.19E-05	6.43E-06
Water	Ion Cr(III)	kg	3.17E-04	3.06E-04	2.87E-04	4.28E-07
Water	Ion Cu	kg	1.22E-03	8.54E-04	8.77E-04	5.70E-06
Water	Ion Hg	kg	1.54E-06	1.53E-06	1.54E-06	1.98E-08
Water	Ion Ni	kg	1.08E-04	8.81E-05	8.18E-05	2.07E-07
Water	Ion Pb	kg	5.07E-04	4.86E-04	5.07E-04	6.09E-06
Water	Ion Zn	kg	1.33E-03	1.32E-03	1.33E-03	1.52E-04
Water	Phosphates	kg	3.47E-03	3.40E-03	3.35E-03	0.00E+00
Water	TOC	kg	3.38E-03	1.65E-03	1.48E-03	6.76E-06
Ancillary product	Burnt lime CaO	kg		4.12E-03	5.88E-03	5.88E-03
Complementary system	EU electricity, medium-voltage	TJ	1.24E-06	9.99E-07	1.80E-08	1.07E-07
Complementary system	Heat from natural gas furnace	TJ	3.02E-06	2.42E-06	2.99E-08	3.75E-07

[a] Including biotic and fossil CO_2.
[b] In TCDD equivalents.

Some selected results for the landfill, MBP, incineration and recycling/production process steps of metals are shown in Table 6.8 (the complete list of emissions and resources is contained in [105]). It can be seen clearly that sanitary landfills emit more CH_4 and NMVOC to the air and ammonia to the water than MBP, and MBP more than the incineration technologies. By contrast, the total amounts of heavy metals released by MBP and the grate technology only differ slightly from the sanitary landfills. The reason is that we assumed that all ingredients of all landfill types (including landfills for the MBP output and combustion residues) are eventually emitted. The small differences in the metal emissions to the water arise due to a few metal emissions to the air (especially concerning MSWI) and the recovery of differing amounts of iron scrap also containing metals other than Fe. In contrast to the conventional technologies, the PECK process has the explicit aim of producing, on the one hand, little polluted mineral product and, on the other hand, fractions with high concentrations of metals that are suitable for recycling. This policy pays off in highly reduced emissions of metals such as Cu^{2+}, Zn^{2+}, and Pb^{2+} to the water.

Concerning the short-term analysis of less than 100 years, sanitary landfills perform best with respect to CO_2, followed by MBP, because these treatment options store carbon in the landfill and, therefore, prevent potential emissions of CO_2 and CH_4 [105]. On the contrary, the thermal technologies immediately transform almost all carbon to CO_2. The short-term metal emissions to the water are only a small fraction of the long-term emissions.

6.3.2 Impact Assessment

The impact assessment was performed with three common European methods: CML'01, Eco-indicator 99, and the Method of Ecological Scarcity (Section 6.2.3). First, we present results concerning the key environmental problems of the treatment methods. Second, we compare these technologies with respect to their environmental and economical performance.

Identification of Hot Spots: Environmentally Relevant Process Steps and Substances

In a first step, we identify environmental 'hot spots' for each technology considered. The goal is to reveal potential areas of improvement for these technologies. Table 6.9 shows the most relevant process steps (including complementary systems) and substances for all technologies and for a selection of impact categories using CML'01. It should be noted that this analysis does not allow any comparison between the technologies, because the total impact potential varies between the technologies. For instance, two identical emissions from technologies A and B might not appear with the same priority in Table 6.9 (one might not be mentioned at all), because technology A emits many other important pollutants, thus the

Table 6.9. Relevant key processes (contributing more than 10% to the total impact potential within one category) and/or key emissions/resources (contributing more than 5% to the total impact potential within one category) using method CML`01 [21] (indefinite time horizon, the original sources for each impact category are quoted in the table). The processes and emissions/resources are listed in order of environmental relevance.

Impact category	Landfill	MBP	Grate incinerator	PECK
Abiotic depletion [41]	Heat production; electricity production; production of pig iron: coal (resource)	Heat production; electricity production; waste collection	Fe-recycling: coal (resource); incineration	Fe-recycling: coal (resource); heat production; incineration
Global warming potential (500 y) [53, 54]	Landfill: CO_2, CH_4 to air; heat production	Incineration, rout, landfill (rout output): CO_2 to air; heat production; electricity production	Incineration: CO_2 to air	Incineration: CO_2 to air
Ozone depletion [98, 99, 100]	CFC R11 and R12 emissions; electricity production	Rout: CFC R11 and R12 emissions; electricity production	Waste collection; incineration; Fe-recycling/production; landfill	Waste collection; Fe-recycling; electricity production; incineration
Human toxicity [55, 56]	Landfill: SbO_4^{2-}, Ba^{2+} and Ni^{2+} to water; electricity production, heat production	Landfill: SbO_4^{2-} and Ba^{2+} to water; electricity production; heat production	Landfill: Ba^{2+}, SbO_4^{2-}, and Ni^{2+} to water; Fe-recycling/production: Cu to air	Fe-recycling: Cu to air, coal generation; incineration: dioxins; landfill: SbO_4^{2-} to water
Freshwater aquatic toxicity [55, 56]	Landfill: Cu^{2+}, Ni^{2+}, and Zn^{2+} to water	Landfill: Cu^{2+}, Ni^{2+}, and Zn^{2+} to water	Landfill: Cu^{2+}, Ni^{2+}, and Zn^{2+} to water	Landfill: Zn^{2+}, Cd^{2+}, and Cu^{2+} to water; Fe-recycling
Marine aquatic toxicity [55, 56]	Landfill: Cu^{2+} and Ni^{2+} to water	Landfill: Cu^{2+} and Ni^{2+} to water	Landfill: Cu^{2+}, Ni^{2+}, and Ba^{2+} to water	Fe-recycling: Coal generation, Cu^{2+} to air, F$^-$ to water; landfill; incineration: F$^-$ to water
Freshwater sedimental toxicity [55, 56]	Landfill: Cu^{2+}, Ni^{2+}, and Zn^{2+} to water	Landfill: Cu^{2+}, Ni^{2+}, and Zn^{2+} to water	Landfill: Cu^{2+}, Ni^{2+}, and Zn^{2+} to water	Landfill: Zn^{2+}, Cd^{2+}, and Cu^{2+} to water; Fe-recycling

Table 6.9 (cont.)

	Landfill	MBP	Grate incinerator	PECK
Marine aquatic toxicity [55, 56]	Landfill: Cu^{2+} and Ni^{2+} to water	Landfill: Cu^{2+} and Ni^{2+} to water	Landfill: Cu^{2+}, Ni^{2+}, and Ba^{2+} to water	Fe-recycling: Coal generation, Cu to air, F^- to water; landfill; incineration: F^- to water
Fresh-water sedimental toxicity [55, 56]	Landfill: Cu^{2+}, Ni^{2+}, and Zn^{2+} to water	Landfill: Cu^{2+}, Ni^{2+}, and Zn^{2+} to water	Landfill: Cu^{2+}, Ni^{2+}, and Zn^{2+} to water	Landfill: Zn2+, Cd2+, and Cu2+ to water; Fe-recycling
Marine sedimental toxicity [55, 56]	Landfill: Cu^{2+} and Ni^{2+} to water	Landfill: Cu^{2+} and Ni^{2+} to water	Landfill: Cu^{2+}, Ni^{2+}, and Ba^{2+} to water	Fe-recycling: Cu to air, coal generation; landfill: Zn^{2+} to water; heat production
Terrestrial ecotoxicity [55, 56]	Landfill: Hg^{2+} to water; electricity production; Zn production: Hg to air	Landfill: Hg^{2+} to water; electricity production; rout/landfill for rout output: Hg to air	Landfill: Hg^{2+} to water; Fe-recycling/production: Cr to air; Zn production: Hg to air	Fe-recycling: Cr and Cu to air; incineration; electricity production
Summer smog [26, 59]	Landfill: NMVOC, CH_4	Electricity production; rout: NMVOC; heat production; waste collection	MSWI: NMVOC; waste collection; Fe-recycling/production: CO	Incineration: NMVOC, NOx to air; waste collection; Fe-recycling: CO to air
Acidification [44]	Electricity production; landfill: NO_x to air; heat production	Electricity production; heat production	Incineration: NO_x to air; Fe-recycling/production; waste collection; landfill: F^- to water	Incineration: NO_x to air; Fe-recycling: SO_x to air; waste collection; electricity production

Table 6.9 (cont.)

| Eutrophi-cation [46, 55] | Landfill: NO_x; electricity and heat production (according to [46] phosphate and ammonia to water) | Electricity and heat production; waste collection, rout / incineration: NH_3 and NO_x to air; (according to [46] phosphate and nitrate to water | MSWI: NO_x to air, NH_3 to air; waste collection; Fe-recycling/ production; landfill | Incineration: NO_x to air, NH_3 to air; waste collection; landfill |

[a] No measurements were available on CFC emissions of sanitary landfills. However, it is known that sanitary landfills accumulate and release CFC to the environment while incinerators destroy them [71, 72]. Therefore, we assumed the emissions of CFC of landfills to be equal to those from MBP.

emission is not relevant *relative* to those.

Several conclusions can be drawn from Table 6.9. First, heavy metal emissions from landfills are important with respect to all toxicity-related categories. Second, the complementary processes from enlarging the system (Table 6.6), in particular the production of energy, seem to be relevant with respect to sanitary landfills and the mechanical-biological treatment. With regard to the thermal technologies, the complementary energy systems play a less important role (they account for less than 11% in all impact categories except for *abiotic depletion*), because the energy-efficiency of these technologies is comparatively high. However, it is notable that the recycling of Fe-scrap recovered from the slag appears as relevant in 8 of the 12 categories considered. Waste collection and transport is only rarely mentioned in Table 6.9, indicating that these processes are not relevant when distances are as short as they are in Switzerland.

In order to find out how important the impact categories are, we defined the energy production of each technology as the functional unit, and compared each treatment technology to average European electricity plants and heat-generating gas furnaces. We considered categories where the waste treatment options scored better than the average energy plants of minor importance. Since the waste treatment technologies also have functions other than energy generation, e.g., the service waste treatment itself, this is a conservative approach.

Sanitary landfills produce less impact than energy plants in the category *abiotic depletion*, MBP in the categories *abiotic depletion* and *acidification*. Concerning the thermal technologies, the treatment technologies are superior to the energy plants in all categories but *global warming* and most *toxicity*-related categories. Therefore, it can be stated that the toxic emissions to water, such as heavy metals and airborne emissions of CO_2, are relevant emissions concerning all technologies if the time horizon considered is indefinite. Concerning sanitary landfill and MBP, emissions of CFC and NMVOC are additionally important, as is CH_4 and possibly ammonia to water for sanitary landfill only.

These results have been confirmed by applying fully aggregating methods such as Eco-indicator 99 and Method of Ecological Scarcity (Table 6.10).

Table 6.10. Environmentally relevant process steps and emissions/resources according to the methods Eco-indicator 99 and Method of Ecological Scarcity (contribution of more than 1% to the total impact potential).

Technology	Relevant processes and emissions/resources according to Eco-indicator 99 (Hierarchist, Egalitarian, or Individualist perspective)	Relevant emissions/resources and processes according to Method of Ecological Scarcity
MSW sanitary landfill	Landfill: Ni^{2+}, Cd^{2+}, and Cu^{2+} to water; CO_2 and CH_4 to air; heat production, and electricity production according to the default Hierarchist perspective. Additionally, Cr^{3+} and Zn^{2+} to water, Cu and Fe as resource, and NMVOC to air concerning the Egalitarian and/or Individualist perspectives.	Landfill: Phosphates and Cu^{2+} to water, NMVOC to air, Hg^{2+} and Zn^{2+} to water, CO_2 to air, Cr^{3+} to water, CH_4 to air, Cd^{2+} and ammonia to water; electricity production; heat production.
Mechanical-biological treatment (MBP)	Landfill (rout output): Ni^{2+}, Cd^{2+}, Cu^{2+}, Zn^{2+} and Cr^{3+} to water; rout/landfill: CO_2 to air; heat production, and electricity production according to the Hierarchist and Egalitarian perspectives. Additionally polychlorobiphenyls to air (rout), Cu and Pb as resource (complementary Cu and Pb production) according to the Individualist.	Phosphates, Cu^{2+}, Hg^{2+}, and Zn^{2+} to water, CO_2 to air, Cr^{3+}, Cd^{2+}, nitrates, and Pb^{2+} to water; electricity production; heat production.
Grate incineration (MSWI)	Landfills (combustion residues): Ni^{2+}, Cd^{2+}, Cu^{2+}, Zn^{2+}, and Cr^{3+} to water; incineration: CO_2 according to the Hierarchist perspective. Additionally coal for Fe-recycling and Cu as resource (complementary Cu production) for the Egalitarian and/or Individualist perspective.	Landfills: phosphates, Cu^{2+}, Hg^{2+}, Zn^{2+}, Cr^{3+}, Cd^{2+}, and Pb^{2+} to water; incineration: CO_2 to air; infrastructure slag landfill.
PECK	Landfill: Cd^{2+} to water; PECK incineration: CO_2, NO_x; heat production; waste collection; Fe-recycling: Cu to air, coal, Fe as resource, Zn to air; electricity production; natural gas for filter ash treatment according to the Hierarchist perspective. In addition Cu as resource (complementary Cu production) for the Egalitarian and/or Individualist perspective.	Incineration: CO_2 and NO_x to air; landfill: Cd^{2+}, Zn^{2+}, Cu^{2+} and Hg^{2+} to water; electricity production; waste collection; infrastructure residue landfill; heat production; Fe-recycling: coal as resource.

The above analysis considers indefinite time horizons, thus accounting for all potential emissions, even if they occur millions of years from now. If only the emissions of the first 100 years were taken into account, heavy metal emissions to water would no longer be considered an important issue (only Hg^{2+} to water according to the Method of Ecological Scarcity). The other emissions listed in Table 6.10 would remain relevant. Since the overall impact potential of all technologies would decrease substantially, further process steps and emissions/resources would contribute, with more than 1%, to the overall impact potential:

- MSW sanitary landfill: Landfill: NO_x to air, Ni and TOC to water, SO_x and N_2O to air; waste collection; infrastructure MSW landfill. If biotic CO_2 were not considered CO_2 emissions would not be relevant during the first 100 years.
- MBP: Waste collection; infrastructure slag landfill; landfill: ammonia to water.
- MSWI: Waste gathering; incineration: NO_x and small particles to air; Fe-recycling/production: Cu to air, Fe as resource, oil from storage EU; slag landfill: diesel consumption of construction machines and energy consumption of buildings; infrastructure slag landfill; production of concrete; infrastructure incineration plant; ancillary products: CaO; production of mineral product: construction sand; infrastructure residue landfill.
- PECK: Infrastructure incineration plant; incineration: small particles to air, Fe-recycling: oil from storage EU, coal as resource; landfill: Ni^{2+} to water, ancillary product: CaO.

In order to identify the relevant impact categories of the short-term analysis, we have defined the functional unit as the energy production (see above) and have only considered the emissions of the first 100 years. Concerning the thermal technologies and the mechanical-biological treatment, the toxicity-related categories are no longer relevant. The reason is that only a small fraction of heavy metals is emitted during the first 100 years, while the major share would be released afterwards and therefore neglected in the short-term analysis. By contrast, *global warming* continues to be important. MBP additionally performs worse than regular energy plants with regard to *ozone depletion, summer smog*, and *eutrophication*. The main reasons for this are the low energy efficiency and the emissions of CFC. Sanitary landfills score worse than average European energy plants in almost all categories. Similar to the long-term analysis, the weak points are the small energy efficiency and emissions such as NMVOC, CFC, and CH_4 to air as well as Ni^{2+} and TOC to water.

Improvement Potentials for Technology Optimization

The analysis above serves to identify several improvement potentials: Above all, the long-term emissions of *heavy metals* deserve further attention. Heavy metals emissions could be prevented, e.g., by recycling the solid residues. Stabilization (vitrification) of heavy metals and finding a 'permanent storage' (as is intended in the case of nuclear waste) can also reduce or delay the impacts of heavy metals. Global warming is another important issue. The most relevant emission in this category is *CO_2*. Technology optimisation can only help little in this regard, and CO_2 sequestration is still futuristic technology. However, the reduction of carbon components in the waste input, e.g., by recycling carbon-containing products, could be a promising strategy. Optimising the *energy efficiency* helps significantly to improve the environmental performance of waste treatment. Concerning the grate and, particularly, PECK technology, another major improvement potential is enhancing the *quality of the Fe-scrap*, e.g., with mechanical pre-treatment. If the scrap can be recycled in an electric arc furnace (which is already done in some cases) instead of a blast furnace, the environmental impact of this process step would be considerably lower.

Technology Comparison

In order to compare the four treatment technologies, we have applied method CML'01 to the inventory results again. The impact potential of the sanitary landfill is set at 100% in all categories and the other technologies are normalized to this value. Note that comparisons are only admitted within the categories but not between them.

Figure 6.6 clearly shows that the incineration of mixed waste is more favourable than direct landfills and the mechanical-biological treatment concerning all impact categories if the method CML'01 is applied. This is primarily due to the poor energy recovery of sanitary landfills and MBP and to emissions such as CH_4, NMVOC, and NO_x to air. MBP scores better than or roughly equal to sanitary landfills in all categories. PECK performs better than grate incineration regarding toxicity. This is the result of PECK's aim to recover heavy metals from the slag

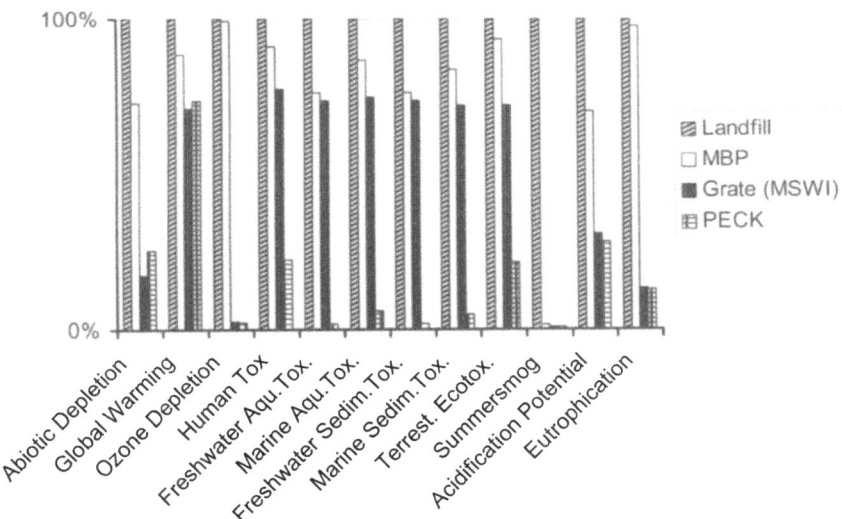

Fig. 6.6. Impact assessment with CML'01 [21] (indefinite time horizon). All values are normalized to the landfill impact potential (=100%). Comparisons between categories are not possible.

and filter ash and, therefore, to prevent metal emissions. In most other categories, PECK performs about equal to the grate technology. Only with respect to the resource oriented categories *abiotic depletion* (CML'01) and *resource surplus energy* (EI 99) is PECK inferior to grate incineration. At first sight this is surprising, because it is the explicit aim of PECK to recover metals and minerals and, thus, to preserve resources. The reason for the poor results in the resource-related categories is the higher energy demand of the PECK technology and the strong weight that many LCIA methods assign to fossil resources in comparison to metal resources.

The application of fully aggregating methods reinforces these results. Figures 6.7 and 6.8 display the LCIA results with Eco-indicator 99 and Method of Ecological Scarcity.

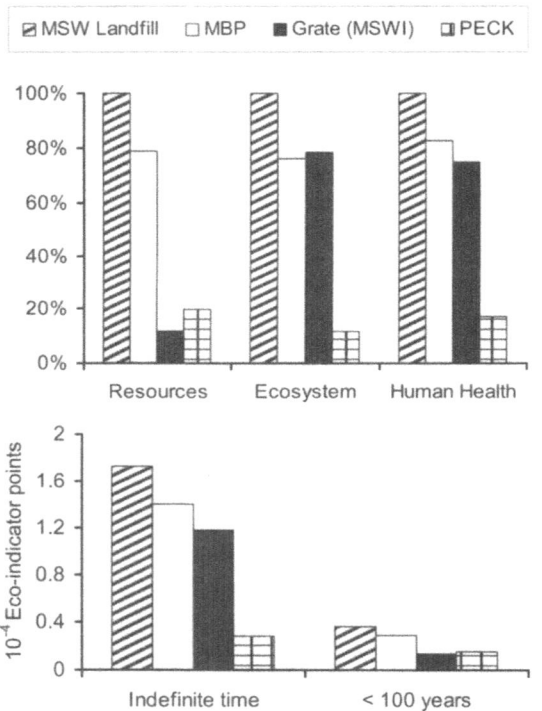

Fig. 6.7. Impact assessment with Eco-indicator 99 (Hierarchist perspective; the results are similar concerning the Egalitarian and Individualist perspective). The above graph shows the partially aggregated results (infinite time horizon). All values are normalized to the landfill impact potential (=100%). The weighted and aggregated results are presented in the graph below (infinite and short-term time horizon).

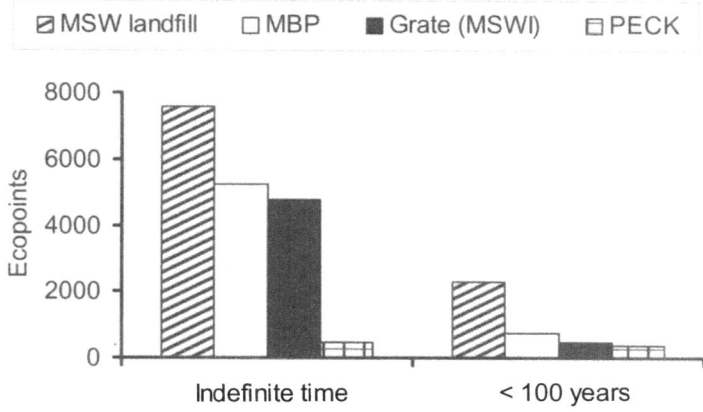

Fig. 6.8. Impact assessment with Method of Ecological Scarcity (indefinite and short-term time horizon).

Figures 6.7 and 6.8 confirm that direct MSW landfills are the ecologically worst treatment option for mixed waste with respect to all damage categories and the overall results. The main reasons are the low energy efficiency and emissions such as CH_4 and NMVOC to the air as well as ammonia to water.

Concerning the impact category *Damage to Mineral and Fossil Resources* (Fig.6.7, above), the thermal technologies score best, because they recover more energy than the other treatment options. PECK is inferior to grate incineration in spite of metal recovery, because it consumes slightly more energy for the recovery of mineral resources. With respect to *Damage to Ecosystem Quality*, MBP and grate incineration perform about equal. The slight differences are due to the Fe-recycling technologies assumed: The iron scrap from MBP was assumed to be re-cycled in an electric arc furnace. This type of recycling produces fewer emissions than blast furnaces, which we assumed to be the technology for Fe-scrap from the thermal treatment processes. PECK is superior to the other technologies, because it recovers metals and thereby prevents their emissions. Concerning *Damage to Human Health* and the aggregated overall results (Fig. 6.7 and 6.8), the ranking is clear: PECK performs best, followed by grate incineration and the mechanical-biological treatment. Sanitary landfills are inferior to all other treatment options.

If only the emissions of the first 100 years were considered, the impact potential would drop considerably concerning all technologies (Fig. 6.7and 6.8). The ranking of the technologies would no longer be as striking. Sanitary landfills would still perform worse than the other technologies with respect to the aggregated results (however, they would score better in the category of global warming). MBP would be slightly inferior to the thermal technologies, mainly because of its low energy efficiency. PECK would no longer score better than grate incineration, because no credit would be given for the avoided long-term emissions of metals. These results show that drawing overall conclusions is wholly dependant on the choice of time frame.

Environmental-Economical Analysis

In this paragraph we will couple the above results from the impact assessment to economic costs. First, we quantify the approximate range of costs per ton MSW for modern treatment plants (Table 6.11). Next, we introduce the concept of Ecological-Economic Efficiency (EEE). Last, we present synthesized results of the ecological and economical assessment.

The costs in Table 6.11 do not include costs for logistics and aftercare of landfills. Logistics account for a large contribution to the overall costs. The cost of waste collection is approximately 72 Euro/t, and the cost of transport 0.1 Euro/tkm plus 12 Euro/t shipping costs [84]. Therefore, logistics will add approximately 85 Euro/t if a distance of 10 km to the treatment plant is assumed. Aftercare of landfills can make sanitary landfills and, to a smaller extent, also other types of landfills much more expensive than indicated in Table 6.11. For instance, the sanitary landfill in Kölliken (Switzerland) [97] needs to be remediated at an estimated cost of between 200 to 235 Mio Euro [18] (571 to 671 Euro/t), because pollutants seep into the groundwater. Concerning the environmental assessment without time restriction, we assumed that no aftercare is performed and that all persistent ingredients of the landfill are emitted (costs of aftercare are therefore not taken into account). With respect to the short-term time horizon, however, costs of aftercare should be taken into account, at least in sensitivity analysis, because remediation could be one reason to justify neglecting all emissions after the first 100 years. Figure 6.9 shows the position of the four treatment technologies in a diagram displaying the environmental impact potential in comparison to economical costs. Relative to grate incineration, sanitary landfills and mechanicalbiological treatment are less costly but environmentally more harmful. PECK is about as costly as grate incineration, but its environmental impact potential is lower.

Table 6.11. Costs of waste treatment (without logistics and future costs for aftercare).

Technology	Costs without logistics[a]	Ref.
MSW sanitary landfill	60 Euro/ton	[84]
Mechanical-biological treatment MBP	80 to 100 Euro/ton MSW	Section 4.1
Grate incineration (MSWI)	100 to 135 Euro/ton MSW	[17]
PECK	104 to 127 Euro[b]	Section 5.2.4

[a] Currency conversion rate from February 2002.

[b] Assuming that the vitrified mineral product can be used or deposited without further costs.

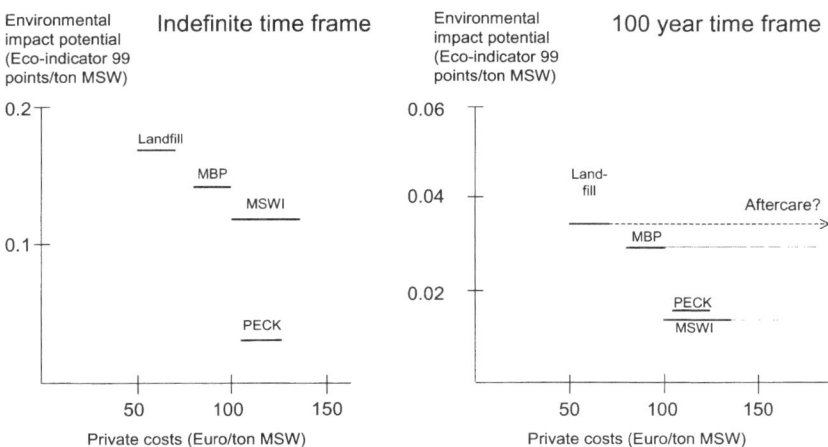

Fig. 6.9. Environmental impact potential (Eco-indicator 99, Hierarchist perspective) and economic costs (without logistics) of the four treatment technologies. The technologies perform better the closer they are located to the left and the bottom in the two graphs

Concerning the short-term contemplation, these two rankings are similar, with the exception of the PECK technology, which performs about equal to grate incineration. However, if we take the cost of aftercare into account (which could justify a temporal cut-off), sanitary landfills could also end up being the worst option concerning costs. In order to better combine the economical and ecological results, we have used the concept of *Ecological-Economic Efficiency (EEE)* [66]. This indicator is defined as the ratio between net ecological benefits (the difference in impact potential of the technologies considered) and the difference in net present value (Equation 6.1). The net present value is the sum of all discounted economic benefits and costs (Equation 6.2). While the EEE quantifies the relation between ecological and economic benefits, it disregards the total volume of an investment which might well be relevant to a given decision.

$$\text{Ecological - economic efficiency (EEE)} = \frac{(\text{NEB}_{\text{TechnologyA}} - \text{NEB}_{\text{TechnologyB}})\left[\text{Point/t MSW}\right]}{(\text{NPV}_{\text{TechnologyA}} - \text{NPV}_{\text{TechnologyB}})\left[\text{Euro/t MSW}\right]} \quad (6.1)$$

where NEB is the net ecological benefit and NPV the net present value.

In order to quantify net ecological benefit, we need to form the difference between the impact potentials of two technologies. For instance, sanitary landfills have an impact potential of 0.172 EI 99 points and grate incineration 0.118 EI 99 points per ton MSW (Hierarchist perspective).

EI 99 Hierarchist	MBP	Grate (MSWI)	PECK	Ecological Scarcity	MBP	Grate (MSWI)	PECK
Landfill	0.0011	0.0009	0.0026	Landfill	78	49	127
MBP		0.0008	0.0043	MBP		17	183
Grate (MSWI)			-0.0596	Grate (MSWI)			-2865

Fig. 6.10. Ecological-economic efficiency (EEE) of choosing one treatment option instead of another. In the tables, each number quantifies the ecological-economical advantage of the technology in the column over the technology in the line. The negative ratio of the PECK technology indicates that both the ecological and economic performance were better. In all other cases, high numbers indicate a high efficiency.

The difference in net ecological benefit is therefore 0.054 EI 99 points/t MSW. The costs given in Table 6.11 are net present costs or the negative of the net present value. The difference in net present value between sanitary landfill and grate incineration is therefore 57,5 Euro/t MSW (average costs). The resulting ecological-economic efficiency of constructing a grate incinerator instead of a sanitary landfill is therefore EEE = 0.0009. In Figure 6.10 we calculate the EEE for all combinations of technologies. Average costs and aggregating LCIA methods were used in the assessment (Eco-indicator 99 and Method of Ecological Scarcity). Concerning Eco-indicator 99, only the Hierarchist perspective is shown. Applying the Egalitarian and Individualist perspectives leads to comparable results. Figure 6.10 shows that PECK is the most efficient ecological-economic alternative to all other treatment technologies. The second efficient alternative to sanitary landfill is MBP, followed by grate incineration.

6.3.3 Interpretation

Identification of Significant Issues

The following parameters have proven to be relevant in the inventory analysis and/or impact assessment (Sections 6.3.1 and 6.3.2):

- Toxicity based impact categories have been identified as a major problem of waste treatment if long-term time horizons are considered. The relevant emissions have a toxic impact on human health and ecosystem health. Global warming is also an important issue. Ozone depletion is relevant concerning the mechanical-biological treatment and sanitary landfill. Acidification and eutrophication are of minor importance concerning sanitary landfill and MBP.

- Heavy metal emissions to the water, such as Cu^{2+}, Ni^{2+}, Cd^{2+}, SbO_4^{2-}, Ba^{2+}, Hg^{2+}, Zn^{2+}, Cr^{3+}, and Pb^{2+} leached from landfills, contribute a large share to the overall impact potential according to all LCIA methods applied. In addition, some emissions to air, such as CO_2 and, to a lesser extent, NO_x, are environmentally relevant. Sanitary landfills (and to a lesser extent MBP) additionally emit significant amounts of CH_4 and NMVOC. Moreover, airborne emissions of Cu are of relevance concerning the recycling of Fe in blast furnaces (this concerns grate and PECK incineration).

- Landfills are the most important process steps with respect to all technologies, if the long-term time perspective is adopted. Complementary processes such as energy production and recycling processes (especially for Fe) also contribute significantly to the total impact potential. In comparison, the biological process (MBP) and the incineration step (grate, PECK, and MBP) only play a minor role, except for in the category *global warming*.

Evaluation of Data Completeness, Sensitivity, and Consistency

The present study has several weaknesses concerning the inventory analysis and impact assessment. These shortcomings should be kept in mind when applying the results.

- The long-term behavior of landfills is uncertain. Since the LCA approach requires a consideration of all emissions from *cradle to grave*, even if uncertainties are significant, we adopted a long-term perspective nevertheless and considered all landfill ingredients as future emissions. Due to the lack of data and to the current inability of LCA to consider emission concentrations in the impact assessment, only emission amounts were taken into account, regardless of their concentrations and temporal occurrence. As a sensitivity analysis, a short-term perspective of less than 100 years was considered. For this time period it is possible to make more reliable predictions based on existing measurements. However, the largest share of emissions is neglected in this approach.

- The fate of landfill emissions in the subsoil has been disregarded. The leachate is assumed to enter the surface water since there are no standard methods available to assess emissions to the groundwater. This simplifying assumption might lead to an underestimation of toxic impact from heavy metals, as has been shown in a case study for MSWI slag landfills [47].

- The impact assessment has been performed without consideration to site-dependent aspects, and this is a simplification. For example, emissions of pollutants at a low level in an urban area can cause environmental impacts different from the same emissions at a high level in rural areas.

- Chloride emissions to the groundwater might be relevant in some places where the groundwater volume is small (Section 2.2). However, the highest concentrations of chloride will occur during the first 100 years, when the drainage system is still working (to a large extent). Therefore, it seems reasonable to assume that most of the chloride that is emitted in high concentrations ends up in the surface water, where it gets diluted sufficiently enough in order not to cause any serious harm.

- Some substances have not been considered. For instance, MSW sanitary land-fills might emit heavy metals to the air during the first 100 years (Section 2.2). However, information about the emission quantity was not available and only Hg emissions to air were considered. Another example is the dioxin contained in the solid incineration outputs (slag and especially filter ash). These emissions might have significant environmental effects that have not been considered in the present study.

- Some metals leave the system through the recovered metal products. For instance, Fe scrap contains other pollutants, such as Cu and Sn. In the current study, we have considered the emissions from the recycling process to air and water. However, most of these pollutants remain in the recycling product itself and the potential impacts during the use and disposal phase of the secondary product were not taken into account. Moreover, the negative effect of these substances on the quality of the recycled material was not considered (e.g., Cu and Sn in Fe scrap diminish the quality of steel [103]).

- It was assumed that Fe scrap from waste incineration needs to be recycled in a blast furnace due to the high oxide content, whereas Fe scrap recovered directly from the waste enters an electric arc furnace. The latter recycling technology produces fewer emissions than the former. Our assumption is debatable because MSWI scrap is currently also treated in electric arc furnaces in some cases, and directly recovered Fe scrap from MSW might be heavily polluted by organic substances so that a thermal pre-treatment might become necessary [103].

- The data quality varies considerably among the technologies considered. The data on PECK is the most uncertain, because this technology has not been implemented as a complete assembly and only some measurements of the solid outputs from trail runs or experimental devices were available (Fig. 6.5). Consequently, many assumptions were made. For example, most emissions to air and to water from the incineration process were assumed to be equal to the emissions of the grate technology. Information on air emissions from the filter ash process – where the sewage sludge additive is burnt – was not available. As an approximation, we assumed that sewage sludge would be fed into the staged incinerator (Fig. 6.5) leading to combustion emissions via that route. The transient product (Fig. 6.5B and C) was neglected, as no information about the composition was available. While it has been shown in trials that the single components of PECK perform well in isolation, there is still a risk that the cross-connected components in the intended PECK assembly might display problems with the recursive nature of the process. Our assumption that the single components perform equally well as stand-alone processes and as connected assembly is therefore debatable.

- Accidental risks such as risks to the working environment, risks of transport accidents, and hygienic risks were neglected in the present work, as is typically done in LCA. However, these risks might be an important factor. For instance, the high temperatures and aggressive gases of the PECK process represent a significant risk for employees and the adjacent population. Concerning the other technologies, MBP bears the highest risks to the working environment, followed by grate incineration and sanitary landfill [84]. Hygienic risks by pathogenic germs seem to be most significant for MBP and sanitary landfill, while this is not a problem for waste incineration [84].

- In a sensitivity analysis, we used Swiss average electricity and heat from an oil furnace as complementary systems for the energy production. The general results of the technology comparison did not change. However, further scenarios that could influence the results are conceivable. The choice of reference systems constitutes a major source of uncertainty [28].

- All emissions of CO_2 were considered irrespective of their origin (biotic or fossil). One could also choose to consider only fossil CO_2. This would, for example, reduce the impact potential in the category of global warming. However, sensitivity calculations showed that our conclusions would not change.

A sensitivity analysis has shown that the results depend to a great extent on the time perspective considered. Long-term emissions contribute the largest share to the overall impact potential. If these emissions are neglected, the ranking of the technologies can no longer be clear. By contrast, the application of different LCIA methods did not influence the results significantly, indicating that the results of the impact assessment are reliable.

Conclusions and Recommendations

The above analysis shows that there are several substantial improvement potentials for the technologies considered. Sanitary landfills and mechanical biological treatment offer inferior performance compared to thermal technologies. The most important factor is the low energy efficiency of these treatment technologies. It can generally be stated that *using the energy content of the waste efficiently* is of great importance when treating waste. This can either be done by *material recycling* (e.g., recycling of aluminum saves 95% of energy compared to primary production [16]) or by *thermal recovery of feedstock energy*.

Long-term releases of heavy metals to the water cause the largest share of impact concerning all treatment options considered. The prevention of such emissions therefore has a high potential for improvement. Promising strategies are *reducing the content of heavy metals in products* (Section 3.1), *recovering heavy metals* (material recycling or PECK technology), or *vitrifying* and thereby stabilizing the heavy metal containing residues (e.g., Section 5.3 or HSR-Process [102]).

The results also showed that the quality of recovered metal products could be improved. For instance, Fe-recycling of iron scrap in a blast furnace produces a larger environmental impact than primary production of pig iron. The reason is that trace metals such as Cu in the Fe-scrap are partially emitted to the air. *Improving the quality of the scrap* is a promising strategy because fewer emissions of trace pollutants would be released in the recycling process due to the smaller content in the scrap input and because cleaner technologies could be used (e.g., electric arc furnace in case of Fe-recycling).

The comparison between the four treatment technologies shows that sanitary landfill is the worst option for mixed waste. This result is independent of the time frame considered and the LCIA method applied. The reasons are higher emissions, especially to the air (e.g., NMVOC, CH_4), and low energy efficiency. If specific waste fractions are considered, some advantages of sanitary landfills can appear applying the short time perspective. One example is the lower amount of CO_2 emissions from landfills compared to those from incineration [33].

The impact score of the mechanical-biological treatment suffers from the low energy-efficiency in comparison to the thermal technologies. The emissions of the MBP process itself are less important with respect to the overall impact, but they play a role for certain impact categories such as ozone depletion.

The potential impacts from the airborne emissions of modern MSWI plants per unit of produced energy are comparable to those of an average energy plant in Europe (except for CO_2). It can therefore be argued that the atmospheric emissions are not a priority issue of waste incineration any longer (under the condition that the energy produced by the waste incinerator replaces average European electricity and heat from gas furnaces).

PECK is favorable to all other treatment options, if long-term time horizons are considered (and if the practical performance meets the targets set). This is due to the recovery of heavy metals on the one hand and a reasonable energy-efficiency on the other hand. The superiority of PECK holds for the aggregated results and all impact categories except for global warming and resource depletion. At first sight, this is surprising as the explicit goal of PECK is the recovery of mineral resources such as Cu and Zn. However, the reason for this is the slightly higher energy use of PECK (for this recovery of minerals) and that most LCIA methods give more weight to the use of fossil resources than mineral resources. However, this deficit in the resource-related impact categories is fully compensated by the higher performance in the toxicity-related categories. Therefore, the energy use of PECK in recovering mineral resources from the solid output is ecologically justified according to the impact assessment methods used.

If only short-term horizons are considered (less than 100 years), the emissions from landfills are very small no matter what technology is applied. Consequently, new technologies such as PECK do not receive credit for preventing long-term emissions. Such technologies might even score worse than the conventional grate incinerator due to their comparatively higher energy use.

Concerning the costs of the technologies, sanitary landfills are the cheapest, followed by the mechanical-biological treatment and the incineration technologies (not considering potential costs of aftercare that could easily reverse the order). Therefore, prioritizing between environmental and economical criteria is necessary when making a decision concerning a new waste treatment plant. In order to facilitate the decision, we applied the concept of ecological-environmental efficiency and compared all technologies with each other. According to our findings,

PECK is the most ecological-economically efficient alternative. However, it must be noted that all data concerning PECK (ecological and economical) are rather uncertain, because this technology has not been implemented at full scale until now. The second best alternative to sanitary landfills would be mechanical-biological treatment, due to its lower costs in comparison to thermal treatment.

Outlook

The significance of heavy metal emissions from landfills and their decisive influence on the overall results suggest that the landfill model deserves further elaboration. Some major shortcomings of the applied model are that the location and the time of emission are disregarded. This has several consequences. First, landfills release pollutants over a very long time as a function of spatial parameters such as infiltration rate and it is not clear whether the same weight should be assigned to emissions occurring at different points in time. Second, emission concentrations are related to the location of the landfill and to a temporal dimension. The consideration of concentrations is very important for the assessment of toxic emissions such as heavy metals, since toxic effects are a function of concentration. A further shortcoming of the above case study is the assumption that all releases from the landfills enter the surface water immediately. This assumption is rather questionable because the technical system, such as the base sealing of the landfill and the drainage system, will fail in the long run. Therefore, it would be more realistic to assume that the emissions seep into the subsoil and eventually reach the groundwater. However, common LCIA methods such as those used in the above case study offer no means to assess the potential impacts of emissions to the groundwater. Initial approaches have been suggested for better modeling of the emission behavior of landfills [49, 88] and to predict and assess the emissions to the groundwater [47, 48, 50], but these approaches are restricted to certain landfill materials and need to be expanded to allow a comparison between different types of landfills.

The above results illustrate that the outcome of the LCA depends on the time frame considered. Different results could be achieved, if even shorter time-horizons than 100 years (e.g., 5 years) were considered. For instance, sanitary landfills might profit from such a strict temporal cut-off because they would get more credit for stored carbon (prevented CO_2-emissions). In combination with a different (cleaner) reference system for energy production, this could make sanitary landfill a favorable option for waste treatment. One of the key questions in LCA of waste treatment processes is, therefore, how future and present emissions should be weighted against each other. This research question will be treated in some depth in the following section.

6.4 Long-Term versus Short-Term Impacts

The above example illustrates that certain environmental decisions involve trade-offs between present and future impacts. Such tradeoffs raise issues of (intergenerational) fairness and equity that are ethical in nature. In general, LCA makes no explicit differentiation between emissions (and, ultimately, impacts and damages) at different points in time. For instance, whether an emission contributes to ozone depletion today or in 200 years is treated equally in LCA. Nevertheless, there are some forms of implicit discounting that are common practice, e.g., temporal system boundaries [30]. Temporal cut-offs are a special case of discounting, with a discount rate of zero for the time horizon considered and of infinity thereafter. Such temporal system boundaries are often proposed for landfill emissions [45, 69]. New thermal waste treatment technologies prevent long-term emissions at the cost of a higher energy use with more immediate impacts. These issues demand a thorough discussion of whether impacts at different points in time should be weighted alike and, therefore, whether discounting (including temporal system boundaries) should be applied in LCA.

6.4.1 The Economical Background of Discounting

In the economic sciences, future costs and benefits are usually discounted to a present value in order to make them comparable to current costs and benefits (cost/benefit analysis) [75]. There are several reasons for valuing one monetary unit of benefit or cost differently at different points of time. For instance, pure time preferences (impatience), the productivity of capital (related to economic growth/decline), and uncertainty/risk perception are all factors that change valuations with time. The way the valuation changes in time is generally known as the discount rate. This rate depends on the factors just mentioned. Thus, in economics, the Net Present Value (NPV) of an investment is calculated as a function of benefits, costs, and the discount rate (Equation 6.2):

$$NPV = \sum_{t=0}^{T} ((B_t - C_t) * \frac{1}{(1 + r)^t}) \tag{6.2}$$

Where B represents the benefits, C the costs, r is the discount rate, and t is a time index. The discount rate of Equation 6.2 is expressed in real terms, net of any changes in the price level. To give an example, a dollar invested today at an interest rate of 5% will have increased to $11.47 in 50 years. Conversely, $100 in 50 years would be worth (would require investing) now $8.72. This latter amount of money represents the NPV.

In economics, the choice of the discount rate is controversially discussed, especially when investments in the public sector are at stake [4, 61, 63]. This discussion involves the question of whether the *private* or the *social* discount rate should be taken. The *private* discount rate can be observed on the financial markets; for instance, a typical value would be between 5 and 7% per year in the European

Union [29]. Many companies, however, calculate using private discount rates greater than 10%. The *social discount rate* is defined as the interest rate at which society is willing to lend money for public projects [75] (Equation 6.3):

$$r = r_{pref} + \eta * r_{gro} \qquad (6.3)$$

Where r_{pref} is the pure rate of time preference (accounting for impatience), r_{gro} the rate of economic growth (growth of the Gross National Product, GNP), and η the negative of the elasticity of the marginal utility of consumption. The marginal utility of consumption decreases with increasing income. For instance, for a family living at the poverty level, an additional Euro income is assumed to have a higher utility than for a millionaire. The elasticity of the marginal utility of consumption is negative, and η therefore has a positive value (η is generally given as between 1 and 3 [4, 75, 80]). The component $\eta * r_{gro}$ thus accounts for the idea that societies will probably be richer in the future (assuming economic growth) and that we should attach less weight to their gains.

In an ideal economy, the social and private discount rate of the financial market should be the same [4, 75]. However, the social discount rate is usually smaller than the private discount rate. Some of the reasons are that private companies have to pay taxes for their benefits and that public projects are thought to be less risky than private projects [75].

6.4.2 Discounting and the Environment: Motivations and Objections

This section discusses a possible transfer of the discounting principle to environmental projects and LCA. As mentioned above, discounting in economics takes into consideration:

1. Changes in the price level (only nominal discount rate);
2. Pure time preference;
3. Productivity of capital and diminishing marginal utility of consumption.
4. Uncertainties.

These four arguments for discounting will be discussed one by one in the following analysis. First, we will generally discuss the justification of these motivations in the context of LCA. In a second step, we suggest possible consequences for the discount rate.

Changes in the Magnitude of Impact

In economics, the nominal discount rate includes changes in the price level. With respect to LCIA, changes in the environment might lead to a change in value of the unit measuring the damage (monetary or not). For instance, an accumulation of heavy metals in the environment might trigger a change in the damage produced by an additional unit of emission. If these changes have not been considered in an earlier phase of the impact assessment, the measuring unit is subject to in-/deflation, because it no longer corresponds to the same magnitude of damage. A 'nominal' discount rate would include such changes in the magnitude of damage. Thus, using a 'real' discount rate requires considering changes in damage magnitude in the impact assessment prior to discounting.

The magnitude of an impact might change, for instance, as a consequence of a changing background concentration of a pollutant or pollutant mix in the environment. The relation between concentrations in the environment and damage is often illustrated in concentration-effect or damage curves. These curves generally have a sigmoid shape. If the background contamination level changes, the position on the damage curve and the slope will change as well. Aggregating LCIA methods use these slopes as weighting factors because they represent the incremental increase of damage as a consequence of an incremental increase in concentration or emission. Therefore, it is important to know the present and future positions on the damage curves in order to estimate the potential impact or damage of the system under study.

The magnitude of damage might also change due to other factors as a function of site and time. Some of those characteristics, to mention a few, are a changing number or distribution of the human population (thereby in/decreasing the number of affected people), a changing sensitivity of ecosystems, an application of remediation technologies in the future, and a change in climatic conditions. Steen [87], however, argues that abatement measures should not be considered in LCA because one of the purposes of LCA is to encourage technology development (which will not be done if this is already taken into account in the damage assessment).

In theory, future changes in the magnitude of damages could be considered in the discount rate. However, it seems more reasonable to model such changes in damage in the impact assessment phase prior to discounting. One reason is that the magnitude of damage is usually not an exponential function of time (an exponential relationship is required when discounting with a constant rate, see Equation 6.2). One way to consider future changes in the magnitude of damage is with a scenario analysis (e.g. [47, 48]).

Discounting Environmental Damages because of Pure Time Preference

In discounting, an important question to be addressed is whether an identical environmental damage can be worth less in the future than today. We will assume that the damage occurs with absolute certainty and that the magnitude of damage is always the same. For instance, assume that two technologies A and B release the same amount of an emission. Technology A releases this emission today, technology B in 100 years. In both cases this emission will have the same impact on the environment. Discounting because of pure time preference would mean that technology B would be preferred to A in spite of the equivalence in impact.

It is well known that in actual decision making people often tend to prefer a present utility to a future utility. Linestone [62] claims that most people have a short planning horizon and that they *are really concerned only with their immediate neighborhood in space and time*. Many other authors and empirical evidence confirm this view (e.g., [1, 60, 73, 83]). People differentiate *according to several kinds of distance or proximity* (geographical, cultural, and temporal). Deciding about whether one cares more about people in the far future than about current people *is a little like deciding whether one cares more about people in one continent than in another, ... or about those with whom one shares history and culture more than those who do not. ...* These *preferences show up in charitable giving, in foreign aid, in immigration policy, and in military intervention* [83]. Discounting because of pure time preference is commonly accepted (though sometimes called irrational), when short-term horizons are concerned. However, in decisions affecting future generations, it is ethically questionable.

There seems to be wide agreement that the welfare of future generations should be a concern to us and that all members of all generations deserve equal treatment, including those not yet born (e.g., [4, 10, 60, 63, 64, 85]). A pure time preference with a positive value implies that future people are not moral objects with equal rights as current people [30]. If LCA wants to meet commonly accepted ethical standards and sustainability criteria, environmental impact harming future generations cannot be subjected to discounting because of pure time preference. According to this line of thinking, we propose to set the pure rate of time preference (r_{pref} in Equation 6.3) to zero. However, it has to be acknowledged that in practice, decision-makers often have and use a rate of pure time preference greater than zero.

Discounting because of Capital Productivity and Diminishing Marginal Utility of Consumption

Discounting because of capital productivity assumes a relationship between environmental damage and economic values. Therefore, we will first discuss if and how monetary equivalents can be found for environmental damage and what objections exist to the monetarization approach. Afterwards, discounting of the monetary equivalent of environmental damage will be discussed assuming that monetarization is accepted.

In economics, especially in cost-benefit analysis, it is common practice (though also controversial) to assign prices to external benefits and costs [75]. These prices are called shadow prices. In environmental economics it is often suggested that a price be assigned to natural assets via taxes [7] or by distributing a limited amount of pollution rights so that the stock market decides upon the price [89]. The objective is to correct limitations of the market, where nature is not perceived as a scarce resource and, as a consequence, not adequately considered in policy and liability decisions [12]. For instance, a coal-fired power plant can have negative health effects, such as asthma, on the population in the neighborhood. The market does not provide a signal that the plant ought to control its air emissions [65]. Assigning a monetary value to these health effects is one approach to put these external costs on par with other costs and to force the power plant to consider them (polluter-pays-principle).

However, the monetarization of human lives or natural assets is often perceived as unethical ([60] and references cited therein). MacLean [64] calls it *morally repugnant* if (human) life is seen as being exchangeable with other utilities. Other authors writing on ethics and especially on economics do not share this opinion [60, 75, 94]. Their main arguments are, first, that natural assets, in fact, do not have a price – but there would probably be a willingness to pay for them if a market existed, and, second, that decisions often require a ranking of preferences between incompatible choices. Indeed, there are many examples where an aggregation of different types of damage is common practice, e.g., liability law or extra payments for high-risk jobs [75]. Also many LCIA methods aggregate damages of different nature and therefore suggest the possibility of tradeoffs [15, 39, 40, 44, 87]. Even if a non-monetarizing method is applied, financial investments are often compared to the environmental benefits (see Environmental-Economical Analysis in Section 6.3.2 and [66]). Therefore, a relation between LCA results and monetary units appears to exist.

The total economic value is composed of the use value, the option value (value for having an option to use something in the future), and the existence value (intrinsic value) [77]. There are various techniques for estimating these external costs; among others, market prices (e.g., of crops which cannot grow any more due to an emission), hedonic prices (e.g., the decrease of house prices as a consequence of increasing noise), travel costs (to visit an area), and the contingent valuation method (surveys of the 'Willingness To Accept' (WTA) or the 'Willingness To Pay' (WTP) of people to prevent or accept a damage, e. g., a reduction in mortality risk) [12, 20, 34, 42, 65, 78, 86]. The WTP/WTA concept is often criticized because the WTP/WTA is smaller in countries where financial resources are scarce than in rich countries ('one dollar one vote'). Therefore, policy actions based on the simple aggregation of these individual measures can have serious distributional implications that need to be considered [12]. In spite of such shortcomings [64, 65, 82, 86], this technique has frequently been used in policy-making [20, 29, 42, 65, 68, 78] and it has widely been discussed in the environmental literature (more than 2000 papers [20]).

From this discussion, we have concluded that there are good reasons and techniques for a monetarization of environmental damages. However, there are groups of people that have ethical objections towards assigning a price to human life and natural assets presuming exchangeability with marketed goods. Of course, these groups of people would also reject any discounting of environmental damage. In the following, we discuss discounting under the conditions that it is accepted to assign monetary values to environmental damages and that a positive view is held towards the market economy.

One reason for discounting in economics is that capital can usually be invested so that it grows in the future. Put in other words, there is a possibility that reducing current consumption and investing the saved resources can increase future consumption. For instance, assume that there are two persons A and B that have the same amount of money available for investment. A puts the money into a saving box and waits for a year. B invests the money in one sack of corn and plants this corn on a field. After paying part of the harvest for rent, B obtains two sacks of corn one year later. It is obvious that A – though not loosing any money - foregoes benefit, because he could have doubled his capital as well instead of merely keeping it. The foregone benefit of A is called opportunity cost. Opportunity cost is one reason why discounting is a must in economics. Another argument for discounting in the context of capital productivity is the diminishing marginal utility of consumption. Economic growth (related to income) causes people to attach less weight to additional gains and, therefore, discounting is needed. Discounting on the basis of capital productivity and diminishing marginal utility of consumption can be done with a positive discount rate in the case of economic growth and with a negative rate in the case of recession (Equation 6.3).

Concerning environmental issues, positive discount rates appear to be inconsistent with the principle of intergenerational fairness, because future damages are considered less serious than current damages [77]. Nevertheless, there are arguments that might justify this type of discounting under certain circumstances. Assume, for example, that monetarization is performed on the basis of prevention or abatements costs. Imagine a landfill from which heavy metals could be extracted and recycled today or in 200 years for the same price of, e.g., 1 Mio Euros, and which will not release any major emissions in the first 200 years. Assuming a discount rate of 2.3%, it would be sufficient to invest 11`000 Euros on the capital market today in order to finance the project in 200 years. It seems logical to say that discounting might be justified in this scenario. However, this approach has a confidence problem. It is difficult to guarantee that the money will not be consumed by intermediate generations. Moreover, abatement costs do not seem to be an adequate measure in an LCA, where damages are valued considering the magnitude of impact or political targets. The two approaches do not seem to match well. On the contrary, the willingness to pay (WTP) and the willingness to accept (WTA) seem to be applicable for LCA purposes. This approach was used for the damage weighting by the EPS method [87] and in the ExternE project [29]. The valuation of some other LCIA methods [15, 40, 44] is based on panel methods or political targets. It seems reasonable to assume that a damage that receives a high weight by a panel (or politicians representing the people) would also receive a high 'price' if the panel would be asked to assign monetary values.

Applying the WTA approach implies that damages can be compensated financially. This is usually assumed in cost-benefit analysis. In its original form, the efficiency criterion of cost-benefit analysis was to *make no change unless the change hurts no one and helps at least someone* [74]. However, this criterion is usually interpreted as *make changes that help some, even at the expense of others, as long as the gainers can fully compensate the loosers* (Kaldor Hicks criterion) [74]. Since the money for compensation of damages occurring in the long-term could be invested on the capital market, discounting could be justified in these cases. The problem with this approach is twofold. First, future generations cannot be asked whether they agree to such compensations; the decision has to be made by current generations who might be prone to biased decisions. For irreversible damages, no option is left to future generations to choose between the natural asset and the compensation. However, such tradeoffs are already practiced in aggregating LCA where irreversible and reversible damages are measured on the same scale (which assumes that tradeoffs are possible). Second, it is not certain that the compensation payment will be passed on by intermediate generations [61]. The compensations would not need to be saved in a separate fund – it could also be 'inherited' in another form. For instance, many economists think that the world economy is growing and that the future will be better off than the present [74]. Therefore, future generations could profit from (and be compensated by) greater welfare. However, this expectation presumes sustainable economic growth, since *environmental destruction holds all the potential for eroding the capital stock of future generations* [76].

These ideas can be summarized as follows: discounting can only be justified if it is believed that future generations are adequately compensated and that they would be satisfied with such compensation.

There are two important consequences for the discount rate: First, according to Rabl [81], the long-term discount rate is unlikely to be higher than the economic growth. Otherwise, any investment would exceed the whole gross national product after a certain time period, which is impossible. Instead, the discount rate would drop as a consequence of an increasing supply of money on the financial markets. However, reinvesting the money constantly could considerably affect the growth of the gross national product.

Second, if income per capita increases, the WTP or WTA is likely to do so as well. Since both the discount rate and the WTP/WTA are related to economic growth, the resulting overall discount rate might therefore be close or equal to zero [10, 80, 81]. It might even lead to the situation of a negative discount rate: Assume that the WTA for an environmental impact increases over time with the social discount rate (the social discount rate reflects society's relative preferences for consumption in the present and future [43]). For illustrative purposes, we set the elasticity of the marginal utility of consumption to -2 ($\eta = 2$, Equation 6.3) and the annual rate of economic growth r_{gro} = annual increase of income = rate of return on investment = 2.3%. From here, the WTA of 100 Euro in year 0 would grow in value to $100 * (1 + r)^t = 100 * (1 + \eta * r_{gro})^t = 100 * (1.046)^t$ in year t. Let us assume that the compensation money can be invested at 2.3% (rate of return on investment), so that after t years the available amount would be $100 * (1 + r_{gro})^t = 100 * (1.023)^t$. This would lead to a negative overall discount rate of -2.3%. Price [80] showed that using common utility functions, the overall discount rate might even go towards negative infinity when $\eta > 1$. This scenario is plausible considering that constant economic growth over a long period of time would lead to a very high income – and perhaps people would in fact demand an extremely high compensation for damages to natural assets or an increase in mortality risk, because the utility of further money would be limited then, whereas natural assets remain scarce. Above a certain level of income, a situation might come up in which no monetary compensation is accepted for any impact to natural assets. These arguments lead to the conclusion that discounting because of the capital productivity could lead to a discount rate close to 0% even if monetarization of damages and discounting of the monetary equivalent is accepted. However, both positive and negative discount rates seem to be possible.

Discounting because of Uncertainties

Risk or uncertainty about future developments might have an influence on the discount rate. There is no doubt that uncertainties exist, especially if time horizons of thousands of years are considered. According to Pearce & Turner [77], the following uncertainties are of relevance:

1. Uncertainty about the existence, magnitude, or quality of damage.
2. Uncertainty about the presence of an individual or society in the future (in the case of long-term emissions)
3. Uncertainty about the preferences of the individual (or society)

There are many ethical objections towards discounting with a discount rate greater than zero because of the uncertainty of the existence of costs. First, uncertainties can be used to justify both positive and negative discount rates [60, 85]. For instance, the population number might grow, the population might get more sensitive to environmental pollution, or environmental impacts unforeseen today may appear in the future thereby increasing the number of potentially affected people. A new remediation technology might be developed, but it might also generate new environmental problems. And even if such a technology could be developed, this does not ethically justify the imposition of risks on the future [85]. *Just because A is better able to deal with B's problems than B is, does not mean that B has the right to impose his problems on A* [85]. Moreover, one of the goals of many LCAs is to prevent environmental harm by identifying key environmental issues and thus stimulate technological development. *If we examined scenarios where future technology would solve all environmental problems we would not get the incitements to develop such technology* [87]. Second, uncertainty is not an ethical justification for rejecting responsibility for future generations [60]. If there is a probability for humankind to exist in the future, and this probability is large, then current generations automatically have the responsibility not to harm future generations [60]. Furthermore, discounting because of the uncertainty of the existence of humankind would encourage a casual attitude to the future, which increases the likelihood of damage to future generations ("self-fulfilling prophecy") [80]. Third, even if the state of the future society differs from today's, it is very likely that fatalities, illnesses, and injuries will still be perceived as damages [60, 80]. Ethicists argue that the preferences of current society should serve as proxy of future generations [85].

From this discussion we conclude that there are severe objections to discounting because of uncertainty about the presence of a society and about the preferences of future societies. Since uncertainty concerning the existence of cost probably does not grow exponentially as a function of time and is difficult to estimate quantitatively [29], we propose considering this latter type of uncertainty with scenario analysis when modeling the magnitude of damage rather than in the 'real' discount rate.

6.4.3 Application of Discounting to LCA of Waste Treatment Technologies

Discounting can only be applied in LCA when information about the temporal occurrence of impacts is available. In Figure 6.11 we show an example, where the predicted impacts of Cd^{2+} emissions from a slag landfill to the groundwater [47] have been discounted with different discount rates.

It can be deduced from Fig 11 that positive discount rates would favor landfills with long-term emissions, as impacts are weighted less the further they occur in the future. For instance, at a discount rate of 1%, the impact from Cd emissions to the groundwater from slag landfills would be roughly 1% of the impact without discounting. All impacts after 1,000 years practically no longer count (less than 0.005% of their value without discounting). At a discount rate of 0.01%, the time horizon considered would increase to 100,000 years. On the contrary, negative discount rate would weigh any type of future impact very high. These results would be more extreme for emissions such as Cu^{2+} that would be emitted for even longer time periods.

The ranking between the treatment technologies depends on the discount rate. Slightly positive discount rates would favor the mechanical-biological treatment and waste incineration. The application of the private discount rate between 5% and 18% would considerably lower the impact score concerning all technologies that include landfills. For instance, the impact of Cd emissions from slag landfills would drop to 0.2%, compared to the impact without discounting (Fig. 6.11); all impacts after 10 to 20 years would virtually be negligible. As a consequence, sanitary landfills could score better than, e.g., grate incineration, because they release most emissions with some amount of temporal retardation, while incineration technologies emit pollutants to the air (e.g. CO_2) immediately. New technologies such as PECK would not score well because they prevent little weighted long-term emissions at the cost of a higher current energy use. By contrast, the application of a negative discount rate would lead to an extremely high impact score of

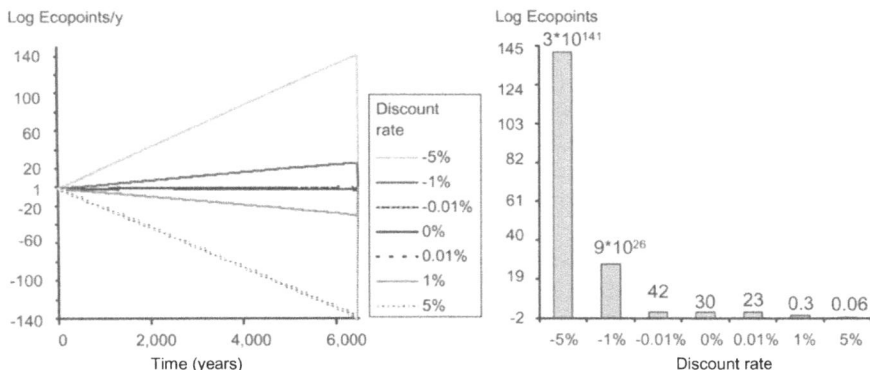

Fig. 6.11. Discounted impact of Cd emissions from slag landfills to the groundwater (Method of Ecological Scarcity, constant background pollution). The left graph displays environmental impact as a function of time, while the right graph presents the results aggregated over indefinite time.

all treatment methods with long-term emissions, so that landfills, mechanical-biological treatment, and grate incineration would be considered very environmentally harmful. However, technologies used for vitrifying the slag and, thereby, prolonging the emission period, would even score worse. Recovering the heavy metals would be the only acceptable way of waste treatment in this case.

While this chapter can provide some assistance for deciding upon the discount rate, the ultimate choice concerning the discount rate remains a value-laden question [30]. However, it can be stated that the long-term discount rate should be close or equal to zero if LCA wants to meet common ethical standards. Therefore, the conclusions drawn in Section 6.3 for the infinite time horizon are important and should be considered, while the short-term analysis is questionable.

6.5 Conclusions

While waste prevention and recycling have been shown to be the most environmental benign choice for many materials [33], this chapter only analyzed MSW treatment technologies. Several conclusions can be drawn:

- LCA is an adequate tool in the analysis of waste treatment processes, because a large number of resources and emissions are considered during the whole life cycle.
- The evaluation of treatment technologies largely depends on the weighting of long-term impacts in comparison to short-term impacts. Our discussion shows that the long-term discount rate should be close or equal to zero if LCA wants to meet common ethical standards.

- Long-term releases of heavy metals to water cause the largest share of impact if an indefinite time horizon is considered, according to the impact assessment methods used. Promising strategies for preventing these emissions are reducing the content of heavy metals in products (Section 3.1), and recovering heavy metals (material recycling or PECK technology). Vitrifying and thereby stabilizing the heavy metal containing residues, or storing them in a 'permanent storage' without contact to water can also reduce or delay the impacts of heavy metals.
- Using the energy content of the waste efficiently is very important when treating waste. This can either be done by material recycling or by thermal recovery of feedstock energy.
- The quality of recovered metal products could be improved. Trace pollutants such as Cu in Fe-scrap cause significant environmental impacts. These pollutants should be separated and possibly recycled.
- Sanitary landfill is the worst treatment option for mixed waste among the technologies considered. This is due to emissions to the air such as NMVOC and CH_4, and a low energy efficiency.
- The mechanical-biological treatment performs better than the sanitary landfill, but its impact score suffers from low energy efficiency in comparison to the thermal technologies.
- The airborne emissions of modern MSWI plants per unit of produced energy are comparable to an average energy plant in Europe. Heavy metal leaching to the water appears to be the most relevant environmental issue.
- PECK performs most favorably compared to all other treatment options if a long-term time perspective is adopted and if the practical performance meets the set targets. The reasons for this are the recovery of heavy metals and a reasonable energy efficiency.
- According to our ecological-economical analysis, PECK performs best, followed by the mechanical-biological treatment. However, it must be noted that all data concerning PECK are uncertain.

References

1. Ahearne JF (2000) Intergenerational Issues Regarding Nuclear Power, Nuclear Waste, and Nuclear Weapons. Risk Analysis 20:763-770
2. Anonymous (1996) Galicia: Threat of Disaster Following Garbage Avalanche. The Environment Digest (news service) 10
3. Anonymous (1997-99) ISO 14040-43. European Committee for Standardization (CEN), Brussels, Belgium
4. Azar C, Sterner T (1996) Discounting and Distributional Considerations in the Context of Global Warming. Ecological Economics 19:169-184
5. Barin I, Igelbüscher A, Zenz F-R (1996) Thermodynamische Analyse der Verfahren zur thermischen Müllentsorgung. R. Berghoff, Essen

6. Baumann H (1998) Life Cycle Assessment and Decision Making - Theories and Practises. PhD thesis, Chamlers University of Technology
7. Baumol WJ, Oates WE (1971) The Use of Standards and Prices of Protection of the Environment. Swedish Journal of Economics 73:42-54
8. Beck A, Bosshart S (1995) Umweltanalyseinstrumente im Vergleich. Diploma thesis. Department of Environmental Sciences, Zurich
9. Belevi H (1998) Environmental Engineering of Municipal Solid Waste Incineration. Verlag der Fachvereine, Dübendorf
10. Birnbacher D (1989) Intergenerationelle Verantwortung oder: Dürfen wir die Zukunft der Menschheit diskontieren? In: Kümmel R, Klawitter J (eds) Umweltschutz und Marktwirtschaft aus der Sicht unterschiedlicher Disziplinen, Würzburg, pp 101-115
11. Björklund A (1998) Environmental System Analysis of Waste Management. Licentiate thesis, KTH
12. Bockstael NE, Freeman AM, Kopp RJ, Portney PR, Smith VK (2000) On Measuring Economic Value. Environmental Science and Technology 34:1384-1389
13. Brunner PH, Zobrist J (1983) Die Müllverbrennung als Quelle von Metallen in der Umwelt. Müll und Abfall 9
14. BUWAL SAftEFaLS (1995) Zusammensetzung der Siedlungsabfälle der Schweiz 1992/93. Schriftenreihe Umwelt. Bundesamt für Umwelt, Wald und Landschaft, Bern
15. BUWAL SAftEFaLS (1998) Bewertung in Ökobilanzen mit der Methode der ökologischen Knappheit. Ökofaktoren 1997. Swiss Agency for the Environment, Forests and Landscape, Bern
16. BUWAL SAftEFaLS (1998) Life Cycle Inventories for Packagings. Schriftenreihe Umwelt. Swiss Agency for the Environment, Forests and Landscape, Bern
17. BUWAL SAftEFaLS (1999) Abfallstatistik 1998. Umwelt-Materialien. Swiss Agency for the Environment, Forests and Landscape, Bern
18. BUWAL SAftEFaLS (2001) Sondermülldeponie Kölliken: Sanierung in Sicht. Swiss Agency for the Environment, Forests and Landscape, Bern
19. BZF (2000) Umwelterklärung 2000 - Bilanzjahr 1999/2000. B.U.S. Zinkrecycling Freiberg GMbH, Freiberg, Germany
20. Carson RT (2000) Contingent Valuation: A User's Guide. Environmental Science and Technology 34:1413-1418
21. Centre of Environmental Science (CML) (2001) Problem Oriented Approach, http://www.leidenuniv.nl/cml/lca2/index.html
22. Clift R, Doig A, Finnveden G (2000) The Application of Life Cycle Assessment to Integrated Solid Waste Management. Trans IchemE 78:279-287
23. CORINAIR (1990) Emission Inventory Guidebook: Group 3 - Combustion in Manufacturing Industry. CORINAIR Coordinated Inventory of Air Emissions, European Environment Agency, European Union
24. CORINAIR (1996) Emission Inventory Guidebook: Group 4 - Production Processes. CORINAIR Coordinated Inventory of Air Emissions, European Environment Agency, European Union
25. De Marchi B, Funtowicz S, Ravet R (1996) Seveso: A paradoxical classic disaster. In: Mitchell JK (ed) The Long Road to Recovery: Community Responses to Industrial Disaster. United Nations University Press, Tokyo - New York - Paris
26. Derwent RG, Jenkin ME, Saunders SM, Pilling MJ (1998) Photochemical Ozone Creation Potentials for Organic Compounds in Northwest Europe calculated with a Master Chemical Mechanism. Atmosperic Environment 32:2429-2441

27. Doka G (2002) Life Cycle Assessment of municipal solid waste incineration using the PECK technology. Report, available on http://www.unite.ch/doka/PECKdoka.pdf, printed version can be ordered from the author wabo@doka.ch. Chemical Engineering Department, ETH Zurich, Zurich

28. Ekvall T, Finnveden G (2001) Allocation in ISO 14041 - A Critical Review. Journal of Cleaner Production 9:197-208

29. European Commission (1995) ExternE: Externalities of Energy. Office for Official Publications of the European Communities, Luxembourg

30. Finnveden G (1997) Valuation Methods Within LCA - Where are the Values? International Journal of LCA 2:163-169

31. Finnveden G (1999) Methodological Aspects of Life Cycle Assessment of Integrated Solid Waste Management Systems. Resources, Conservation and Recycling 26:173-187

32. Finnveden G (2000) On the Limitations of Life Cycle Assessment and Environmental Systems Analysis Tools in General. International Journal of Life Cycle Assessment 5:229-238

33. Finnveden G, Johansson J, Lind P, Moberg A (2000) Life Cycle Assessment of Energy from Solid Waste. Stockholm University, fms 137, FOA-B--00-00622-222--SE, www.fms.ecology.su.se, Stockholm

34. Fischhoff B (2000) Informed Consent for Eliciting Environmental Values. Environmental Science and Technology 34:1439-1444

35. Fliedner A (1999) Organic Waste Treatment in Biocells. A Computer Based Modelling Approach in the Context of Environmental Systems Analysis. Masters thesis, Royal Institute of Technology, Stockholm

36. Frischknecht R (1997) Goal and Scope Definition and Inventory Analysis. In: Udo de Haes HA, Wrisberg N (eds) Life Cycle Assessment: State of the Art and Research Needs. Eco-informa Press, Bayreuth

37. Frischknecht R, Bollens U, Bosshart S, Ciot M, Ciseri L, Doka G, Dones R, Gantner U, Hirschier R, Martin A (1996) Ökoinventare von Energiesystemen. Grundlage für den Vergleich von Energiesystemen und den Einbezug von Energiesystemen in Ökobilanzen für die Schweiz. 3. Auflage. ETH Zurich and Paul Scherrer Institut Villigen / Würenlingen, Switzerland

38. Fugleberg S (1999) Finnish Expert Report on Best Available Techniques in Zinc Production. Finnish Environment Institute, Helsinki

39. Goedkoop M (1995) The Eco-indicator 95 - Weighting Method for Environmental Effects that damage Ecosystems or Human Health on a European Scale. Novem, Amersfoort

40. Goedkoop M, Spriensma R, Müller-Wenk R, Hofstetter P, Köllner T, Mettier T, Braunschweig A, Frischknecht R, van de Meent D, Rikken M, Breure T, Heijungs R, Lindeijer E, Sas H (1999) The Eco-indicator 99: A Damage Oriented Method for Life Cycle Impact Assessment. Methodology Report. Pré Consultants, Amersfoort

41. Guinée J.B. (ed.) (2001) Life Cycle Assessment an operational guide to the ISO standard, Kluwer Academic Publishers, Dordrecht/London/Boston

42. Hammitt JK (2000) Valuing Mortality Risk: Theory and Practice. Environmental Science and Technology 34:1396-1400

43. Hammitt JK, Harvey CM (2000) Equity, Efficiency, Uncertainty, and the Mitigation of Global Climate Change. Risk Analysis 20:851 - 860

44. Hauschild M, Wenzel H (1997) Environmental Assessment of Products; Scientific Background. Chapman & Hall, London
45. Heijungs R, Huppes G (1999) Inventory Modelling in the Dutch LCA Methodology Project. In: Guinée J (ed) Danish-Dutch Workshop on LCA Methods, Leiden (NL)
46. Heijungs R, Guinee J, Huppes G, Lankreijer RM, Udo de Haes HA, Wegener-Sleeswijk A (1992) Environmental Life Cycle Assessment of Products. CML, Leiden
47. Hellweg S (2000) Time- and Site-Dependent Life-Cycle Assessment of Thermal Waste Treatment Processes. PhD thesis, www.dissertation.de, ETH Zurich
48. Hellweg S, Hofstetter TB, Hungerbühler K Time-Dependent Life-Cycle Assessment of Emissions from Slag Landfills with the Help of Scenario Analysis. Submitted to Journal of Cleaner Production
49. Hellweg S, Hofstetter TB, Hungerbühler K (2002) Modeling Waste Incineration for Life Cycle Inventory Analysis in Switzerland. Environmental Modeling and Assessment 6:219-235
50. Hellweg S, Fischer U, Hofstetter TB, Hungerbühler K Site-Dependent Fate Assessment: Transport of Heavy Metals in Soil. Submitted to Journal of Cleaner Production
51. Höffken E, Hammer R, Schicks H, Ullrich W (1988) Chemische Zusammensetzung und Verwendung von aufbereitetem Schrott. Stahl und Eisen 108:801-806
52. Hofstetter P (1998) Perspectives in Life Cycle Impact Assessment: A Structured Approach to combine Models of the Technosphere, Ecosphere and Valuesphere. Kluwers Academic Publishers
53. Houghton JT, Meira Filho LG, Callander BA, Harris N, Kattenberg A, Maskell K (eds) (1995) Climate change 1995. The Science of Climate Change; Contribution of WGI to the second assessment report of the intergovernmental panel on climate change. Cambridge University Press, Cambridge
54. Houghton JT, Meira Filho LG, Bruce J, Lee H, Callander BA, Haites E, Harris N, Maskell K (eds) (1994) Climate change 1994. Radiative Forcing of Climate Change: An Evaluation of the IPCC IS92 Emissions Scenarios. Cambridge University Press, Cambridge
55. Huijbregts MAJ (1999) Priority Assessment of Toxic Substances in the Frame of LCA. University of Amsterdam (NL), Amsterdam
56. Huijbregts MAJ (2000) Priority Assessment of Toxic Substances in the Frame of LCA. Time Horizon Dependency of Toxic Potentials calculated with the Multi-media Fate, Exposure and Effects Model USES-LCA. Institute for Biodiversity and Ecosystem Dynamics, University of Amsterdam, Amsterdam (NL)
57. IPCC (2000) Best Available Techniques Reference Document on the Production of Iron and Steel. Integrated Pollution Prevention and Control (IPPC), European Commission Directorate-General Joint Research Centre
58. IPCC (2000) Reference Document on Best Available Techniques in the Non Ferrous Metals Industries. Integrated Pollution Prevention and Control (IPPC), European Commission Directorate-General Joint Research Centre
59. Jenkin ME, Hayman GD (1999) Photochemical Ozone Creation Potentials for Oxygenated Volatile Organic Compounds: Sensitivity to Variations in Kinetic and Mechanistic Parameters. Atmospheric Environment 33:1775-1293
60. Leist A (1996) Ökologische Ethik II: Gerechtigkeit, Ökonomie, Politik. In: Nida-Rümelin J (ed) Angewandte Ethik. Die Bereichsethiken und ihre theoretische Fundierung. Kröner, Stuttgart, pp 386-457

61. Lind RC (1990) Reassessing the Government's Discount Rate Policy in Light of New Theory and Data in a World Economy with a High Degree of Capital Mobility. Journal of Environmental Economics and Management 18:S8-S28
62. Linestone HA (1973) On Discounting the Future. Technological Forecasting and Social Change 4:335-338
63. Livingstone I, Tribe M (1995) Projects with Long Time Horizons: Their Economic Appraisal and the Discount Rate. Project Appraisal 10:66-76
64. MacLean DE (1990) Comparing Values in Environmental Policies: Moral Issues and Moral Arguments. In: Hammond PB, Coppock R (eds) Valuing Health Risks, Costs, and Benefits for Environmental Decision Making. National Academy Press, Washington, pp 83-106
65. Matthews HS, Lave LB (2000) Applications of Environmental Valuation for Determining Externality Costs. Environmental Science and Technology 34:1384-1389
66. Meier M (1997) Eco-Efficiency Evaluation of Waste Gas Purification Systems in the Chemical Industry. PhD thesis, ETH Zurich
67. Morf L, Ritter E, Brunner PH (1997) Güter- und Stoffbilanz der MVA Wels. Technische Universität Wien, Wien
68. Mourato S, Ozdemiroglu E, Foster V (2000) Evaluating Health and Environmental Impacts of Pesticide Use: Implications for the Design of Ecolabels and Pesticide Taxes. Environmental Science and Technology 34:1456-1461
69. Nielsen PH, Hauschild M (1998) Product Specific Emissions from Municipal Solid Waste Landfills; Part I: Landfill Model. International Journal of LCA 3:158-168
70. Norgate TE, Rankin WJ (2000) Life Cycle Assessment of Copper and Nickel Production. In: Minerals C (ed) International Conference on Minerals Processing and Extractive Metallurgy. Minprex 2000, Clayton, Victoria, Australia
71. Obernosterer RW (1994) Flüchtige Hologenkohlenwasserstoffe FCKW, CKW, Halone; Stoffflussanalyse Oesterreich, Technische Universität
72. Obernosterer RW, Brunner PH (1997) Construction Wastes as the Main Future Source for CFC-Emissions. In: Bringezu S, M. F-K, Kleijn R, Palm V (eds) Regional and National Material Flow Accounting: From Paradigm to Practice of Sustainability, Leiden
73. Okrent D (1999) On Intergenerational Equity and Its Clash with Intragenerational Equity and on the Need for Policies to guide the Regulation of Disposal of Wastes and Other Activities Posing Very Long-Term Risks. Risk Analysis 19:877-900
74. Page T (1977) Conservation and Economic Efficiency. John Hopkins University Press, Baltimore
75. Pearce DW (1983) Cost-Benefit Analyses, 2 edn. The Macmillan Press LTD, London
76. Pearce DW (1983) Ethics, Economics and the Environment. International Journal of Environmental Studies 21:1-3
77. Pearce DW, Turner R (1990) Economics of Natural Resources and the Environment. Harvester Wheatsheaf, London
78. Pearce DW, Seccombe-Hett T (2000) Economic Valuation and Environmental Decision-Making in Europe. Environmental Science and Technology 34:1419-1425
79. Pré Consultants (1999) SimaPro. Amersfoort, The Netherlands
80. Price C (2000) Discounting Compensation for Injuries. Risk Analysis 20:839 - 849
81. Rabl A (1996) Discounting of Long-Term Costs: What would Future Generations prefer us to do? Ecological Economics 17:137-145
82. Sagoff M (2000) Environmental Economics and the Conflation of Value and Benefit. Environmental Science and Technology 34:1426-1431

83. Schelling TC (2000) Intergenerational and International Discounting. Risk Analysis 20:833 - 837

84. Schmidt I, Kicherer A, Zwahr H, Ludwigshafen A (2001) Ökoeffizienz-Analyse Restmüllentsorgung, Mechanisch-biologische Abfallbehandlung, Müllverbrennung sowie Deponie. BASF, Ludwigshafen (D)

85. Shrader-Frechette (2000) Duties to Future Generations, Proxy Consent, Intra- and Intergenerational Equity: The Case of Nuclear Waste. Risk Analysis 20:771-778

86. Spash CL (2000) Multiple Value Expression in Contingent Valuation: Economics and Ethics. Environmental Science and Technology 34:1433-1437

87. Steen B (1999) A Systematic Approach to Environmental Priority Strategies in Product Development (EPS). Version 2000 - General System Characteristics. Centre for Environmental Assessment of Products and Material System; Chalmers University of Technology, Stockholm, Sweden

88. Sundqvist J-O, Finnveden G, Stripple H, Albertsson A-C, Karlsson S, Berendson J, Höglund L-O (1997) Life Cycle Assessment and Solid Waste - Stage 2. AFR. Swedish Environmental Protection Agency, Stockholm

89. Tietenberg T (1990) Economic Instruments for Environmental Regulation. Oxford Review of Economic Policy 6

90. Tillman A-M (1999) Significance of Decision-Making for LCA Methodology. Environmental Impact Assessment Review 20:113-123

91. Traber D (2001) Petrology, Geochemistry and Leaching Behaviour of Glassy Residues of Municipal Solid Waste Incineration and their Use as Secondary Raw Material. PhD thesis, Bern University

92. Udo de Haes HA, Jolliet O, Finnveden G, Hauschild M, Krewitt W, Müller-Wenk R (1999) Best Available Practice Regarding Impact Categories and Category Indicators in Life Cycle Impact Assessment; Background Document for the Second Working Group on Life Cycle Impact Assessment of SETAC-Europe. Part 1 and 2. International Journal of Life Cycle Assessment 4:66-74 and 167-174

93. Udo de Haes HA, Jolliet O, Finnveden G, Goedkoop M, Hauschild M, Hertwich EG, Hofstetter P, Klöpffer W, Krewitt W, Lindeijer EW, Müller-Wenk R, Olsen SI, Pennington DW, Potting J, Steen B (2002 (forthcoming)) Towards Best Practice in Life Cycle Impact Assessment, Pensacola, Florida

94. van Beukering P, Ooosterhuis F, Spaninks F (1998) Economic Valuation in Life Cycle Assessment. Applied to Recycling and Solid Waste Management. Working Paper. Institute for Environmental Studies. Vrije Universität, Amsterdam

95. Wallmann R (1999) Ökologische Bewertung der mechanisch-biologischen Restabfallbehandlung und der Müllverbrennung auf Basis von Energie- und Schadgasbilanzen. Arbeitskreis für die Nutzbarmachung von Siedlungsabfaellen e.V. (ANS), Stuttgart

96. Weidema B (1998) Application Typologies for Life Cycle Assessment. Int. J. LCA 3:237-240

97. Wenger C, Jordi B (2001) Eine giftige Erbschaft als Lehrstück. Umwelt (BUWAL, SAEFL), 2, ISSN 1424-7186, Bern

98. WMO (World Meteorological Organisation) (1992) Scientific Assessment of Ozone Depletion: 1991. Global Ozone Research and Monitoring Project. Geneva

99. WMO (World Meteorological Organisation) (1995) Scientific Assessment of Ozone Depletion: 1994. Global Ozone Research and Monitoring Project. Geneva

100. WMO (World Meteorological Organisation) (1999) Scientific Assessment of Ozone Depletion: 1998. Global Ozone Research and Monitoring Project. Geneva

101. Wrisberg N, Udo de Haes HA, Triebswetter U, Eder P, Clift R (2002) Analytical Tools for Environmental Design and Management in a Systems Perspective. Kluwer Academic Publishers, Dordrecht/London/Boston

102. Zeltner C (1998) Petrologische Evaluation der thermischen Behandlung von Siedlungsabfällen über Schmelzprozesse. PhD thesis, ETH Zurich

103. Zeltner C (2001) Personal communication

104. Zimmermann P, Doka G, Huber F, Labhardt A, Menard M (1996) Ökoinventare von Entsorgungsprozessen, Grundlagen zur Integration der Entsorgung in Ökobilanzen. ESU-Reihe. Institut für Energietechnik, ETH Zurich, Zürich

105. Hellweg S, Doka G, Finnveden G, Hungerbühler K. (2002) Electronic Appendix to Chapter 6, http://msw.web.psi.ch

7 Assessing and Improving Social Compatibility

Vicente Carabias, Herbert Winistörfer, Walter Joos, with contributions by
Christopher Rootes, Ortwin Renn, Elke Schneider, Hans Kastenholz, and
Richard V. Anthony

"People start pollution, people can stop it!" This slogan, coined for an anti-
pollution campaign in the USA, is convincing at first sight. The preceding
chapters of this book have shown that there is a broad range of options for dealing
with solid wastes in an environmentally and economically acceptable way and that
we indeed have the tools to limit waste related pollution to an acceptable level.
There is, however, a broad gap between knowledge about environmental issues,
on the one hand, and taking action to change the situation, on the other.
Overcoming this gap is one of the central challenges in bringing environmental
issues from the domain of analysis and scientific understanding to the reality of
every-day life.

We all contribute to the ever-increasing masses of solid waste that have to be
dealt with. The average consumers do, however, not perceive the result of their
consumption as a problem, as long as it is regularly removed from their sight. By
definition, waste is material that people want to get rid of, so it is perfectly
understandable that they will not want to accept it when it comes back *en masse* to
their neighborhood in the form of incinerators, landfills or large recycling plants,
with real or perceived risks that are imposed on them by others. The resulting
NIMBY ("Not in my backyard!") attitude is very common concerning
installations for the treatment or deposition of waste. This is particularly
prominent if incineration is concerned. Although in the long-term incineration is

objectively a much better option than landfilling (Chapter 6), it is very often perceived as too much of a risk for local pollution and therefore feared by people in the neighborhood. It seems that people prefer to live with the long-term environmental burden of a landfill rather than with the risk of local emissions from an incinerator (which have become near zero with the air pollution control systems in place in modern installations). The infamous NIMBY attitude has a temporal dimension: "Not In My Life-Time!"

Planners, administrators, developers of environmental technology, environmental authorities, and operators of waste treatment plants have a range of tools for predicting trends in quantity and quality of solid waste for treatment, recycling or disposal and a choice of well proven solutions for handling the quantities and qualities in an environmentally acceptable way. Planning procedures are time consuming and require specialist know-how. The case studies from Switzerland, England, Germany and the USA in the following chapter exemplify that, conspicuously often, the planners forget to involve the affected public in the decision-making process at an early stage, which can result in greater or lesser articulate local opposition and campaigning against the choices favored by the specialists. It seems that the social acceptance of waste installations is the least predictable, though very often *the* decisive issue in establishing locations for waste installations. Achieving acceptance of large-scale waste management operations requires an open dialogue between all stakeholders. It is of central importance that public participation begins at the earliest possible time. The examples given in this chapter clearly show that the dialogue will not work, or even be contra-productive, if it starts after important decisions have been made.

One step in the direction of a better handling of the social aspects in a waste management project might be the development of a suitable assessment tool, such as the Social Compatibility Analysis (SCA) described below in section 7.2. SCA allows for an advanced identification of those aspects of a given technology or project that may give rise to social and/or political controversy in implementation.

Waste management is about the magic of making waste disappear, be it by transforming it into something useful, such as recycled products or energy, or by burying it. While it is clear that the waste management specialists are the ones who are best informed about the economic and ecological advantages and disadvantages of different options, they should bear in mind that decisions concerning the realization of a project should not be made on technical grounds alone and that decisions have to be supported by those who are potentially affected. Ultimately, what is called for is a democratic process of decision-making concerning the available options, which are covered in the previous chapters of this book.

In the future, acceptance for new or improved technologies should be optimized by starting the social acceptance dialogue at the very beginning of technical evolution, i.e. in the research and development phase, when new concepts are first conceived.

7.1 An Introduction to "Social Compatibility"

Sustainability encompasses an ecological, a social and an economic dimension. New strategies and new ways of solving problems in all areas of society not only influence the ecological and economic environment, but also the social environment. If the social consequences of such strategies and solutions are neglected, their successful implementation may be at risk due to a lack of public or social acceptance.

Skorupinski & Ott [49] suppose that factual *acceptance* of technological risks may be due to various (subjective) reasons. The range of possible reasons reaches from pure ignorance or fear of jeopardizing employment to free and informed consent. To talk of *acceptability* on the other hand presupposes an evaluation with objective criteria (Section 7.2), which renders possible a differentiation between acceptable and unacceptable outcomes. This is also the case with the *social compatibility* of planning and measures, especially of any type of infrastructure project which is not simply a question of acceptance but, moreover, of acceptability. Analogous to the natural environment, there is also a social environment based on concrete social realities in which planning, measures and projects all play a role [8]. These social realities are illustrated with the criteria for social compatibility in Table 7.1 of Section 7.2.

Staub-Bernasconi [51] perceives social compatibility as an essential building block for a *multi-dimensional conception of environmental compatibility*. She maintains that social compatibility asks for the "intended and unintended, positive and negative influences of social systems (e.g. economics, politics) on fulfilling people's needs and on their ability to live together. In this regard the emphasis is placed on *psychological needs* (elimination of fears, psychological security relative to violence, traffic etc), *social needs* (social integration, social recognition and fair interaction and exchange) and *cultural needs* (questions of purpose and meaning, issues of cultural and sub-cultural identity)." In the 70's, there was already an intensive movement towards greater *social awareness. Social reports, social accounting and a systematic structure of social indicators* resulted from this movement. Research focused on the methodology of assessing the social consequences of technology, i.e. "technology assessment". It was deemed necessary to research the influence that technological changes exerted not only on economic and social developments, but also particularly on people and their social needs [8]. Social compatibility went hand in hand with three other criteria: *economic, environmental and international compatibility.*

Promoters of the *pluralistic value approach* considered social compatibility a necessity for taking into account the prevalent social values within technological developments. One way to do this is to guarantee the ability to overcome potential conflicts via the democratic decision-making processes. Advocates of the *future oriented acceptability approach* [32, 51] defined social compatibility as the social implication of a certain technological choice in accord with social order and development (i.e. constitutional state, separation of powers, democracy, multi-party system, a federal system, local democracy, constitutional rights, etc.). This approach is also called *constitutional compatibility* or *normative social compatibility*. "A technology, which is compatible with the constitution and with the environment, is therefore also socially compatible" [49]. According to Eichener [19] there are at least four other concepts of social compatibility in addition to the normative one. They are as follows:

- empirical social compatibility
- distributive social compatibility
- procedural social compatibility
- consensus oriented social compatibility.

Empirical social compatibility seeks to determine the way the affected social groups/stakeholders understand social compatibility by means of empirical social research. When such research is carried out through the use of standardized representative surveys targeted at those groups immediately affected, certain difficulties come to light. In addition to the problem of method (especially in the operational field), there is the problem of an inadequately informed target group. This means that, especially in the case of prospective evaluations, future scenarios of project development become a controversial issue. Nevertheless, the problem of diverging interests can be largely solved by accepting the majority decision, whilst also seeking to address the problem of protecting the interests of minority groups.

According to the theory of *distributive social compatibility*, a technology is socially compatible when its application and costs are "fairly" distributed, and where the social costs of the technology are, firstly, minimized, and secondly, born by those who actually profit the most from the technology. In this way, for example, the need for preventative measures (e.g. qualification measures) or compensatory measures (e.g. employment protection measures) can be deduced, and should be proportionate to the productivity gained by employing new technology.

According to the concept of *procedural social compatibility*, a technology is socially compatible when those affected (the stakeholders) are adequately involved in its implementation. This comprises an adequate participation qualification as well as opportunities to participate in the planning, decision-making and implementation processes. This might include, if necessary, changing internal, operational decision-making structures. As with the concept of empirical social compatibility, the burden of definition is assigned to those directly affected, who, despite extensive consulting, further education and qualification, are confronted with the problem of limited information concerning the direct and indirect consequences of the technology.

Finally, according to the term *consensus oriented social compatibility*, the more socially compatible a technology is, the wider the consensus for it will be and, thus, the lower the opposition to it. However, this concept might be opposed by maintaining that it unleashes an innovation-unfriendly "consensus totalisation". Furthermore, the consensus principle would always lead to a minimal solution that is based on the smallest common denominator.

Skorupinski [48] proposes, on the other hand, to understand social compatibility as democratic compatibility: Therefore a development is socially compatible, if all the affected stakeholder in a common discourse do principally agree with a taken decision.

With Eichener [19] we can state that a socio-technological system is deemed more socially compatible

- the more normative criteria it corresponds to (e.g. scientific),
- the more it reflects the subjective interests of the affected group and/or is accepted by its representatives,
- the more it reflects a fair distribution of the use and costs of the technology,
- the more the affected groups participate in its implementation,
- the more it is accepted by all the affected groups.

In short, there are essentially two dimensions to the term *social compatibility*; the *subjective*, which calls for acceptance, i.e. whether the population and the organisations directly affected consider a measure as positive, and the *objective*, which calls for changes to a social environment based on concrete social criteria (such as discrimination, education & training, impact on inhabited areas, income distribution, information / communication, participation, transparency, risks for the population). As a result, social compatibility can be either empirically assessed by means of a survey, or by employing objective criteria like those already mentioned.

Social compatibility, as one of the three pillars of sustainability, the others being environmental and economic sustainability, constitutes an increasingly important criterion for the assessment of waste management strategies, concepts, planning and projects.

7.2 The Tool "Social Compatibility Analysis SCA"

Since the Brundtland Report [5] and the 1992 Rio Conference defined the social aspect as the third dimension of the concept of sustainability, along with the ecological and economic dimensions, interest in *social compatibility* and the assessment of social aspects has been rekindled [27]. The integration of the social dimension into the processes of decision making, planning and problem solving, requires an innovative and interdisciplinary approach. There are already tested and approved methods and instruments for evaluating the ecological and economic dimensions of activities, projects, developments, products and organisations in terms of their *sustainability*. However, such methods and instruments are conspicuously lacking when it comes to the *social dimension*.

In order to include social compatibility in planning processes, a tool similar to that of the environmental impact assessment (Section 5.1) has been developed. In contrast to the subjective evaluation method with which the acceptance of those directly affected is estimated, our proposed method defines objective criteria

(Table 7.1) according to which social compatibility is evaluated. Therefore, the objects under assessment are the intended as well as the unintended, the positive as well as the negative impacts on the fulfilment of human needs and the cohabitation of human beings, with particular attention given to psychological, social and cultural needs [8, 51].

Detailed research work has been carried out concerning the social compatibility of waste management policies and projects in Switzerland. Results obtained from a Delphi Expert-Questioning concerning the social aspects of waste management in Switzerland [10] led to the development of a tool for recording and evaluating social compatibility. This tool, the *Social Compatibility Analysis SCA* [55], is based on the ABC method embodied in business administration and already applied in the assessment of environmental impacts [23]. The central concept of the Social Compatibility Analysis is based on a general set of objective criteria for social compatibility (Table 7.1), which, in respect to specific assessment problems, can be adapted. Within the waste management field, for example, a specific criteria catalogue has been compiled and used for a specific project in Thun, Switzerland (Section 7.4.1). The SCA-tool is applicable in the analysis or assessment of many systems, such as plannings, projects, processes, products, organisations. The user of the SCA-tool divides the system into a number of subsystems. A product, for example, could be divided into subsystems by choosing the life cycle phases preproduction, production, use and disposal. The user then chooses several of the above mentioned evaluation criteria and assigns all subsystems to classes A (highly relevant social problems), B (of medium relevance) or C (of low relevance) for all the chosen criteria. The *Social Compatibility Analysis SCA* is a derivative of the ABC-Analysis, which is based on the premise that a small number of A components contributes in a major way to a problem, whilst a much greater number of C components contributes in only a minor way. All those components of medium relevance are assigned to class B. Assignment to classes A and B are explained and commented on (Table 7.2). Since A components are highly relevant to a problem, they are the main focus for improvement. The results achieved through this method reflect the semi-quantitative assessment of the user. If more than one stakeholder is involved, the result of the evaluation may represent the consensus of a user group.

The *SCA* has already been successfully implemented in several projects; for example, in defining a regional concept for sewage sludge disposal and for planning a large thermal waste disposal plant (Table 7.2). SCA-results with reference to the construction of the above mentioned thermal waste disposal plant have, for example, shown that parallel to the decision making process within legally defined parameters, appropriate participation processes must also be initiated from beginning of the planning phase to avoid acceptance problems. Additionally, the application of the SCA-tool has indicated that the *SCA* is a particularly valuable tool when the social dimension of a project is concerned, when the clarification of various stakeholder assessments is needed, or when sets of solutions are to be negotiated [12].

Table 7.1. General Criteria for Social Compatibility

Criteria	Explanation & examples
Participation	Participation opportunities for affected parties
Information & Communication	Understanding the decision-making process; transparent information policy
Compliance with social laws	Adherence to constitutional rights and principles; adherence to social rights
Occupation	Quantity (Employment level in Industry and Business; I&B); Quality (Required educational standard in I&B); etc.
Education & Training	Availability, accessibility and quality of educational facilities and information channels (nursery, elementary, public, private, and other educational institutions)
Material well-being	Level of income; distribution of wealth; social security (senior citizen provision, medical provision, unemployment benefits, etc.); cost of living; etc.
Working conditions in developing regions	Exploitative work practices, such as child and forced labour; working with dangerous substances
Protection against discrimination	Gender; racial; religious; disability; lifestyle; etc.
Protection of minorities	Minorities in democratic decision-making processes
Public security	Presence of law & order representatives; surveillance and lighting of public utilities; guaranteed safe disposal; etc.
Protection against health hazards	Evaluation of toxicology; evaluation of radiation; damage duration; etc.
Protection against nuisance	Noise; odor; vibration; landscape; etc.
Living space	Quantity (encroachment or destruction of living space); quality (proportion of green/recreational areas, view, sunlight, social infrastructure, rent regulations, opportunities for development, etc.)
Handling of risks for the population	Prevention; damage limitations; liability/insurance; communication; etc.
Winner – loser symmetry	Between Individuals, regions, countries, continents (first and third world), present and future generations (maintenance of decision opportunities)

An assessment of social compatibility using the tool *SCA* is decision-support oriented and shows what the user or user group considers the weaknesses of a project to be. The various subjective assessments of different stakeholder groups concerning the social consequences of a problem can be compared and clearly documented [55]. The result is subjective in that the evaluation criteria and the categorisation into A, B or C are dependent upon the user. On this basis, comparisons between different problem areas are not possible. However, comparisons with assessment periods within a problem/project completed earlier are possible if the individual classifications were clearly defined and substantiated from the very beginning. Similar to an ecological life cycle assessment (LCA, see Chapter 5), the *SCA* is not only suitable for comparing several variants, but also for performing an in depth analyses of one particular aspect under evaluation, with the aim of identifying specific areas for potential improvement.

Table 7.2. Social Compatibility Analysis SCA: An example of application in the case of a waste incineration plant. SC = Social Compatibility [A = SC is minimal, and there is a great need for further improvement (measures), B = SC is limited, and there is a need for further measures; **C** = SC is evident, and further action is not required].

Criteria	Need & Basic decision	Planning phase	Construc- tion phase	Operational phase		Deconstruc- tion phase
				Operatio nal processes	Materia l flows	
Participation	A[a]	A[b]	C	A[c]	C	C
Information & Communication	A[d]	A[e]	B[f]	A[g]	B[h]	C
Compliance with social laws	B[i]	B[j]	C	C	C	C
...	...	...	...	...	...	...

[a] The population directly affected by the plant has no legal means of being involved in the decision making process (the actual need for the plant in the first place, and the decision to build it). Appropriate voluntary instruments for citizen participation should be undertaken as required.
[b] The population directly affected by the plant has no legal means of influencing either the choice of plant location or the type of technology. Canton regulations governing the construction of any new facility require that the canton eventually initiate a plebiscite concerning the project. Appropriate voluntary instruments for citizen participation should be undertaken as required.
[c] The population directly affected by the plant has no means of influencing the operational phase of the plant. The formation of a "watch-dog" group which would include parties from each stakeholder group is yet to be examined.
[d] ...

Value conflicts are often exposed both between the criteria of a specific group and between the groups themselves. One example is the balance between the social compatibility of employment (working places) and costs for the public (in the case of a public employer).

It is both sensible and necessary to record and evaluate the social aspects with respect to any given project, in as much as this approach would best serve the public interest. Authorities, business sectors, and other associations should take it upon themselves to implement a social compatibility analysis within legally defined parameters [55]. This instrument could be used to find consensus oriented solutions to conflicts, if the social dimension of a subject for assessment has to be structured, and if the diverging assessments of different stakeholder groups are to be made visible. It would then facilitate the negotiation of the various solution approaches. The *SCA* is therefore suitable for use in participatory processes such as acceptance dialogues as a means of visualizing different evaluations and standpoints of various interest groups, thus providing a common basis for discussion and solution finding.

7.3 Instruments for Improving Social Compatibility

Joos et al. [26, 27] show that decision transparency, interregional cooperation, information policy, and public participation are important factors with regard to the public acceptance of waste management. The results and recommendations of a Delphi Expert-Questioning on the development of waste management in Switzerland [10] can be utilized in order to improve the social compatibility of waste management. The experts considered the *public acceptance dialogue* (in the sense of participatory processes indicated in Section 7.3.3) to be an appropriate means of ensuring the social compatibility of new waste treatment plants and of arriving at a consensus. Furthermore, the instruments for assessing social compatibility (i.e. the *SCA* in Section 7.2) should be implemented in the planning processes as well. *Policy approaches* (Seciton 7.3.1) such as the waste avoidance strategy are supported by most of the questioned experts. A majority of the experts is also of the opinion that environmentally responsible behaviour can be promoted above all by means of *financial incentive systems* (Section 7.3.2). These primary environmental measures influence (indirectly) the social compatibility of waste management as well.

In short, there are different levels on which to take action in the improvement of social compatibility [27]: there is the strategic, governmental level which sets rules, the economic level which gives incentives to certain behaviour, and the participative level which improves public acceptance in actual projects. These three levels will be discussed in the following sections 7.3.1 through 7.3.3.

7.3.1 Policy Approaches

Effective government policy does not depend on one area only. Pollution prevention has many dimensions, and the evaluation of possibilities over the life cycle of a product or service could reveal that a variety of interventions are necessary to achieve the desired environmental objectives. Thus, governments should take a systematic approach, using a variety of instruments. What the most suitable combination of economic, regulatory and information policies is will have to be determined by each country individually, although attention must be paid to the international effects (for example, in relation to trade issues) of national policies, especially concerning products. As was stated in Section 7.2, decisions concerning ecological effects often have social consequences. Therefore, interventions towards more ecological solutions often contribute to more social compatibility as well.

Economically, the generation of waste is a waste of natural resources or raw materials; however, as the digestive system of the human body suggests, the generation of waste cannot be avoided altogether. Similarly, economic activity, by its very nature, cannot avoid generating waste [30]. Waste generation can be influenced, however (see e.g. Section 3.1.6). One way to do this is through *cleaner technology*, which encompasses the objective of waste minimisation. *Individuals* can also influence the reduction of waste generation. Examples include the use of returnable rather than disposable bottles, the purchase of more durable or less energy-consuming products, and a change in the general consumer mentality, which dictates the purchasing of new things rather than reusing old items etc.

There is another element of waste prevention policy which is often overlooked. It involves the *reduction of heavy metals and other undesirable substances in products* in order to avoid their presence in the waste stream at a later stage. Frequently, such heavy metals or other undesirable substances qualify as "hazardous" waste, thus making their recycling, recovery or safe disposal more expensive [30]. Several steps have been taken to reduce the presence of undesirable substances in products. For example, the European Directive 91/157EWG and its amendment 98/101/EG regulate the use of heavy metals in batteries.

Of high relevance is the existence of a *supportive framework of environmental and social regulations* to create a strong demand for the development and diffusion of cleaner technologies. Companies lack the essential motivation to install even the most cost-effective pollution and waste reduction technologies if there is no societal demand (generally expressed in environmental standards and regulation of polluting sources) for pollution prevention and control. Additionally, while there may be cost savings in terms of energy and materials in the long run, businesses which expend substantial capital on pollution prevention, may be placed at an economic disadvantage (at least at short run) vis-à-vis competitors, which do not have to meet environmental requirements [34].

Despite its implementation shortcomings, most experts of the Delphi Questioning welcomed the *"avoid-reduce-reuse-recycle-dispose" strategy* [10]. They were especially critical of the attempts to avoid and reduce waste to date, deeming them insufficient. They therefore called for *more information and financial incentives.* Such measures could - due to consumer behaviour - put more pressure on producers to take further steps in avoiding and reducing waste resulting from production processes.

However, the willingness of the population to reduce the amount of waste by reducing consumption was given a low rating by the experts. It should ultimately be *the task of the producer/provider to satisfy consumer needs with minimal use of resources.* Furthermore, there is a trend towards the *consumption of more environmentally and socially friendly products.* It therefore goes without saying that "well-being through avoidance" [33] can only be achieved by changes in consumption patterns, production methods and product type: efficiency (i.e. towards less material intensive production) and sufficiency (i.e. towards less consumption) strategies must be concurrently adopted.

In order to redirect patterns of consumption, it is necessary to do *scientific research in the field of using rather than owning goods*, that is, into duration of the use of goods. The aim would be to encourage manufacturing industries to *increase the longevity of their products*, to *reduce the need for repairs* and to *use resources more efficiently*. Concerted efforts to avoid waste lead to a reduction in both the flow and consumption of materials. This ultimately means a decrease in the production and consumption of goods [31].

Most experts regard the achievement of a *closed cycle (or zero waste) economy* (Section 3.1) – where used goods are almost totally disassembled and their parts, or at least their materials, reused or recycled – to be desirable yet unrealistic [26]. Both "globalisation", i.e. increased distances between place of production and place of consumption, thus, increased availability, as well as low priced raw materials are seen as obstacles to this goal. Higher priority should be given to making products more durable, i.e. striving for *product longevity*.

A 70% majority of experts agree that there should be a legal stipulation for *products to display their environmental impact*. Apart from the experts representing the economic sector, all other sectors involved in the Delphi Questioning agreed with this statement [27]. Therefore, it would be *obligatory to assess the environmental impact of products*. Eco-labelling programmes, with the express purpose of identifying and promoting products whose producers have taken the most action to incorporate environmental objectives in the constitution of the product and in the production process, have been established in several countries [34]. Mechanisms which give consumers access to environmental information about products (and stimulate producers to assess and produce such information) are fundamental for any successful programme in order to promote cleaner technologies and waste minimisation.

The importance of *public awareness* has already been noted in connection with product information. Another aspect of public awareness is "the right to know" about the releasing of contaminants into air, water, and land by sources of pollution and waste. There are many ways to foster appropriate educational approaches at all levels related to the full variety of pollution prevention / waste minimisation actions; for example, adequate informational and educational programmes by both public schools and non-governmental organisations [34].

7.3.2 Financial Incentives

The purpose of *economic instruments* used in the environmental field is to influence decision-making and behaviour by firms and individuals. As compared to direct regulations, economic instruments allow actors the freedom to respond to the stimulus in ways they themselves decide are most beneficial. Moreover, if a certain environmental goal is to be attained, economic instruments will, at least in theory, promote the most cost-effective behaviour [34].

A majority of the Delphi Questioning experts [10] is of the opinion that environmentally responsible behaviour can be promoted above all by means of financial incentive systems. The effectiveness of such incentive systems was clearly confirmed in the expert questioning. To encourage an increase in the return of valuable waste materials and/or problematic waste-products such as batteries, *deposit/return charges* usually provide a more effective solution than integrating disposal charges into the actual price. However, administrative and logistical expenditure involved in both enforcing a deposit system and preventing its abuse, should be in direct proportion to the desired ecological use.

Table 7.3. Applying financial incentives to encourage the separation of different types of waste (the percentage of experts in favor of financial incentives is indicated in brackets).

Deposit/return charges for	Disposal charges made in advance for
• reusable PET bottles (88%)	• paper (63%)
• batteries (69%)	• cardboard (60%)
• aluminium cans (60%)	• used tyres (57%)
• recyclable PET-bottles (56%)	• electrical appliances (56%)

In various contexts experts emphasize the effectiveness of *financial incentives* as well as *voluntary agreements among producers*, which are both seen to be superior to command-and-control tools and (non-binding) appeals and exhortations. Voluntary agreements can be a very effective means of promoting cleaner technologies (including product improvements) and pollution prevention in general. They often work faster and with more flexibility than regulatory programmes, and thus help to promote innovative and cost-effective approaches to cutting pollution and waste. Voluntary approaches also serve a very important function in building trust and credibility among the private, governmental and public sectors [34]. It is hoped that the authorities do not exclusively focus on the economy when seeking cooperation, but also involve other social groups in this process. The seeking of cooperation can therefore act as a catalyst for increasing public awareness towards sustainability [6].

The majority of experts proposed two primary measures of equal worth for the improvement of careful use of goods and products. Once again, *financial incentives, ecological taxes and full cost calculation* were cited in this context. Ecological taxes would render goods with high environmental impacts more expensive and people might use them with more care as a consequence. In addition to economic measures, however, experts also called for *increased public relations work*, for enhanced *(environmental) education/training*, as well as for the promotion of *existing good examples*, which would illustrate similar value changes towards more ecological behaviour. Methods of dealing with goods and products in a responsible manner should also be taught in ecology lessons at school.

Some waste incineration plants have high total costs for certain types of waste services. More transparent cost calculations would reveal reduction potential in the total costs of waste incineration plants. Many experts felt that savings could be made in the personnel and logistical areas, as well as in the cost of supra-regional collaboration. Waste disposal from built-up areas should not be first and foremost profit orientated. Rather, guaranteed safe disposal should take precedence [11]. The high investment costs for waste incinerators require a permanent flow of waste over the lifetime of the incinerator – generally about thirty years – to operate economically. This stifles innovation and makes the recurrence of reuse, recycling or other technologies impossible, even where such innovations could be economically advantageous [30].

7.3.3 Participatory Processes

More and more decision makers and affected parties engaged in solving environmental problems are recognizing that traditional decision making strategies are insufficient. Often heavily influenced by scientific analysis and judgement, these kinds of decisions are vulnerable to two major critiques. First, they do not give proper consideration to the affected interests and therefore suffer from a lack of public acceptance. Second, they rely almost exclusively on systematic observations and general theories while neglecting the local and applied knowledge of the people most familiar with the problem. These decisions might produce outcomes that are incompetent, irrelevant, or simply non-functional. Citizen involvement in decision making, i.e. participatory processes, has been widely acknowledged as a potential partial solution to these problems [39, 40]. Public participation can be achieved through forums for exchange which should facilitate communication between government, citizens, stakeholders, and businesses regarding a specific decision or problem with the aim of reaching a consensual solution. These forums can include activities such as public hearings and meetings, focus groups, surveys, citizen advisory committees, referendums and initiatives, round tables, mediations, negotiations, and other such models.

Motivation for Public Participation

In participatory democratic theory, public involvement is morally and functionally integral to the emergence and the sustenance of the two central values of democracy: popular sovereignty and political equity. It is generally accepted that democracy is the outcome of an agreement among people who establish a sovereignty based upon popular and mutual consent. All power within the sovereignty is allocated through this agreement. A democracy's functional value or capacity is measured by the soundness of the decisions reached in light of the needs of the community and by the scope of public participation in reaching them. Rousseau argued that a sovereignty is composed of all citizens and requires input via public involvement to determine legitimate objectives. In his reasoning, public participation is justified out of necessity: citizens must engage in political affairs to keep the state alive, for only through interaction can the general will emerge from the plurality of particular wills [40].

Table 7.4. Individual or group arguments for and against participation in mediation-procedures; an example of the participatory processes [13, 36]

Arguments against participating	Arguments for participating
<ul><li>Fear of losing in the participation procedure</li><li>Fear that weaknesses will be sought out</li><li>Fear of taking on responsibility</li><li>Fear of being involved (loosing legitimacy for opposition) and being afraid of a role change</li><li>Insecurities concerning new or unknown procedures</li><li>Expenditure of additional time and resources</li><li>Insecurity concerning trustworthiness and deception</li><li>Party politics</li><li>Shortage of personnel and lacking know how</li></ul>	<ul><li>Public declaration of need</li><li>Access to early and improved information</li><li>Simple access to experts and planners</li><li>Avoiding the neglect of specific local interests</li><li>Possibility of participating and influencing the process from the outset (of search processes, contextual influence, for example)</li><li>Improved basis for decisions</li><li>Potential to find most beneficial solution for the involved communities</li><li>Transparency and openness / fair procedure</li><li>Improved positions for negotiation (balancing responsibility & positions of control)</li><li>Discussions concerning content rather than legalities</li></ul>

Functional analytic justifications for public participation are based on the contribution of participatory processes to the social system's need to maintain itself [40]. It is the explicit aim of participatory processes to increase the acceptance of planning processes that might otherwise be exposed to resistance of various origins. Additionally, it should achieve political acceptance and thereby increase the likelihood of successful legal procedures. One way to achieve this is to gain acceptance among the general public. This requires, for example, the balancing of public and particular interests in order to increase the likelihood of the implementation of a certain project. Empirical evidence shows that an interested public can be well informed and won over through a step by step discussion of the situation, and an open presentation of the facts. Central to this process is an information policy that not only does not avoid addressing the problems and risks of the planning, but also shows potential solutions and is open for public discussion [25].

Who Takes the Risks in Technological Developments?

Taking a risk is characterized by knowledge of the circumstances and by its voluntary nature. This is definitely not the case if, as in certain technologies like those used in waste treatment plants, the decisions for technological developments are made by a small number of people, and many others are then confronted with the consequences. In fact, the situation can be described in terms of the overall potential of the consequences transforming a risk into a danger for the individual. The individual is exposed to these consequences without having voluntarily made the decision concerning this risk. The problem of accountability and responsibility takes a completely new form. Whereas taking a risk is bound up with responsibility for the consequences, undesired consequences of technological developments are often not directly caused by those confronted with the consequences. People are willing to suffer harm if they feel it is justified or if it serves other goals. At the same time, they may reject even the slightest chance of being hurt if they feel the risk is imposed on them or violates their attitudes and values [37]. The perspective of those who are confronted with danger is being transformed into a potential of protest which is helping to make authorities decide in favor of more secure technological options [50].

Siting Waste Facilities

Controversy concerning facility location for the disposal and treatment of both hazardous and domestic waste is widespread in all industrialised countries. Proponents (both public and private) of these facilities point to the need for them and to their technical suitability, while opponents (local communities and environmental groups) emphasize the risks on health and the environment associated with the substances to be disposed of or treated by the process [15].

The causes of the NIMBY[1] syndrome have been enquired into in policy analysis, risk analysis and sociological literature. While everybody is aware that waste is generated every day by every human being and that economic activity is not possible without the generation of waste, the subject of waste management, treatment and disposal is such a taboo in our society that the whole discussion is often reduced to the question of where the waste installation should be located. Arguments against local waste facility sites are usually biased. Indeed, with regard to environmental and consumer protection aspects – transport, emissions, odour, noise, encroachment on landscape, etc. – there is no difference between a production plant and a waste installation; yet the production plant meets much less resistance from residents [30]. The most accurate studies argue that it is not only the characteristics of the projects (their technical suitability, economic efficiency,

---

[1] The NIMBY (Not In My Back Yard!) labels refers to fervent local citizen opposition to siting proposals or land-use activities with potential adverse impacts. Broad public support for environmental values, dread of unknown and uncontrollable risks to personal health, and the sudden increase in publicly available information have been posited as explanations for this reaction [41].

relation to undesired and therefore suspect waste materials, etc.) but also, or even more so, the features of the decision making process that are crucial in determining the acceptability and feasibility of waste facilities projects.

Possible remedies for the *NIMBY* syndrome have been suggested and some of them, such as the introduction of information and participatory procedures in the decision making process, have been implemented. Claus & Wiedemann [13] have therefore set up basic rules for a *fair assessment of location for waste disposal facilities:*

1. Initiate and implement a widely encompassing participatory process
2. Reach consensual agreement that the status quo is no longer a solution
3. Aim for consensus
4. Build trust
5. Search for the solution that is most appropriate to the problem
6. Guarantee the highest possible security
7. Make known all potentially negative impacts of the plant
8. Offer compensation for the impacts to the municipality of the location
9. Negotiate the applicable conditions for a plant and potential influences on the operation
10. Search for acceptable locations in a voluntary process
11. Approach the evaluation of the location as a contest
12. Ensure that the spatial distribution of public infrastructures is balanced
13. Set an adequate time table
14. Keep your options open

The main reason for local opposition to potentially environmentally threatening plants and technology is the fear of potential breakdown and subsequent damage which may carry uncertain consequences for health, the environment and the standard of living within the community. In general, such reservations are based on previous experience of technical breakdowns which are often due to budget cuts, negligent behaviour or accidents. Another reason for citizen protest is the fear of material loss in case a plant should be built in their vicinity. Many plants, such as waste treatment plants, offer few new employment possibilities, and therefore generate minimal additional income for the affected community. The local population may also oppose a project because they expect unpleasant odours, noise or other encroachments on their quality of life.

Table 7.5. Location based advantages and disadvantages of waste treatment plants [24]

Advantages	Disadvantages
• Improvement on former methods of waste disposal	• Negative impact on landscape
	• Topographic requirement
• Workplaces	• Emissions
• Business tax	• Negative feeling
• Cheaper energy?	• Potential dangers
• Political advantages?	• Additional traffic
• Compensation measures	• Community image / attractiveness
• Location advantage	

Finally, citizens or communities also oppose such projects because they consider it unjust if their particular community has to bear both the burden and the risk of an entire region's waste management. The inhabitants of rural communities are particularly opposed to the idea of having the waste of industrialized regions that are already economically advantaged on their doorstep (town-country-conflicts). However, the expected cost-benefit-ratio is not the only reason why people may oppose a project. Governmental decision making processes are often the targets of criticism as well. The existing channels open to public participation are, in many cases, not perceived as open and fair. Social acceptance of any policy is closely linked to the idea that the procedure by which a decision is made is fair [38]. Any public participation must meet the criteria of fairness and competence [38]. Competence includes the fact that a guaranteed safe disposal of waste must be given.

An integral waste management planning procedure for a waste treatment plant should therefore consist of the following [25]:

- An assessment of the need for a waste treatment plant that is based on all applicable reduction and recycling options.
- An assessment of the present situation that clearly indicates the extent to which waste management planning also influences present production and consumption habits (on a regional or even national level).
- An assessment of technical and organisational potential in order to reduce the risks in relation to the interests of the waste producers (consumer as well as industry). Waste treatment plants that can be planned without high costs are virtually impossible. Plants that increase the cost of waste management are contrary to the interests of local industries. This field of tension has to be debated in a public manner.
- Therefore it is important to call for public forums/events where all interested groups and stakeholders are present and are allowed to express their concerns. It is important to clarify the various conflicting interests and to debate their relevance in public.
- It is important to choose a method that allows the problem to be tackled step by step, and that is tolerant of mistakes. Planners have to include multiple

perceptions of risk and to search for adequate solutions. If that proves impossible, conflicts of interests must remain transparent.

- Individual fears – rationally justified or not – should be taken seriously. All fears have to be taken into consideration and to be treated openly.

The aims of the implementation of a participatory method have to be clearly stated from the beginning. This is important in order to achieve the acceptance of the procedure itself. Furthermore, the existence of a clear basis simplifies future evaluation.

The procedures can be defined as follows [see 20]:

---

1. Initiation phase

- Kick-off by an initiator
- Find an agent of change or a mediator (neutral, independent person/party)
- Clarify questions of finance (who will finance the process: the authorities, a public/private company, other interested party?)

2. Preparation phase

- Draw up a conflict analysis (relevant parties, matters of conflict, procedure, etc.)
- Choose participants for negotiations
- Define range of flexibility
- Co-operative search for information

3. Search for conflict-solutions

- Focus on interests, not on positions
- Find win-win solutions
- Split the conflict into manageable parts that can be resolved independently

4. Transition phase/Implementation

- Documentation and communication of the results
- Place the participating parties under obligation
- Define procedures for potential future conflicts

The participants should reach a consensus about what procedures will be implemented. It is important that all parties agree to this common basis. Then, an institution or a person should be selected who is responsible for the preparation and the proceedings of the mediation process. The qualifications and responsibilities of this institution, inclusive of financial compensation, have to be clearly stated. There is usually a general agreement about who is responsible for informing the public and the media. There should be an assessment of potential controversies and of potential areas of conflict. Subsequently, potential methods of addressing and solving these conflicts should be designed [7].

The following table gives an overview of the potential applications of participatory processes and other approaches for the improvement of social compatibility within waste management:

Table 7.6. Potential application of participatory processes

Dimension	Class	Description of class	Type of participatory process
Time	Long term	Long range planning (> 5 years)	Round table, "future-workshop"
	Mid range	Planning to be implemented in the near future	Participation, citizen panels
	Short term	Short term planning, implementation phases	Participation, mediation, round table, SCA
Space	Local	Participation in local communities and neighbourhoods	Participation, citizen panels, round tables, SCA
	Regional	Regions, Cantons, extra-local spaces	Round table, SCA
	National	(Inter-)nationally applicable procedures	Round table, SCA
Type of conflict	Open	Conflict is openly dealt with; agreement seems infeasible	Mediation, SCA
	Latent	Conflicts are visible, clear fronts yet to be established	Round table, SCA
	No conflict	No conflicts expected / visible	Participation, SCA

SCA = Social Compatibility Analysis; Participation = other participatory processes

A *round table* is a systematically subdivided process that is guided by a neutral moderator. If needed, external experts are invited to testify and to aid the process with additional information. Every group of stakeholders is to be represented by the same number of persons (independent of its political or economic position). Among other points of import concerning representation, this also visually reinforces the norm of equal rights [4]. It is, though, possible to have different forms of representation for stakeholders in the various groups regarding specific issues, as well as concerning spatial topics, potential political influences (input of recommendations, preparing decisions, entering a caveat etc.), and time frame (for a limited time or as an institutional right).

The philosophy behind *citizen panels* is that citizens, experts, and stakeholders can resolve environmental conflicts through their respective expertise. Stakeholders are valuable sources of concerns and criteria in evaluating options, since, as the name indicates, their particular interests are at stake. Experts are needed for providing technical data and making explicit the relations between options and impacts. Citizens must live with the consequences, and are therefore the best judges in evaluating the decision options. The intention of citizen panels is to bring these three perspectives together in a productive fashion [38].

Mediation is a process of negotiation in which a wide variety of conflicting parties seek a fair balance concerning their respective interests. This process is led by a neutral agent. The search for an ecologically and socially compatible solution is central to this process and the solution is sought through negotiations that are based on fairness, openness and information. A reduction of mistrust can be achieved by an open declaration of the differences concerning interests and perspectives. Some typical perspectives are as follows: Planners, for example, see themselves as victims of planning insecurities. It is not possible for them to foresee the decisions the courts will make. Affected persons, on the other hand, see themselves as having to bear the biggest burden in that they have to contribute more than their fair share to the well being of the general public. Some interested parties are looking for the cheapest way to dispose of waste, other parties think that waste can, in principle, be prevented. *Processes of mediation* offer a forum for discussing the various points of view and serve as a means of developing alternatives.

The procedures of the already mentioned conflict analysis will differ depending upon the stage of the conflict in which mediation is initiated, as well as the agreed duties of the mediator. If the conflict over the plans or the location of a plant continues for a long time, it might be more beneficial to identify the matters of dispute along with the relevant parties, than to initiate a new planning and mediation process. Nevertheless, even at an advanced stage of conflict, the focus should remain on interests rather than on interested parties.

It can be very advantageous to design charts which indicate the respective parties, their specific interests, the points of conflict, and potential options for their solution. The following questions should be answered by the mediator during this phase [20]:

---

1. Parties/Stakeholder-Groups

- What are the most important parties and who are their spokespersons?
- Are all affected interests equally represented in the process?
- Are all parties willing to find a consensual solution to the conflict?
- Are the parties willing and able to collaborate?

2. Matter of dispute

- What aspects best characterize the conflict? Is the point of contention due to differing interests or differing values or both?
- In what way can problems be best defined?
- What are the core elements of the dispute?
- What are the less relevant matters of the dispute?

- Which matters of dispute are negotiable?
- What are the core interests of the respective parties?
- What interests do the respective parties have in common?
- What positions do the parties take in their negotiation?
- What options for conflict resolution exist?

3. Procedures

- What is the position of the respective parties think concerning the implementation of consensual conflict resolution strategies?
- What are the potential benefits to the parties resulting from this process?
- What structural forces will influence the conflict solution (time framework, legal activities, financial constraints)?
- What obstacles have to be overcome in the process?
- Are there parties with specific experience in consensual conflict resolution?
- What are the chances of success in the implementation of the procedures?

In addition to the conflict analysis, there should be preliminary talks involving the different parties in order to evaluate expectations concerning the mediation. This will aid in determining a party's willingness to participate as well as their preconditions concerning the contents and procedures. It is thus as much a task of motivation as a task for the transmission of trust into the planned procedure [13].

In order to reach a "win-win-solution" in which all parties benefit, all participants have to be able to accept the interests of potential adversaries. Instead of searching for the one solution, the search for a variety of alternatives – even in smaller details – must have priority. A process of collective brainstorming offers initial access to the complexities of this approach and can also be helpful in other difficult phases. Renn et al. present additional requirements for a successful participatory process in Section 7.4.3.

Why do Some Campaigns Against Waste Facilities Succeed Where Others Fail?

Christopher Rootes

Most recent attempts to explain the pattern of successes and failures of campaigns against waste facilities such as incinerators and landfills have focused upon the ways the issues are framed by the opponents and the proponents of a particular development, and on the language used by them in the course of campaigns.

The development in recent years of *ecological modernization (EM)* – "a discourse that recognizes the structural character of the environmental problems but none the less assumes that existing political, economic, and social institutions can internalize … care for the environment" [21] – has become "the most credible way of "talking Green" in spheres of environmental policy-making" [22]. If indeed EM has become the hegemonic discourse, we might expect that the success of proposals for new waste facilities such as incinerators and of the campaigns against them will be determined, at least in part, by the degree of skill with which the developers on the one hand and the campaigners on the other employ that discourse.

Waste authorities and corporations have adapted quickly. Mass burn incineration without energy recovery is now regarded as a mere "technological fix" to the waste problem, but incineration with energy recovery, suitably relabeled "energy-from-waste", is enthusiastically embraced as the ecologically modern alternative.

For local communities opposed to the siting of waste facilities, however, there are a number of strategic dilemmas involved in attempts to raise the level of campaigns beyond the local and immediate, and the discourse of ecological modernization is less clearly advantageous. Research in the US has concluded that NIMBY protests are more likely to succeed if they can reframe the issues as wider issues of environmental management capable of appealing to a broader public [53]. However, whilst the adoption of a universalizing discourse may be a necessary condition if opposition to a facility is to be made respectable in the eyes of and to recruit the support of non-local actors, it may appear abstract and over-complex to local lay people who may be more easily and more intensely mobilized by NIMBY campaigns that focus on particular local concerns and values. Good universal arguments often fail, and even good universalist decisions may have locally unpalatable outcomes. Intense NIMBY campaigns sometimes succeed even where good universal arguments are weak or absent.

Another dilemma arises if the baton of opposition is passed from local communities to more institutionally privileged actors such as lawyers and local councillors. Although such actors may more expertly make arguments that are consistent with institutionalized rules, procedures and principles, reliance upon them – and the relocation of contention to courtrooms and council chambers – may hasten the demobilization of the local community, with the resultant loss of the "people power" that is a critical resource when campaigners are attempting to influence political decisions, which decisions on waste facility siting inevitably are.

Nor are "consensus-building", deliberative and inclusive procedures necessarily advantageous to local communities. Typically, they involve protracted inquiries and consultations that tax the limited resources of locals and exacerbate the difficulties of campaigners in maintaining the commitment and momentum of those local people who are personally ill-resourced to participate effectively in formal consultative processes. Not only are resources and people tied up in consultation not available for community organizing, but the evident disparity between the resources at the disposal of governmental authorities and corporations on the one hand and local communities on the other may itself be demoralizing to local campaigners. Not infrequently, elaborate consultative procedures produce neither the outcomes local campaigners desire, nor the legitimization of waste facility siting that authorities seek [35, 41]. Not surprisingly, local campaigners are often cynical about such exercises and may, on strategic grounds, be wise to refuse to participate in them.

The recently fashionable concerns of social scientists with discourse have often been accompanied by neglect of the structures of power in communities and between communities, governments and corporations. There is, however, evidence that in the US, waste incinerators were deliberately sited in areas where least resistance was expected: in smaller, rural communities with poor and poorly educated populations employed in resource-extractive industries or agriculture [14]. Other US research, however, found that static characteristics of communities were not strongly determinant of the success of local campaigns against waste incinerators [53].

Research in England suggests a more complicated picture [45]. If, in the US, prior networks were not an important determinant of successful resistance to incinerator proposals [54], in England personal and political networks in communities have been crucial to the success of local campaigns, and help to explain why village communities have often been more successful in resisting waste facilities even on "brownfield" sites than have their counterparts in towns. At the formal, institutional level, villages are distinguished from towns by their possession of a lower tier of local government – the Parish Council – which may act as a nucleus of participatory local organization. The social character of villages also appears to be important. In England, as elsewhere in western Europe, villages are often socially mixed communities in which local interaction is relatively intense as well as socially diverse. Villages, moreover, are the chosen homes of many people who work (or have worked) elsewhere and who are embedded in networks of expertise that can be brought to bear in local campaigns. Suburbs and inner urban areas, by contrast, are often more socially homogeneous and have relatively less resourceful populations.

These observations suggest that we should at least qualify US findings that the properties of communities and the existence of prior networks are less determinant of the success of local campaigns than are the skill, imagination and energy of campaigners [53, 54]. Local environmental action is the contingent product of a volatile cocktail of structural and conjunctural constraints and opportunities, and the actions and inaction of the individual citizens who are confronted by and, sometimes, attempt to surmount constraints and to create opportunities [44]. The efforts and ingenuity of campaigners are undoubtedly important, but the outcomes of contention over waste facilities seem generally to depend less upon the qualities of campaigns than upon *political opportunity structures* that campaigners are powerless to change.

Campaigners' adoption of a universalizing discourse is no guarantee of success if the political context is unpropitious. Indeed, consideration of a large number of cases suggests that the ways in which the issues are framed by campaigners are relatively unimportant to the chances of success of campaigns. Much more important are political opportunities, both those that are genuinely structural (such as local authority boundaries) and those that are more strictly conjunctural (such as the alliance structures within and between local councils), as well as the policy imperatives of local and national governments, especially as these change over time[2].

For local communities, the defeat of proposals for waste facility siting may be the only kind of success that matters, but even failed campaigns may nevertheless succeed in heightening local awareness of waste issues and in building community organization and solidarity that educates and empowers citizens and survives as the basis for future political resistance. Nothing is more likely to get people thinking about waste than the prospect of an incinerator on their doorstep, and nothing is more likely to mobilize apolitical citizens into the political process than a grassroots campaign.

Citizens with an interest in ensuring the adoption of the least damaging waste management strategies are generally recommended to become involved in the decision-making process at the very outset, when local, regional and national waste management plans are at the consultation stage. The major shortcoming of this recommendation is that, for most people most of the time, waste is an invisible issue. As a result, however hard authorities try to interest and involve the public, they generally fail so long as the issue remains at the abstract level of policy. Most people only become aware of waste as an issue when an incinerator or landfill is proposed for their backyard.

It is, then, all very well (and doubtless right) to recommend that people concern themselves with waste at the very outset, but the general public will not do so, and environmental groups generally do not have the resources to do so. The solution may lie in encouraging precisely the kind of involvement planning authorities and waste corporations generally fear most. If, from the outset, governments (national or local) were to fund environmental groups to mobilize critical arguments (and perhaps the public), there would be a greater likelihood that a reasonable range of perspectives would be introduced into the planning process from the beginning, and that the outcomes would command wider public acceptance.

7.4 Siting Waste Facilities: Case Studies

Siting a new waste treatment plant often triggers fierce opposition among the local population. Concerning the social compatibility of the siting procedures of waste facilities, the decision-making process is of major importance. The following case studies illustrate several decision-making approaches and their consequences in different countries.

---

[2] The importance of the national policy climate to choices of waste strategy and the outcomes of local mobilizations is evident in the US: whereas in the 1980s most incinerator projects were realised with little opposition, very few new incinerators have been commissioned and many proposed projects have failed to gain approval since the increased federal intervention in waste policy and the development of the national anti-incineration movement in the 1980s, [45, 54].

7.4.1 Local Opposition against a Waste Disposal Plant in Switzerland

In times of decreasing technological acceptance and increasing risk factors within society, environmental technology, especially in the area of waste treatment/disposal facilities, has fallen under the banner of increased public scrutiny. In many cases, such as in a project planned for the Swiss city of Thun, local citizens have demonstrated fierce opposition to the construction of new waste treatment plants. The opposition against the planned waste treatment plant in Thun was organised during 1996 after local citizens had been informed by the plant operator that their city had already been chosen as the intended site (Table 7.7).

Table 7.7. History of the intended waste treatment plant in Thun, Switzerland (adapted from [56]).

Date	Fact
1993	Evaluation of 31 possible locations by authorized experts [3] resulting in the utility company's decision on Thun.
1994	Utility company decision concerning appropriate technology of Siemens AG for waste treatment.
1995	Working within the cantonal regulations for planning permission (this restricts the participation of the local community).
2/1996	Participation of the local citizens concerning the cantonal regulations. The public is invited to express their ideas via a pre-formatted reply card.
11/1996	Public presentation of cantonal building regulations. The building application submitted by the utility company provokes numerous protests / objections.
2/1997	Petition against the plant submitted to the cantonal government.
5/1997	Cantonal government grants planning permission. Protesters register an official complaint with the administrative court.
5/1998	The administrative court rejects the complaint.
6/1998	The plant manager disengages Siemens AG due to serious technical problems in the pilot plant in Germany. Instead, a conventional waste-to-energy technology with less burning capacity is chosen.

The initial public reaction was that in 1997 the affected neighbours in the vicinity of the chosen location signed a petition against the waste facility location. The cantonal administration rejected this petition on the grounds that waste treatment is of cantonal interest, whereas the opposition was a minority who did not see the need for such a waste facility in the region of Thun. Nevertheless, a broad opposition against the intended waste treatment plant arose in 1997, in which local environmental & consumer organisations as well as affected communities went as a group to the administrative court in Berne.

Round Table on Waste Management, Thun, 1998/1999

The Round Table for Waste Disposal in the Thun region was established at the end of 1998. After the collapse of the thermal waste treatment plant project by Siemens AG (Schwelbrennanlage SBA), SBA Thun, it was hoped that this "Round Table" would bring about a consensus of opinion among the stakeholder groups concerning the future strategy of waste disposal in the region. They had learned from prior error: Opposition groups (local environmental and consumer organizations; political parties; affected communities) and the operator of the proposed waste treatment plant Thun, the plant manager AVAG[3], had decided to take into account the need for participatory processes. The Department of Building and Energy, Canton Bern, as well as the City Council of Thun were likewise represented in this way. Representatives from the Swiss Agency for the Environment, Forests and Landscape (SAEFL), and from cantonal environmental departments had also been invited to the Round Table as experts.

Table 7.8. Interests and expectations of opposing parties participating at the Round Table.

Party in opposition to waste treatment plant	Interest at the Round Table	Expected outcome
Activist Group: "Clean Air Thun" (AGL, "Aktion Gesunde Luft Thun")	Initiating the investigation into alternative projects.	Waste prevention, alternative scenario (AGL/FLT, Freie Liste Thun).
Free list Thun (political party FLT)	Working in a constructive environment against an undesirable waste treatment policy.	Waste prevention, alternative scenario (AGL/FLT), find alternative to the principle means of waste disposal: incineration.
Bernstr.-Leist (local inhabitants initiative)	Protecting the interests of local inhabitants.	Re-consider the location Thun.
Political community Hilterfingen	Protecting the interests of local inhabitants.	No waste treatment plant in Thun.
IG Velo Thun (Lobby group of cyclists)	Improving air quality; open and honest dialogue.	Waste treatment in an economically and ecologically viable way. Aim: the reduction of waste quantity.
Lerchenfeld-Leist (local inhabitants initiative)	Protecting the interests of local inhabitants.	Re-considering location, investigation into alternative methods of waste treatment.
Pro Regio Thun	Improving quality of life as well as the potential for tourism of tourism in the region of Thun.	Investigation of alternative waste treatment strategies. No waste treatment plant close to the city.

---

[3]   AVAG = AG für Abfallverwertung, waste disposal company of the Thun region.

Our research team had been requested – after substantial co-initiation of the Round Table – to make a scientific study of this participatory process [9]. There had been certain indications that, out of the often difficult and complex situation which exists between the planners/operators and opponents of a waste disposal plant, some constructive negotiations had emerged.

The main demands made by different stakeholder groups regarding the waste treatment plant are summarized in Table 7.8.

There was absolute opposition by some to the specific location of the planned plant. Based on an evaluation of the location, the utility company (AVAG) saw no need for a re-assessment of the location. The question therefore remained as to whether the location issue was still a point of discussion. Furthermore, waste plans using alternative waste treatment techniques had also been requested.

The choice of location seemed to be an emotional, not a technical question. In an attempt to counter emotional factors by information, the local audience had to be informed as to why it was necessary to deliver the waste to this densely populated area. What followed was a discussion about the specific location.

Overview of the Round Table's Process

The initiatives of both the future utility (AVAG: installation of the support group "Waste treatment plant Thun"), as well as the main opposing organisation (Pro Regio Thun: talks at a round table) were indicators of a positive new orientation which finally led to the "Round Table of Waste Management Thun".

AVAG insisted that a person with strong ties to the region, yet not connected to the utilities company, moderate the process. However, the opposition soon began to suspect that there were personal interests behind this regional involvement. The opposition objected to the person chosen for moderation on several counts. He was the manager of a technology company in a near by village and was therefore considered to be involved in the local economy. Furthermore, being in the same political party as the responsible local councillor of Thun (who was, at the same time, also a member of the board of AVAG), he was able to reject a political intervention asking for a moratorium of the specific technology proposed for the planned waste treatment plant.

Nevertheless, after the first meeting with the moderator, in which he categorically insisted on a neutral and consensus oriented solution finding process, those in opposition signalled their willingness to participate in the process of a round table under the direction of the designated moderator.

Principal questions could thus be resolved relatively early in the process: basic questions had to be considered outside the meeting room in a private environment. During Round Table discussions, factual restraints or prejudices could not be determined. Despite the general consensus, different interpretations were evident. The parties in support of the planning process interpreted this as an invitation to continue the planning process in order not to lose federal subsidies. They insisted that this behaviour would not prejudice the building of a waste treatment plant.

Fig. 7.1. Fear that "SBA Thun will eat the city of Thun" [2].

The opposing parties, however, interpreted this continued planning as a total prejudice looking for a new waste treatment plant. Therefore, they called for a halt to the treatment plant planning process in favour of an alternative waste treatment plan.

The differing commitments of the various stakeholder groups and their insistence on a limited scope of action repeatedly showed that consensus building would not be easy. However, during the meetings of the Round Table there was generally a good atmosphere. This positive climate was due, at least in part, to the efficient leadership of the moderator. The climate was disturbed by an occasional emotional outburst, followed by corresponding interventions of the moderator.

The Round Table lost its credibility when the canton rejected the opposition's request for a feasibility study of (future oriented) alternative waste treatment concepts. The opposing parties then had only two options remaining: they could either choose legislative action and pursue their wishes by registering their protest with the regional court, or they could hope for a change in the basic facts giving evidence that there is no need for a further waste treatment plant.

The opposition was afraid that a waste treatment plant would be imposed on the city of Thun, a solution that was clearly against the interest of the majority of the city inhabitants. It was also said that the local inhabitants were never really informed about the consequences of the cantonal plant construction plans. In the public's view, the Round Table held the function of searching for acceptable solutions for the whole of the region. Local municipalities in the vicinity that supported the construction of the plant were said to be acting according to the NIMBY principle (Not In My Back Yard). A successful application of the participatory process should have included all the interests of the local, regional as well as the supra-regional inhabitants.

According to the moderator, one of the aims of the Round Table was to uncover new facts regarding waste management conditions in the region of Thun and regarding the intended waste treatment plant and to bring these factors into the discussion. However, it was clear from the beginning that these new facts could be interpreted in various ways. Another crucial factor was time. The Round Table could have been much more effective if the cantonal authorities had accepted a new investigation into the feasibility of the location and of alternative waste treatment. Furthermore, the moderator did not understand why the location had not been re-considered. In his personal view, this Round Table served no real purpose if the decision of the cantonal government had already been made.

All participants agreed upon the reason for the failure of the Round Table: its initiation occurred 2-3 years too late. At the time the Round Table was initiated, the various stakeholder positions had already been established and were virtually unchangeable. The relevant and important decisions had already been made.

Comparison between Theory and Practice in the Case of the Waste Treatment Plant in Thun

Local citizen opposition to waste treatment plants is – according to experts – mainly based on the fear that odour emissions, reduced air quality and increasing traffic will adversely affect their quality of life. If such fears are not taken seriously by authorities, planners and utility companies, and if citizens are not informed in an appropriate way, but simply confronted with finalized projects, the basis for trust between the various stakeholders could be seriously impeded. According to a majority of experts, these situations could be avoided by actively involving the local community at an early stage in the proceedings.

The scientific accompaniment and the analysis of the Round Table Waste Management Thun have given a chance to take a critical look at the processes involved, and to compare fact to theory in the participatory processes in hopes of establishing some guidelines for the future [9]. The insights gleaned from this comparison should lead to recommendations that might be taken into consideration in an eventual re-establishment of a participatory process or for similar cases.

The initial establishment of the Round Table Waste Management Thun was based on less than optimal preconditions: the positions of the various stakeholders were virtually set and beyond discussion. The local utility AVAG was only willing to initiate the participatory process when put under pressure by the local municipality despite hopeless conditions for the opposition, as well as by a legal recommendation.

The predetermined positions of the stakeholders, which effectively provided the setting for an escalated conflict for which the stakeholders were unable to find any solution (the conditions for the implementation of the plant having already been established), called for a process of mediation instead of a round table. Mediation would also have been more suitable as its results are usually agreements that can be implemented step by step by the participants in conflict. Special consideration has to be given to the circumstances and the selection of the relevant participatory process [4].

Although AVAG wanted to involve the various stakeholders, it nevertheless chose the moderator. This initially led to a certain level of mistrust, and to the implicit doubt concerning the neutrality of the moderator. This mistrust continued despite the repeated insistence on the part of the moderator concerning his neutrality and independence. One of the main opposing parties (Pro Regio Thun) continued to doubt the moderator up to the point of actually withdrawing from the Round Table. They claimed that the moderator had not behaved neutrally, and had shown his biased position by lobbying for the proponents in several public events. In addition, they claimed that the moderator chose to repeatedly and categorically ignore their various propositions. Pro Regio Thun insisted that a moderator should assume a more neutral role as an independent agent (Pro Regio Thun, March 31st 1999).

The search for *scopes of action* and *innovative solutions*, as well as the *facilitation of fair dialogue* were deemed the core functions of an independent agent. The moderator was not expected to come up with direct solutions himself.

In retrospect, it is clear that the moderator was not fully accepted by some of the opposing parties. It is always being difficult to find an ideal candidate to suit all the various respective needs. Therefore, it is recommended that - especially in the case of larger negotiations - a team of moderators who share the task of moderating is installed.

In addition to its late establishment, what was missing from the beginning was a clearly structured procedure for the Round Table Waste Management Thun. The Round Table immediately started with a presentation of the points of view of the participants and their aims without having previously clarified the purpose of the Round Table, or the way in which it should proceed. Theory (Section 7.3.3.) says that it is important as a first step to clarify the procedures with the consensus of all participants. It should also be clear who is leading the process. At this point, an institution or a person should be selected to be responsible for the preparation and the implementation of the participatory process. The competences and duties of this institution, regarding financial aspects as well, should be clearly stated at the outset. In addition, there is usually a general agreement about who is responsible for informing the public and the media. There has to be an assessment of potential controversies and lines of conflict. Building upon this assessment, there has to be an evaluation of potential solutions on how to handle the controversial areas.

Before starting the process, the moderator held interviews with the main parties of potential conflict, such as AVAG and Pro Regio Thun. However, there was no deeper analysis of conflict such as was suggested in Section 7.3.3. It is very helpful to design charts that show the respective parties and their specific interests as well as the points of conflict, plus the potential options for their solution.

It soon became clear that an important party in the conflict resolution was missing. AVAG clearly stated that it was only the contractor of the canton of Berne, and that it only led the project on behalf of the canton. Despite this, the cantonal authorities were only invited at a rather late stage in the process, and were only represented by the secretary of the respective department. When selecting the people to participate in the procedure, it is important to include those people in the mediation process who have power in the effective procedure. Close ties between the informal and the formal procedures improve the implementation of the results of the environmental mediation and ensure that the people in power are part of both processes, are familiar with the points of discussion and are willing to learn from the results of the mediation process. This improves the implementation of the results of mediation.

During the process of setting the procedure and rules for the Round Table (such as follows: do not form prejudices, no single party has power to make decisions for the Round Table, clarify waste management strategies, evaluation of location and relevant facts, the moderator is exclusively entitled to contact the media), those in opposition were not happy with several of the stated rules. Some of this discontent came up again when the rules were broken. One of the major shortcomings of the Round Table was its failure to establish binding rules. The Round Table had absolutely no power in influencing the decisions for the planning process of the waste treatment plant in Thun. Therefore, the opposing groups saw it as a "fig leaf". The binding rules must be clearly communicated and agreed upon. The more the binding rules are relevant for all participating parties, the more motivated people are and thus the results are bound to be more substantial.

In this case, the disillusioned moderator deemed it necessary state the following: "According to the clarification of the aims of all the stakeholders, some of the opposing parties are fundamentally questioning the present location, as well as the overall need for a new waste treatment plant. Therefore, it will be very difficult to find a consensus." Certain decisions had been made and were hardly debatable. As a result, emotional outbursts replaced the exchange of well researched information. This led to an erosion of trust concerning fair media coverage with all stakeholders. Those in opposition said that: "All sides did investigate environmental compatibility but nobody dared to question the fundamentally important question of social compatibility concerning this project." A neutral social compatibility analysis of the waste treatment plant project, as well as of the various potential locations of the plant, parallel to a thorough environmental impact assessment is therefore strongly recommended.

All participants should be informed prior to the participatory process, that there is only a chance of success if they are open to new solutions, and if they are ready to build consensus. Concerning these points, the relevant questions are:

- Are there any options that are compatible with all points of view?
- Is there the possibility of dividing the parts of the problem that can be resolved consensually from those parts that are potentially conflictual in a way that the first can be implemented step by step without impeding the progress of the others?
- Are compromises possible for conflictual parts of the problem?

If these things were taken into account, it is not understandable why the location was excluded from re-assessment. The opposing parties mainly questioned the location. Therefore, the unwillingness of AVAG, as well as the canton, to re-assess the location evaluation of 1993 severely damaged the chances of finding a consensual solution. The Round Table might have reached a greater level of success if the canton had approved a re-evaluation of the location (Fig. 7.2).

The trust of concerned local people in the efficiency of the Round Table was further damaged by the building planner's rejection of an independent assessment of the optimisation of waste management of the region. The local people had the impression that the Round Table only served as a means of information exchange, and as a new forum to communicate widely known points of view.

In order to reach "win-win-solutions" it takes the willingness of all participants to listen and to want to understand the arguments of the other parties, and to concede their value. Instead of searching for a direct solution, many alternatives, even individual details, must be assessed. Brainstorming offers an initial point of entry to the complexities of this condition, and might also be helpful in difficult phases.

A clearly established and stated time plan would have allowed all stakeholders to assess their costs from the beginning, and to accept or reject them. Time has to be allotted for the treatment of a problem without any unnecessary pressure. Instead, much time was lost in discussions, and those in opposition had the impression that this was a tactic to abuse the pledge of secrecy in order to maintain their silence. The whole project of the waste treatment plant in Thun was driven by time pressure. The implicit aim of the canton as well as of AVAG was a

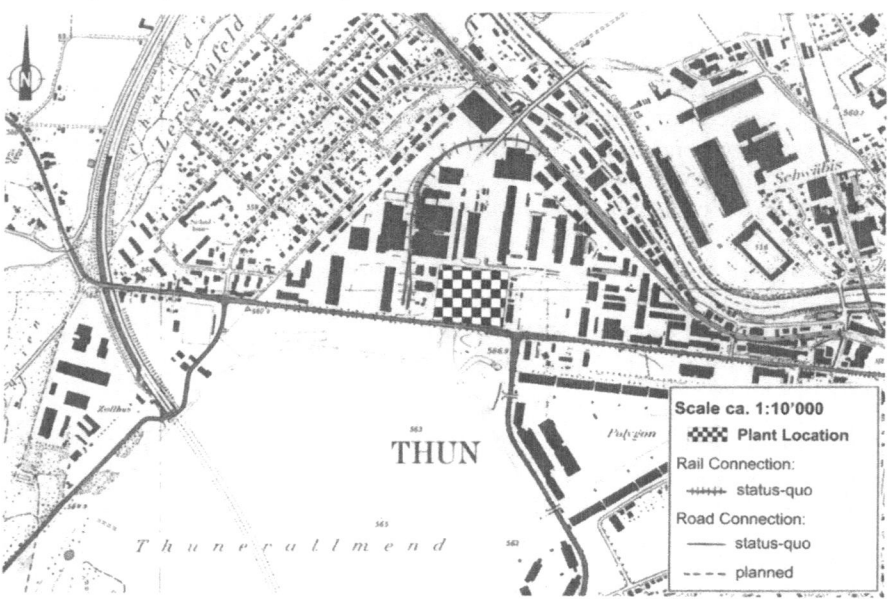

Fig. 7.2. Planned location of the waste treatment plant in the city of Thun: The closest residential area is Lerchenfeld, at a distance of 500 m (AVAG, 1993, [3]).

maximization of the potential federal subsidies. They therefore did not find time to adequately evaluate the opposing parties' statements.

Relevant information must be mobilised and offered to all participants. All parties must be willing to learn from this information and to take part in a discussion about it, and then to re-evaluate their respective positions accordingly. It is to the merit of the moderator that all participants are adequately informed and that new facts are brought to the discussion table. From his point of view, the main aim of the round table is to start a discussion between all relevant stakeholders. In a follow-up discussion, the aim should be to develop strategies for the sustainable development of local waste management.

The parties in opposition called for a feasibility study (Table 7.9), which was intended to indicate alternative waste management scenarios and to assess their potential for realisation. They could not understand why the incinerator received so much support among officials and experts when, to the citizens, the options of recycling, reducing, and reusing waste were left unexplored, let alone implemented [38]. They called for an expert group to survey alternative waste management ways in which they should have been represented. Nevertheless, neither AVAG nor the canton supported the actualization of the feasibility study for alternative waste management.

Table 7.9. Points of reference for the requested feasibility study by the opposing parties.

Points to be considered	Content to be considered
Geographical area	Location, accessibility, region represented by AVAG
Social Compatibility	Security, health, noise, potential mental effects
Process engineering	Increasing the rate of waste avoidance, sorting out, recycling, reducing the volume of burning
Ecological	Increasing the substitution of fossil fuel, CO_2 accounting, reduction of harmful substances
Economic	Overall cost efficiency, alternative use of the planned location, ensuring the care for long term security of the landfill site
Future Orientation	Assessment of future process engineering potential, supra-regional impact

Legal Assessment of the Case "Waste Treatment Plant Thun"

The participatory process should be implemented in the planning procedure as early as possible. This would help to reduce time pressure and, thus, not impede research into alternative possibilities. It is essential to clarify where there is room for discussion and the implementation of alternatives, and where there is no room at the beginning of the process. The broader the options, the higher the chances of

having an open discussion that might lead to a fair solution, but also the more difficult the discussion will be. If there is no consensus in the beginning about the procedures, then there will most likely be no consensual result forthcoming.
Points that should have been discussed are as follows [47]:

- The *assessment of the need*: If there is a legitimate need for burning capacity (according to the criteria agreed upon), then the plant itself cannot be questioned. This was precisely the unresolved point right up to the end in Thun. The question was asked within the framework of the participatory process, but during the process potential alternative assessments were only presented and never really discussed. If such a fundamental question is asked, it has to be resolved before entering into the participatory process.

- The *assessment of the location*: A discourse is taken up too late if its only purpose is to assess a predetermined location, and the project as such cannot be reassessed. If the potential of the participatory process is only to fine tune a few modalities, then the whole process starts far too late to really have the function of a forum of discourse. The participatory process should have started at the time of the evaluation of a potential location. The evaluation that took place in 1993 is plausible and comprehensible, but the basic criteria and their valuation are not the only ones possible. Those are the questions that are legitimately controversial and therefore implicitly sensitive to public collaboration.

Should conflicts between the stakeholders (e.g. waste management utilities, cantonal authorities, federal authorities and local opposition groups) result in court action, there is the threat of drawn out and costly law suits. Constructive, fact oriented discussion between the different stakeholder groups becomes virtually impossible. This is due to the confrontational nature of the basic structures of the procedures. On the one hand there are the planners, on the other there are the registered protesters, and finally, there is a yes/no decision by the relevant authority, which leaves no room for negotiation. The Thun example clearly shows that the legally prescribed opportunities for participation in Switzerland are unsuitable for reaching generally acceptable decisions.

Moreover, it is important to note that participatory processes are not intended to serve the purpose of finding acceptance for predetermined decisions. Rather, they should help in the search for generally acceptable solutions for all the stakeholders, *before* any decisions are actually made.

7.4.2 Local Mobilizations against Waste Incinerators in England

Christopher Rootes

The UK consigns a higher proportion of its waste to landfill than any EU state except Ireland. However, the EC landfill directive and the increasing likelihood that the UK will fail to meet its targets for the reduction of the amount of waste going to landfill have stimulated both the national government and local waste authorities urgently to consider alternatives. Nowhere has this urgency been greater than in Kent, almost all of whose waste is currently landfilled, much of it in neighbouring Essex under contracts which are about to expire.

Kent County Council (KCC), as the responsible waste authority, identified incineration with energy recovery as a major part of its alternative strategy and encouraged Kent Enviropower (KE), a consortium of British and multi-national engineering companies, to build and operate a large waste incinerator. An initial proposal to site it near a cement works in the village of Halling encountered fierce local opposition, as did another proposal to build a waste incinerator on the site of an existing power station at Kingsnorth. Both Halling and Kingsnorth are in an already polluted and economically depressed area which, under local government reorganization, was to be excised from Kent to create the new Medway Towns Authority. The prospect of Kent burning its waste in Medway raised questions of democracy to which the new Labour government could be expected to pay close attention, especially as all three Medway MPs elected in 1997 were Labour.

Siting a Waste Incineration Plant: Arguments for the New Site and First Reactions from Local Residents

Both the Halling and Kingsnorth proposals were abandoned and, just before Christmas 1997, it was announced that KE was applying to build an incinerator to burn 500'000 tonnes per annum of non-hazardous municipal and industrial solid waste at Allington Quarry, near Maidstone (Fig. 7.3). The attractions of the Allington site to an incinerator developer are obvious: bordered by a small commercial estate and overlooking a paper mill whose chimneys emit large volumes of steam, it is almost directly accessible by motorway, rail and river.

This, however, is less important than the fact that the 24 hectare site constitutes a major part of the narrow "strategic gap" of undeveloped land that separates the boroughs of Maidstone to the east and Tonbridge and Malling to the west. In planning terms, this "strategic gap" has been sacrosanct, zealously defended against a succession of would-be developers, and enshrined in the Kent Structure Plan, the document which guides all land use development in the County.

It was, therefore, an enormous shock to local residents to be told that this site had been proposed for the construction of a massive waste-to-energy plant. KE had identified Allington as a potential incinerator site in 1993 and had approached

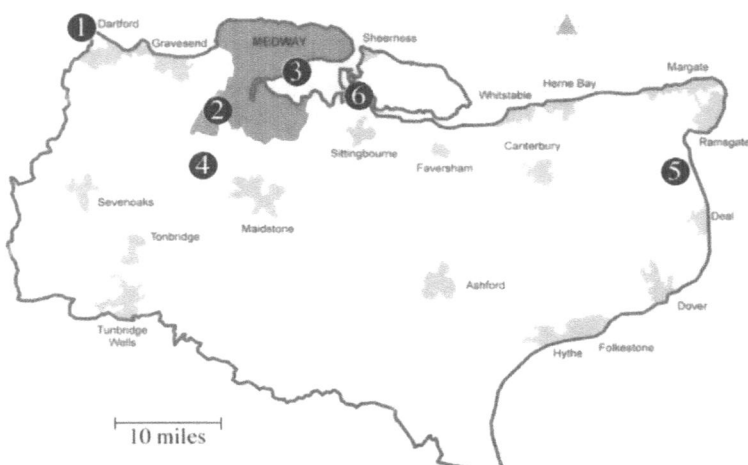

Fig. 7.3. Sites of proposed incinerators in Kent and Medway. Key to proposed incinerator sites: 1. Belvedere (South east London), 2. Halling, 3. Kingsnorth, 4. Allington Quarry, 5. Richborough, 7. Ridham Dock

the quarry operator, but incineration did not at that time fit with the strategy of the quarry operator's waste management subsidiary which was simply to use inert waste to fill the holes left by quarrying.

Several factors combined to change this. Firstly, the UK government's new landfill tax reduced the amount of builders' waste sent to landfill, with the result that Allington Quarry would take 40 to 50 years to fill rather than the twenty originally anticipated, and so became a less commercially attractive operation for the landfill company. Second, the Inspector's comments on the new Kent Waste Local Plan criticised the method by which KCC had selected the five brownfield sites designated as potential incinerator sites and said they should have no greater weight than any others which met the appropriate criteria. At about the same time, a consultant's analysis of potential sites identified Allington Quarry as the best waste incinerator site in the County – it met all seven criteria and was of sufficient size. The proximity principle also favoured Allington: it was well-placed to serve all four of the largest towns in mid and west Kent.

Opposition against the Location of a Waste Incineration Plant

The immediate opposition to the Allington proposal came from the Maidstone Green Party and the quickly formed Maidstone Incinerator Action Group, modelled on the existing Halling Incinerator Action Group. A series of public meetings attracted large numbers of concerned residents, and all the local district Councillors announced their opposition to the incinerator.

The campaign was at first dominated by the residents' action group, and energetically conducted both by new recruits and by a leading veteran of the Halling and Kingsnorth campaigns. Public meetings were held, highly professional presentations were made to Councillors, and Council meetings were lobbied. At this stage, the tactics of objectors were to question the claims made for the safety, efficiency and environmental acceptability of incineration and to argue that more environmentally friendly alternatives were not being considered. To counter KCC's scepticism about the prospects for a dramatic increase in reduction, reuse and recycling to reduce waste growth, a strong case was made for the use of industrial-scale anaerobic digesters, and composting campaigners came from other parts of Kent to assist in establishing a community recycling and composting scheme for the area.

Although the campaign quickly secured the support of Maidstone and Tonbrige and Malling Borough Councillors, it appeared to make no impression either on KCC officers or the members of the majority Conservative group on KCC. As the campaign dragged on, personal differences within the residents' action group diminished its effectiveness, and the main burden of opposition was assumed by the two Borough Councils. Within KCC, the most prominent opposition came from two Councillors from the opposition Liberal Democrat group. Although persuaded by the arguments against incineration, for tactical reasons they couched their opposition to the proposal mainly in land-use planning terms. The Council officers' response was to dismiss as minimal the likely impact of traffic associated with the plant, and to argue that the overwhelming need for new waste disposal facilities as an alternative to landfill overrode normal land-use planning constraints.

Over 3'000 letters objecting to the proposal were received by KCC and strong objections were lodged by all the neighbouring District and Parish Councils as well as the local water supply company which feared contamination of a main aquifer below the site. Nevertheless, the majority of KCC's Planning Committee (all Conservatives and none from the immediately affected area) voted in November 1999 to approve the proposal. Because the development was in breach of the County Structure Plan it was automatically referred to the Secretary of State for the Environment who promptly announced that he would not intervene, explaining that it was policy to over-rule local authorities' decisions only where they were in serious conflict with planning guidelines and /or had wider planning implications. Maidstone and Tonbridge and Malling Borough Councils applied unsuccessfully to the High Court for judicial review of the decision, and although the Environment Agency (EA) had by November 2001 still not issued a certificate permitting the operation of the proposed plant, KCC is confident that it will do so, that construction will commence in 2002, and that the incinerator will be operating in 2004.

Several Things are Interesting about this Case

The proposal was to build what the developers proclaimed to be a "state of the art" waste-to-energy plant based on a fluidised bed incinerator of a kind already operating in urban locations in Sweden. Objectors suggested that the contents of the waste stream were likely to be different (and potentially more toxic) in

England than in Sweden where environmental awareness among the public is greater. They also seized on the fact that the incinerator proposed for Allington would be much larger than the Swedish plants and so would be not merely "state of the art", but an experiment- a view shared by those in the waste industry who favour the more common "mass burn" process or other methods of waste disposal. Yet Councillors and officers dismissed anaerobic digestion – the alternative proposed by objectors – as "an untried technology", despite the presentation of evidence of its successful use in the US.

Scientifically educated objectors and environmental groups raised the point of health risks allegedly associated with incineration, especially those posed by dioxins and furans. At the key public meeting, KE's project engineer did not deny those risks but described them as ultimately unquantifiable because scientific knowledge is never complete. He reiterated the company's assurance that the waste industry was uniquely highly regulated and that the plant would conform to the highest EU standards and would be closely monitored by the EA.

Either because arguments about possible health hazards appeared too abstract and hypothetical or, in view of the claim and counter-claim between developers, council officers and campaigners, perhaps simply too complicated and contested, members of the public generally focused on things they could understand – smell, dust, vermin, traffic, or the effect upon property values. There was little evidence among the public of an awareness of new risks associated with advanced technology. But nor does it appear that the arguments, good or bad, about the proposed technology, the risks of incineration, or the loss of amenity to the local population, had any influence on the decisions of either the waste authority or the government.

Although the Allington campaign was less locally intense than those at other nearby sites where incinerator proposals were withdrawn, the decisive differences between them did not lie in the way the issues were framed. The universalist anti-incineration arguments presented in the course of the Allington campaign had been honed and made more comprehensive and sophisticated as a result of the earlier campaigns, and Councillors had little option but to focus on grounds of objection consistent with planning laws. The crucial difference was in the political structural contexts within which the campaigns were conducted.

If local authority boundaries did much to defeat proposals for incinerators at Kingsnorth and Halling, they had the opposite effect for Allington. Allington is on the edge of the redrawn County of Kent. The districts in Kent most affected by the incinerator are represented by Liberal Democrat councillors, but the majority of the Councillors who took decisions on the incinerator were Conservatives representing voters living up to forty miles away in east, west and north Kent. Many of these Councillors were being strongly lobbied by local campaigners – and the Council for the Protection of Rural England – against landfill operations in their own areas. Thus they had no interest in rejecting a proposal that appeared to offer a solution to a pressing problem and whose undesirable effects would not be felt by their supporters. It was relatively easy for Kent's Conservative Councillors to represent the waste to be burned at Allington as a problem arising in Maidstone and deserving of disposal there, despite KCC officers' briefings that made it clear that although the household waste would be drawn from the nearest four boroughs, the 300'000 tonnes of commercial and industrial waste might be drawn from a much wider area.[4]

Timing was also a Factor

In the years that it had taken to find a suitable site for an incinerator, Kent's need for new facilities to deal with an ever increasing quantity of waste was becoming ever more urgent. The publication of the government's strategy document, *A Way With Waste*, in 1999 and the Environment Minister's evident frustration at slow progress toward recycling targets meant that the government was increasingly resigned to the construction of a large number of incinerators as a major contribution toward meeting targets for the reduction of landfill. Thus local urgency combined with a lack of national government resolve to resist incineration, and the battle against the Allington incinerator was lost without even a public inquiry.

KCC has contracted to deliver a minimum tonnage of waste to Allington, but recently increased government targets for recycling threaten to reduce the volume of waste available for incineration and so make it more likely that KCC's projections will prove over-optimistic. Having committed itself to Allington, KCC thus has a strong interest in rejecting possible rival incinerator proposals. In 2000, deploying arguments its members and officers had rejected when they were made by opponents of the Allington scheme, KCC rejected proposals by the French-owned waste corporation, SITA, to build waste incinerators at Richborough, near Sandwich, and Ridham, near Sittingbourne.

KCC officers recommended refusal of the Ridham application because of uncertainty about local air quality and effects on wildlife. Yet English Nature concluded that possible emissions from the incinerator would not adversely affect wildlife, and the EA commented that "even with the most pessimistic assumptions, National Air Quality Guidelines would probably not be breached." KCC officers put the most negative construction on this, emphasising the word "probably" in order to highlight the possible risk, and Councillors obliged, declaring that KCC "should not do anything that puts at risk the health of the people of Kent." Yet both Allington and Kingsnorth were recommended for approval, and Councillors were disposed to approve them, when the EA had not received a first stage application and so was unable even to comment on possible environmental hazards. The precautionary principle, it appears, was invoked only when it was convenient to do so. A final irony is that SITA, having appealed against KCC's rejection of its planning application for Ridham, has secured a

---

[4] Only much later (in 2001) did it emerge that, in order for KCC to supply the incinerator with the contracted tonnage of waste, it would be necessary to truck domestic waste in from more distant parts of Kent.

public inquiry[5]. Developers, it seems, have rights to public hearings that the affected citizenry and their elected representatives do not.

Political Contention over Incineration in England

The saga of contention over incinerator siting in Kent has parallels elsewhere in England [45]. Hampshire County Council has over-ridden local opposition to approve two waste incinerators, whereas in Portsmouth, recently excised from Hampshire to become a new unitary local authority, Councillors rejected officers' recommendation to approve an incinerator application. The political context of planning decisions appears to be all-important. In geographically large counties such as Kent and Hampshire, it is much easier to get a majority of Councillors to take locally unpopular decisions than it is in geographically compact authorities like Medway and Portsmouth in which, as in other new unitary authorities in England, Councillors have "no place to hide" from unpopular decisions and so have rejected incineration as a means of waste disposal.

If political structures and the imperatives of national policy have been more important than the quality of arguments and local campaigns in determining decisions on the siting of waste incinerators in Kent and Hampshire, the case of Essex suggests that even in large counties different outcomes may be possible. There a consortium of borough councils, building on the experience of their successful collective resistance to a County plan to increase house-building, formed to ensure that the Essex Waste Local Plan did not recommend incineration without full consideration of alternatives. Essex, however, had the advantage of being late to develop a Waste Plan and so could draw on the experiences of Hampshire and Kent.

Conclusions

Nevertheless, collective resistance to incineration will not of itself solve the increasing problem of waste. As the experience of another shire county – Surrey – demonstrates, even an exemplary process of public consultation may fail not only to identify any publicly or politically acceptable waste disposal sites but also to persuade citizens to accept responsibility for the waste they produce. So long as

---

[5] SITA has appealed principally on the grounds that KCC has wrongly claimed that there is no need for additional incineration capacity in Kent. SITA asserts that the fluidised bed (FB) technology proposed for Allington is inappropriate for the incineration of municipal solid waste, that it will be more costly than mass burn (MB), and that it will not deliver the promised contribution to recycling targets, not least because, by comparison with MB, FB produces much larger quantities of toxic fly ash (>14% of the tonnage of waste cf 4% for MB) which must be treated as special waste for which there is no licensed disposal facility in Kent. SITA further contends that, with the withdrawal of Kvaerner from the KE consortium, the developers of Allington have never built or operated an incinerator and will, in view of the problems with FB technology, be incapable of delivering the promised capacity at the contracted price.

recycling rates remain low and the production of waste continues to increase, political contention over incineration in England seems likely to grow.

Despite assurances that permitted emissions present no risk to human health, residents show no signs of being prepared to welcome waste incinerators into their surroundings. Yet contention has until now largely been focused upon the particular sites designated for incinerators and only belatedly, and often as a by-product of local campaigns of resistance, have more general issues of waste management strategy been raised. Local campaigners have, as a result, been largely at the mercy of the caprice of local circumstances.

Recently, however, national political and environmental groups have been increasingly willing to take up the issue. During 2001, Greenpeace, an organization that had not previously been active on municipal waste issues in Britain, launched a campaign against incineration with a characteristically spectacular protest against the Edmonton incinerator in London, and in February 2002 Greenpeace activists occupied the "flagship" SELCHP incinerator in south-east London. In both cases, the protests were designed to highlight the health risks posed by the emission of dioxins, both incinerators having been at least briefly in breach of permitted emission levels, and to encourage consideration of alternative modes of waste disposal such as recycling and composting. The opposition Conservative Party has called for a moratorium on the building of new waste incinerators, its environment spokesman has called on the government to investigate Greenpeace's claims about dioxin emissions, and Liberal Democrat Councillors and Green members of the London Assembly have supported the Greenpeace actions. There are indications, then, that the political context may be changing and that the contention over waste management may shift from battles over siting decisions to debates about sustainable strategies that bring issues of consumption and civic responsibility back to centre stage.

7.4.3 Public Participation for a Waste Management Plan in Germany

Ortwin Renn, Elke Schneider, and Hans Kastenholz

The Controversy of Solid Waste Management in Germany

In Germany, special administrative entities named "Gebietskörperschaften" (a county or a regional district) are responsible for waste collection, treatment and disposal. But in a densely populated country like Germany it becomes more and more difficult to allocate new sites for waste disposal. Together with increased efforts for recycling and other strategies for waste reduction, a law [1] has been established defining specific standards for all disposable material after the year 2005. To comply with these legal requirements, all counties or cities will have to upgrade their waste treatment system by the year 2005 in order to meet these standards. While the legal setting does not explicitly call for a certain technical option of waste treatment, in practice incineration is given priority because at the moment it is the only treatment technology which meets the official criteria for disposal. But critics question the scientific validity of the criteria and doubt that the law itself will survive in the longer run. Both lines of criticism are discussed widely and this causes a great deal of uncertainty in the public debate on waste management planning.

The Structure of Participation

The official process of waste management planning now in use in the Northern Black Forest Region was a response to the previously mentioned legal situation in Germany. After a history of various planning efforts and systems implemented in this field, in 1993 three rural counties and the City of Pforzheim formed a cooperation to seek a common solution on the regional level for their waste problems. For this task a special planning organization (P.A.N.[6]) was established and supported by the counties and the City. An engineering consultant was hired to provide the decision-making committees with the necessary technical information. When P.A.N. representatives learned about the Stuttgart Center of Technology Assessment's (directed by the author) experience with conducting public participation in a structured manner, they decided to implement such a program in the upcoming planning process and began collaborating with the Center [overviews in 43, 46].

In the official process of decision making, the task of developing the waste management concept was divided into three consecutive decision phases, each of them setting the necessary framework for the following phase. The first step consisted of an estimation of capacity requirements for waste management for the year 2005, determining the minimum and maximum capacity needed for the regional concept. Based on the results of this phase, in the second phase the appropriate treatment technologies (type of facility) had to be selected in order to define the specific selection criteria for the third step, i.e., the siting of the facilities. The results of the participation program had to be available according to this time frame and the discussions were to proceed parallel to the topics of the official phases. Fig. 7.4 illustrates the involvement scheme that was developed for this case, based on a conflict analysis prior to the start of the participation program (adopted from [46]). The decision-making methods changed during the three phases. In phases II and III the organizers of the Center applied the value-tree and MAU decision analysis [18, 28, 29, 42] methods for structuring the decision making process. The first phase (waste prognosis) did not call for such a procedure, as forecasting is an evolutionary procedure.

---

[6] P.A.N. GmbH = Gesellschaft zur Planung der Restabfallbehandlung in der Region Nordschwarzwald, Deutschland (Society for Planning the Disposal of Residues in the Northern Black Forest Region, Germany)

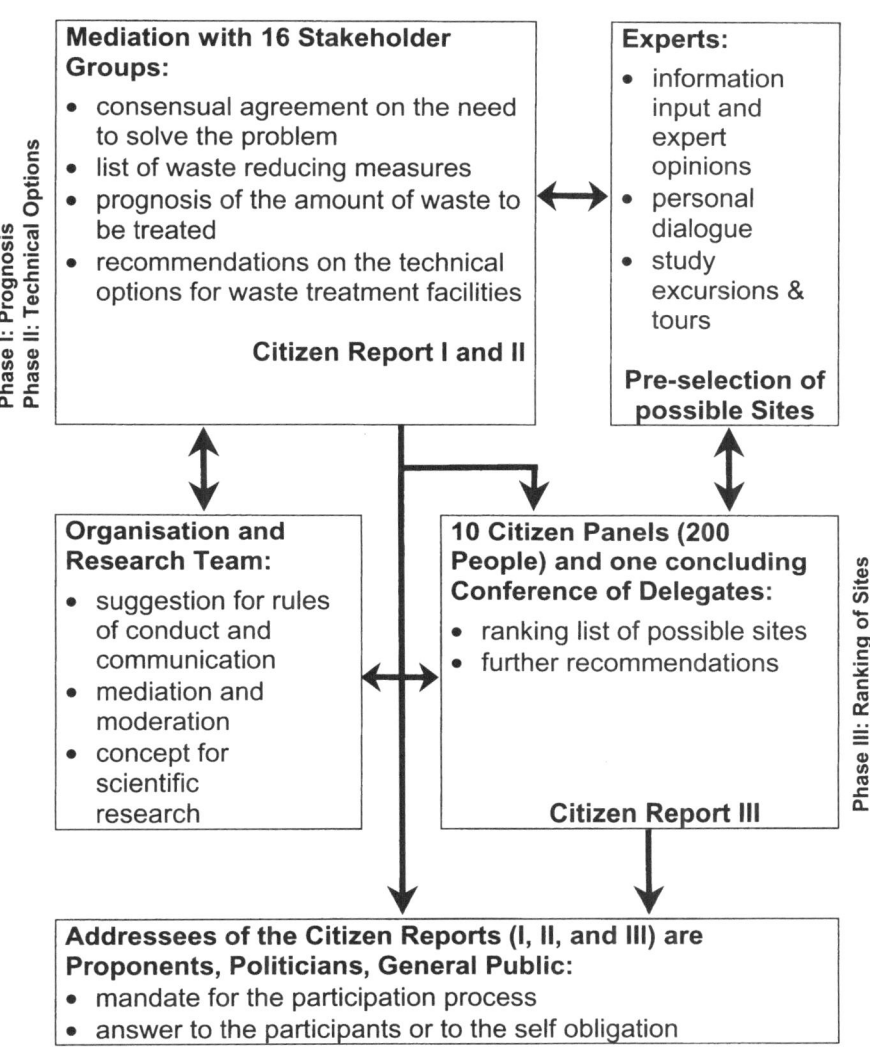

Fig. 7.4. Applied Model for the Project: Macro-structure

In phases I and II stakeholder groups convened in a series of consensus conferences, first developing a waste capacity estimation for 2005 and then ranking possible technical options for waste treatment. After these tasks were completed and the political decision for a combination concept for the region was made, the third phase of site selection was initiated. The task was to locate one central incinerator and two biomechanical treatment plants in the region. Sixteen communities had been identified in a preliminary suitability study by the consultant as potential sites for waste treatment plants, some of them being suitable for both basic methods. In this phase we applied a modified version of the concept of planning cells [16, 17]. A random selection of approximately 200 inhabitants from the potential site regions was conducted. The 200 randomly selected citizens were assigned to one of 10 parallel working citizen panels, each consisting of the same number of representatives from each potential site community.

The panels were given the task of determining the most suitable sites among these sixteen. Four of the panels focused on the siting of the incinerator and the other six developed criteria for siting the biomechanical treatment plants. They developed site selection criteria and ranked all the sites, considering social, political, ecological, and economic impacts as well as equity issues including benefit-sharing packages. The team again used value-tree analysis and a modified MAU-procedure to reach a consensus among the participants.

The decision making process included:

- construction of a value-tree in one brainstorming and several discussion sessions, establishing consensus on the values and their hierarchical structure;
- construction of a catalogue of criteria which could be used as benchmarks for collecting and processing information for each potential site on each criterion;
- evaluation of the criteria;
- judging and ranking of the options relative to each other according to their performance profiles with regard to the different criteria;
- discussing the results and compiling a final document.

Each of the ten panels reached a unanimous conclusion with respect to the ranking list. In the end, every group elected three delegates who were to meet in a special conference with the objective of composing one common suggestion for a combination method incorporating both treatment technologies in the optimal way. Finally each panel was given the opportunity to comment on the result of the conference of delegates and all suggestions were included in the citizen report. Fig. 7.5 shows the applied set of methodologies in each of the three phases (adopted from [46]).

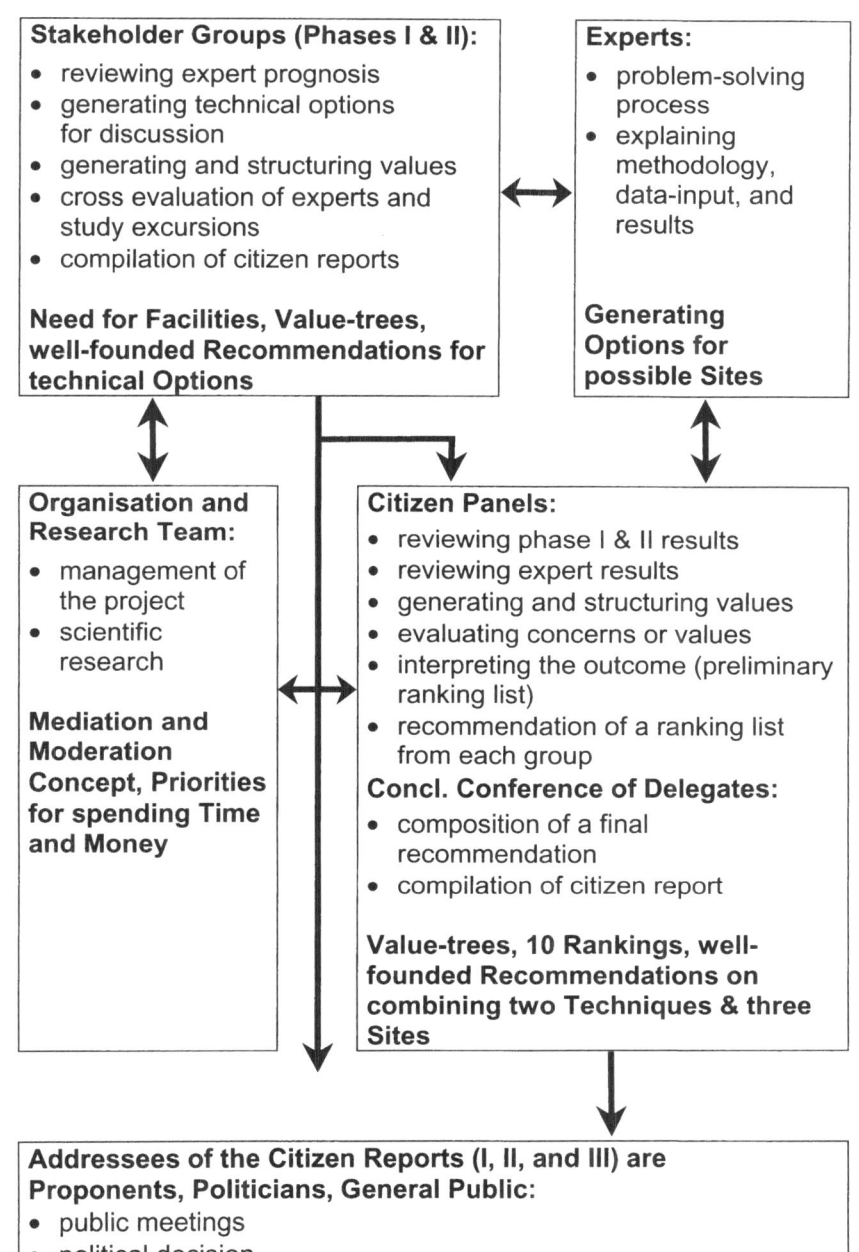

Stakeholder Groups (Phases I & II):
- reviewing expert prognosis
- generating technical options for discussion
- generating and structuring values
- cross evaluation of experts and study excursions
- compilation of citizen reports

Need for Facilities, Value-trees, well-founded Recommendations for technical Options

Experts:
- problem-solving process
- explaining methodology, data-input, and results

Generating Options for possible Sites

Organisation and Research Team:
- management of the project
- scientific research

Mediation and Moderation Concept, Priorities for spending Time and Money

Citizen Panels:
- reviewing phase I & II results
- reviewing expert results
- generating and structuring values
- evaluating concerns or values
- interpreting the outcome (preliminary ranking list)
- recommendation of a ranking list from each group

Concl. Conference of Delegates:
- composition of a final recommendation
- compilation of citizen report

Value-trees, 10 Rankings, well-founded Recommendations on combining two Techniques & three Sites

Addressees of the Citizen Reports (I, II, and III) are Proponents, Politicians, General Public:
- public meetings
- political decision

Fig. 7.5. Applied Methodologies within the Project: Micro-structure

The Results

Phases I and II resulted in a documentation of substantiating arguments in favor of the four alternative solutions suggested by the participants of the Round Table. The discursive process was successful in producing a common document in which the potential risks and benefits of all four options were described. This document was signed by all participants and was later used as important input for the political decision making body (County parliaments and City Council of Pforzheim). It certainly validated the willingness of all parties (even the most radical representatives) to accept a equally representative and well balanced procedure when making these decisions. They all agreed that each option had specific advantages and disadvantages and that it was legitimate to assign trade-off potentials between them. Fundamentalist positions were gradually phased out during the deliberation process. Purely strategic reasoning was replaced by an exchange of constructive arguments, evaluations, and interpretations.

For the policy makers in the community and county parliaments, this processing of complex information and the structuring of many potential options in four feasible and, in their own rationale, optimized policy alternatives provided the needed input to make a substantiated and legitimate decision. All parliaments voted for alternative 2 (pre-handling, drying and processing of waste before incinerating the remaining dry waste), guided by the discussion of the Round Table. Although not all participants liked the final vote, they accepted that the appointed decision maker made a deliberate choice among equally feasible and defendable alternatives on the basis of legitimate political preferences.

In phase III, the 10 citizen panels were asked to evaluate various potential sites for up to three biochemical plants and one incinerator. Before we assembled the panels we conducted a survey among the participants asking them about their attitudes toward risk and general perceptions, as well as their expectations. Similar to the participatory experiences that our team had in Switzerland in 1992 [41], almost 80% of the citizens were convinced that their hometown was not suitable for any type of waste disposal facility [52]. More than 60% believed that incinerators were associated with high risks for human health and the environment, 40% believed this was also true for the biochemical plants. One third of the citizens were very skeptical about the prospect of finding a common solution to the waste problem in the area, another 40% were relatively skeptical, 20% were fairly optimistic and 6% very optimistic.

The survey was repeated in the middle of the participation procedure (before making the final decision) and several months after the procedure. The survey results demonstrated that the participants gained confidence in the participation process [52]. Almost 65% were convinced that they could reach a consensual decision when they were asked in the second survey. They also showed more openness towards offering their hometown (40% still thought their town was not suitable, 42% though it may be suitable, and 5% thought it was definitely suitable, the remaining 3% voiced no opinion). Of special interest are the results concerning individual perceptions of the participation process itself: More than

85% of the respondents were convinced that they have learned a lot during the deliberations. They emphasized that they found the procedure to be fair with respect to the different positions as well as to the populations of the different site regions [43]. They also recommended the procedure for similar future disputes (74%).

The last indicator of a substantial change due to the deliberation process was the consequent decisions. All ten groups reached a consensus on their final verdict. Nine groups provided a prioritized list for siting one of the planned waste disposal facilities. The four groups looking for a location to host an incinerator all agreed that the densely populated city of Pforzheim was the most suitable site. The reasons were as follows:

- waste heat could be used for district heating;
- the solution was financially very attractive;
- the waste incineration plant was close to those people who generate the waste (fairness issue);
- the waste incineration plant would not impose risks on the local population that were higher than other risks to which they were already exposed;
- the waste incineration plant was close to the town hall and the city administration which would provide an additional assurance that the regulatory agencies within the administration would commit themselves to thorough inspections (since they would be the first victims if they failed to do so).

These explicitly articulated reasons provide sufficient evidence that the original fear of exaggerated risks was overcome during the process of deliberation. In contrast to the representatives of phase II, the citizens were not committed beforehand and thus willing to change their previous beliefs and even attitudes. They had nothing to lose by examining each argument and decided on the basis of the evidence provided whether the argument was valid or invalid or whether it should be modified. Creating sufficient dialectical space provided the participants with ample opportunities to reach a consensus or at least a compromise.

Conclusions

Involving citizens in the decision making process requires careful planning, thoughtful preparation, and the flexibility to change procedures on the demand of the affected constituencies. A cooperative discourse aims at integrating public input prior to the final decision. It is meant to address public concerns, to collect local knowledge, and to exchange arguments among the various stakeholder groups. Such a pre-decisional discourse can only succeed if the following requirements are met:

1. *A clear mandate for the discourse participants:* What are the topics of discussion? What results are expected of them?
2. *A clear understanding of the options and permissible outcomes of such a process:* If, for example, the site for a risk producing facility is already chosen,

the discourse can only focus on issues such as choice of technology, emission control, and compensation.

3. *A predefined time table:* It is necessary to allocate sufficient time for all the deliberations, but a clear schedule including deadlines is required to make the discourse effective and product-oriented.

4. *A mutual understanding of how the results of the discourse will be integrated into the decision making process of the regulatory agency:* As a pre-decisional tool the recommendations cannot serve as binding requests. Rather, they should be regarded as consultancy reports similar to the scientific consultants who articulate technical recommendations to the legitimate public authorities.

The experiences from our project together with other case studies in this field clearly show, that citizens are capable of making prudent recommendations when given the opportunity to deliberate over a complex and uncertain planning task with an ambiguous outcome.

7.4.4 Public Influence on Siting Waste Facilities in California, USA

Richard V. Anthony

The public demand for efficient management of resources postulates a zero waste strategy. Most public opposition to disposal facilities siting have to do with the improper disposal of wasted resources. The argument of consumer convenience is used in favor of a landfill or incinerator. However, neighbors to such facilities view such a potential condition as a threat to their property value and quality of life. Many of the facility siting issues of environmental justice movements are based on quality of life and property values.

The following case studies will discuss several successful siting attempts for landfill and waste to energy plants. The success in each case is an example of the perseverance of a government agency and/or its private sector contractor in establishing a long-term disposal system. Each siting had opposition and local issues of environmental impacts.

Siting Municipal Solid Waste Landfills

The California County of Fresno successfully sited 100 years of landfill capacity in the late 1980s. The County is located in the middle of the San Joaquin Valley in the California Central Valley. Much of the world's food supply is grown in this region, which can be seen from the moon. On the east is the Sierra Nevada Mountain Range, and 150 miles west is the California Coastal Mountain Range. The Sierra Nevada receives abundant snowfall and, through the use of dams and irrigation canals, provides water for agriculture. The Valley has a huge sole source ground water aquifer.

The lack of long-term landfill capacity was the subject of the County Waste Management Plan in 1978. The plan called for the evaluation of several sites in the northeastern part of the County. The consideration of these sites for waste management plants roused the anger of the adjacent ranchers and farmers, as the eastern sites were closer to the mountains and the water supply than areas further west where the ground water was deeper.

The Solid Waste Authority allotted the task of choosing potential sites to the Farm Bureau. The Farm Bureau is a non-governmental organization (NGO) representing the agriculture community. As they had been vocal about the planned sites, it made sense to offer them an opportunity to identify sites that met with their standards. The sites they chose were on the west side of the valley on marginal farmland where the ground water was deep. The selected site was an expansion of an existing site.

California law required an environmental impact report. A Citizens Advisory Committee reviewed the report and recommended double liners and other environmental mitigations. There was citizen opposition from neighboring regions.

This use prevented potential farmland from producing goods and created a waste disposal site for the regions waste. The number of jobs required for running the plant are far fewer than what source separating the material and selling it to commodity and compost markets would have required.

The California County of San Diego spent millions of dollars in identifying and performing research for environmental impact reports to locate a northern county landfill site. San Diego County is the sixth largest county in population in the United States. In the 1990's, the County Solid Waste Division controlled all the landfill sites in the region except for the one used by the City of San Diego. The County provided disposal for 18 cities and 1.2 million people.

The County and the City of San Diego had faced rapid growth at that time. The suburbs enveloped rural areas; sewer treatment plants and landfills as well as agricultural uses became public nuisances. Waste to Energy as an alternative was rejected in the City of San Diego by ballot. The demand from the local citizens was that the material ought to be recycled.

A technical advisory committee located the landfill sites scientifically. Size and proximity to the highway had high points. The best canyons seemed to be near water flows. Neighbors near and down grade from the potential sites became concerned about ground water contamination. Neighbors from existing sites began to organize and protest existing operations. The environmental impact report was forced to answer thousands of questions raised by citizens at public meetings and through letters.

The site could never satisfy the need to create jobs and industry as only a few are created by landfill. The county and, subsequently, the promoter weren't able to generate substantial interest for the construction industry, as the infrastructure consists of road work and dirt moving. The landfill site for North County has yet to be permitted.

Siting Waste to Energy Plants

In the late 1980s, *Fresno County* negotiated with a private entity to provide a waste to energy mass burn system and a transfer station to take residual to the new lined regional landfill. Concern with the need for a regional landfill and its possible impacts on ground water led the planning agency to negotiate for volume reduction and steam and energy production. The site was placed in an industrial area near a rail spur and adjacent to a steam run cotton oil plant. The Sierra Club, a local NGO with the mission of environmental advocacy, and the local citizens committee supported the project.

The question of flow control and recycling became the issue of protest. Paper recyclers were concerned that potential recyclable fibers were being committed long term to the financing of the plant. Even with a payment from the utility for the alternatively produced energy, there was a need for a long-term tip fee.

Another issue was the liability for the disposal of the ash and other hazardous waste that would be generated at the facility. The ability to meet the air pollution requirements in an area deemed by the air pollution control board as non-attainment was never tested, as the environmental impact report on this facility was never filed. Although it was identified as an economic disagreement, the contract was never signed due to risk the private company was going to have to take if their technology could not meet the air standards.

The technology and its promises of steam and energy weren't enough to bring additional funds to the project and provide it with the support it needed. The subsidized financing of the landfill made land filling seem like a better alternative. The lack of support from the local recycling community was a final blow.

San Diego County, like Fresno, initiated a waste to energy process while looking for a new landfill site. The election in the City of San Diego banning waste to energy plants impacted the county's planned facility. The City of San Marcos, the host city, voted for the plant. The ensuing controversy led the developer to include the community in the design of the facility. The final design was a materials processing facility making a refuse-derived fuel to be burned in a suspended fire dedicated boiler. The plant would recover 450 tons of the 2000-ton per day stream. The chief opposition came from local lawyers who lived in a nearby suburb.

The inability to get an air permit for the burners within the time allotted by the public bond sale forced the County to split the project and build the materials recovery facility first. This was done. The burners were never permitted and the waste to energy portion of the facility was never built.

The materials recovery facility created about 250 jobs recovering 450 tons of recycled materials per day. The plant was eventually shut down as cities found subsidized landfill was less expensive.

Conclusions

The following can be expected when siting disposal facilities.

- Siting landfills and incinerators will bring out anger in a community.
- The closer the citizen lives to the facility the more intense the indignation.
- Involving locals in the siting process allows all the issues to be discussed.
- Some areas are more appropriate than others.

Those in positions of authority within contemporary waste management programs must address the following questions:

- "What is the problem and what are the alternatives?" If the problem is one of lack of capacity for permanent disposal of unwanted discards, what are the alternatives? If thermo-destruction and land burial raise issues of air, water and land pollution, is there another alternative? Can these materials go back as recyclables and compostables?
- Is there an environmental fatal flaw? Will the process create new environmental problems? Will the long-term costs of resource depletion impact the short-term savings? Does the public really believe that there will be no environmental impact?
- Will it create jobs and industry? How many jobs and how much material feedstock will be created by the system choice?

The general public is suspect of wasting. The idea of wasting resources doesn't comply with "common sense" thinking. The older generation has memories of public dumps where valuable parts and products could be found. In a time of increasing population and depleting resources, landfills and incinerators are not popular.

The lack of suitable answers to the questions; "What is the problem and what are the alternatives?" "Is there an environmental fatal flaw?" "Will it create jobs and industry?"; will create controversy from local citizens looking for a way to stop an unwanted facility.

The old method of volume reduction and land filling does not answer the questions of alternatives, environment and jobs. Subsidizing wasting and supporting a throw-away society is frustrating for the general public and defies their common sense.

The public needs to know that the community authorities are the caretakers of the resources in the community. The quality of the air they breathe, the water they drink and the general quality of life is entrusted to local and regional government. The choices for the disposal of unwanted discards must follow a method of highest and best use and the systems for collecting and transporting these discards must be directed to recycling and composting facilities.

7.4.5 Conclusions of the Case Studies

Upon comparison of the presented case studies, it becomes evident that a certain *level of institutionalized participative policy making* seems to be an important factor for successful waste management planning. In practice, this means that situations in which no participative procedure is defined for siting a waste treatment facility (see UK case study, Section 7.4.2), or where the existing procedure is too exclusive due to legal and/or bureaucratic terms, or in which participation is implemented too late (see Swiss case study, Section 7.4.1), seem to be a very unlikely candidates for a successful outcome. In consequence, NIMBY reactions could be detected in the UK and in the Swiss cases, whereas in the German case (Section 7.4.3), *public participation* was implemented *from the beginning* for the site selection procedure in order to achieve general consent for the facility site.

The public does indeed have something to contribute to the planning process. The rationality of public input depends, however, on the procedure of involvement. Given a *conducive and supportive structure for discourse*, citizens are capable of understanding and processing risk-related information and of articulating well-reasoned recommendations. The discourse models are an attempt to design a procedure that allows citizens to take advantage of their full potential and they include the professional knowledge and expertise necessary to *make substantiated and democratically legitimate decisions*.

However, the chance for finding a satisfying solution depends upon the amount of *room for maneuver or flexibility*. From the start, the framework of the Swiss round table procedure was very limiting . Despite the various pledges, there were options that could not be discussed and the time pressure increased from day to day. There was not enough *time and flexibility* to build a consensus. The process clearly showed that reduced flexibility also reduces a willingness to compromise, and can thus contribute to the failure of a round table. What was missing in the Swiss case – a clearly structured participative procedure with an appropriate and efficient time plan – was present in the German case (Section 7.4.3).

It should be evident that people's present and future needs must be taken seriously by the administration. The responsible administrative body for waste management as well as the waste facility manager should consider the *public* as a waste management *client*. This is illustrated in the Californian cases, where the authorities are presented as "the *caretakers* of the resources in the community" (Section 7.4.4). In contrast to the US cases, the European cases (Sections 7.4.1-7.4.3) show another relationship between the administration and the public: The public is often regarded as a troublemaker and, therefore, not adequately involved in the waste planning procedures.

In all cases, there was a demand for the evaluation of *more environmentally friendly alternatives to waste incineration*. Whilst in the German case *measures for waste reducing* were listed parallel to the preselection of possible waste facility sites, in the Swiss as well as in the UK cases the officials and the experts did not support the options of recycling, reducing, and reusing waste. In one of the Californian cases a materials recovery facility was built, because the waste incineration plant was not permitted.

The increasing production of waste and the on-going debate over incineration call for various measures to improve social compatibility. Policy approaches (see Section 7.3.1) should consider the public demands, financial incentives (Section 7.3.2) should help to reduce the production of waste, and participatory processes (Section 7.3.1) could address public concerns in order to get a more socially compatible waste management.

7.5 Conclusions

Significant progress has been made in recent years in reducing the environmental impact of waste treatment. Waste management is, however, still far from meeting the criteria of sustainability. In particular, the social aspects of public waste management [see 26] have been addressed to only a small extent by scientists, planners and decision makers until now.

Recording and assessing social aspects is vital when it comes to matters of public interest. Authorities, business sectors, and associations could, for example, conduct social compatibility analyses (Section 7.2) to find consensus oriented solutions to conflicts, thus helping to structure the social dimension of a subject for assessment and to reveal the diverging assessments of different stakeholder groups. It hereby facilitates the negotiation of various solution approaches.

Planning and decision making processes must be substantially strengthened by involving stakeholders in the population as well as environmental organisations in the *participatory processes* (Section 7.3.3). Inter-regional waste management co-ordination and financial incentive systems promoting waste avoidance and recycling should operate in tandem with participation measures.

Sustainability can only be achieved by working within the framework of a long-term, social-educational process. This requires both effective communication and creativity. Participation and a democratic process must therefore be strengthened on both national and international levels (the concept must be based on the public having the *right to participate* in decision-making). This requires the availability of corresponding instruments of action, which improve administrative transparency and ensure the efficient involvement of the (affected) public. These elements are also demanded by *Local Agenda 21* processes.

The general public, important interest groups and various authorities should be fully integrated into the decision-making process. Consensus seeking is therefore an essential step on the road to arriving at a decision. This fact works significantly in favor of moves towards sustainable development. Nevertheless, participation mechanisms, above all when affected minorities are concerned, have to be increasingly strengthened. Networks addressing the problems of sustainable development (i.e. participatory processes) must focus on include a wide spectrum of participants covering many fields of interest (e.g. Federal Government, regional and local councils, business and industry, organisations), both general and scientific, as well as local interest groups and public initiatives.

Participatory processes create the chance for all possible interests to be taken into consideration, and by actively involving those affected, not only help to encourage more informed decisions, but also promise solutions which anticipate increased acceptance from the parties concerned. That is why participatory processes should, as they are not as of yet a legal requirement, at least be recommended to the court as additional processes.

At the same time it is very important to ensure that schools and universities promote environmental awareness. Because of their future orientation, *practically oriented research* programmes which will be able to assess developments in the field of waste management on a regular basis, (such as those in Universities of Applied Science) will assume a key role in this respect. As independent institutions they are best equipped to make objective comparisons between, and assessments of various waste treatment or recycling options, or to perform in depth studies of waste avoidance strategies. As with the issue of waste disposal, the matters of value transference and sustainable development should be well integrated into the educational programmes.

On the basis of our research, within a Delphi Expert-Questioning [10], the following *criteria* must be applied *for assuring the social compatibility of a waste management* programme:

- Accessibility of information, transparency of decision-making and decision-execution.
- Ensuring of participation rights for the affected public (within a direct-democratic framework); here the interests of future generations have to be considered.
- Individual and collective interests, as well as local and regional interests, must be weighed against each other, and where possible reconciled. Where conflicting, mediation processes or other conflict resolution methods must be carried out.
- Basic life opportunities (work, recreation, risk avoidance, and needs for food, water and warmth) must be justly available to all.

The environmentally responsible behaviour of people must be promoted above all by means of financial incentive and educational systems. A process of participative decision making on the siting of a waste facility shows substantial difference to other decision making processes: The different stakeholders have many varied interests. The public, for example, is interested in the choice of waste treatment technology as well as in the siting of facilities, the (potentially harmful) emissions emanating from such installations, the transport working in conjunction with the installations, and the costs which they have to cover as consumers or as tax-payers. On the contrary, the administrative body is primarily interested in having enough capacity for secure waste disposal – independent from the locality of the facility site. Therefore, starting with stakeholder-involvement from the earliest stages of waste planning is strongly recommended. Even as early as the political decision-making on waste management concepts at local, regional and national level, where basic decisions will be made, it is advantageous to carry out participatory processes. In short, sustainable waste management should ensure social compatibility by considering the interests of all the affected stakeholders. Approaches to assess, evaluate and improve social compatibility within waste management have been presented and should be put into practice.

References

1. Verordnung über die umweltverträgliche Ablagerung von Siedlungsabfällen vom 20. Februar 2001 (BGBl. I S. 305). Germany
2. Anonymous (5 5, 1998) Einsprachen gegen SBA Thun abgewiesen. Thuner Tagblatt, Thun
3. AVAG (1993) Kehrichtverbrennungsanlage KVA AVAG. Bericht zur Standortevaluation. AG für Abfallverwertung (AVAG), Zurich, p 77
4. Beckmann J, Keck G (1999) Beteiligungsverfahren in Theorie und Anwendung. Akademie für Technikfolgenabschätzung in Baden-Württemberg, Stuttgart
5. Brundtland G, al e (1987) Our Common Future. University Press, Oxford
6. Brunner U (6 13/14, 1998) Staatliche Rechtsetzung contra Selbstregulierung. Neue Ansätze im Umweltschutzrecht mit Vorbildcharakter. Neue Zürcher Zeitung, Zurich
7. Büchel D (1996) Konsensus-Konferenz und verwandte Methoden in der Schweiz. Schweizerischer Wissenschaftsrat, Bern
8. Bückmann W (1982) Aspekte der Sozialverträglichkeitsprüfung. IfZ GmbH, Berlin
9. Carabias V, Winistörfer H, Joos W (2000) Runder Tisch Abfallentsorgung Thun 1998/99. Zürcher Hochschule Winterthur, Winterthur
10. Carabias V, Winistörfer H, Joos W, Stücheli A (1999a) Delphi-Befragung 1997/98. Zürcher Hochschule Winterthur, Winterthur
11. Carabias V, Joos W, Seiler H-J, Winistörfer H (1999b) Legal dimensions of Social Compatibility of Public Waste Management in Switzerland. In: Boucquey N (ed) European Consumer Law and Waste Management. Centre de droit de la consommation. Université Catholique de Louvain, Louvain-la-Neuve, pp 326-340
12. Carabias-Hütter V, Winistörfer H (2001) Tools Needed for Sustainability Evaluation: the Social Compatibility Analysis (SCA). In: CISA Environmental Sanitary Engineering Centre (ed) Sardinia 2001 - Eighth International Waste Management and Landfill Symposium, S. Margherita di Pula, Cagliari, Italy
13. Claus F, Wiedemann PM (1994) Umweltkonflikte. Vermittlungsverfahren zu ihrer Lösung. Eberhard Blottner Verlag, Taunusstein
14. Cole LR, Foster SR (2001) From the Ground Up: Environmental Racism and the Rise of the Environmental Justice Movement. New York University Press, New York London, p 3
15. Dente B, Fareri P, Ligteringen J (1998) The Waste and the Backyard. Kluwer Academic Publishers, Dordrecht
16. Dienel PC (1989) Contributing to Social Decision Methodology: Citizen Reports on Technological Projects. In: Vlek C, Cvetkovich G (eds) Social Decision Methodology for Technological Projects. Kluwer Academic Press, Dordrecht, pp 133-150
17. Dienel PC, Renn O (1995) Planning Cells: A Gate to "Fractal" Mediation. In: Renn O, Webler T, Wiedemann P (eds) Fairness and Competence in Citizen Participation.

Evaluating New Models for Environmental Discourse. Kluwer, Dordrecht Boston, pp 117-140

18. Edwards W (1977) How to Use Multiattribute Utility Measurement for Social Decision Making. IEEE Transactions on Systems, Man, and Cybernetics SMC-7:326-340

19. Eichener V, Mai M (1993) Sozialverträgliche Technik - Gestaltung und Bewertung. Deutscher Universitäts Verlag, Wiesbaden

20. Gassner H, Holznagel B, Lahl U (1992) Mediation: Verhandlungen als Mittel der Konsensfindung bei Umweltstreitigkeiten. Economica Verlag, Bonn

21. Hajer MA (1995a) The Politics of Environmental Discourse: Ecological Modernization and the Policy Process. Clarendon Press, Oxford, p 25

22. Hajer MA (1995b) The Politics of Environmental Discourse: Ecological Modernization and the Policy Process. Clarendon Press, Oxford, p 30

23. Hallay H, Pfriem R (1992) Öko-Controlling: Umweltschutz im mittelständischen Unternehmen. Campus, Frankfurt

24. Hauber (1989) Wege zur Erhöhung der Akzeptanz von Abfallentsorgungsanlagen. Müll und Abfall 1/89

25. Herbold R (2001) Mediation. Internal Report. Universität Bielefeld

26. Joos W, Carabias V, Winistörfer H, Stücheli A (1999a) Social Aspects of Public Waste Management in Switzerland. Waste Management. International Journal of Integrated Waste Management, Science & Technology 19:417-425

27. Joos W, Carabias V, Winistörfer H, Stücheli A (1999b) Ansätze zur Erhebung, Bewertung und Verbesserung der Sozialverträglichkeit in der Abfallwirtschaft. In: Stucki S, et al. (eds) Forschung für eine nachhaltige Abfallwirtschaft. PSI, Villigen

28. Keeney RL, Renn O, von Winterfeldt D (1987) Structuring West Germany's Energy Objectives. Energy Policy 15:352-362

29. Keeney RL, Renn O, von Winterfeldt D, Kotte U (1984) Die Wertbaumanalyse. In: Renn O (ed) Technik und Sozialer Wandel. HTV Edition, München

30. Krämer L (1999) Community Waste Management and Consumer Rights - Theory and Practice. In: Boucquey N (ed) European Consumer Law and Waste Management. Centre de droit de la consommation. Université Catholique de Louvain, Louvain-la-Neuve, pp 32-56

31. Looss A, Katz C (1995) Abfallvermeidung. Strategien, Instrumente und Bewertungskriterien. Erich Schmidt Verlag, Berlin

32. Meier B (1988) Sozialverträglichkeit, Deutung und Kritik einer neuen Leitidee. Deutscher Instituts Verlag, Köln

33. Müller M, Hennicke P (1994) Wohlstand durch Vermeiden. Mit der Ökologie aus der Krise. Wissenschaftliche Buchgesellschaft, Darmstadt

34. OECD (1997) Cleaner Production and Waste Minimisation in OECD and Dynamic Non-member Economies. OECD, Paris

35. Petts J (1995) Waste Management Strategy Development: A Case Study of Community Involvement and Consensus-Building in Hampshire. Journal of Environmental Planning and Management 38:519-536

36. Petts J (1997) The Public-Expert Interface in Local Waste Management Decisions: Expertise, Credibility and Process. Public understand. Sci. 6:359-381

37. Renn O (1998) The Role of Risk Communication and Public Dialogue for Improving Risk Management. Risk Decision and Policy 3:5-30

38. Renn O, Webler T (1992) Anticipating Conflicts: Public Participation in Managing the Solid Waste Crisis. GAIA 1:84-94
39. Renn O, et al. (1997) Discursive Methods in Environmental Decision Making. Business Strategy and the Environment 6:218-231
40. Renn O, Webler T, Wiedemann P (1995) Fairness and Competence in Citizen Participation. Evaluating Models for Environmental Discourse. Kluwer Academic Publishers, Dordrecht
41. Renn O, Webler T, Kastenholz H (1998b) Procedural and Substantive Fairness in Landfill Siting: a Swiss Case Study. In: Löfstedt R, Frewer L (eds) The Earthscan Reader in Risk and Modern Society. Earthscan, London, pp 253-270
42. Renn O, Kastenholz H, Schild P, Wilhelm U (1998a) Abfallpolitik im kooperativen Diskurs. Bürgerbeteiligung bei der Standortsuche für eine Deponie im Kanton Aargau. Hochschulverlag AG an der ETH, Zurich
43. Renn O, Schrimpf M, Büttner T, Carius R, Köberle S, Oppermann B, Schneider E, Zöller K (1999) Abfallwirtschaft 2005. Bürger planen ein regionales Abfallkonzept. Nomos, Baden-Baden
44. Rootes C (1997) Shaping Collective Action: Structure, Contingency and Knowledge. In: Edmondson R (ed) The Political Context of Collective Action. Routledge, London New York, pp 81-104
45. Rootes C (2001) Discourse, Opportunity or Structure? The Development and Outcomes of Local Mobilisations Against Waste Incinerators in England. ECPR Joint Sessions, April 6-11, 2001, Grenoble
46. Schneider E, Oppermann B, Renn O (1998) Experiences from Germany: Application of a Structured Model of Public Participation in Waste Management Planning. Interact. Journal of Public Participation 4:63-72
47. Seiler HJ (1999) Fallbeispiel KVA Thun - Demokratiedefizit? Untersuchung von rechtlichen Aspekten im Rahmen des SPPU. Report. IP Abfall, Münsingen
48. Skorupinski B (1996) Gentechnik für die Schädlingsbekämpfung - eine ethische Bewertung der Freisetzung gentechnisch veränderter Organismen in der Landwirtschaft. Enke, Stuttgart
49. Skorupinski B, Ott K (2000) Technikfolgenabschätzung und Ethik. vdf, Zurich
50. Skorupinski B, Ott K (2002) Technology Assessment and Ethics - Determining a Relationship in Theory and Practice. Poiesis & Praxis, International Journal of Technology Assessment and Ethics of Science 2
51. Staub-Bernasconi S (1991) Sozialverträglichkeit - Bausteine auf dem Weg zu einer mehrdimensionalen Konzeption von Umweltverträglichkeit. Schweizerischer Wissenschaftsrat, Bern
52. Vorwerk V, Kämper E (1997) Evaluation der 3. Phase des Bürgerbeteiligungsverfahrens in der Region Nordschwarzwald. Working Report No. 70. Akademie für Technikfolgenabschätzung, Stuttgart
53. Walsh E, Warland R, Clayton Smith D (1993) Backyard NIMBYS and Incinerator Sitings: Implications for Social Movement Theory. Social Problems 40
54. Walsh E, Warland R, Clayton Smith D (1997) Don't burn it here: Grassroots Challenges to Trash Incineration, No. 1. Penn State Univ. Press, University Park PA
55. Winistörfer H (1999) Sozialverträglichkeit. Ein pragmatischer Ansatz für deren Bewertung. Tagung IP Abfall & SIGA/ASS, Baden
56. Winistörfer H, Carabias V (12 22, 1998) Wie sozial verträglich ist unsere Abfallwirtschaft? Neue Zürcher Zeitung, Zurich

8 Towards Sustainable Waste Management

Regula Winzeler, Peter Hofer, and Leo Morf, with a contribution by Zhao Youcai

Sustainable development, as originally defined by the Brundtland Commission, aims to ensure that "... it meets the needs of the present generations without compromising the ability of further generations to meet their own needs." The second part of this definition of sustainability, relating to future generations, is particularly important in the field addressed by this book. While, at least in industrialized countries, waste management practices have developed to a stage where direct impacts on the environment are significantly reduced, it is the long-term environmental problems that still need to be addressed. Making sure that waste deposits will not pose a hazard for coming generations is as much of an open problem as making sure that resources will not be depleted at a rate which cannot be sustained. Waste management or, rather, the management of material flows within an economic system is therefore an important area for testing sustainable solutions.

Reducing the specific material throughput of the economy substantially and at the same time finding ways of approaching high recycling efficiency is a long-term guideline for sustainable development. Working towards this goal will mean integrating waste management as a part of the management of all material resources, i.e. production, consumption, recycling, treatment and final disposal (Integrated Waste Management, IWM).

The following chapter attempts an outline of possible roadmaps towards sustainable solutions in waste management. A roadmap to sustainability requires a well-balanced and holistic consideration of ecological, economical and social objectives, some of which are conflicting. Perfecting the environmental performance of waste management operations, such as incinerators with high recovery rates for secondary raw materials towards true "zero waste" standards, may be technically feasible, but will incur high costs which could even result in an ecological backfire if the population is not willing to pay these costs. High disposal costs will encourage citizens to choose illegal ways of disposing of their waste, which, in turn, will result in more pollution: estimates by the Swiss federal Office of the Environment have shown that emissions of dioxins from illegally burned waste exceed the ones from incinerators.

With a set of well-defined targets for sustainable integrated resource management, the authors of the following chapter have created tools which allow a systematic analysis of possible future scenarios and an assessment of policies for leading society towards these targets. The authors present five possibilities of future waste management systems as scenarios resulting from specific policy options for which an analysis of the target system has been carried out. In an interdisciplinary effort the authors formalized the analysis of all aspects of waste management. It is clear that an optimization of resource management is, at the same time, a local issue in as far as waste disposal is a very localized issue (considering local economic boundary conditions, consumption patterns, cultural attitudes, etc.) and a global issue, where part of the resource consumption is concerned. The framework provided by this analysis allows an assessment of possible policy decisions with respect to sustainability targets for, e.g. the Swiss economy. The findings of the Swiss analysis have been utilized by a Chinese colleague (Zhao Youcai) who, in section 8.3.5, applies the target system of the Swiss group to the situation in an emerging economy.

The scenarios emphasizing reduced resource use and improvements at the front end (scenarios termed "better product" and "fewer products, new values") show more sustainability-supporting elements than those scenarios which rely on classical waste management instruments and technologies that focus on the end of the pipe. The question of "how to get there from here..." is, by way of illustration, exemplified in the attached case studies.

Section 8.3.3 analyses the potential of eco-taxing for achieving a transition towards better products. Eco-taxing, i.e. taxing of energy or specific materials instead of taxing labour, has been discussed as a global steering instrument to create incentives for reducing large material flows, e.g. the flow of CO_2 into the atmosphere due to the use of fossil energy. The analysis shows that the introduction of an energy or CO_2 tax, as it is discussed in various countries, would have little noticeable impact on the overall reduction of MSW and uncertain influence on the concentrations of hazardous substances in MSW. In order to be relevant for a transition to better products with respect to recycling and/or long term deposition, specific ecological taxes will have to be applied for the steering of specific elements with particularly high hazard potential, such as cadmium. The relevance of waste management in steering the flows of cadmium is also confirmed by the material flow analysis which was carried out for the chemical elements C, N, Cl and Cd. (section 8.3.2).

The problematic materials in MSW in a long-term perspective, are metallic parts and materials that have a highly complex composition and, at the same time, a short life cycle. Optimizing the chemical composition of complex materials, such as electronic devices, with respect to sustainable resource management is a formidable task. Electronic hardware, e.g. in computers, have a short use life, which is not determined by the actual wear of the equipment, but by the rate of innovation underlying its development. Strategies for designing electronic equipment which take the full life cycle into consideration need to be aware of the fact that for such products, the essential parts cannot be designed for longevity. Section 8.3.4 analyses the situation of computer hardware from this point of view with the conclusion that there is scope for improving the design if one considers that only a minor fraction (by weight!) of a computer is actually sensitive to fast innovation cycles.

The analysis of this chapter shows clearly that there is no simple recipe that can make the transition to sustainable resource management happen. We will have to rely on a number of technologies, organizational strategies, and policy decisions working towards a set of targets which are in accord with and aim towards sustainable development. The development towards a more sustainable management of material resources is not a straightforward process, but a tortuous path which must consider all aspects of sustainability at any given time.

8.1 Sustainability of Waste Management and Treatment

8.1.1 Definition of "Sustainable Development"

The Brundtland Commission in 1987 and the Rio Conference in 1992 have defined guidelines for a *sustainable development* of humans and environment. Humanity has "to ensure that it meets the needs of the present generations without compromising the ability of future generations to meet their own needs" [32, p.8]. The three main objectives of a sustainable development are "the protection of man and environment", "economic compatibility" and "social compatibility". They are connected to each other in a dynamic triangle and interdependent; they are of equal significance and are to be respected equally.

Many countries have committed themselves to include the term "sustainable development" in their political agenda and to orientate the national economy to sustainable criteria.

8.1.2 Definition of Integrated Waste Management

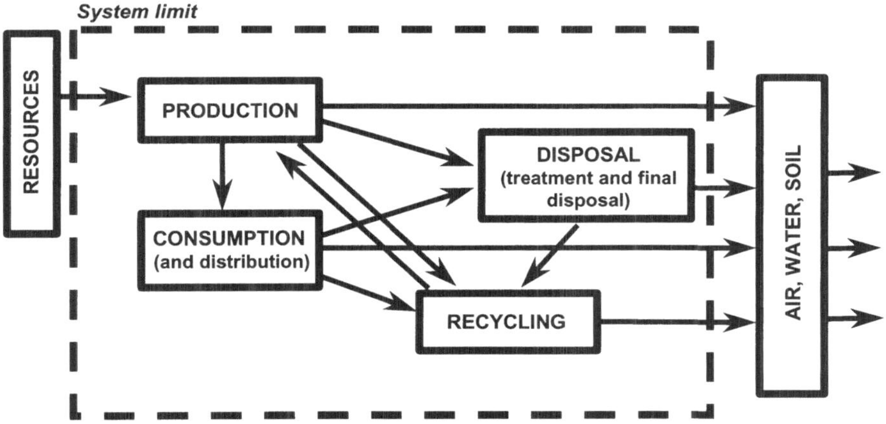

Fig. 8.1. Integrated Waste Management. The rectangles show the processes, the arrows mark the flows of materials and goods within, into and out of the system

In this chapter Integrated Waste Management (IWM) is used to describe "*Waste management in its entirety*". It takes into consideration the whole life cycle of products, from exploring and quarrying resources, through the production and trade of goods, to the disposal of waste - "from cradle to grave". Therefore the processes of production, consumption, recycling and disposal are to be taken into account (Fig. 8.1). Ratios and characteristics of production and consumption should be considered, as this makes it possible to determine the required information about the amount, the quality and the composition of the waste to be treated at an early stage (*early recognition*). Therefore, an optimal directing of disposal and recycling concerns the entire national economy, of which waste management is a part [10, 13]. Though waste management ("disposal" and "recycling") cannot direct or determine the processes of production and consumption, it can give recommendations for the attention of producers and consumers, their associations and the authorities.

8.1.3 Target System: Goals for a Sustainable Waste Management

Steering national economy and waste management necessitates a definition of targets and/or goals, that is, statements about the intended measurable results [27]. Therefore, the standards for a future sustainable waste management have had to be set, *targets* have had to be defined and put into a *target system* according to a hierarchical order. The principle target refers to the term "sustainable development" in the sense of the Brundtland definition. Additionally a new systematic structure and subdivision into three levels has been applied to the target system (see below). The system is consequently oriented towards waste management.

This target system supplements recent documentation from the Swiss authorities concerning targets for waste management [13]. The target system for a sustainable waste management provides new targets mainly in the field of "economic compatibility" and "social compatibility".

The target system was developed in a series of several steps. Current laws, published concepts and guidelines on sustainable development in Switzerland, in addition to other pertinent scientific literature, were studied with regard to waste management. A target system was proposed, discussed and developed during several workshops with an interdisciplinary team consisting of natural scientists, economists and social scientists from academia, public administration and private industry. The target system was eventually confirmed by the team. Initially developed by this team of Swiss experts in response to the situation of waste management in Switzerland in 1997, the target system is based on the specific development of waste management in this country. Experts from other countries would, no doubt, define different sustainable target systems which would pertain specifically to the development of waste management within their own countries. While scientific and technical targets might be generally acknowledged, economic and, particularly, social targets are expected to differ. Therefore, the target system shown in the following tables should be recognized as one among many potential propositions of targets for sustainable waste management.

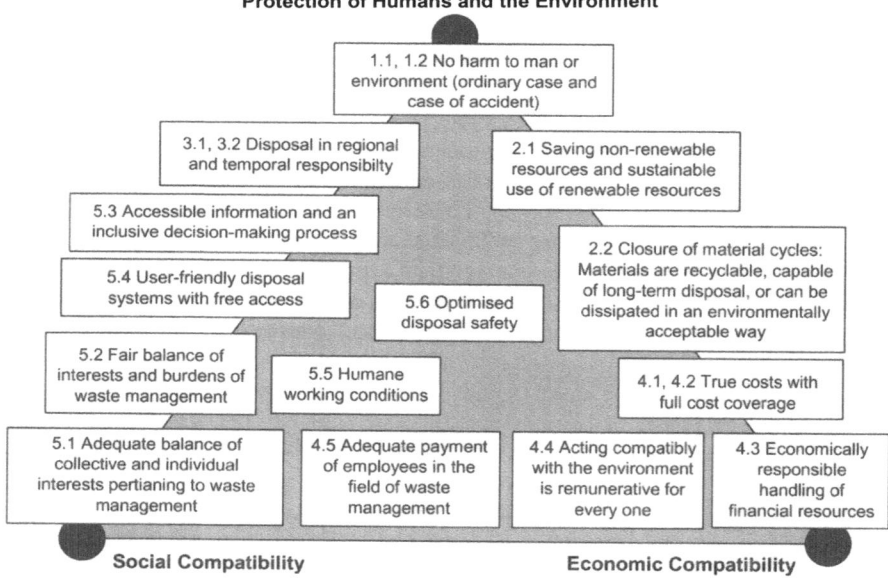

Fig. 8.2. The target system of a sustainable waste management: This illustrates the position of each target within the "dynamic triangle", the *Overview of Targets* (below) describes the target system of a sustainable waste management in detail. The targets are long-term models which should be fulfilled over the course of several generations

Overview of Targets

Main target 1: Protection of man and environment

- Sub-target 1.1: *No harm to man or environment in ordinary case*
- Sub-target 1.2: *No harm to man or environment in the case of an accident*
    Waste management is to cause no long-term damage, neither in ordinary case nor by accident (earthquakes, big fires, incidents in industries). Anthropogenic material flows are to be lower than the geogenic (natural) material flows or lie within the oscillation of the natural flows.

Main target 2: Conservation of resources

- Sub-target 2.1: *Saving non-renewable resources and sustainable use of renewable resources*
    The resources (material, products / goods, energy, space, information) are to be optimally used. Renewable resources are to be used in a sustainable way: they are to be exploited and used at no greater rate than the rate at which they renew themselves ("Harvesting rates do not exceed regeneration rates" [17, p. 271]). Non-renewable resources are to be used in an economical and thrifty manner and, if possible, substituted with renewable resources. The efficiency in production is to be increased, so that resources are saved.
- Sub-target 2.2: *Closure of material cycles:*
    Material cycles are to be closed in an ecologically sensitive and economically reasonable way. Materials should either be recyclable, capable of long-term disposal [11], or they can be dissipated in an environmentally sustainable way (e.g. CO_2 in the atmosphere, chlorine in rivers). The aim is to define an adequate final sink for each substance. It is important that the limits of the absorbing capacity of the ecosystems are taken into consideration ("Waste emissions do not exceed the assimilative capacity of the local environment" [17, p. 271]).

Main target 3: Waste problems are to be solved here and now

- Sub-target 3.1: *Disposal in regional responsibility*
- Sub-target 3.2: *Disposal in temporal responsibility*
    Disposal in regional and temporal responsibility of a region means that no waste, especially no hazardous waste is left to accumulate over a span of years for the care of future generations. Interregional and international solutions are included, if the disposal paths of waste to be treated are under control and if an interregional compensation of interests plays an active role.

Main target 4: Waste management is to be organized in an economically acceptable way

- Sub-target 4.1: *True costs with full cost coverage: Internalization of external costs for damages to environment and human health*
- Sub-target 4.2: *True costs with full cost coverage: Internalization of external costs for risks (accidents, aftercare for landfills)*
    The full costs of waste management should be distributed according to the "polluter-pays" principle. Full cost coverage is appropriate here. Therefore, the costs incurred through damage to environment and human health are to be integrated into the overall

calculation of costs. All costs caused by risks of waste management should be integrated. These might be, for instance, costs caused by accidents or costs for aftercare of landfills (recultivation, off-gas treatment, cleaning of seepage water from landfill, etc).

- Sub-target 4.3: *Economically responsible handling of financial resources*
  The responsible handling of money means the efficient use of financial resources and striving for a good cost-benefit ratio. The resulting costs of waste management should be both economically and politically acceptable.

- Sub-target 4.4: *Acting compatibly with the environment is remunerative for every one*
  A system of financial and non-financial incentives (e.g. taxes, deposits, social respect) makes environmentally friendly behaviour remunerative for every one. Harmful action to the environment is prosecuted.

- Sub-target 4.5: Adequate payment of employees in the field of waste management
  Payments should be dependent on the functional specification and not on gender and social status. For fulltime employment, payment should be at least as high as the legally fixed subsistence levels.

Main target 5: Waste management is to be socially compatible

- Sub-target 5.1: *Adequate balance of collective and individual interests pertaining to waste management*
  The balance of the long-term collective interests of the society and the individual interests should be optimized. The changes necessary for achieving this balance should be such that nobody's personal situation changes for the worse.

- Sub-target 5.2: *Fair balance of interests and burdens of waste management*
  There should be a fair balance in the interests and burdens of waste management on various social levels as well as within different regions (central and marginal regions).

- Sub-target 5.3: *Accessible information and an inclusive decision-making process*
  Information on waste management, be it general or case specific, should be accessible for any involved person and should provide the necessary basis for well-reasoned participation. Participation is to be, in principle, possible for anyone concerned, yet the measure of participation has to be determined for each case individually.

- Sub-target 5.4: *User-friendly disposal systems with free access*
  Waste management and disposal systems are to be user-friendly: The disposal systems, like disposal plants and the decentralized collection of recyclables, should be easy to understand and the disposal paths clear and simple. There is to be free access for everybody.

- Sub-target 5.5: *Humane working conditions*
  Working conditions within the field of waste management are to be humane, include a high level of occupational safety, provide hygienic workplaces and fulfillment.

Sub-target 5.6: *Optimized disposal safety*
  Adequate capacities for disposal of goods are to be guaranteed. This means that accumulated waste is to be disposed of within a reasonable time period and in an environmentally acceptable way. Disposal capacities are not maximized but optimized: reserve capacity of disposal plants should differ only slightly from the effective need.

The target system shows different aspects of sustainable waste management and takes the three main objectives, protection of man and environment, economic compatibility and social compatibility into equal consideration.

The above-shown target system of sustainable waste management contains several *target conflicts*. An obvious example is the conflict between "Closure of material cycles" (Sub-target 2.2) and "Economically responsible handling of financial resources" (Sub-target 4.3). Closing material cycles can be very expensive, for instance, if recycled materials must compete with cheap primary resources; or if expensive, energy-intensive technologies or processes are necessary for the recycling of substances. These and other conflict areas were recognized, though not eliminated. The definition of sustainable development itself contains conflicts that, in fact, originate in the definition. Open and constructive discussion concerning the conflicting targets and possible reconciliation must take place for each case individually, with particular and equally proportional attention given to the objectives of protection of man and environment, economic compatibility and social compatibility.

8.1.4 Comparison of the Sustainability Targets with Current Governmental Targets and Established Law in Switzerland

The Overview of Targets shows the target system of a sustainable waste management compared with the targets of the Guidelines on Swiss Waste Management [13], of the Concept of Swiss Waste Management [15] and other strategy-papers written by the Swiss authorities (Strategy of Federal Government [14], Strategy of Waste Management [16]), as well as several laws (Protection of Environment Act [3], Technical Ordinance about Waste [6], Accidence Ordinance [7], Federal Labour Act [1] and Ordinance 3 of Federal Accidence Insurance Act [2]). The comparison facilitates the regulation of whether the targets are appropriate (*control of targets*) and whether the three main objectives - protection of man and environment, economic compatibility and social compatibility - are given equal consideration. Neither the pursued targets nor the effectiveness of already implemented measures, however, is treated here.

From the comparison between current Swiss targets and the above defined sustainable targets of waste management, the following can be concluded:

- Many of the current targets in Switzerland are in line with sustainability. If one considers the targets integrally (*integral consideration*) there are gaps: Current targets in the field *substances / environment / technique* are complete. Contrarily, in the field "economy / economic efficiency" there are some gaps, and even more in the field "society / policy / authorities".
- Targets conflicting with sustainability were not identified.

Table 8.1. Comparison of the target system of a sustainable waste management with the current targets in Switzerland (part 1)

Targets of a sustainable waste management (see *Overview of Targets*)	Current targets of waste management according to strategies of Swiss authorities	Rate of sustainability*	Deficits of current targets
SUBSTANCES / ENVIRONMENT / TECHNIQUE (Main targets 1-3)			
Sub-target 1.1: No harm to man or environment in ordinary case	Protection of man and environment according to the precaution principle [3, Article 1, 6, Article 1]. The Guidelines on Swiss Waste Management [13] assign top priority to the precaution principle.	●●	No deficits in targets.
Sub-target 1.2: No harm to man or environment in the case of an accident	Protection of people and environment from serious damage due to accidents [7, Article 1].	●●	No deficits in targets.
Sub-target 2.1: Saving non-renewable resources and sustainable use of renewable resources	Saving resources like materials, products / goods, energy, space (especially landfill volumes) [15].	●●	No deficits in targets.
Sub-target 2.2: Closure of material cycles: Materials are to be recyclable, capable of long-term disposal, or can be dissipated in an environmentally acceptable way.	Waste must be recycled in an environmentally benign way, disposal systems cause only two different classes of materials: substances are either recyclable or capable of long-term disposal [13].	●●	In addition, substances neither recyclable nor capable of long-term disposal should be dissipated in an environmentally acceptable way.
Sub-target 3.1: Disposal in regional responsibility	Disposal of product/material within the country, in co-operation with other countries [13].	●●	No deficits in targets.
Sub-target 3.2: Disposal in temporal responsibility	Disposal systems are only to produce residues capable of long-term disposal (apart from recyclable substances), meaning that the residues cause only emissions that are environmentally acceptable over hundreds of years [13].	●●	No deficits in targets.

* The rate of sustainability measures the sustainability while comparing the current targets in Switzerland with the sustainable targets: ●● = sustainable, ●○ = partly sustainable, ○○ = not sustainable

Table 8.2. Comparison of the target system of a sustainable waste management with the current targets in Switzerland (part 2)

Targets of a sustainable waste management (see *Overview of Targets*)	Current targets of waste management according to strategies of Swiss authorities	Rate of sustainability*	Deficits of current targets
ECONOMY / ECONOMIC EFFICIENCY (Main target 4)			
Sub-target 4.1: True costs with full cost coverage: Internalization of external costs for damages to the environment and human health	Fees for waste treatment and long-term disposal are to be cost- and risk-adequate, including fees for foreseeable or potential subsequent costs [13].	●○	The actual targets do not consider the internalization of external costs.
Sub-target 4.2: True costs with full cost coverage: Internalization of external costs for risks (accidents, aftercare for landfills)	Fees for waste treatment and long-term disposal have to be cost- and risk-adequate, including fees for foreseeable or potential subsequent costs [13].	●○	The actual targets do not consider the internalization of external costs.
Sub-target 4.3: Economically responsible handling of financial resources	Recycling must function in accordance with business management in long-term consideration [13]. Recycling has to be remunerative on existing markets.	●○	Economically responsible handling of money is not an explicit target, yet an implicit one in terms of cost effectiveness
Sub-target 4.4: Acting compatibly with the environment is remunerative for everyone	Tax incentives are to be considered [13]. The ecological tax reform is identified as a target [14].	●○	The demand for tax incentives contains tools that make environmentally friendly behavior remunerative. However, the target itself is not named.
Sub-target 4.5: Adequate payment of employees working in the field of waste management	No target formulated.	○○	Target is missing.

* The rate of sustainability measures the sustainability while comparing the current targets in Switzerland with the sustainable targets: ●● = sustainable, ●○ = partly sustainable, ○○ = not sustainable

Table 8.3. Comparison of the target system of a sustainable waste management with the current targets in Switzerland (part 3)

Targets of a sustainable waste management (see *Overview of Targets*)	Current targets of waste management according to strategies of Swiss authorities	Rate of sustainability*	Deficits of current targets
SOCIETY / POLICY / AUTHORITIES (Main target 5)			
Sub-target 5.1: Adequate balance of collective and individual interests pertaining to waste management	No target formulated.	○○	Target is missing.
Sub-target 5.2: Fair balance of interests and burdens of waste management	Equally distributed incineration plants [16].	●○	The target currently only concerns municipal solid waste.
Sub-target 5.3: Accessible information and inclusive decision-making process	No target formulated.	○○	Target is missing.
Sub-target 5.4: User-friendly and freely accessible disposal systems with free access	No target formulated.	○○	Target is missing.
Sub-target 5.5: Humane working conditions	Employers must take measures in favor of employee health and safety [1, Article 6]). Hygienic workplaces and high occupational safety [2; 8, Article 2] must be realized.	●●	No deficits in targets.
Sub-target 5.6: Optimized disposal safety	Optimizing the incineration capacity [16].	●○	The optimized disposal safety is concentrated on combustible solid waste.

* The rate of sustainability measures the sustainability while comparing the current targets in Switzerland with the sustainable targets: ●● = sustainable, ●○ = partly sustainable, ○○ = not sustainable

- The analysis of the current targets shows that targets in the field of natural sciences are generally more clearly defined than targets concerning social aspects. Social aspects in waste management are hardly identified.
- Another general deficit is the isolated consideration of waste management instead of considering IWM, as part of the national economy (Fig. 8.1).

Although not investigated in detail, it is obvious that some measures have already been realized without any defined targets. This is because these targets are self-evident or their achievement was almost complete and therefore not considered to be urgent. By contrast, some targets are only expected to be reached with great difficulty and in the distant future, like, for instance, sub-targets 4.1 and 4.2: True costs with full cost coverage.

8.1.5 The Target System as Planning Instrument for Authorities

The target system is a tool for guiding waste management towards a sustainable development. It helps national and/or regional authorities in defining their targets on waste management. Two regional government administrations in Switzerland, canton Thurgau and canton Zurich, have used this tool as a *planning instrument*. They based their discussions on the above-shown system. They modified it, solved the target conflicts and established priorities. They created their own target system, co-ordinated with the current situation and problems in their respective cantons. The discussions helped them to determine their strategies in waste management.

8.2 Scenarios for a Future Waste Management and Treatment

8.2.1 Terminology and Methods

While section 8.1, through the target system, illustrates potential directions to be taken, in this section the possibilities of how to achieve sustainability in waste management are investigated.

The terms used in this work are defined as follows:
- Visions are subjective wishes or draft images and conceptualizations of future states that might or might not be realistic.
- Scenarios are objectively conceivable and possible future states which can be described. Scenarios include the methods of how to reach the defined future states [20, 31].

- A strategy is the sum of all necessary measures and practices for the achievement of targets [27].
- Measures and procedures are the specific instructions for the achievement of targets.

Future Workshops: Method of Scenario Building

An interdisciplinary team of natural scientists, economists and social scientists from academia, public administration and private industry has developed several utopia and visions for a sustainable waste management in Switzerland. These utopia and visions emerged in *"future workshops"* [24]. The aim of this method is to collect ideas and visions about the future, without bothering about the constraints of the current situation. Out of many different utopia, visions and conceptualizations about the future, five conceivable scenarios of future waste management in Switzerland were defined and described in detail, including a description of short-, middle- and long-term measures needed to implement the scenarios according to the *method of scenario building*. Investigations of the influence of an ecological tax reform were added.

Method of Scenario Evaluation

- The sub-targets of the target system of a sustainable waste management described in the *Overview of Targets* (8.1.3) were translated into measurable or certifiable criteria regarding material, technical, ecological, economical and social aspects (section 8.2.3).

For each of the scenarios, there was an evaluation of whether the criteria had a positive (supporting), a negative (inhibiting), or neutral effect on the development towards sustainability. The evaluation of scenarios and their identified sustainability-supporting elements provided the basis for building strategies, as well as concrete measures and procedures to be realized, see section 8.4).

8.2.2 Description of Scenarios

Table 8.4 provides an overview of the five scenarios developed. These scenarios will each be individually described in the following text. In addition to the five scenarios, the influence of an ecological tax reform on waste management was analyzed separately in section 8.3.

Table **8.4.** Overview of five scenarios of future waste management in Switzerland

Scenarios	Intention	Optimized field
1: Waste management: Central co-ordination	• Mandatory co-ordination of waste management by government	Structures of organization
2: Waste management: Free market	• Intensification of free market in waste management	Structures of economic system
3: Better products	• Ecologically improved products	Quality of products, production process
4: Fewer products, new values	• Less consumption (frugality) • Leasing and shared use of items/products instead of ownership (strategy of sufficiency) • Change in awareness towards co-operative action, compatible with the environment	Quantity of products, consumption process
5: Simplified waste logistics	• Simplified logistics in collection and disposal of solid waste • High-tech disposal: Thermal treatment of waste with residues capable of long-term disposal • High level of user-friendliness	Logistics and process technology

The scenarios are conceivably possible and wishful future states. They cover a certain variety of possible developments, in 20 to 100 years from now, but are not claiming to be complete. Four requirements were defined:

• Switzerland will remain a democratic and constitutional state.
• Adequate disposal capacity will be guaranteed.
• The "polluter-pays" principle will be applicable.
• An ecological tax reform will be implemented.

According to the method of developing scenarios in workshops, the proposed scenarios are rather positive, idealized pictures of waste management in the future ("wishful thinking"). They might be criticized for being unrealistic, yet they are innovative. The aim of this method is not to find the best scenario, but rather – as mentioned above – to find elements that will lead development towards more sustainability. Scenarios and their evaluation are, thus, instruments used to gain more information about how to achieve sustainability. This method of scenario evaluation is not intended for making long-term, future predictions.

Scenario 1: "Waste Management: Central Co-ordination"

The aim of Scenario 1 is to achieve optimal structures of organization with a focus on recycling and disposal processes (Fig. 8.1).

• Centralized organization and control of waste management is performed by a central disposal-company. The state acts as surveillance authority. Municipal solid waste is detailed to disposal plants such as incineration plants, landfills and recycling plants. Collection of waste, separate collection of recyclables and recycling processes are uniform and nationwide.

- Incineration prices of all disposal plants are uniformly assessed and harmonized. Differences in effective costs (mainly different fix costs of the plants) are balanced (Solidarity principle).
- Central data collection acts as the basis for improved average utilization of disposal plants and for minimal use of reserve capacities. The goal is that the total capacity of disposal plants differs only slightly from the effective need. Directing the waste flows with balancing under- and over-capacities of disposal plants and with the aim of an optimized disposal safety is performed.
- Collection of recyclables (glass, metals, paper etc) is supported through deposits for increased return rates. Pre-paid disposal fees are used for financing expendable and costly disposal technologies.

Scenario 2: "Waste Management: Free Market"

Like Scenario 1, Scenario 2 focuses on the processes of disposal and recycling. While Scenario 1 optimizes the organizational structures, the aim of Scenario 2 is a waste management regulated by free market.

- Free trade of disposal services and waste takes place. Waste and recycling goods can be traded in regular shopping centers, markets, grocery stores etc.
- Market requirements and legal environmental standards are defined by the state and controlled by the authorities. The correct treatment of waste and recyclables is established in an official disposal manual.
- Operation of disposal plants and disposal and recycling industries through private companies (privatization of current facilities).
- Agents of waste management enterprises are *disposal plants*, such as incinerations, landfills and recycling plants, as well as *disposal companies* such as haulers, and distributors. Disposal plants, on the one hand, are licensed through the authorities and are committed to respecting the legal environmental standards. Disposal companies, on the other hand, receive waste and look after the correct treatment or recycling according to the official disposal manual. In principle, anybody can operate a disposal company, but a license is required for special waste categories such as, for instance, hazardous waste.
- Public waste collection systems are abolished. A system of private delivery or privately organized collection system can develop on the free disposal market.

Scenario 3: "Better Products"

The main focus of Scenario 3 is optimizing the production process for ecologically improved products.

- Environmental policy changes from ecological micro steering with primarily end-of-pipe measures to ecological macro steering with comprehensive regulation of energy, material and waste, regional planning, traffic, risk and hazard potentials.

- Introduction of a tax on hazardous substances such as heavy metals (cadmium, for example).
- Material and energy optimization of products throughout their life cycle (and therefore consideration of the processes production, consumption, recycling and disposal).
- Use of few and unproblematic chemical substances in the production of goods, as far as possible (limited to carbon, silicon, hydrogen, oxygen, nitrogen, iron, aluminum, calcium and sodium, for instance).
- Compliance with environmental standards according to the principle of the best available technology, meaning technically possible and economically acceptable.
- Producer responsibility for providing environment-friendly production and disposal.
- Material accounting on the regional and industrial level. Substance accounting on the level of key industries important for specific substances.

Scenario 4: "Less Products, New Values"

Scenario 4 focuses mainly on the processes of consumption, though is also inclusive of production. The principle is one of reducing the quantity of products/production through the development of a new lifestyle guided by co-operative, environmentally responsible behavior.

- Change in awareness towards co-operative instead of self-seeking behavior, increase in collaborative values and increased awareness of the environment.
- Standard of living compatible with sustainability targets. Growing importance of non-materialistic welfare as opposed to the current materialistic and affluence-oriented standard of living (strategy of sufficiency).
- A system of financial and non-financial incentives (such as social acceptance) supports ecological and social action (honoring, sanctioning).
- Tax on non-renewable resources with high taxation of harmful substances.
- New using forms: Purchase of using rights instead of goods (Leasing and shared use of things instead of owning them); increase of the use intensity and duration of goods through durability and repair-friendliness.
- Responsibility of producers for producing high-quality goods, for declaring the ingredients, and for disposing goods after use in an environmentally acceptable way.

Scenario 5: "Simplified Waste Logistics"

Scenario 5 concentrates on optimized disposal processes and simplifies the waste logistics.

- Thermal treatment of as many waste categories as possible; treatment of combustion residues is improved (high-tech disposal) and expanded.
- Simplified waste logistics without separating waste at the source. This increases the user-friendliness and convenience. Communal collection accepts all categories of waste. A separate or decentralized collection of recyclables, organized by communal authorities, does not exist. A separate collection organized by private initiative is not excluded.
- Separate collection only for construction waste and sewage sludge. Decentralized composting of green waste (waste from gardening, agriculture etc), paper and glass takes place through private initiative only.
- Recycling is focused on internal industrial processes and on metal recovery from the treatment of combustion residues (see Chapter 4).

8.2.3 Evaluation of Scenarios

The evaluation of the scenarios was undertaken by the members of the above named workshop team (section 8.2.1). Each evaluator had consequently to review all scenarios in evaluating the criteria, comparing the situations described in the scenarios with the situations in 1997. Some examples of criteria have been:

- Substance flows from the waste management system (Fig. 8.1) into the environment: e.g., carbon, nitrogen, chlorine, cadmium and organic compounds
- Number of waste treatment plants with significant risk potential
- Amount of waste recycled, illegally disposed, or deposited in MSW landfills
- Relation of disposal costs to gross national product
- Existence of financial incentive systems
- Distribution of waste disposal plants
- Relation of disposal costs to the controllable net income of the individual
- Relation of waste related workplaces of high quality and security to the total number of workplaces

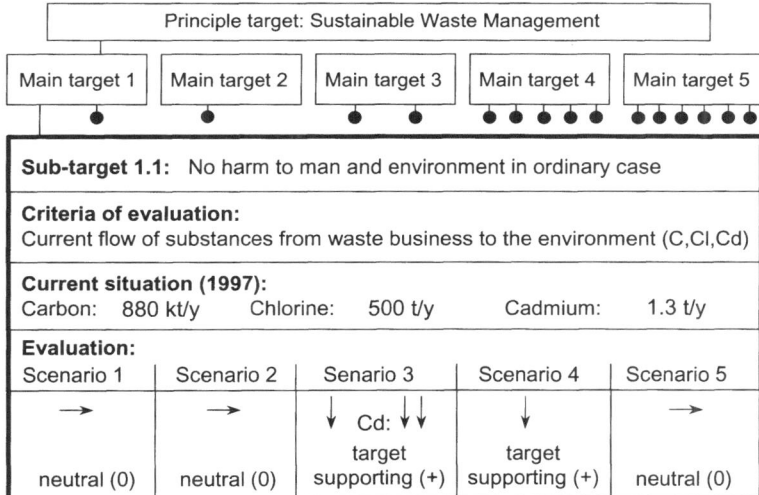

Fig. 8.3. Example for the methodology for evaluating scenarios

Tables 8.5 to 8.7 show the results of the evaluation of the five scenarios in detail. The arrows indicate whether the criteria - compared with the situation in 1997 – is increasing (↑), decreasing (↓), or not changing (→). The elements are sustainability-supporting (+), sustainability-inhibiting (-) or neutral (0). Two signs mean that the evaluation within the team was controversial.

The evaluation by the 15 members of the workshop team can be summarized as follows:

- Scenario 1 ("Central co-ordination") and Scenario 2 ("Free market") with changing structures in organization and economic system show comparatively few elements supporting sustainability. In the field of incineration and landfill with high investments and long-term responsibility (disposal safety, aftercare for landfills), "Central co-ordination" is a sustainable option. On the other hand, in fields with lower investments, such as a decentralized collection of recyclables and recycling, "Free market" (Scenario 2) – combined with government-issued requirements – is a promising strategy. "Central co-ordination" and "Free market" are middle-term options.
- Improved production in Scenario 3 ("Better products") and the change in practice towards less consumption and co-operative behavior, compatible with the environment in Scenario 4 ("Fewer products, new values") are sustainable options, which however take a long time for implementation. "Leasing and shared use of goods instead of owning them", which means increasing the use intensity of goods through durability, repair-friendliness and shared use of goods, is also a sustainable method. Scenario 4 ("Fewer products, new values"), with its clear social focus, does not contribute more to the achievement of social targets

Table 8.5. Results of the evaluation of the five scenarios for a future waste management in Switzerland (Part 1)

SUBSTANCES / ENVIRONMENT / TECHNIQUE (Main targets 1-3)					
Sub-target 1.1: No harm to man or environment in ordinary case	**Criterion 1**: Anthropogenic substance flows from the waste management system (Fig. 8.1) to the environment (to the final storage, to the water, to the air), for example, for the elements carbon, nitrogen, chlorine and the organic compounds				

Result of evaluation*:	Scenario 1	Scenario 2	Scenario 3	Scenario 4	Scenario 5
Criterion 1:	→ 0	→ 0	↓ +	↓ +	→↓** 0+

Sub-target 1.2: No harm to man or environment in case of an accident	**Criterion 1**: Number of waste treatment plants with a significant risk potential				

Result of evaluation*:	Scenario 1	Scenario 2	Scenario 3	Scenario 4	Scenario 5
Criterion 1:	→↓ 0+	↑ -	↓ +	↓ +	→↑ 0-

Sub-target 2.1: Saving non-renewable resources and sustainable use of renewable resources	**Criterion 1**: Amount of non-renewable resources (materials) used in Switzerland **Criterion 2**: Total energy consumption in waste management in Switzerland and comparison with the gross energy consumption in Switzerland				

Result of evaluation*:	Scenario 1	Scenario 2	Scenario 3	Scenario 4	Scenario 5
Criterion 1:	→↓ 0+	→↓ 0+	↓ +	↓↓ +	→↓ 0+
Criterion 2:	→ 0	→ 0	↓ +	↓ +	↑ -

Sub-target 2.2: Closure of material cycles	**Criterion 1**: The amount of waste as part of the total amount of waste, which is recycled, potentially capable of long-term disposal or can be dissipated in an environmentally acceptable way (%)				

Result of evaluation*:	Scenario 1	Scenario 2	Scenario 3	Scenario 4	Scenario 5
Criterion 1:	↑ +	→↑ 0+	↑↑ +	↑↑ +	↓↑ -+

Sub-target 3.1: Disposal in regional responsibility	**Criterion 1**: Part of the solid and fluid waste accumulated in a region, is disposed in regional, ecological and economic responsibility and in respect to an over-regional compensation of interests				

Result of evaluation*	Scenario 1	Scenario 2	Scenario 3	Scenario 4	Scenario 5
Criterion 1:	→↑ 0+	↓ -	→↑ 0+	↑ +	→↑ 0+

Sub-target 3.2: Disposal in temporal responsibility	**Criterion 1**: Amount of waste deposited in MSW landfills **Criterion 2**: Amount of waste, illegally disposed **Criterion 3**: Amount of recycled waste with hazardous substances				

Result of evaluation*:	Scenario 1	Scenario 2	Scenario 3	Scenario 4	Scenario 5
Criterion 1:	↓ +	↓ +	↓ +	↓↓ +	↓↓↑ +-
Criterion 2:	→↑ 0-	↑→ -0	→ 0	↓ +	→↓ 0+
Criterion 3:	→ 0	↑ -	↓ +	↓ +	↓ +

- Comparison of criteria with the current situation in 1997: Criteria are ↑: increasing, ↓: decreasing, →: not changing. Element is +: sustainability-supporting, -: sustainability-inhibiting, 0: neutral.
- ** →: carbon, nitrogen, chlorine, ↓: cadmium and organic compounds

Table 8.6. Results of the evaluation of the five scenarios for future waste management in Switzerland (Part 2)

ECONOMY / ECONOMIC EFFICIENCY (Main target 4)				

Sub-target 4.1:
True costs with full cost coverage (1)

Criterion 1: The chances of achieving/implementing true costs with full cost coverage for running costs (operation, maintenance, staff) and capital costs in IWM

Criterion 2: The chances of achieving/implementing true costs with full cost coverage for damage to the environment and human health in IWM

Result of evaluation*:	Scenario 1	Scenario 2	Scenario 3	Scenario 4	Scenario 5
Criterion 1:	↑→ +0	↑→ +0	→ 0	↑ +	→ 0
Criterion 2:	↑→ +0	↓ -	↑ +	↑ +	→ 0

Sub-target 4.2:
True costs with full cost coverage (2)

Criterion 1: The chances of achieving/implementing true costs with full cost coverage for accidents in waste management

Criterion 2: The chances of achieving/implementing true costs with full cost coverage for aftercare of landfills in waste management

Result of evaluation*:	Scenario 1	Scenario 2	Scenario 3	Scenario 4	Scenario 5
Criterion 1:	↑→ +0	→↓ 0-	↑ +	↑ +	↑→↓ +0-
Criterion 2:	↑ +	↓ -	↑ +	↑ +	↑↓ +-

Sub-target 4.3:
Economically responsible handling of financial resources

Criterion 1: Relation of disposal costs to gross national product

Criterion 2: Relation of developmental and production costs to gross national product

Result of evaluation*:	Scenario 1	Scenario 2	Scenario 3	Scenario 4	Scenario 5
Criterion 1:	→↓ 0+	↓ +	↓ +	↓ +	↑ -
Criterion 2:	→ 0	→ 0	↑ -	↑→ 0-	→ 0

Sub-target 4.4:
Acting compatibly with the environment is remunerative for every one

Criterion 1: Existence of financial and non-financial incentive systems which make acting compatibly with the environment remunerative for every one

Result of evaluation*:	Scenario 1	Scenario 2	Scenario 3	Scenario 4	Scenario 5
Criterion 1:	→ 0	→↑ 0+	↑ +	↑ +	↓ -

Sub-target 4.5:
Adequate payment of employees in the field of waste management

Criterion 1: Standardized level of payment for jobs in waste management

Result of evaluation*:	Scenario 1	Scenario 2	Scenario 3	Scenario 4	Scenario 5
Criterion 1:	→ 0	↓ -	→ 0	→ 0	↑ +

* Comparison of criteria with the situation in 1997: Criteria are ↑: increasing, ↓: decreasing, →: not changing. Element is +: sustainability-supporting, -: sustainability-inhibiting, 0: neutral.

Table 8.7. Results of the evaluation of the five scenarios for future waste management in Switzerland (Part 3)

SOCIETY / POLICY / AUTHORITIES (Main target 5)				
Sub-target 5.1: Adequate balance of collective and individual interests pertaining to waste management	**Criterion 1:** Adequate co-ordination of long-term collective interests for a sustainable development of waste management and individual interests, with the aim of reaching the Pareto-equilibrium			

Result of evaluation*:	Scenario 1	Scenario 2	Scenario 3	Scenario 4	Scenario 5
Criterion 1:	→ 0	→↓ 0-	↑ +	↑↑ +	→ 0

| **Sub-target 5.2:** Fair balance of interests and burdens of waste management | **Criterion 1:** Relation of disposal costs to the controllable net income of the individual, level of income is taken into account **Criterion 2:** Distribution of waste disposal plants in Switzerland, in coeval consideration of the amount of waste | | | |

Result of evaluation*	Scenario 1	Scenario 2	Scenario 3	Scenario 4	Scenario 5
Criterion 1:	↓→ +0	↓↑ +-	↓ +	↓ +	↑ -
Criterion 2:	↑ +	↓ -	→ 0	→ 0	↓→ -0

| **Sub-target 5.3:** Accessible information and inclusive decision-making process | **Criterion 1:** Accessibility of credible data and information concerning waste flows and emissions **Criterion 2:** Accessibility of information on costs **Criterion 3:** Accessibility of information on product content (declaration of products) **Criterion 4:** Possibility for participation in discussion and decision-making in waste management questions | | | |

Result of evaluation*:	Scenario 1	Scenario 2	Scenario 3	Scenario 4	Scenario 5
Criterion 1:	↑ +	↓ -	↑ +	→ 0	↑ +
Criterion 2:	↑ +	↓ -	→ 0	↑ +	↑ +
Criterion 3:	→ 0	↑→ +0	↑ +	↑ +	→↓ 0-
Criterion 4:	→ 0	→ 0	→ 0	→ 0	→ 0

| **Sub-target 5.4:** User-friendly and disposal systems with free access | **Criterion 1:** Required time per annum per household for waste disposal **Criterion 2:** User-friendliness of disposal paths and waste treatment plants **Criterion 3:** Possibility of free choice of disposal path and waste treatment plant | | | |

Result of evaluation*:	Scenario 1	Scenario 2	Scenario 3	Scenario 4	Scenario 5
Criterion 1:	→ 0	↑↓ -+	→↑ 0-	→↑ 0-	↓ +
Criterion 2:	↑ +	↓ -	→ 0	↓ -	↑ +
Criterion 3:	↓ -	↑ +	→ 0	→ 0	↓ -

| **Sub-target 5.5:** Humane working conditions | **Criterion 1:** Number of workplaces in waste management **Criterion 2:** Relation of waste related workplaces of high quality and security to the total number of workplaces (measuring the relation of sick- or accident-days to the total number of workdays) | | | |

Result of evaluation*:	Scenario 1	Scenario 2	Scenario 3	Scenario 4	Scenario 5
Criterion 1:	→ 0	→ 0	↑ +	↓ -	↓ -
Criterion 2:	→ 0	↑ -	↓ +	↓ +	↓ +

Table 8.7. (cont.)

Sub-target 5.6: Optimized disposal safety	**Criterion 1:** Guaranty of basic waste disposal, everybody has the possibility to dispose waste legally and correctly. **Criterion 2:** Relation of the existing middle- and long-term agents in waste management to the amount of waste **Criterion 3:** Difference between the reserve capacity for the disposal plants and the effective need

Result of evaluation*:	Scenario 1	Scenario 2	Scenario 3	Scenario 4	Scenario 5
Criterion 1:	→ 0	↓ -	→ 0	→ 0	→ 0
Criterion 2:	→ 0	↓ -	→ 0	→ 0	↑ +
Criterion 3:	↓ +	↓↑ +-	→ 0	→ 0	↓ +

* Comparison of criteria with the current situation in 1997: Criteria are ↑: increasing, ↓: decreasing, →: not changing. Element is +: sustainability-supporting, -: sustainability-inhibiting, 0: neutral.

than the elements of the other scenarios do: The sub-targets containing clear social aspects, such as "User-friendly disposal system with free access" (Sub-target 5.4) or "Humane working conditions" (Sub-target 5.5) were not all determined to have a sustainability supporting (+) effect in Scenario 4. The conclusion is that, not only social, but also ecological and economic elements of scenarios can equally contribute to "social compatibility".

• The success of Scenario 5 ("Simplified waste logistics") depends mainly on whether and to what extent residues can be made capable of long-term disposal or recyclable with high-tech incineration processes. Another question is whether there is a market for recyclables such as metals recovered from filter ash and slag (see Chapter 4). Under current economic conditions and considering that newly developed technologies are lacking practical operational experience, Scenario 5 is a middle-term option for sustainability.

The evaluation of scenarios was based on the knowledge and experiences of the workshop team, and the results represent the view of these experts only. Even though the evaluation was controversial in some points and is, of course, subjectively determined by the evaluators, it provides a sufficient variety of sustainability-supporting elements, providing material for recommendations for sustainable action (section 8.4).

8.3 Case studies

8.3.1 Aim of the Case Studies

While an integral focus on waste management was selected in the previous sections for work with targets and scenarios, in selected fields, case studies were intended for finding/developing more information about specific aspects. The subjects of the four case studies are as follows:

- Material flow analysis can provide fundamental knowledge for developing a sustainable waste management. As an example the first case study shows the results of a material flow analysis for Cadmium in Switzerland.
- In the second case study [26], Scenario 3 ("Better products") and the possibilities of material and energy optimization of products in production, recycling and disposal were investigated in the field of electrical and electronic scrap.
- In the third case study [19], the influence of an ecological tax reform and other market instruments for waste management in general and in specific relation to the five defined scenarios were investigated.
- The fourth case study focuses on Chinese waste management. The target system defined in section 8.1.2 was tested and adapted to the development of waste management in China.

8.3.2 Material Flow Analysis as a Base for a Sustainable Waste Management

The aim of this case study [18] was to determine material flows in Switzerland for the following four substances: carbon (as carrier of energy and nutrients), cadmium (as toxic element applied used in consumer goods), chlorine and nitrogen (as essential nutrients and potential air polluters). Material flow analysis (substance flow analysis) is a method to establish the material balances of the inputs, outputs and stocks of human settlements ("the anthroposphere"), and contribute to our understanding of the material flows and material accumulations in human-made systems. This analysis is essential in assessing the relative importance of the material flows and the options available for influencing these flows as well as the relative importance of different processes in the entire system (e.g. waste management).

The material balances were established for the nineteen-eighties and the nineteen-nineties in the last century. They are based on ascertained data. Based on existing know how with potential sources, the partitioning and the fate of the selected substances as well as the metabolism of the entire system investigated are described. In comparing the corresponding material balances, the investigations focused especially primarily on measures with regard to defined targets in waste management.

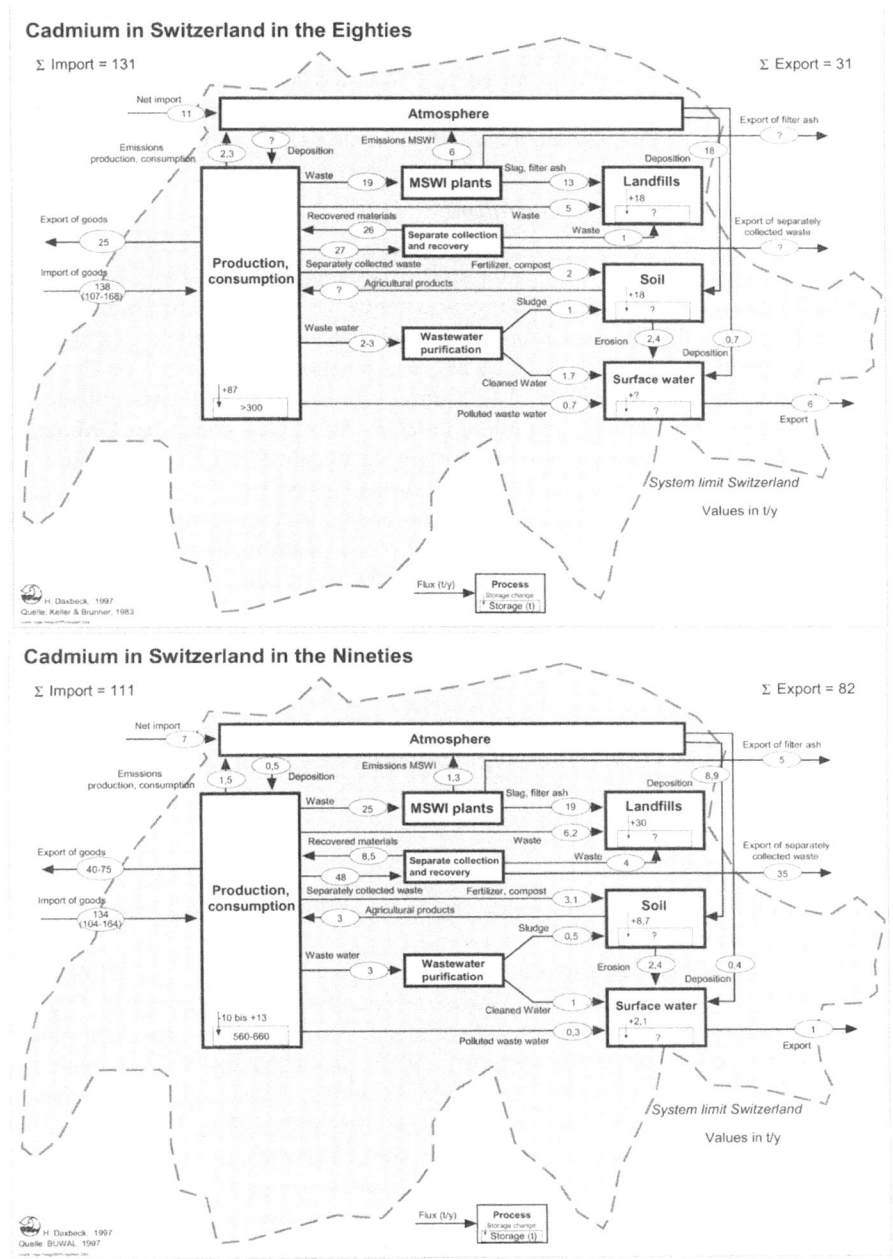

Fig. 8.4. Material flow analysis of cadmium in the eighties and nineties in Switzerland [18]; MSWI: Municipal Solid Waste Incinerator

In this summary of the case study we focus on one substance only: Cadmium. With this example it is demonstrated, how the methodology of material flow analysis can be used as a useful instrument for early recognition (section 8.1.2), for the control of material fluxes, for priority settings and for the selection of efficient technical and organizational measures in waste management.

Material Flow Analysis of Cadmium

As can be seen in Fig. 8.4, consumption of cadmium decreased by 3% from the eighties to the nineties. The remaining flows into the environment are massively decreased by 40% for the same time period. Solid waste management as a concentration and retention process, holds a primary position of influence on the cadmium metabolism in Switzerland. The cadmium flows into incineration plants increased by 30% from the eighties to the nineties. The Swiss Clean Air Ordinance [4] determines the maximally permissible emission concentration of cadmium in waste gas to be 0.1 mg/m^3. Thus the airborne emissions from incineration plants have been reduced very successfully by almost 80%. An additional indirect effect is a 50% reduction in the atmospheric deposition of cadmium onto ground-soil.

The Ordinance on Material [5] has effected the reduction of cadmium flows in many goods by up to 90%. However, at the same time, the cadmium use in nickel-cadmium accumulators has increased. Accumulators are not regulated by the Ordinance of Material [5] and they are responsible for almost 70% of the current load.

The legal regulation has so far only caused a displacement of the problem, rather than solving it (*problem shifting*), for the entire load of cadmium has hardly changed.

Conclusions

The following main conclusions out of the material flow analysis for the four substances can be summarized:

1. Waste management as part of the whole national economy: Using material flow analysis changes the way a system is perceived and/or analyzed. The focus is not limited to waste management alone. It includes the substances coming in and going out of the system of waste management itself: It includes flows "from the cradle to the grave." Research has shown that waste management is very often not the key process for controlling material flows in the anthroposphere. For cadmium (and similarly for chlorine), waste management is a key process to control material flows to the environment. Nitrogen, for instance, is clearly more influenced by agriculture, waste management play a minor role. Fossil fuels are, at 90%, the most important carbon-substrate, which means that the main flows of carbon are not controlled by waste management activities, too. Nevertheless, waste management has a - in part, very efficient - retention function, for instance, with respect to emissions of cadmium, organic compounds or hydrochloride. Because of these reasons, it is of great impor-

tance that the whole national economy, of which waste management is a part, be taken into consideration [10, 13].

2. Efficiency of the legal regulations: The efficiency of legal regulations was high, when either technical and/or technological solutions or alternatives were already available and proven in practice, or if they were later developed (due to legal regulations). For example, through the installation of flue gas cleaning systems, the emissions of cadmium could be reduced by factor 5, and the emissions of hydrochloride decreased by a factor of almost 10. Legal regulations to avoid or prohibit substances in certain consumer products were also rather successful, but have not been able to solve the full problem related to the concerned substance. A displacement of the problem was often the result (*problem shifting*). Until the end of the eighties, selective solutions had been applied to solve problems day by day. The result of these measures was the shifting of a problem into other environmental compartments, like from air to soil or from soil to groundwater, etc. (*problem shifting*). Legal regulation of substances (Ordinance of material [5]) was applied to reduce the consumption of hazardous substances in certain applications (e.g. cadmium as stabilizer in plastics). This type of very specific regulation related to a certain application often just displaces the problem. This is mainly due to an inelastic nature in economic production and to the use of many substances (e.g. cadmium, chlorine), a variety of possible intentional applications, and the possible occurrence of the substances as a contaminant in other products (cadmium in phosphate, heavy metals in recycling products). The aim of organizational measures in the future is not merely to formulate restrictions and prohibitions. More system-approached solution finding is to be applied. The control of waste management processes is rather performed through a mixture of system-based knowledge, market instruments, voluntary measures and responsibility on behalf of the producers and consumers. The future will show how quickly and efficiently these measures will take effect.

3. Equal requirements for disposal and recycling processes: The avoidance concepts used until now have not been able to affect a clear reduction of the used materials and were therefore not very successful. In the anthroposphere, huge stocks of materials and substances have been building up and are still increasing. These stocks have to be used / recycled or disposed of in the future. Materials and substances are increasingly infiltrated in recycling processes. Therefore, it is necessary that for these processes, the requirements regarding environmental protection are as strict as for disposal processes. The necessary legal regulations have to be defined.

4. Achievements of the material flow analysis as a planning instrument: As shown with the four substances carbon, cadmium, chlorine and nitrogen, the material flow analysis is a suitable instrument for early recognition, for the control of material fluxes, for priority setting and for the selection of efficient technical and organizational measures in waste management. Due to material flow analysis, future resource potentials, future restraints of resources or environmental impacts, caused by the current handling of goods, materials, substances and

waste, can be recognized. It is possible to prove the efficiency of technical and organizational measures with regard to the defined targets in waste management. Material flow analysis is, therefore, a necessary planning instrument, which provides important decision-making material in waste management. Without knowledge of the documentation and steering of material flows, an optimized waste management, regarding optimal use of resources and long-dated protection of environment, is not possible.

8.3.3 Scenario "Better Products" – The Example of the Personal Computer

This case study [26] investigates the possibilities of improving products, which is the main focus of the Scenario "Better products", in the field of electronics (section 8.2.2). This case study focuses on recycling processes as well as on the design and production of electrical and electronic appliances. Products, devices and materials as well as recycling processes are analyzed how recycling and design/production can be optimized. The example of the personal computer (PC) is used here to illustrate the practical application of this tool.

Electrical and Electronic Appliances

Collection and recycling of electrical and electronic scrap is well organized in Switzerland. Electrical and electronic scrap can be returned for recycling to all stores selling appliances. Recycling of devices and materials is provided by a few private organizations. Legally, this field is directed through the "Act on the Return, Take back and Disposal of Electrical and Electronic Appliances" (VREG) [9].

The amount of electrical and electronic scrap in Switzerland in 1998 was 90'000 t [23, 30]. The rate of recycling of electrical and electronic appliances was relatively high (58'500 t, 65%), 50% (27'250 t) of this amount was iron scrap, which can be recycled easily. 25% of the appliances (22'500 t) were exported and used abroad as second-hand products. 10% (9'000 t) were treated in incinerators.

The highest turnover of products in 1998 was in the branch of "Home appliances" (40%, 62'000 t), followed by "Electronic data processing (EDP) and office automation" (36%, 56'500 t). The highest total amount of circuit board is found in "EDP and office automation" at 2'600 t (65%). This is also the group with the highest increase potential within the next five years (+40%). Considering the environmental importance of the circuit board and the high yearly turnover, optimizing recycling processes need to focus on the product group "EDP and office automation".

The Portfolio-Grid-Method

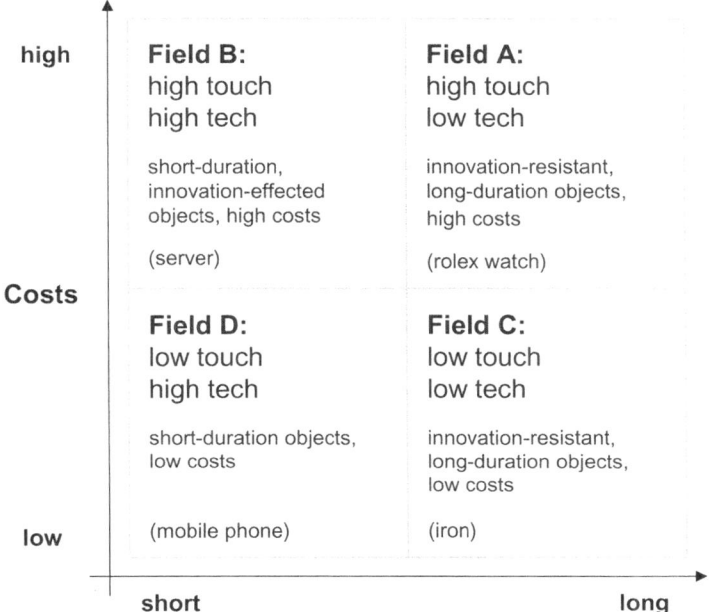

Innovation resistance: period of innovation cycle

Fig. 8.5. Portfolio-grid for evaluating products with the axes "innovation resistance" (represented through "innovation cycle") and "costs". The four fields A, B, C and D are characterized by high/low innovation cycles and high/low costs

The portfolio-grid method, well known in financial branches, is adapted and used in the waste management field of electrical and electronic scrap to describe products, devices and materials as well as recycling processes and, in turn, to generate information on how recycling and design/production might be optimized.

"Innovation resistance" and "Costs" of products were recognized to be the main influence factors in the using-system of electrical and electronic appliances [26]. Each product is a composition of different units such as devices or materials. The durability of a product is, on the whole, not determined through the natural age limit of the entire product, but through the one unit with the lowest innovation resistance and the shortest innovation cycle (in number per year) respectively. A product containing highly innovation-dependant devices will quickly become market-uncompetitive, if not obsolete, due to innovations and new developments - its use life is limited by these innovations (*innovation determined use life*). The term "innovation resistance" is therefore a measure for the longevity of a product or device.

In the application of the portfolio-grid, by plugging a product into a field, it can be determined for each product, whether or not it is *expensive*, and whether it will be *innovation resistant* and, therefore, has a potentially high degree of longevity, or, by contrast, soon to be made obsolete by innovations (*Portfolio-grid method*). A positioning within the portfolio-grid is generally applicable for products, devices, materials, substances and recycling processes. This method allows for the description of them, and thereby facilitates a determination of ideal points of departure for optimization.

Application to Electronic Products

First the portfolio-grid method is applied to the level of products. The example of the PC in Fig. 8.6 (left) shows its position within field D as a product with rather low costs and short innovation cycles, while typewriters, on the contrary, are cheaper but have a longer use-life.
Fields D and C along the axis "Innovation cycle" in Fig. 8.6 (right) show that there are equal quantities of long-lived (43% by weight) and short-lived (47%) products currently on the market. The values in fields B and D along the axis "Costs" indicate that there are more low-cost (47%) than high-cost products currently on the market (8%).

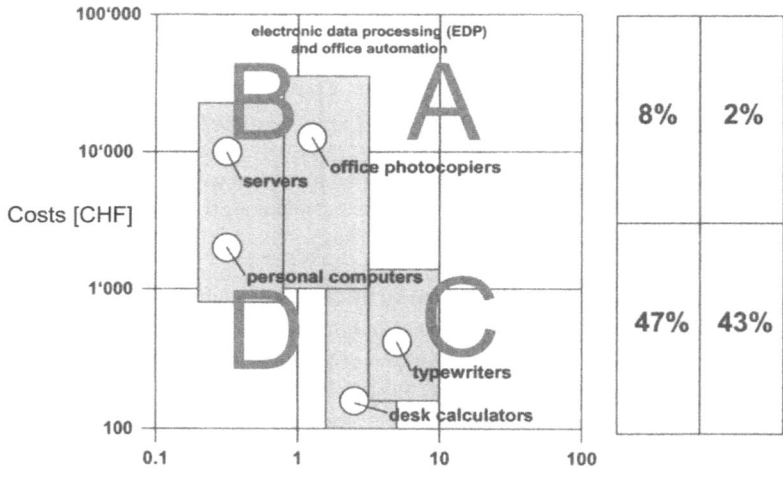

Fig. 8.6. Level of products: Position of the Personal Computer, belonging to the group of "EDP and office automation", within the portfolio-grid, Switzerland 1998 (left); and rate of percent by weight for all electrical and electronic appliances, Switzerland 1998 (right). Grey areas indicate common variations of data. White circles indicate the respective "center of gravity" of product group

Two main trends can be seen in Fig. 8.6.: A trend towards more innovation-effected or short-lived products and a trend towards low-cost products. Increasing pressure from the global market encourages more innovation. The use phase and the innovation cycles of products decrease. At the same time, the production of low-cost goods is encouraged by a principle of profit maximization and the ever-changing "needs" or demands of the consumer. Another trend is that the product groups grow together, and more "intelligent" products, which can perform several functions at the same time, come onto the market. The development of mobile phones which can be used for calling, writing short messages, surfing the world wide web, organizing one's schedule, etc. illustrates this trend.

Application to Electronic Devices

The innovation resistance of the different units can vary greatly, as the example of the PC in Fig. 8.7 shows. The processor, the hard disc and random access memory, the board and the expansion card are subjected to a condition of fast innovation; actually, all below one year. The processor of a PC is both the most costly and the most innovation-effected device. Distribution of quantity (in percentage of weight) shows a different picture: 75 to 90% of the devices in a PC are "old" technologies with high innovation resistance.

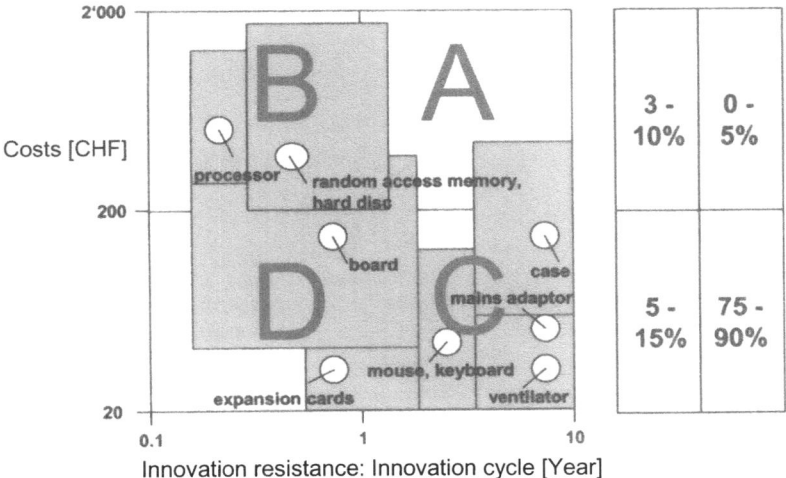

Fig. 8.7. Level of devices: Distribution of main devices of a Personal Computer within the portfolio-grid and stated in percent of weight; price level in 1998. Grey areas indicate common variations of data. White circles indicate the respective "center of gravity" of product group

It turns out that one has to focus on design and production as well as on the recycling processes for the devices to avoid a heavy increase in non-recyclable electrical and electronic waste. Inversely proportional is the distribution of the investments in Fig. 8.7: 80% of the turnover is apportioned to innovation-effected devices, only 20% to the other parts.

Application to Materials and Substances

Still using the PC as an example, the level of materials and substances is investigated (Fig. 8.8). The axis "Costs" here represents the costs of waste disposal (below the value zero) and the profit of recyclables sold as secondary resources (above the value zero)
The segmentation into the two fields A and D is obvious: Recycling is only profitable for materials that can be separated with simple, known technologies. Classification of materials into constant and unmixed fractions is an important requirement for recycling. However, only the metallic fractions can be handled easily in this way. The separation of compounds such as plastic compounds and circuit board is difficult or even impossible. This further illustrates the importance of product design. At the point of design and production, the producer must already be thinking of how the devices and materials can be recycled and how the recycling logistics can be simplified.

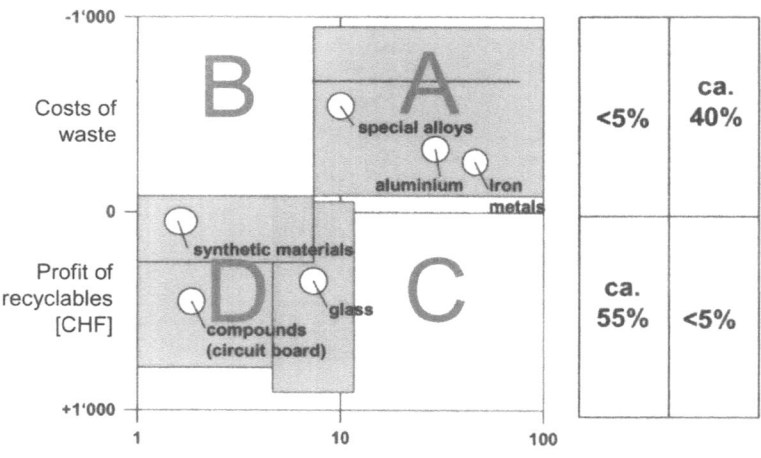

Fig. 8.8. Level of materials: Distribution of materials of a Personal Computer within the portfolio-grid, price level in 1998, stated in percentage of weight, reference year: 1991. Grey areas indicate common variations of data. White circles indicate the respective "center of gravity" of product group

Application to Recycling Processes

An additional application of the portfolio-grid is in the evaluation of the recycling processes and the material fractions arising from the rebuilding and separation processes of appliances respectively. This evaluation provides a distribution similar to the one shown in Fig. 8.8. Some fractions, mostly the compounds and hazardous substances, are highly innovation-effected and the recycling is costly. Metals and glass, on the other hand, are long-lasting fractions and can be sold as secondary resources.

Points of Action for Optimization in Production, Consumption and Recycling

Using the described portfolio-grid method, it is possible to characterize products, devices, materials and substances, as well as processes, in order to identify, subsequently, ideal points of action for optimization in production, consumption and recycling. The positioning of a specific product within the portfolio-grid will help to determine the possibilities and decisions on action to be taken, as well as the possible points in the lifecycle of a product, where the action is taken. The influence on the scope of action was qualitatively evaluated for each quadrant of the portfolio-grid regarding the following three aspects (Table 8.8):

- Technical aspects such as technology and know how,
- Logistic aspects such as spatial and time co-ordination, and
- Social aspects such as behavior and independence.

Table 8.8. Points and scope of action, in co-ordination with the positioning of a specific product or device, etc., in the portfolio-grid (t: technical aspects, l: logistic aspects, s: social aspects; +: increasing, -: decreasing, 0: no effect).

Point of action	Resulting scope of action											
	Quadrant A			Quadrant B			Quadrant C			Quadrant D		
	t	l	s	t	l	s	t	l	s	t	l	s
Production												
Procurement of raw and other material	+	+	+	+	0	0	0	0	0	-	-	-
Fabrication of devices and products	+	+	+	+	-	0	0	0	0	-	-	-
Research and development	+	+	+	+	-	-	0	0	0	-	-	-
Conception of appliances	+	+	+	0	-	0	0	0	0	-	-	0
Organization of distribution	+	+	+	0	-	0	0	0	0	-	-	-
Organization of services and attendance	+	+	+	-	-	-	+	0	0	-	-	-
Process of fabrication	+	+	+	0	-	0	+	0	0	-	-	0
Import	+	+	+	0	-	0	0	0	0	0	-	-

Table 8.8. (cont.)

Point of action	Quadrant A			Quadrant B			Quadrant C			Quadrant D		
	t	l	s	t	l	s	t	l	s	t	l	s
Evaluation												
Conceptual requirements	+	+	+	0	+	+	0	0	0	-	-	-
Purchase decision	0	0	0	0	0	0	0	0	0	0	0	0
Developmental possibilities	+	0	+	0	0	-	0	0	+	-	0	-
Operational safety	0	0	0	0	0	0	0	0	0	0	0	0
Use												
Method of use	+	0	0	0	0	0	0	0	0	-	-	-
Intensity of use	+	0	0	0	0	0	+	0	0	-	-	-
Marketing decisions	+	0	+	0	0	0	0	-	0	-	-	-
Reparation (this does not mean *repairs*)	+	+	+	0	+	0	0	0	0	-	-	-
Services and maintenance	+	+	0	0	+	0	0	-	0	-	-	-
Operational effort	+	+	+	-	0	0	0	0	0	0	0	-
Changing needs	+	0	0	+	0	0	0	0	0	-	0	0
Recycling												
Recycling of products	+	0	+	+	-	0	0	+	+	-	+	0
Recycling of devices	+	+	+	0	0	0	0	0	0	-	-	-
Recycling of material	+	0	0	0	0	0	+	+	+	-	-	-
Recycling of substances	0	0	0	0	0	0	0	0	0	-	0	0
Technology and know-how of recycling	0	0	0	-	0	0	0	0	0	-	0	0
Distribution and logistics	0	0	0	-	0	0	+	+	+	-	-	-

Conclusions for the Scenario "Better Products"

The developmental potential towards sustainable recycling is found in the area of products/devices and of materials/substances. The following points have to be considered and put into practice on the level of products and devices:

- A scattered distribution of devices in various fields of the portfolio-grid should be avoided.
- Innovation-effected and low-cost devices must be considered in the process of product design, such that they can be replaced without problems.
- A simple disassembly of devices must be possible.

On the level materials and substances:

- Materials should be standardized.
- The diversity of materials should be decreased.
- Composite materials should be avoided.
- Hazardous substances should be avoided.

Table 8.9 illustrates the characterization and suitable waste management strategies for each of the four field-types of the portfolio-grid.

One must ascertain the suitable strategy of recycling method for each product and device in a separate evaluation process in order to find the most suitable point of action for optimizations in production, consumption and recycling. The recognizable trend moves in the direction of innovation-effected, short-living and low-cost products (same power and capacity for less money in the PC example). The portfolio-grid is a method which facilitates the ascertaining of points of action for optimization and, at the same time, helps to detect which devices must be altered for greater longevity in order for the entire product to achieve greater longevity.

Table 8.9. Characterization of the four field-types and the suitable waste management strategies

Field-types	Waste management strategies
Field A: High touch, low tech Innovation-resistant, long-duration objects, high costs	• Increasing life-duration • Recycling of products with differentiated recycling logistics
Field B: High touch, high tech Short-duration, innovation-effected objects, high costs	• Modular construction, constructional separation of innovation-effected devises • Recycling of devices and products (quality assurance of the re-used devices is important)
Field C: Low touch, low tech Innovation-resistant, long-duration objects, low costs	• Increasing life-duration • Recycling of products and materials • Eventual recycling of substances
Field D: Low touch, high tech Short-duration objects, low costs	• Modular construction, constructional separation of innovation-effected devices • Recycling of devices • Recycling of materials and substances, incineration (thermal recycling)

8.3.4 The Influence of an Ecological Tax Reform on Waste Management

In several European countries, an ecological tax reform or elements of market instruments are being discussed or are implemented. The influence of an ecological tax reform on political economics and especially on waste management in Switzerland is investigated as well as the consequences of market instruments in the five described scenarios [28]. The term "market instruments", used in this case study to mean various types of taxes and fees, however, excludes other economic instruments, such as tradable emission licenses.

Three Levels of Economic Steering of Material Flows

Market instruments in the field of waste management can serve the purpose of financing or steering. Current taxes in waste management simultaneously have financing as well as steering effects, to lesser and greater extents. The environmental market instruments can be divided into three steering levels:

1. *Basic or global steering on the level of goods*: Concerns the flows of goods; the quantity of the flows is decisive
2. *Steering of substances:* Concerns the substances causing problems in waste disposal
3. *Steering of processes*: Concerns separation processes, reusing processes, recycling processes, as well as detoxification and disposal processes

These three instruments will be discussed one by one in the following text.

Global Steering on the Level of Goods

Global steering can be performed by implementing taxes on resources like energy or on emissions like CO_2 or SO_2. A tax on fossil fuels has been proposed in Switzerland. The tax rate being discussed [22] is one which will increase by 2% for electricity and 3.5% for fossil fuels every year over a period of ten years. After ten years, tax rates will be at 13% and 32% respectively. At this stage, demand for energy should be reduced by 8%, and taxes account for 3.3 billion CHF. These tax rates are based on the model of ETRT (Ecological Tax Reform Tax reduction) [22]. Main intentions of the model ETRT are as follows:

- Tax rates are raised on all fossil fuels and on electrical energy.
- Tax rates are implemented step by step and gradually increased (see above).
- Tax revenues are used for reducing the associated employer outlay.

Part of the model ETRT are cushion measures, which limit the net load through the raised tax to a maximum of one percent of the gross production value of the respective enterprises or branches.

The effects of model ETRT on the quantity of waste were estimated in this case study [28] for the waste-relevant branches as follows: It was assumed, that the change in the gross production value in the waste-relevant branches has a direct effect on the amount of waste. Therefore, the change of inland consumption had to be calculated first. Afterwards the waste fractions were assigned to the selected branches, if possible, and the change of waste quantity was determined.

In addition to the energy industry, which is the central concern, the estimations in this case study show that only 13 other economical branches are affected substantially by an increase of more than 0.5% in the gross production costs. Seven out of these 13 branches are waste-relevant: textiles, fine paper and cardboard, non-metallic minerals, chemistry (plastic, rubber), metallurgy, machine and vehicle building and electrical, electronic and optical devices. Associated employer outlay will decrease by 6.4% on average, which corresponds to 1.03% of the total wage costs.

The effect of an energy tax on the amount of solid waste is very small within the ten years considered. The amount of solid waste will decrease – in comparison with the development of a reference model without an ecological tax reform and with zero growth [22] – by 32'000 tons per year. This corresponds to 0.66% of the amount of solid waste within the seven branches, respectively 0.5% of all municipal solid waste plus the waste of industrial and small business enterprises. (In fact, the amount of solid waste will increase for plastic and rubber, machine- and vehicle-building, electric and electronic appliances industries, because the decrease of wage costs will outweigh the increase of energy costs. In all other branches the amount will decrease).

As a result of the energy tax, full costs of collecting and transporting municipal solid waste will increase slightly (by 1.4%). This will make recycling and disposal slightly more costly, but will not endanger them. In the long run, tax-caused innovation processes will emerge and result in a decreasing amount of waste. However, the influence on the quality of waste is uncertain and could even be negative with new high-technical materials containing hazardous substances.

In summary, global taxes at the rates shown above have only a small influence on the reduction of solid waste. For the same reason, they contribute little to sustainability in waste management. In order to really support sustainability in waste management, tax rates would have to be substantially higher. At the same time, supporting instruments should be implemented in order to inhibit a reduction in waste quality through new problematic substances. Global steering with high tax rates (which would be higher than the ones investigated in this case study) appears to be an important instrument for reducing the entire volume of material flows, but is not suitable for specific use on problem substances.

The Level of Substances: Substance Tax on Cadmium

Cadmium and other heavy metals belong to the group of most important problem substances in the waste management in Switzerland today (Chapter 5 and [12, 21, 29]). Therefore, the effect of a tax on cadmium, in order to reduce its use and emissions, was examined [28].

In the nineteen-nineties, more than 90% of all cadmium employed in Switzerland was contained in nickel-cadmium-accumulators, pigments and stabilizers used for the production of synthetic materials, solders and alloys (section 2.2 and [18]). The cadmium tax should therefore focus on these products. Tax is determined by the amount of cadmium used in a particular product. The tax is raised on the import of cadmium or products containing it. One condition for a successful implementation of the tax is – as a supporting measure – the legal obligation to declare the cadmium upon its import at the border. The tax rate should be high enough to induce a substitution of cadmium by other substances.

Nickel-cadmium-accumulators are responsible for 50% of current cadmium flows [18]. The investigations show that a tax rate of between 10 and 30% of the current price of nickel-cadmium-accumulators may induce a substitution by nickel-metalhydride-cells. The latter have additional positive attributes, such as no memory-effect, decreasing prices, etc. This makes them more attractive for consumers and could help in the facilitation of the substitution. With respect to the other applications of cadmium, it is more difficult to find substitutes. It is possible that the substitution of cadmium (e.g. in stabilizers by lead [25]) will bring about worse product attributes and causes additional costs. The longevity of the products without cadmium could also decrease. But with the development of adequate technologies and new products, these additional costs may most probably be reduced after some years. The cadmium tax on nickel-cadmium-accumulators is a pragmatic approach in the reduction of cadmium with low organizational effort, and does therefore not solve the entire cadmium problem.

The implementation of a specific tax on cadmium or other substances, in combination with the foreseeable and gradual increase of tax rates will encourage industries to look for substitutes or alternative materials with the same attributes. They would have a chance to reduce the costs of adaptation. Additional measures would be necessary in order to avoid substitution by more rather than less problematic substances.

The Level of Processes: Recycling of Electrical and Electronic Waste

The influence of market instruments on selected recycling processes in the field of electric and electronic appliances was examined [28]. The current situation concerning electrical and electronic scrap is described in section 8.3.2 and in [26].

According to a plan of sustainable waste management, electrical and electronic scrap should be avoided as much as possible, and the accumulated scrap should be recycled or disposed of with a low emissions process. Reducing and avoiding them can be achieved through an extension of use life and through the reuse of appliances or devices. Extending the use life of electrical and electronic appliances is not easy. The use period is often limited by innovations (section 8.3.2). The longevity of a computer, for instance, is determined by the one element which is first made obsolete due to new developments and innovations. The longevity of the whole product therefore depends on the element with the shortest life (*innovation determined use-life*).

In order to break with this dependence, the following requirements for product development and design should be implemented:

- Standardization of devices which are not technology- or marketing-relevant and which are therefore not easily made obsolete by innovations and have a longer use life.
- The development of module design which will allow them to be exchanged or repaired separately.
- Declaration of composition of the devices

Being able to influence the industrial product design and various associated processes (such as marketing, distribution, selling and repair logistics) is currently still a difficult task, because the suitable instruments as well as the necessary information are lacking. The risk of errors with economic consequences is high.

More effective is the reuse of appliances and repair of damaged appliances. This can clearly help in the reduction of the amount of elements to be disposed of. Repair services should be supported and expanded, but can hardly be influenced by the current instruments of waste management. Therefore, the efforts should focus on instruments which can influence the recycling processes and the ecological disposal or volume of electronic scrap. In Switzerland, the VREG [9] already regulates the return and disposal of electrical and electronic appliances to a great extent. In addition to that, the following steering instruments are to be implemented:

• Tax on electrical and electronic scrap, in order to make the disposal of appliances which can be recycled (e.g. metalliferous components, picture tubes etc), more costly. Tax objects would be recyclable appliances to be brought to the incineration plant, the amount of the tax possibly varying according to the degree of hazardous potential of the contained substances. Recyclers of electrical and electronic appliances are responsible to pay the tax.

• Obligation to declare materials: Recycling is easier, if the composition of the materials and components of appliances is declared and the declaration is coordinated on an international level.

In the field of process steering, the recycling and disposal processes are important. With the VREG [9] in the area of electrical and electronic scrap, the introduction of a tax has been proposed on non-recycled appliances, which has a positive influence on the recycling logistics. In other fields, the requirements must first be enforced, in order to prevent negative effects which could follow.

Conclusions: Market Instruments and their Contribution to a Sustainable Waste Management

Three levels of economic steering have been considered for the use as market instruments. While global steering with high tax rates (which would be higher than the ones examined in this case study) is an effective instrument for reducing material flows altogether, it is less suitable for the specific reduction of problem materials. The steering of materials and substances through a tax on problematic substances is very effective, when the specific substance is to be reduced, and not entirely eliminated. Steering of processes can be used with recycling, reuse or disposal processes (with the effect of reducing waste amounts and eliminating problematic substances), but is not suitable for directly influencing the production processes.

Table 8.10 shows an overview over the discussed market instruments and their influences on sustainability in waste management.

Table 8.10. Overview of market instruments for a sustainable waste management (++++ strong effect, +++ clear effect, ++ supporting effect, + weak effect, 0 no effect/neutral, - negative effect, -- clear negative effect), slightly changed according to [28].

Effect on	Energy-/CO_2-tax with refund of yield	Tax on materials and substances (Cd, Pb, Hg, Cu, VOC, etc)	Pre-paid disposal fee and pre-paid recycling fee	Tax on landfill	Tax on waste, disposal bag fee	Basic fee on disposal for households and industries	Deposit
Dematerializing, net product, reduction of energy- and resource-use	+++	++	+	0	0	0	++
Reduction of CO_2-emissions	+++	+	+	+++	+	0	++
Reduction of air pollution	+++	++	+	+++	+	0	+
Innovation effects: New eco-friendly products and processes	++++	++++	++	0	0	0	0
Reduction of problem materials, amelioration of waste quality	-	++++	+++	0	++	0	+
Incentive for separate collection of recyclables	+	+[a]	++++	+	++++	0	++++
Incentive for reuse and recycling	+	+[a]	+++	+	++	0	++++
Financing of recycling logistics	0	+++	++++	0	+++	+++	0
Financing of disposal logistics	0	+++	++++	++++	++++	++++	0
Realization of the "polluter-pays" principle	++++	++++	+++	+	++++	+	++++
Socially acceptable distribution of burdens in waste management	+, +++[b]	0	0	-	-	--	

[a] For the taxed materials and substances: ++++.
[b] Depending on how the tax revenues are refunded.

According to Table 8.10, conclusions about market instruments with an effect towards sustainability in waste management are as follows:

- A high global tax such as an energy tax or a CO_2-tax with a refund is an important instrument for a sustainable economy and for sustainable waste management as a part of the national economy (Fig. 8.1). On waste management alone, its effects are low.
- Specific substance taxes on problematic, hazardous materials such as heavy metals are important instruments as long as they do not facilitate problem shifting. Additionally, a stock of materials in consumer products as well as the level of emissions in the environment must be taken into consideration.
- Pre-paid disposal fees, disposal-bag fees and deposits are necessary reinforcement measures in the closing of material cycles and detoxification through recycling and reuse.
- A tax on recyclable waste can facilitate the improvement of recycling processes and product development.

All these instruments can only be effective if accompanied by the applicable laws and effective legal enforcement.

In the following, various market instruments appropriate for the development of the desired state of the five scenarios are described. The market instruments can be effective tools for a sustainable economy and waste management, but cannot solve problems related to specific critical substances in waste. Therefore, the energy- and CO_2-tax (referred to as ecological tax reform in the present work) is an important condition in all of the five scenarios (section 8.2.2), despite the fact that its effect on the waste flows is low.

Scenario 1 ("Central co-ordination") and Scenario 2 ("Free market") focus on different structures of organization and different economic systems. Market instruments such as taxes on substances (avoidance of problem materials) and financing instruments are to be implemented by the following:

- Main instruments are pre-paid disposal fees, disposal bag fees, taxes for financing waste logistics as well as for the separation of waste (steering instruments)
- Supporting instruments: Taxes on materials, pre-paid disposal fees and pre-paid recycling fees (steering instruments)

Scenario 3 ("Better products") and Scenario 4 ("Fewer products, new values") focus on the quality and quantity of products. Suitable market instruments are as follows:

- Main instruments: A system of high taxes on materials and problem substances in order to facilitate the reduction of hazardous substances (Scenario 3).
- Supporting instruments: Disposal bag fees, tax on waste in addition to pre-paid disposal fees and pre-paid recycling fees to help finance the recycling logistics and separation and recycling processes (steering instrument)

Scenario 5 ("Simplified waste logistics") focuses on high-tech disposal with incineration for many waste categories. The residues should be suited for long-term disposal or for use as resources. The market instruments are as follows:

- Main instruments: Disposal bag fees, tax on waste and additional basic fees on waste.
- Supporting instruments for all hazardous substances which are not sufficiently controllable

8.3.5 Development Towards Sustainable Waste Management in China

Zhao Youcai

Sustainable Development Targets in Shanghai

Since 1978, when China adopted a policy of greater exposure to the outside world, Shanghai industry has always insisted that economic development be as important as environmental protection. It has attempted to increase GDP and production efficiency without increasing waste volume. During the rapid industrial development after 1995, the level of major pollutants should not rise above what they were in 1995. In the 21st century, the emission of pollution will be greatly reduced in comparison with conditions in 1995. The governmental document "National Total Discharged Amount Control Plan of Major Pollutants During Ninth Five Years Plan" specified that the production of industrial solid waste be restricted to below 12.30 million tons and all hazardous waste be treated properly in 2000. Nevertheless, this objective has only partly been fulfilled, as observed in 2001. Healthcare waste and much hazardous waste is still dumped or stored onsite by the generators.

In terms of environmental sanitation, the current total length of road in Shanghai that should be cleaned on a daily basis is 1518 km, with a total area of 26.31 million m². Approximately 40% of the roads are classified as First Class Roads and thus cleaned by machines. The remaining roads are manually cleaned. Because of large-scale construction in Shanghai, sanitation work becomes much more demanding.

In road sanitation, all refuse is collected in a mixed state. The government is striving for separation and individual collection for various component groups. The primary concern is to find a method of treating refuse. As part of the primary objectives, 95% of all refuse are to be properly treated and the treatment method of refuse is to change from the current landfill method to a combination of incineration and landfill treatment. In 2003, two incineration plants will be built, one in East Shanghai and one in West Shanghai. The changes in refuse transportation will soon be completed. An inland transportation system for the incineration plants will gradually be established.

Classified collection is to be performed. In 2010, the following objectives are to be met: all refuse will be properly treated through incineration and sanitary landfill, all road cleaning will be mechanized, and all refuse transportation will be loop-closed and containerized.

Four measures have been proposed:

1. Prevention should be combined with treatment. Prevention is more important than treatment. New technology should be utilized and disseminated. The energy recovery should be optimized.
2. The strategic regulation of industrial structures should be prompted in order to optimize resource distribution.
3. The 107 major enterprises that generate the most pollutants in Shanghai should be monitored and supervised in order to urge them to reach the national standards.
4. Management and statute must be strengthened. In order to meet the objective of "Zero Discharge of Hazardous Waste" as determined in "The Notice of Shanghai Government for Implementation of the State Council Decision on Environmental Protection", it is necessary to strengthen the management of industrial hazardous waste and hospital hazardous wastes, and to promote the license system for the disposal of hazardous waste. The Shanghai Government has decided to construct a hazardous waste landfill with a total capacity of 10,000t/y in Jiading and a incineration plant with a capacity of 19,000t/y in Jinsun.

Sustainability Targets and their Implementation in Chinese Law

Table 8.11 compares the target system for a sustainable waste management (section 8.1.3) to current Chinese laws and formulated targets.

Table 8.11. Comparison of the target system of a sustainable waste management with the current formulated targets in China.

Targets of a sustainable waste management (see *Overview of Targets)*	Current targets of waste management according to Chinese governmental authorities	Rate of sustain- ability*	Deficits of current targets
SUBSTANCES / ENVIRONMENT / TECHNIQUE (Main targets 1-3)			
Sub-target 1.1: No harm to man or environment in ordinary cases	Protection of man and environment is one of China's fundamental policies as claimed in the Constitution of the People's Republic of China and The Solid Waste Pollution Prevention and Control Law.	●●	No deficits in targets.

Table 8.11. (cont.)

Targets of a sustainable waste management (see *Overview of Targets*)	Current targets of waste management according to Chinese governmental authorities	Rate of sustainability*	Deficits of current targets
SUBSTANCES / ENVIRONMENT / TECHNIQUE (Main targets 1-3)			
Sub-target 1.2: No harm to man or environment due to accidents	No targets formulated. Such accidents often occur, but the relevant parties always react slowly.	○○	Target is missing.
Sub-target 2.1: Saving non-renewable resources and sustainable use of renewable resources	8 cities, including Shanghai, Beijing, Shenzhen, etc., were selected to sort and separate waste at the source in 2000.	●○	Too early to conclude at this point.
Sub-target 2.2: Closure of material cycles: Materials are recyclable, capable of long-term disposal, or can be dissipated in an environmentally acceptable way.	Waste prevention, reuse and recycling are encouraged by law and provisional regulations. However, landfill is still the primary method of disposal for MSW. Composting is on the decline while incineration is increasing, particularly in large cities.	●●	No deficits in targets. Landfill and incineration are two methods for MSW treatment. In general, landfill is preferred to incineration.
Sub-target 3.1: Disposal in regional responsibility	Disposal within the region of generation. No transfer of waste across provinces is allowed before official permits have been obtained by all relevant parties..	●○	No deficits in targets. However, the targets have not been fully implemented thus far. Several small-scale sanitary landfills may be constructed within a limited area based on the administration divisions in a middle scale city. Also, lack of capital investment in the region may delay the establishment of required facilities.
Sub-target 3.2: Disposal in temporal responsibility	Landfill is permitted as the primary disposal method for MSW. The generators should treat their own hazardous waste, such as healthcare and industrial waste, using landfill and incineration.	●○	Leachate generated at landfills may cause environmental pollution. Many companies are reluctant to treat the waste they generate due to lack of money or environmental awareness.

Table 8.11. (cont.)

Targets of a sustainable waste management (see *Overview of Targets*)	Current targets of waste management according to Chinese governmental authorities	Rate of sustain- ability*	Deficits of current targets
"ECONOMY / ECONOMIC EFFICIENCY" (Main target 4)			
Sub-target 4.1: True costs with full cost coverage: Internalization of external costs for damages on environment and human health	Target partly formulated.	● ○	Insufficient facilities available for safe treatment of MSW due to a lack of investment. Local government is the only provider of investment.
Sub-target 4.2: True costs with full cost coverage: Internalization of external costs for risks (accidents, aftercare for landfills)	Target partly formulated.	○ ○	
Sub-target 4.3: Economically responsible handling of financial resources	Recycling within company or waste exchange among companies is emphasised. Recycling must be remunerative on an existing market.	● ○	Benefits from recycling is still a principle for all actors.
Sub-target 4.4: Acting compatibly with the environment is remunerative for every one	Tax incentives are to be considered. Ecological tax reform has been identified as a target.	● ○	The demand for tax incentives contains tools that make action remunerative.
Sub-target 4.5: Adequate payment of employees working in waste management	No target formulated.	○ ○	Demand is missing.
SOCIETY / POLICY / AUTHORITIES (Main target 5)			
Sub-target 5.1: Adequate co-ordination of collective and individual interests on waste management	No target formulated.	○ ○	Demand is missing.
Sub-target 5.2: Equally balanced interests and burdens of waste management	No target formulated.	● ○	
Sub-target 5.3: Accessible information for possible participation in discussion and decision	No target formulated.	○ ○	Demand is missing.
Sub-target 5.4: User-friendly and freely accessible disposal systems	No target formulated.	○ ○	Demand is missing.

Table 8.11. (cont.)

Targets of a sustainable waste management (see *Overview of Targets)*	Current targets of waste management according to Chinese governmental authorities	Rate of sustainability*	Deficits of current targets
Sub-target 5.5: Humane working conditions	Employers must take measures in the protection employee health. Hygienic workplaces and high occupational safety must be implemented.	●●	No deficits in targets.
Sub-target 5.6: Optimized disposal safety	Landfill as primary option, incineration to be considered when landfill is unavailable.	●○	Optimized disposal safety is to concentrate on lower investment and easier management.

* The rate of sustainability measures the sustainability while comparing the current targets in Shanghai with the sustainable targets: ●● = sustainable, ●○ = partly sustainable, ○○ = not sustainable

8.4 Recommendations for Development Towards Sustainability

The following recommendations for the development towards more sustainability in waste management are based on the (qualitative) results of the evaluation in the five scenarios (section 8.2.) and the conclusions of the case studies (section 8.3.). The recommendations combine well-known facts with new findings.

First, the conditions for a sustainable waste management (section 8.4.1) will be defined; second, strategies that help to achieve sustainability (section 8.4.2) will be proposed; and third, current courses of action to be taken will be proposed.

The scenario evaluation, performed by the team of experts, and the identified sustainability-supporting elements have acted as the basis for developing new strategies and measures / procedures. The results are representative of the knowledge and experiences of the panel of experts that constitute the workshop team. The following strategies are, therefore, not exclusive of other possibilities. Other possible scenarios and strategies, such as the *life-cycle management* developed in Germany, were not investigated.

8.4.1 Conditions for a Sustainable Waste Management

The following conditions are indispensable for development towards sustainability.

- A system of sustainable waste management must take into consideration the whole life cycle of products, from exploring and quarrying resources, through the production of goods, to the disposal of waste ("from cradle to grave"). Therefore the processes "production", "consumption", "recycling" and "disposal" are to be regarded (Fig. 8.1).
- The targets of a sustainable waste management (see Overview of Tables, 8.1.3) are to be discussed on both national and regional levels. The target conflicts should be solved, in so far as this is possible, and priorities established.
- The targets and strategies must be communicated to the public. The development towards sustainability must be accompanied by a corresponding information and communication policy. Implementation of these factors will increase the chances of success.
- The legal conditions must be clearly defined and implemented accordingly. Consequently, the conditions should be predictable over reasonable periods of time for authorities, industries and private persons, and are therefore deemed fair.
- A quality control system must be installed in order to control the effect of measures taken on the national and regional level.
- Within IWM, "Central co-ordination" and "Free market" are both recognized by the federal government to be target-supporting structures of organization and economic systems. They are to be implemented in the appropriate fields.
- Market instruments, such as an ecological tax reform, as well as incentives, such as deposits, etc, are recognized as target supporting elements and are therefore increasingly implemented.

8.4.2 Strategies, Measures and Procedures

The development towards sustainability in waste management should be aligned with the following strategies. This selection is neither complete, nor exclusive, and additional strategies can be added.

1. *Liberalization with clear conditions*: IWM should be liberalized in selected fields and under clear conditions. This means, on one hand, liberalizing the market concerning, for instance, the collection and recycling of recyclables. The public can profit from the innovative ability of private industries, without endangering disposal guaranties. On the other hand, it means forced co-ordination, for instance, in the field of incineration or landfills. The investments are high and cannot be amortized within a few years. The costs should be distributed according to the "polluter-pays" principle.

2. *Optimized balance between avoidance, recycling and disposal*: An optimized balance between avoidance, recycling and disposal must be established, in principle, for any single substance. Priority must be put on substances harmful to humans and the environment or put on important resources. Sustainability-supporting elements, such as the following, must be taken into account:

- increasing the use intensity of goods through longevity, repair-friendliness and shared use of goods (e.g. car sharing),
- supporting recycling in suitable fields and as far as is ecologically sensible and economically acceptable,
- decreasing the amount of solid waste by stimulating a change in awareness towards less consumption and co-operative behavior, and
- minimizing disposal costs, but ensuring a high environmental protection level.

3. *Exclusive discharge of waste suitable for long-term disposal in landfills*: Landfilling is reserved for waste or substances which do not cause harm to man and environment over a period of hundreds of years [11]. This is a middle-term option that might be reached through the following:

- supporting recycling in suitable fields and as far as ecologically sensible and economically acceptable,
- mandatory incineration of all burnable waste, as far as they are not recyclable
- high-tech incineration should produce residues which are either suitable for long-term disposal or recyclable.

For more information about recently developed treatment-technologies producing residues that are either similar to rocks and therefore capable for long-term disposal or fully recyclable because they are similar to ores, see Chapter 4.

4. *Increased responsibility of producers*: Increase of responsibility of producers concerns the following:

material and energy optimization of products in production, consumption, recycling and disposal, in combination with the following:

- avoidance of environmental impact,
- material and substance accounting in industries and small enterprises, or
- transparency through declaration of products.

Material and substance accounting is to be limited to substances with important material flows or to problematic substances such as heavy metals and toxic substances.

5. *Increased responsibility of consumers*: Changing awareness towards co-operative behavior and less consumption (frugality) is a target-viable but long-term option. An increase of use intensity in goods is another sustainability-supporting element. Transparency through declaration of materials is necessary for consumers to take responsibility. Material accounting in industrial and small enterprises is a necessary condition.

6. *High transparency and user-friendliness*: High transparency and user-friendliness means simple and clear disposal systems and paths. Materials must be declared. Sustainable action must be remunerative for every one through a system of financial and non-financial incentives. An open-exchange based policy of information and communication is a necessary condition.

The Guidelines on Swiss Waste Management [13] and the Swiss Protection of Environment Act [3] aim to achieve a waste management not harmful to the environment. This goal should be reached through the strategies "avoidance", "reduction", "recycling" and "disposal" with the proposed priority [3, Article 30, 13]. This waste *hierarchy* needs to be questioned though, since sustainability does not only strive for the "protection of man and environment", but also for "economic compatibility" and "social compatibility". Therefore, the present priorities have to be adjusted accordingly. A problem substance in a composted material, for instance, may be better concentrated trough the incineration process (in order to be disposed in an environmentally capable way) instead of uncontrollably downcycling it in material cycles.

The increase of responsibility on behalf of producers and consumers has been a sustainability-supporting element in several of the evaluated scenarios. The responsibilities in the field of "disposal" and "recycling" in Switzerland today are clearly set forth in the Swiss Protection of Environment Act [3, Article 31]. The responsibility of producers and consumers in the field of "production" and "consumption" is currently still regulated insufficiently.

The strategies should be discussed and specified for each country, with reference to the current situation. As an example, the strategy "Increased responsibility of producers" and the suitable measures and procedures for concerned parties are described below:

Strategy: Increase of the Responsibility of Producers (Example)

Measure: Material and energy optimizing of products
The responsibility of producers must be increased by obliging them to optimize the products within IWM, which takes into consideration the whole life cycle of products and therefore the processes "production", "consumption", "recycling" and "disposal".

Procedures:

- Introduction of an ecological tax reform as global steering for the energy optimization of products.
- Introduction of a tax on problematic substances, for instance, the heavy metal cadmium, as a necessary condition for substance (material) optimization of goods. The basis for determining the tax would be the amount of cadmium contained in a product, such as nickel-cadmium-accumulators. Tax subjects are the importer and a small circle of concerned actors, which would thus allow for low administration costs. Degree of taxation: For nickel-cadmium accumulators

there already exits technically and economically suitable substitutes. The relatively small increase in the price of between 10 and 30% can already cause a reduction of the cadmium flows of 50% [28].

- Implementation of environmental impact assessment on products. With this assessment the production of a specific good can be evaluated by the authorities and optimized for more environment-friendly production processes. The LCA could be basis of this assessment (see Chapter 6).

Support of recycling within selected fields with the goal of using more recyclables as resources in production processes. The degree of sensibility and supporting value must be assessed for each cycle.

Responsible actors and concerned people:
Democratically elected representatives must introduce the chosen instruments. Importers and producers are the responsible actors.

8.5 Summary and Conclusions

Sustainable waste management considers the *whole life cycle of products* from the cradle to the grave. Therefore, the processes "production", "consumption", "recycling" and "disposal" are to be regarded within IWM (Fig. 8.1).

A *target system for sustainable waste management* was developed regarding the three main objectives of a sustainable development; "protection of man and environment", "economic compatibility" and "social compatibility". The comparison of the current targets in Switzerland with the described sustainable targets illustrated that there are currently gaps in the targets (section 8.1).

Five *scenarios* for a future waste management in Switzerland were developed in future workshops and evaluated using criteria based on the system of sustainable targets (section 8.2). Based on the *evaluation of scenarios*, several recommendations for the achievement of sustainability in waste management were defined (strategies, measures/procedures, section 8.4).

In four case studies special aspects concerning targets, scenarios and the steering of sustainable waste management were investigated in detail:

- A suitable instrument for controlling sustainable waste management is the *material flow analysis* (section 8.3.2, [18]). As shown for the substances carbon, cadmium, chlorine and nitrogen, the material flow analysis is a suitable instrument for early recognition, for the control of material fluxes, for control of targets and priority setting, as well as for the selection of efficient technical and organizational measures in waste management.
- Possibilities for *making better products* were investigated in the field of electrical and electronic appliances (section 8.3.3, [26]). The *portfolio-grid method*, well known in the financial sector, was adapted and used to characterize products, devices and materials and to generate evidence on how recycling and de-

sign/production can be optimized. The example of the personal computer illustrates, in greater detail, the application of this tool.

• Three main steering levels have been considered for the use of *market instruments* in a sustainable waste management (section 8.3.4, [28]). Global steering, the steering of materials and substances, and steering of processes were investigated for three examples: Global tax on electricity and fossil energy, substance tax on cadmium, and recycling of electrical and electronic waste.

• One case study has investigated the current situation in China concerning target settings in waste management (section 8.3.5). The study compares current targets in China with the proposed sustainable target system in section 8.1, and focuses on the current implementation of *sustainable targets in China*.

Based on the present study, the main conclusions are as follows:

• The term of sustainability in waste management was substantiated on the basis of set targets, mostly found in literature, for the three main objectives of "protection of man and environment", "economic compatibility" and "social compatibility". While the technical and scientific targets are principally accepted, there is not yet full agreement about the economic and social targets. The *target system* is a tool for guiding waste management towards sustainable development. It helps authorities in defining their targets on waste management. The proposed system of sustainable targets (section 8.1) has been established by a team of experts and reflects their views and experiences, which in turn were determined by the specific development in Switzerland. The target system proposed here is intended as a base for structured target setting, testing, adaptation, and priority establishing elsewhere (planning instrument).

• The method of *development and evaluation of scenarios* for a future waste management (section 8.2) described here does not take as its starting point the current situation, but is based on idealized future states in 20 to 100 years. The method is meant to act as a proposition for finding strategies towards more sustainability. Though some possible strategies are described, the particular *strategies* as well as the *measures* and *procedures* must be defined through each operator referring to the specific case.

• *Market instruments* are discussed widely, but only sparingly implemented. They should be applied according to their possible effect: Global steering is an instrument for reducing material flows altogether, but is less suitable for the specific reduction of problematic materials; it only has an explicit effect at high tax rates. The steering of materials and substances through a tax on hazardous substances can be very effective if the specific substance should be reduced, but not eliminated entirely; problem shifting must be avoided. Steering of processes can be used with recycling, reuse or disposal processes, but is not suitable for directly influencing the production processes.

• The method of *material flow analysis*, used in one case study (8.3.2) is a planning tool, which provides good decision support in waste management. With this instrument it can be seen whether or not measures for problem solving in waste management were efficient in the past. The tool is well established in science and increasingly used in practice. *Material and substance* accounting (focused on key substances which are either hazardous , or important resources) should be implemented and used more systematically in public administration, industry and small enterprises.

References

1. Bundesgesetz über die Arbeit in Industrie, Gewerbe und Handel (Arbeitsgesetz, ArG) vom 13. März 1964
2. Bundesgesetz über die Unfallversicherung (Unfallversicherungsgesetz, UVG) vom 20. März 1981 (mit seitherigen Änderungen)
3. Bundesgesetz über den Umweltschutz (Umweltschutzgesetz, USG) vom 7. Oktober 1983 (Stand am 21. Dezember 1999)
4. Luftreinhalte-Verordung (LRV) vom 16. Dezember 1985 (Stand am 28. März 2000)
5. Verordnung über umweltgefährdende Stoffe (Stoffverordnung, StoV) vom 9. Juni 1986 (Stand am 31. Juli 2001)
6. Technische Verordnung über Abfälle (TVA) vom 10. Dezember 1990 (Stand am 28. März 2000)
7. Störfallverordnung (StFV) vom 27. Februar 1991 (Stand am 28. März 2000)
8. Verordnung 3 zum Arbeitsgesetz (ArGV 3) vom 18. August 1993
9. Verordnung über die Rückgabe, die Rücknahme und die Entsorgung elektrischer und elektronischer Geräte (VREG) vom 14. Januar 1998 (Stand am 20. März 2000)
10. Baccini P, Bächler M, Brunner PH, Henseler G (1985) Von der Entsorgung zum Stoffhaushalt: Die Steuerung anthropogener Stoffflüsse als interdisziplinäre Aufgabe. Müll und Abfall 4/85
11. Brunner PH (1992) Wo stehen wir auf dem Weg zur "Endlagerqualität"? Oesterreichische Wasserwirtschaft 9/10
12. Brunner PH, Zobrist J (1983) Die Müllverbrennung als Quelle von Metallen in der Umwelt. Müll und Abfall 9/83
13. Bundesamt für Umweltschutz (1986) Leitbild für die schweizerische Abfallwirtschaft, ausgearbeitet von der Eidgenössischen Kommission für Abfallwirtschaft. Schriftenreihe Umweltschutz 51
14. Bundesrat (1997) Nachhaltige Entwicklung in der Schweiz: Strategie. Bern
15. BUWAL (1992) Abfallkonzept für die Schweiz, Ziele, Massnahmen, Wirkung. Schriftenreihe Umwelt 173
16. BUWAL (1999) Abfallstrategie 2000: Die Entsorgung der brennbaren Abfälle nach dem Ablagerungsverbot am 1. Januar 2000. Pressekonferenz von Vertretern des Bundesamtes für Umwelt, Wald und Landschaft am 29. Januar 1999
17. Daly E, Townsend KN (1993) Valuing the Earth, Economics, Ecology, Ethics. The MIT Press, Cambridge, Massachusetts London
18. Daxbeck H, Morf L, Brunner PH (1998) Stoffflussanalysen als Grundlagen für eine ressourcenorientierte Abfallwirtschaft. Technische Universität Wien, Institut für Wassergüte und Abfallwirtschaft, Wien
19. Duden (1989) Deutsches Universalwörterbuch A-Z, 2nd edn. Dudenverlag, Mannheim Wien Zurich, p 1681

20. Gausemeier J, Fink A, Schlake O (1995) Szenario-Management: Planen und Führen mit Szenarien. Hanser, München

21. Hellweg S, (2000) Time- and Site-Dependent Life-Cycle Assessment of Thermal Waste Treatment Processes. PhD thesis, ETH Zurich

22. INFRAS (1999) Soziale und räumliche Verteilungswirkungen von Energieabgaben. Eidgenössische Materialzentrale, Bern

23. Jung U (1995) Elektronikschrott-Recyclingkonzepte im Vergleich. Wanderer Verlag, Heere

24. Jungk R, Müllert NR (1997) Zukunftswerkstätten, Mit Phantasie gegen Routine und Resignation. Heyne, München

25. Klepper G, Michaelis P, Mahlau G (1995) Industrial Metabolism, a Case Study of the Economics of Cadmium Control. Mohr, Tübingen, p 75

26. Müller M, Huber D (2000) Einfluss der Verwertungsprozesse auf die Dynamik der Abfallwirtschaft. Eine Übersicht über Umfeld, Rahmenbedingungen und Einflussfaktoren. Baden

27. Mutafoff A, Glatz I (2001) Ziele vereinbaren und Strategien realisieren. Erfolgsfaktoren der Unternehmens- und Mitarbeiterführung. Verlag Moderne Industrie, Landsberg/Lech

28. Ott W, Kälin R (2000) Marktwirtschaftliche Instrumente für eine nachhaltigere Schweiz. Müll und Abfall 9/00:547-556

29. SAFEL (Swiss Agency for the Environment Forests and Landscape) Cadmium. Schriftenreihe Umwelt 295

30. SWICO Kommission Umwelt (1999) Tätigkeitsbericht, April.

31. von Reibnitz U (1992) Szenario-Technik. Instrumente für die unternehmerische und persönliche Erfolgsplanung, 2 edn. Gabler, Wiesbaden

32. World Commission on Environment and Development (1987) Our Common Future. University Press, Oxford

9 Concluding Remarks

Christian Ludwig and Samuel Stucki

Waste Management: a Local Issue with Global Consequences

San Diego, Kyoto, Shanghai, Helsinki, Amsterdam, Zürich... - Experts from many different countries, including plant operators, consultants, university professors and scientists have contributed to this book. A colorful think tank was assembled in order to summarize current knowledge on waste management from different perspectives.

The various contributions in this book clearly reflect the regional background of the individual authors. By no means can it be said that there is a unified global agreement on how to manage the waste problems of the future. This is primarily due to the various perceptions of the problem of waste. However, all agree that waste management and the treatment of waste is indispensable for any urban society. Dumping waste, i.e. landfilling without any further precautions, is not a valid choice for any society. Today, fast developing regions with inadequate waste management systems are confronted with serious urban pollution and health hazards. Developed regions are faced with inherited landfill burdens to the local (water, soil, and air pollutants) as well as the global environment (greenhouse gases). There is a broad consensus among experts that waste streams have to be reduced and landfills containing untreated MSW should be avoided.

The ecological objectives of waste management are twofold: to reduce the impact of pollutant emissions from waste into the environment to levels the ecosystem can reasonably and non-detrimentally manage, and to reduce resource consumption to a rate which is compatible with the regeneration of renewables, or the substitution of non-renewables. Sustainable waste management should try to minimize overall impacts on nature and humans, be it by recycling, recovering or appropriate treatment for safe disposal. The challenge waste management is faced with is to find ways of reaching, in a socially and economically compatible way, the ecological goal of attaining and maintaining well-balanced material cycles now and for the indefinite future.

It is important that consensus is reached on an international level for the objectives concerning sustainable waste management. Waste treatment and management practices, however, have to be optimized on a regional basis, according to regional conditions. There are large differences in the composition and heating value of MSW, depending on the level of local income. Treatment technologies need primarily to be adjusted to the properties of MSW and their likely development. Moreover, it is clear that the best available technologies may not be affordable solutions for the majority of regions in the world. Regarding this unavoidable fact, biological treatment methods are well adapted to MSW with high contents of food wastes, have a high potential as low cost alternatives to thermal treatment methods, and are therefore particularly well suited for developing countries.

Thermal Treatment: A Contribution to Sustainable Waste Management

The choice of treatment technologies, particularly concerning the application of incineration, has been a controversial issue in international discussions. Thermal treatment technologies have been introduced mainly in Europe and Japan, where the population density is so high that the MSW problem was primarily a problem of landfill space and sanitation. The first incinerators had no, or merely rudimentary flue gas cleaning systems. Today, the air pollution control systems are so efficient that incinerators have become a true sink, rather than a source of pollutants. The residues of conventional incinerators can be deposited safely. However, long-term behavior with respect to heavy metal leaching cannot be predicted with high confidence. Because of their high content of toxic metals and their high leachability, untreated residues such as filter ash cannot be used in civil engineering applications.

It has been shown that incineration can be further improved to generate better products with substantially lower heavy metal concentrations. The optimization of the current incineration technology is very cost effective for improving today's products. Heavy metals, recovered from incinerator ash can be refined by metallurgical processes. However, there are limited opportunities for reusing the purified bottom ash as substitute materials in, for example, cement production. Therefore, it is necessary to further improve their pozzolanic or hydraulic properties.

Vitrified products have been found to be very resistant against leaching, even if the heavy metal concentrations are not necessarily lower than in conventional incineration residues. Whereas these glassy products are used as building materials in some countries, others prefer a 'no regret' policy and plan to deposit these materials in suitable long-term landfills in order to reduce dissipation of their potentially toxic contents into the environment.

Efficient Material and Energy Recovery

Waste is seen in part as a renewable energy source because of its biomass content. Energy recovery from waste, therefore, can substantially contribute to the attainment of CO_2 reduction targets.

Landfilling has been proposed as a viable mid-term method for carbon sequestration. However, even if installations for recovering landfill gases are applied, there are still emissions of the potent greenhouse gas methane, which offset the carbon storage achieved. Moreover, the efficiency of sanitary landfills for containing pollutants is rather poor and toxic substances are still emitted into air, soil and water. Organics, such as plastics, may have a long residence time in a sanitary landfill; bio-chemical transformations will occur very slowly, therefore the problems will be left to later generations. If efficient energy recovery is utilized, thermal processes can very efficiently reduce overall greenhouse gas emissions. Incineration strongly supports an environmental policy which postulates that those who generate the waste problems should solve them, here and now.

In life cycle assessment (LCA) of MSW treatment technologies the results largely depend on how long-term impacts are valued in comparison to short-term impacts. If long-term impacts are ignored or discounted, then landfilling turns out to be the best option; if, however, environmental burdens for later generations are valued the same as for the present generation, treatment technologies with a high degree of separation of metals score best. If LCA wants to meet common ethical standards, the long-term view should be taken.

Energy is an important resource which is highly rated in LCA. Energy recovery is a must for organic materials, which are difficult to recycle. There are options for the improvement of high-grade energy (i.e. electricity) yield from incinerators. Producing refuse derived fuel (RDF) by mechanical sorting of MSW is an attractive alternative to mass-burning in incinerators: efficient energy conversion for power or for thermal production processes is possible. However, there are drawbacks which need to be considered: Fly ash from RDF combustion contains highly soluble heavy metal compounds. Fossil fuel substitution by waste derived fuels should only be considered if there is a clear overall environmental benefit.

Mechanical-biological treatment performs better than sanitary landfills, but its impact score suffers from low energy efficiency in comparison to thermal treatment technologies.

As already stated, exploiting the energy content of waste is very important when treating waste. Material recycling can have an important energy savings aspect as well; this can be seen, for example, in the recycling of energy intensive products such as aluminum containers.

Socially and Economically Acceptable Solutions

The best eco-technical solution is not sustainable if it cannot be transferred to the market. It has become increasingly clear that technical feasibility alone by no means guarantees the practicability of a proposed solution. Social and economic acceptance ultimately determines whether or not an ecological improvement will eventually be brought from the laboratory into the real world.

The assessment of social aspects has often been neglected in waste management, especially in the planning and building of large installations. Social compatibility analysis could be a valuable tool for pinpointing potential areas of conflict and for introducing guided acceptance dialogues. One area of application for this instrument is in the consensus-oriented process for conflict management, where, e.g., the social dimension of an object to be assessed must be structured, divergent opinions of different groups need to be clarified and possible solutions must be discussed.

Flexibility for Future Options and Developments

The results of recent R&D into thermal, mechanical, and biological separation processes for achieving substantial redirection of material flows in waste treatment facilities have been presented in this book and show that technical solutions are feasible.

In the long-term, new ways of product design and production processes will eventually lead to a material ecology that avoids the wasting of important material resources as well as polluting. There is, however, no alternative to end-of-pipe technologies in the short- to medium-term.

Waste management requires large-scale investments to be made. Large investments (e.g. incinerators) require careful planning, because once they are completed, a given policy is firmly established. Accordingly any change in policy takes a long time. This might also be one of the problems with public acceptance. Large-scale investments block the way for radical change within the following 20 to 40 years! Changes are also required at the level of consumption. People need to become aware of the fact that the rate of consumption we have in the industrialized countries is not sustainable.

The qualities of recovered materials, be it as secondary raw materials or as fuel, need further improvement as well: raw materials with respect to their heavy metal content and characteristics as construction material, fuels with respect to their heavy metal content and fuel properties. Clever combinations of all options - recycling, biological, mechanical and thermal treatment technologies - will be needed to increase current efficiencies in order to save energy and material resources, and to lower the environmental impacts.

Research and development for the further advancement of waste technologies and management options should be continued. There is a need to find better ways of integrally assessing social, economic and ecological criteria in waste management. Clearly, in the long-term, design for recycling and recovery should become mandatory in product development. At the same time, waste streams should increasingly be treated as an ingredient for production processes, be it as a material or an energy resource. A change in paradigm is requested and required; the perspective of waste as a nuisance to be gotten rid of must be transformed into that of waste as a resource of a prospering economy.

Abbreviations

A	Austria
ABB	ASEA Brown Boveri AG
ABB AshArc	"name of ash melting process"
ABB Deglor	"name of vitrification process"
AbfAblV	Abfallablagerungsverordnung, (D)
ACLSM	Active Control Landfill and Stabilization Method
AD	anaerobic digestion
AGL	Aktion Gesunde Luft Thun (CH)
AOX	adsorbable organic halogens
APC	air pollution control
ARA	American Automotive Recyclers Association
ASR	automotive shredder residues
ASTM	American Society for Testing and Materials
AT_4	"respiration coefficient: amount of oxygen consumed within 96 hours"
AVAG	"AG für Abfallverwertung, waste disposal company of the Thun region (CH)"
AVI	see GDA/AVI
AVR	"Dutch MSWI plant"
BA	bottom ash
BCR	"official reference material of FA of European Community Bureau of Reference"
BFB	bubbling fluidized bed
BimSchV	"German legal requirements concerning air pollution control"
BLA	"Dutch Air Emissions from Waste Incineration Order"
BOD	biological oxygen demand
BTU	British Thermal Unit (BTU)
BUWAL	Bundesamt für Umwelt, Wald und Landschaft, (CH)
C	carbon
CBR	California Bearing Ratio
CE	consumer electronics
CEM	"standard of portland cement"
CEN	European. Committee on normalisation
CFB	circulating fluidized bed
CFC	chloro-fluoro-carbon

CFD	computational fluid dynamics
CH	Switzerland
CH_4	methane
CHF	Swiss Franc, Swiss currency
CML	"a life-cycle impact assessment (LCIA) method"
CO	carbon monoxide
CO_2	carbon dioxide
COD	Chemical Oxygen Demand
CombS	combined system
ConvS	conventional system
CRT	cathode ray tube
CSH	calcium-silicate-hydrates
CSR	continuously stirred reactor
CTU	CT Umwelttechnik
D	Germany
DDT	Dichlor diphenyl trichlor ethane, "biocide"
DEV-S4	"German leaching procedure"
DIN	Deutsche Industrie Norm
DIP	De-Inked-Pulp
DNA	deoxyribonucleic acid
DSD	Dual System Deutschland
EA	Environment Agency
ECO	economiser [tech.]
$E_{CRS,\vartheta}$	temperature dependent energy exchange ratio
EDIP	"a life-cycle impact assessment (LCIA) method"
EDP	electronic data processing
EDTA	ethylene dimethyl tetraamin acid
EEE	electrical and electronic equipment
EEE	Ecological-Economic Efficiency
EFA	electrostatic filter ash
ELV	end-of-life-vehicles
EM	ecological modernization
EMPA	"Swiss governmental Material Testing and Research Institute"
EN	Euro Norm
EO	explorable ores
EPS	"a life-cycle impact assessment (LCIA) method"
EPS	extracellular polymeric substances
ESP	electrostatic precipitator
EU	European Union
EXAFS	Extended X-ray Absorption Fine Structure Spectroscopy
FA	fly ash, filter ash
FB	fluidised bed
FC	filter cake
FLT	Freie Liste Thun, (CH)

FLU	"APC residue from MSWI plant Lucerne, (CH)"
FLUAPUR	"name of filter ash treatment process"
FNU	"APC residue from MSWI plant Niederurnen, (CH)"
GB_{21}	"gas formation within 21 days"
GDA/AVI	"Amsterdam Municipal Waste Processing Department"
GDP	Gross Domestic Product of a country
GHG	Green House Gas
GNP	Gross National Product
HCB	Hexachlorobenzene
HCN	Hydrocyanic acid
HDPE	high density polyethylene
HE-WTE	High Efficiency Waste-to-Energy
HHW	household hazardous waste
HLW	nuclear high-level waste
h_n	net calorific value
HS	Blast furnace slag
HT	high temperature treatment technologies
I&B	Industry and Business
i.s.s.d/i.s.s.dry	under standard conditions, dry
ICP-OES	Inductively Coupled Plasma Optical Emission Spectrometer
IEA	International Energy Agency
IG	Interessen Gemeinschaft
IP-Waste	Integrated Project Waste
IR	infrared
ISO	International Organisation of Standardisation
ISWA	International Solid Waste Association
IT	industrial technology
IWM	Integrated Waste Management
JLT46	"Japanese leaching test method for soil"
KCC	Kent County Council
KE	Kent Enviropower
L/S	liquid/solid ratio
LCA	Life-Cycle Assessment
LCD	liquid crystal displays
LCI	Life-Cycle Inventory
LCIA	Life-Cycle Impact Assessment
LHV	lower heating value
l_{min}	minimum air requirement
LSLP	laboratory-scale leaching plant
MBP	mechanical-biological pre-treatment (of MSW)
MFR	material processing facility
MSW	municipal solid waste
MSWI	municipal solid waste incinerator
Mtoe	million tons of oil equivalent

MW	municipal solid waste
MWI	municipal waste incineration
MWTE	"public utility in the Netherlands"
MWWTS	Municipal waste water treatment sludge
NEB	net ecological benefit
NGO	non-governmental organization
NH_3	ammonia
NIMBY	"Not in my backyard!"
NL	The Netherlands
NMVOC	non-methane volatile organic compounds
NOx	nitrogen oxide
NPV	net present value
O&M	Operation and Maintenance
OECD	Organization for Economic Co-operation and Development
PAH	polycyclic aromatic hydrocarbons
PAN	„Gesellschaft zur Planung der Restabfallbehandlung, (D)"
PBB	polybrominated biphenyls
PBDD/F	polybrominated dibenzo dioxins and furans
PBDE	polybrominated diphenylethers
PC	personal computer
PCB	polychlorinated biphenyls
PCDD/F	polychlorinated dibenzo dioxins and furans)
PDF	"a life-cycle impact assessment (LCIA) method"
Pe	Peclet Number
PE	Polyethylen
PECK	„Paul Scherrer Institut - Eberhard Recycling - CT Umwelttechnik – Küpat"
PET	polyethylene terephthalate
PF	primary (regular) fuel
PFR	plug flow reactor
pH	log H_3O^+ concentration
PLFA	phospholipid fatty acids
PS	power station
PS	Polystyrole
PSA	pressure swing adsorption
PTFE	Teflon
PVC	polyvinyl chloride
QRR	quartz-glass tube
QZ	Quartz
R&D	Research and Development
RAL-GZ 724	"quality certificate"
RBRC	Rechargeable Battery Recycling Corporation
RDF	refuse derived fuel
REF	recovered fuel

RES	renewable energy sources
RESH	residues from shredding
RR park	resource recovery park
RRRF	Robbins Resource Recovery Facility
SAEFL	Swiss Agency for the Environment, Forests and Landscape
SBA	Siemens-Schwelbrennanlage
SC	Social Compatibility
SCA	Social Compatibility Analysis
SCR	selective catalytic reduction
SCW	Supercritical water
SCWO	supercritical water oxidation
SEM	Scanning electron micrographs
SEM	scanning electron microscope
SF	substitute fuel
SFA	Fly ash
SI	saturation index
SIA	Schweizerischer Ingenieur- und Architektur-Verband
SLF	shredder light fraction, fluff
SNCR	selective non-catalytic reduction
SO_2	sulfur dioxide
SPPE	Swiss Priority Program Environment
SWICO	Swiss business association of information, communication and organisation technologies
TC	Transfer coefficients
TCDD	Tetrachlor dibenzo dioxin „Seveso-Dioxin"
TCE	trichloroethene
TDS	Thermo-Desorption-Spectroscopy
TEES	Thermochemical Environmental Energy System
TEQ	TCDD equivalent
TOC	Total Organic Carbon
TOW	Swiss Technical Ordinance on Waste (see also TVA)
TV	television set
TVA	Schweizerische Technische Verordnung über Abfälle (see also TOW)
TWR	transpiring wall reactor
UCPTE	"a life-cycle impact assessment (LCIA) method"
UK	United Kingdom
US	United States of America
VFA	Volatile fatty acids
v_{min}	minimum amount of flue gas
VREG	"Act on Return, Take back and Disposal of Electrical and Electronic Appliances"
VS	Vergasen/Verbrennen-Schmelzen
VTT	"finnish national research centre"

WEEE	Electrical and Electronic Equipment
WTA	"Willingness To Accept"
WtE	Waste to Energy
WTP	"Willingness To Pay"
WWTP	Waste Water Treatment Plant
ϑ	temperature
χ	concentration
λ	stoichiometric ratio
ω	frequency
ξ	concentration (mass related)
Ψ	concentration (volume-related)

Index